FUNCTIONAL ANALYSIS

Princeton Lectures in Analysis

I Fourier Analysis: An Introduction

II Complex Analysis

III Real Analysis: Measure Theory, Integration, and Hilbert Spaces

IV Functional Analysis: Introduction to Further Topics in Analysis

Princeton Lectures in Analysis

IV

FUNCTIONAL ANALYSIS

Introduction to Further Topics in Analysis

Elias M. Stein

&

Rami Shakarchi

PRINCETON UNIVERSITY PRESS
PRINCETON AND OXFORD

Published by Princeton University Press
41 William Street, Princeton, New Jersey 08540

In the United Kingdom: Princeton University Press
6 Oxford Street, Woodstock, Oxfordshire, OX20 1TW

Library of Congress Cataloging-in-Publication Data is available

ISBN 978-0-691-11387-6

British Library Cataloging-in-Publication Data is available

This book has been composed in LaTeX

The publisher would like to acknowledge the authors of this volume for providing the camera-ready copy from which this book was printed.

Printed on acid-free paper. ∞

press.princeton.edu

Printed in the United States of America

10 9 8 7 6 5 4

To my grandchildren

Carolyn, Alison, Jason

E.M.S.

To my parents

Mohamed & Mireille

and my brother

Karim

R.S.

Foreword

Beginning in the spring of 2000, a series of four one-semester courses were taught at Princeton University whose purpose was to present, in an integrated manner, the core areas of analysis. The objective was to make plain the organic unity that exists between the various parts of the subject, and to illustrate the wide applicability of ideas of analysis to other fields of mathematics and science. The present series of books is an elaboration of the lectures that were given.

While there are a number of excellent texts dealing with individual parts of what we cover, our exposition aims at a different goal: presenting the various sub-areas of analysis not as separate disciplines, but rather as highly interconnected. It is our view that seeing these relations and their resulting synergies will motivate the reader to attain a better understanding of the subject as a whole. With this outcome in mind, we have concentrated on the main ideas and theorems that have shaped the field (sometimes sacrificing a more systematic approach), and we have been sensitive to the historical order in which the logic of the subject developed.

We have organized our exposition into four volumes, each reflecting the material covered in a semester. Their contents may be broadly summarized as follows:

I. Fourier series and integrals.

II. Complex analysis.

III. Measure theory, Lebesgue integration, and Hilbert spaces.

IV. A selection of further topics, including functional analysis, distributions, and elements of probability theory.

However, this listing does not by itself give a complete picture of the many interconnections that are presented, nor of the applications to other branches that are highlighted. To give a few examples: the elements of (finite) Fourier series studied in Book I, which lead to Dirichlet characters, and from there to the infinitude of primes in an arithmetic progression; the X-ray and Radon transforms, which arise in a number of problems in Book I, and reappear in Book III to play an important role in understanding Besicovitch-like sets in two and three dimensions; Fatou's theorem, which guarantees the existence of boundary values of bounded holomorphic functions in the disc, and whose proof relies on ideas developed in each of the first three books; and the theta function, which first occurs in Book I in the solution of the heat equation, and is then used in Book II to find the number of ways an integer can be represented as the sum of two or four squares, and in the analytic continuation of the zeta function.

A few further words about the books and the courses on which they were based. These courses where given at a rather intensive pace, with 48 lecture-hours a semester. The weekly problem sets played an indispensable part, and as a result exercises and problems have a similarly important role in our books. Each chapter has a series of "Exercises" that are tied directly to the text, and while some are easy, others may require more effort. However, the substantial number of hints that are given should enable the reader to attack most exercises. There are also more involved and challenging "Problems"; the ones that are most difficult, or go beyond the scope of the text, are marked with an asterisk.

Despite the substantial connections that exist between the different volumes, enough overlapping material has been provided so that each of the first three books requires only minimal prerequisites: acquaintance with elementary topics in analysis such as limits, series, differentiable functions, and Riemann integration, together with some exposure to linear algebra. This makes these books accessible to students interested in such diverse disciplines as mathematics, physics, engineering, and finance, at both the undergraduate and graduate level.

It is with great pleasure that we express our appreciation to all who have aided in this enterprise. We are particularly grateful to the students who participated in the four courses. Their continuing interest, enthusiasm, and dedication provided the encouragement that made this project possible. We also wish to thank Adrian Banner and José Luis Rodrigo for their special help in running the courses, and their efforts to see that the students got the most from each class. In addition, Adrian Banner also made valuable suggestions that are incorporated in the text.

We wish also to record a note of special thanks for the following individuals: Charles Fefferman, who taught the first week (successfully launching the whole project!); Paul Hagelstein, who in addition to reading part of the manuscript taught several weeks of one of the courses, and has since taken over the teaching of the second round of the series; and Daniel Levine, who gave valuable help in proofreading. Last but not least, our thanks go to Gerree Pecht, for her consummate skill in typesetting and for the time and energy she spent in the preparation of all aspects of the lectures, such as transparencies, notes, and the manuscript.

We are also happy to acknowledge our indebtedness for the support we received from the 250th Anniversary Fund of Princeton University, and the National Science Foundation's VIGRE program.

Elias M. Stein

Rami Shakarchi

Princeton, New Jersey

August 2002

As with the previous volumes, we are happy to record our great debt to Daniel Levine. The final version of this book has been much improved because of his help. He read the entire manuscript with great care and made valuable suggestions that have been incorporated in the text. We also wish to take this opportunity to thank Hart Smith and Polam Yung for proofreading parts of the book.

May 2011

Contents

Preface to Book IV

Functional analysis, as generally understood, brought with it a change of focus from the study of functions on everyday geometric spaces such as $\mathbb{R}$, $\mathbb{R}^d$, etc., to the analysis of abstract infinite-dimensional spaces, for example, functions spaces and Banach spaces. As such it established a key framework for the development of modern analysis.

Our first goal in this volume is to present the basic ideas of this theory, with particular emphasis on their connection to harmonic analysis. A second objective is to provide an introduction to some further topics to which any serious student of analysis ought to be exposed: probability theory, several complex variables and oscillatory integrals. Our choice of these subjects is guided, in the first instance, by their intrinsic interest. Moreover, these topics complement and extend ideas in the previous books in this series, and they serve our overarching goal of making plain the organic unity that exists between the various parts of analysis.

Underlying this unity is the role of Fourier analysis in its interrelation with partial differential equations, complex analysis, and number theory. It is also exemplified by some of the specific questions that arose initially in the previous volumes and that are taken up again here: namely, the Dirichlet problem, ultimately treated by Brownian motion; the Radon transform, with its connection to Besicovitch sets; nowhere differentiable functions; and some problems in number theory, now formulated as distributions of lattice points. We hope that this choice of material will not only provide a broader view of analysis, but will also inspire the reader to pursue the further study of this subject.

1 L^p Spaces and Banach Spaces

In this work the assumption of quadratic integrability will be replaced by the integrability of $|f(x)|^p$. The analysis of these function classes will shed a particular light on the real and apparent advantages of the exponent 2; one can also expect that it will provide essential material for an axiomatic study of function spaces.

F. Riesz, 1910

At present I propose above all to gather results about linear operators defined in certain general spaces, notably those that will here be called *spaces of type (B)...*

S. Banach, 1932

Function spaces, in particular L^p spaces, play a central role in many questions in analysis. The special importance of L^p spaces may be said to derive from the fact that they offer a partial but useful generalization of the fundamental L^2 space of square integrable functions.

In order of logical simplicity, the space L^1 comes first since it occurs already in the description of functions integrable in the Lebesgue sense. Connected to it via duality is the L^∞ space of bounded functions, whose supremum norm carries over from the more familiar space of continuous functions. Of independent interest is the L^2 space, whose origins are tied up with basic issues in Fourier analysis. The intermediate L^p spaces are in this sense an artifice, although of a most inspired and fortuitous kind. That this is the case will be illustrated by results in the next and succeeding chapters.

In this chapter we will concentrate on the basic structural facts about the L^p spaces. Here part of the theory, in particular the study of their linear functionals, is best formulated in the more general context of Banach spaces. An incidental benefit of this more abstract view-point is that it leads us to the surprising discovery of a finitely additive measure on *all* subsets, consistent with Lebesgue measure.

1 L^p spaces

Throughout this chapter $(X, \mathcal{F}, \mu)$ denotes a σ-finite measure space: X denotes the underlying space, $\mathcal{F}$ the σ-algebra of measurable sets, and μ the measure. If $1 \leq p < \infty$, the space $L^p(X, \mathcal{F}, \mu)$ consists of all complex-valued measurable functions on X that satisfy

$$\int_X |f(x)|^p \, d\mu(x) < \infty. \tag{1}$$

To simplify the notation, we write $L^p(X, \mu)$, or $L^p(X)$, or simply L^p when the underlying measure space has been specified. Then, if $f \in L^p(X, \mathcal{F}, \mu)$ we define the L^p **norm** of f by

$$\|f\|_{L^p(X,\mathcal{F},\mu)} = \left(\int_X |f(x)|^p \, d\mu(x) \right)^{1/p}.$$

We also abbreviate this to $\|f\|_{L^p(X)}$, $\|f\|_{L^p}$, or $\|f\|_p$.

When $p = 1$ the space $L^1(X, \mathcal{F}, \mu)$ consists of all integrable functions on X, and we have shown in Chapter 6 of Book III, that L^1 together with $\| \cdot \|_{L^1}$ is a complete normed vector space. Also, the case $p = 2$ warrants special attention: it is a Hilbert space.

We note here that we encounter the same technical point that we already discussed in Book III. The problem is that $\|f\|_{L^p} = 0$ does not imply that $f = 0$, but merely $f = 0$ almost everywhere (for the measure μ). Therefore, the precise definition of L^p requires introducing the equivalence relation, in which f and g are equivalent if $f = g$ a.e. Then, L^p consists of all equivalence classes of functions which satisfy (1). However, in practice there is little risk of error by thinking of elements in L^p as functions rather than equivalence classes of functions.

The following are some common examples of L^p spaces.

(a) The case $X = \mathbb{R}^d$ and μ equals Lebesgue measure is often used in practice. There, we have

$$\|f\|_{L^p} = \left(\int_{\mathbb{R}^d} |f(x)|^p \, dx \right)^{1/p}.$$

(b) Also, one can take $X = \mathbb{Z}$, and μ equal to the counting measure. Then, we get the "discrete" version of the L^p spaces. Measurable functions are simply sequences $f = \{a_n\}_{n \in \mathbb{Z}}$ of complex numbers, and

$$\|f\|_{L^p} = \left(\sum_{n=-\infty}^{\infty} |a_n|^p \right)^{1/p} .$$

When $p = 2$, we recover the familiar sequence space $\ell^2(\mathbb{Z})$.

The spaces L^p are examples of normed vector spaces. The basic property satisfied by the norm is the triangle inequality, which we shall prove shortly.

The range of p which is of interest in most applications is $1 \leq p < \infty$, and later also $p = \infty$. There are at least two reasons why we restrict our attention to these values of p: when $0 < p < 1$, the function $\|\cdot\|_{L^p}$ does not satisfy the triangle inequality, and moreover, for such p, the space L^p has no non-trivial bounded linear functionals.[1] (See Exercise 2.)

When $p = 1$ the norm $\|\cdot\|_{L^1}$ satisfies the triangle inequality, and L^1 is a complete normed vector space. When $p = 2$, this result continues to hold, although one needs the Cauchy-Schwarz inequality to prove it. In the same way, for $1 \leq p < \infty$ the proof of the triangle inequality relies on a generalized version of the Cauchy-Schwarz inequality. This is Hölder's inequality, which is also the key in the duality of the L^p spaces, as we will see in Section 4.

1.1 The Hölder and Minkowski inequalities

If the two exponents p and q satisfy $1 \leq p, q \leq \infty$, and the relation

$$\frac{1}{p} + \frac{1}{q} = 1$$

holds, we say that p and q are **conjugate** or **dual** exponents. Here, we use the convention $1/\infty = 0$. Later, we shall sometimes use p' to denote the conjugate exponent of p. Note that $p = 2$ is self-dual, that is, $p = q = 2$; also $p = 1, \infty$ corresponds to $q = \infty, 1$ respectively.

Theorem 1.1 (Hölder) *Suppose* $1 < p < \infty$ *and* $1 < q < \infty$ *are conjugate exponents. If* $f \in L^p$ *and* $g \in L^q$, *then* $fg \in L^1$ *and*

$$\|fg\|_{L^1} \leq \|f\|_{L^p} \|g\|_{L^q}.$$

Note. Once we have defined L^∞ (see Section 2) the corresponding inequality for the exponents 1 and ∞ will be seen to be essentially trivial.

---

[1] We will define what we mean by a bounded linear functional later in the chapter.

The proof of the theorem relies on a simple generalized form of the arithmetic-geometric mean inequality: if $A, B \geq 0$, and $0 \leq \theta \leq 1$, then

$$A^\theta B^{1-\theta} \leq \theta A + (1-\theta)B. \tag{2}$$

Note that when $\theta = 1/2$, the inequality (2) states the familiar fact that the geometric mean of two numbers is majorized by their arithmetic mean.

To establish (2), we observe first that we may assume $B \neq 0$, and replacing A by AB, we see that it suffices to prove that $A^\theta \leq \theta A + (1-\theta)$. If we let $f(x) = x^\theta - \theta x - (1-\theta)$, then $f'(x) = \theta(x^{\theta-1} - 1)$. Thus $f(x)$ increases when $0 \leq x \leq 1$ and decreases when $1 \leq x$, and we see that the continuous function f attains a maximum at $x = 1$, where $f(1) = 0$. Therefore $f(A) \leq 0$, as desired.

To prove Hölder's inequality we argue as follows. If either $\|f\|_{L^p} = 0$ or $\|f\|_{L^q} = 0$, then $fg = 0$ a.e. and the inequality is obviously verified. Therefore, we may assume that neither of these norms vanish, and after replacing f by $f/\|f\|_{L^p}$ and g by $g/\|g\|_{L^q}$, we may further assume that $\|f\|_{L^p} = \|g\|_{L^q} = 1$. We now need to prove that $\|fg\|_{L^1} \leq 1$.

If we set $A = |f(x)|^p$, $B = |g(x)|^q$, and $\theta = 1/p$ so that $1 - \theta = 1/q$, then (2) gives

$$|f(x)g(x)| \leq \frac{1}{p}|f(x)|^p + \frac{1}{q}|g(x)|^q.$$

Integrating this inequality yields $\|fg\|_{L^1} \leq 1$, and the proof of the Hölder inequality is complete.

For the case when the equality $\|fg\|_{L^1} = \|f\|_{L^p}\|g\|_{L^q}$ holds, see Exercise 3.

We are now ready to prove the triangle inequality for the L^p norm.

Theorem 1.2 (Minkowski) *If $1 \leq p < \infty$ and $f, g \in L^p$, then $f + g \in L^p$ and $\|f+g\|_{L^p} \leq \|f\|_{L^p} + \|g\|_{L^p}$.*

Proof. The case $p = 1$ is obtained by integrating $|f(x) + g(x)| \leq |f(x)| + |g(x)|$. When $p > 1$, we may begin by verifying that $f + g \in L^p$, when both f and g belong to L^p. Indeed,

$$|f(x) + g(x)|^p \leq 2^p(|f(x)|^p + |g(x)|^p),$$

as can be seen by considering separately the cases $|f(x)| \leq |g(x)|$ and $|g(x)| \leq |f(x)|$. Next we note that

$$|f(x) + g(x)|^p \leq |f(x)|\,|f(x) + g(x)|^{p-1} + |g(x)|\,|f(x) + g(x)|^{p-1}.$$

If q denotes the conjugate exponent of p, then $(p-1)q = p$, so we see that $(f+g)^{p-1}$ belongs to L^q, and therefore Hölder's inequality applied to the two terms on the right-hand side of the above inequality gives

$$\|f+g\|_{L^p}^p \leq \|f\|_{L^p}\|(f+g)^{p-1}\|_{L^q} + \|g\|_{L^p}\|(f+g)^{p-1}\|_{L^q}. \tag{3}$$

However, using once again $(p-1)q = p$, we get

$$\|(f+g)^{p-1}\|_{L^q} = \|f+g\|_{L^p}^{p/q}.$$

From (3), since $p - p/q = 1$, and because we may suppose that $\|f+g\|_{L^p} > 0$, we find

$$\|f+g\|_{L^p} \leq \|f\|_{L^p} + \|g\|_{L^p},$$

so the proof is finished.

1.2 Completeness of L^p

The triangle inequality makes L^p into a metric space with distance $d(f,g) = \|f-g\|_{L^p}$. The basic analytic fact is that L^p is **complete** in the sense that every Cauchy sequence in the norm $\|\cdot\|_{L^p}$ converges to an element *in* L^p.

Taking limits is a necessity in many problems, and the L^p spaces would be of little use if they were not complete. Fortunately, like L^1 and L^2, the general L^p space does satisfy this desirable property.

Theorem 1.3 *The space $L^p(X, \mathcal{F}, \mu)$ is complete in the norm $\|\cdot\|_{L^p}$.*

Proof. The argument is essentially the same as for L^1 (or L^2); see Section 2, Chapter 2 and Section 1, Chapter 4 in Book III. Let $\{f_n\}_{n=1}^{\infty}$ be a Cauchy sequence in L^p, and consider a subsequence $\{f_{n_k}\}_{k=1}^{\infty}$ of $\{f_n\}$ with the following property $\|f_{n_{k+1}} - f_{n_k}\|_{L^p} \leq 2^{-k}$ for all $k \geq 1$. We now consider the series whose convergence will be seen below

$$f(x) = f_{n_1}(x) + \sum_{k=1}^{\infty}(f_{n_{k+1}}(x) - f_{n_k}(x))$$

and

$$g(x) = |f_{n_1}(x)| + \sum_{k=1}^{\infty}|f_{n_{k+1}}(x) - f_{n_k}(x)|,$$

and the corresponding partial sums

$$S_K(f)(x) = f_{n_1}(x) + \sum_{k=1}^{K}(f_{n_{k+1}}(x) - f_{n_k}(x))$$

and

$$S_K(g)(x) = |f_{n_1}(x)| + \sum_{k=1}^{K}|f_{n_{k+1}}(x) - f_{n_k}(x)|.$$

The triangle inequality for L^p implies

$$\begin{aligned}\|S_K(g)\|_{L^p} &\leq \|f_{n_1}\|_{L^p} + \sum_{k=1}^{K}\|f_{n_{k+1}} - f_{n_k}\|_{L^p} \\ &\leq \|f_{n_1}\|_{L^p} + \sum_{k=1}^{K}2^{-k}.\end{aligned}$$

Letting K tend to infinity, and applying the monotone convergence theorem proves that $\int g^p < \infty$, and therefore the series defining g, and hence the series defining f converges almost everywhere, and $f \in L^p$.

We now show that f is the desired limit of the sequence $\{f_n\}$. Since (by construction of the telescopic series) the $(K-1)^{\text{th}}$ partial sum of this series is precisely f_{n_K}, we find that

$$f_{n_K}(x) \to f(x) \quad \text{a.e. } x.$$

To prove that $f_{n_K} \to f$ in L^p as well, we first observe that

$$\begin{aligned}|f(x) - S_K(f)(x)|^p &\leq [2\max(|f(x)|, |S_K(f)(x)|)]^p \\ &\leq 2^p|f(x)|^p + 2^p|S_K(f)(x)|^p \\ &\leq 2^{p+1}|g(x)|^p,\end{aligned}$$

for all K. Then, we may apply the dominated convergence theorem to get $\|f_{n_K} - f\|_{L^p} \to 0$ as K tends to infinity.

Finally, the last step of the proof consists of recalling that $\{f_n\}$ is Cauchy. Given $\epsilon > 0$, there exists N so that for all $n, m > N$ we have $\|f_n - f_m\|_{L^p} < \epsilon/2$. If n_K is chosen so that $n_K > N$, and $\|f_{n_K} - f\|_{L^p} < \epsilon/2$, then the triangle inequality implies

$$\|f_n - f\|_{L^p} \leq \|f_n - f_{n_K}\|_{L^p} + \|f_{n_K} - f\|_{L^p} < \epsilon$$

whenever $n > N$. This concludes the proof of the theorem.

1.3 Further remarks

We begin by looking at some possible inclusion relations between the various L^p spaces. The matter is simple if the underlying space has finite measure.

Proposition 1.4 *If X has finite positive measure, and $p_0 \leq p_1$, then $L^{p_1}(X) \subset L^{p_0}(X)$ and*

$$\frac{1}{\mu(X)^{1/p_0}} \|f\|_{L^{p_0}} \leq \frac{1}{\mu(X)^{1/p_1}} \|f\|_{L^{p_1}}.$$

We may assume that $p_1 > p_0$. Suppose $f \in L^{p_1}$, and set $F = |f|^{p_0}$, $G = 1$, $p = p_1/p_0 > 1$, and $1/p + 1/q = 1$, in Hölder's inequality applied to F and G. This yields

$$\|f\|_{L^{p_0}}^{p_0} \leq \left(\int |f|^{p_1} \right)^{p_0/p_1} \cdot \mu(X)^{1-p_0/p_1}.$$

In particular, we find that $\|f\|_{L^{p_0}} < \infty$. Moreover, by taking the p_0^{th} root of both sides of the above equation, we find that the inequality in the proposition holds.

However, as is easily seen, such inclusion does not hold when X has infinite measure. (See Exercise 1). Yet, in an interesting special case the opposite inclusion does hold.

Proposition 1.5 *If $X = \mathbb{Z}$ is equipped with counting measure, then the reverse inclusion holds, namely $L^{p_0}(\mathbb{Z}) \subset L^{p_1}(\mathbb{Z})$ if $p_0 \leq p_1$. Moreover, $\|f\|_{L^{p_1}} \leq \|f\|_{L^{p_0}}$.*

Indeed, if $f = \{f(n)\}_{n \in \mathbb{Z}}$, then $\sum |f(n)|^{p_0} = \|f\|_{L^{p_0}}^{p_0}$, and $\sup_n |f(n)| \leq \|f\|_{L^{p_0}}$. However

$$\begin{aligned} \sum |f(n)|^{p_1} &= \sum |f(n)|^{p_0} |f(n)|^{p_1 - p_0} \\ &\leq (\sup_n |f(n)|)^{p_1 - p_0} \|f\|_{L^{p_0}}^{p_0} \\ &\leq \|f\|_{L^{p_0}}^{p_1}. \end{aligned}$$

Thus $\|f\|_{L^{p_1}} \leq \|f\|_{L^{p_0}}$.

2 The case $p = \infty$

Finally, we also consider the limiting case $p = \infty$. The space L^∞ will be defined as all functions that are "essentially bounded" in the following sense. We take the space $L^\infty(X, \mathcal{F}, \mu)$ to consist of all (equivalence classes of) measurable functions on X, so that there exists a positive number $0 < M < \infty$, with

$$|f(x)| \leq M \qquad \text{a.e. } x.$$

Then, we define $\|f\|_{L^\infty(X,\mathcal{F},\mu)}$ to be the infimum of all possible values M satisfying the above inequality. The quantity $\|f\|_{L^\infty}$ is sometimes called the **essential-supremum** of f.

We note that with this definition, we have $|f(x)| \leq \|f\|_{L^\infty}$ for a.e. x. Indeed, if $E = \{x : |f(x)| > \|f\|_{L^\infty}\}$, and $E_n = \{x : |f(x)| > \|f\|_{L^\infty} + 1/n\}$, then we have $\mu(E_n) = 0$, and $E = \bigcup E_n$, hence $\mu(E) = 0$.

Theorem 2.1 *The vector space L^∞ equipped with $\|\cdot\|_{L^\infty}$ is a complete vector space.*

This assertion is easy to verify and is left to the reader. Moreover, Hölder's inequality continues to hold for values of p and q in the larger range $1 \leq p, q \leq \infty$, once we take $p = 1$ and $q = \infty$ as conjugate exponents, as we mentioned before.

The fact that L^∞ is a limiting case of L^p when p tends to ∞ can be understood as follows.

Proposition 2.2 *Suppose $f \in L^\infty$ is supported on a set of finite measure. Then $f \in L^p$ for all $p < \infty$, and*

$$\|f\|_{L^p} \to \|f\|_{L^\infty} \qquad \text{as } p \to \infty.$$

Proof. Let E be a measurable subset of X with $\mu(E) < \infty$, and so that f vanishes in the complement of E. If $\mu(E) = 0$, then $\|f\|_{L^\infty} = \|f\|_{L^p} = 0$ and there is nothing to prove. Otherwise

$$\|f\|_{L^p} = \left(\int_E |f(x)|^p \, d\mu\right)^{1/p} \leq \left(\int_E \|f\|_{L^\infty}^p \, d\mu\right)^{1/p} \leq \|f\|_{L^\infty}\mu(E)^{1/p}.$$

Since $\mu(E)^{1/p} \to 1$ as $p \to \infty$, we find that $\limsup_{p\to\infty} \|f\|_{L^p} \leq \|f\|_{L^\infty}$.

On the other hand, given $\epsilon > 0$, we have

$$\mu(\{x : \ |f(x)| \geq \|f\|_{L^\infty} - \epsilon\}) \geq \delta \qquad \text{for some } \delta > 0,$$

hence

$$\int_X |f|^p \, d\mu \geq \delta(\|f\|_{L^\infty} - \epsilon)^p.$$

Therefore $\liminf_{p\to\infty} \|f\|_{L^p} \geq \|f\|_{L^\infty} - \epsilon$, and since ϵ is arbitrary, we have $\liminf_{p\to\infty} \|f\|_{L^p} \geq \|f\|_{L^\infty}$. Hence the limit $\lim_{p\to\infty} \|f\|_{L^p}$ exists, and equals $\|f\|_{L^\infty}$.

3 Banach spaces

We introduce here a general notion which encompasses the L^p spaces as specific examples.

First, a **normed vector space** consists of an underlying vector space V over a field of scalars (the real or complex numbers), together with a **norm** $\|\cdot\| : V \to \mathbb{R}^+$ that satisfies:

- $\|v\| = 0$ if and only if $v = 0$.
- $\|\alpha v\| = |\alpha| \, \|v\|$, whenever α is a scalar and $v \in V$.
- $\|v + w\| \le \|v\| + \|w\|$ for all $v, w \in V$.

The space V is said to be **complete** if whenever $\{v_n\}$ is a Cauchy sequence in V, that is, $\|v_n - v_m\| \to 0$ as $n, m \to \infty$, then there exists a $v \in V$ such that $\|v_n - v\| \to 0$ as $n \to \infty$.

A complete normed vector space is called a **Banach space**. Here again, we stress the importance of the fact that Cauchy sequences converge to a limit in the space itself, hence the space is "closed" under limiting operations.

3.1 Examples

The real numbers $\mathbb{R}$ with the usual absolute value form an initial example of a Banach space. Other easy examples are $\mathbb{R}^d$, with the Euclidean norm, and more generally a Hilbert space with its norm given in terms of its inner product.

Several further relevant examples are as follows:

EXAMPLE 1. The family of L^p spaces with $1 \le p \le \infty$ which we have just introduced are also important examples of Banach spaces (Theorem 1.3 and Theorem 2.1). Incidentally, L^2 is the only Hilbert space in the family L^p, where $1 \le p \le \infty$ (Exercise 25) and this in part accounts for the special flavor of the analysis carried out in L^2 as opposed to L^1 or more generally L^p for $p \ne 2$.

Finally, observe that since the triangle inequality fails in general when $0 < p < 1$, $\|\cdot\|_{L^p}$ is not a norm on L^p for this range of p, hence it is not a Banach space.

EXAMPLE 2. Another example of a Banach space is $C([0,1])$, or more generally $C(X)$ with X a compact set in a metric space, as will be defined in Section 7. By definition, $C(X)$ is the vector space of continuous functions on X equipped with the sup-norm $\|f\| = \sup_{x\in X} |f(x)|$. Completeness is guaranteed by the fact that the uniform limit of a sequence of continuous functions is also continuous.

EXAMPLE 3. Two further examples are important in various applications. The first is the space $\Lambda^\alpha(\mathbb{R})$ of all bounded functions on $\mathbb{R}$ which satisfy a **Hölder (or Lipschitz) condition of exponent** α with $0 < \alpha \leq 1$, that is,

$$\sup_{t_1 \neq t_2} \frac{|f(t_1) - f(t_2)|}{|t_1 - t_2|^\alpha} < \infty.$$

Observe that f is then necessarily continuous; also the only interesting case is when $\alpha \leq 1$, since a function which satisfies a Hölder condition of exponent α with $\alpha > 1$ is constant.[2]

More generally, this space can be defined on $\mathbb{R}^d$; it consists of continuous functions f equipped with the norm

$$\|f\|_{\Lambda^\alpha(\mathbb{R}^d)} = \sup_{x\in\mathbb{R}^d} |f(x)| + \sup_{x\neq y} \frac{|f(x) - f(y)|}{|x - y|^\alpha}.$$

With this norm, $\Lambda^\alpha(\mathbb{R}^d)$ is a Banach space (see also Exercise 29).

EXAMPLE 4. A function $f \in L^p(\mathbb{R}^d)$ is said to have **weak derivatives** in L^p up to order k, if for every multi-index $\alpha = (\alpha_1, \ldots, \alpha_d)$ with $|\alpha| = \alpha_1 + \cdots + \alpha_d \leq k$, there is a $g_\alpha \in L^p$ with

$$\int_{\mathbb{R}^d} g_\alpha(x)\varphi(x)\, dx = (-1)^{|\alpha|} \int_{\mathbb{R}^d} f(x)\partial_x^\alpha \varphi(x)\, dx \tag{4}$$

for all smooth functions φ that have compact support in $\mathbb{R}^d$. Here, we use the multi-index notation

$$\partial_x^\alpha = \left(\frac{\partial}{\partial x}\right)^\alpha = \left(\frac{\partial}{\partial x_1}\right)^{\alpha_1} \cdots \left(\frac{\partial}{\partial x_d}\right)^{\alpha_d}.$$

Clearly, the functions g_α (when they exist) are unique, and we also write $\partial_x^\alpha f = g_\alpha$. This definition arises from the relationship (4) which holds whenever f is itself smooth, and g equals the usual derivative $\partial_x^\alpha f$, as follows from an integration by parts (see also Section 3.1, Chapter 5 in Book III).

[2]We have already encountered this space in Book I, Chapter 2 and Book III, Chapter 7.

The space $L^p_k(\mathbb{R}^d)$ is the subspace of $L^p(\mathbb{R}^d)$ of all functions that have weak derivatives up to order k. (The concept of weak derivatives will reappear in Chapter 3 in the setting of derivatives in the sense of distributions.) This space is usually referred to as a **Sobolev space**. A norm that turns $L^p_k(\mathbb{R}^d)$ into a Banach space is

$$\|f\|_{L^p_k(\mathbb{R}^d)} = \sum_{|\alpha|\leq k} \|\partial_x^\alpha f\|_{L^p(\mathbb{R}^d)}.$$

EXAMPLE 5. In the case $p = 2$, we note in the above example that an L^2 function f belongs to $L^2_k(\mathbb{R}^d)$ if and only if $(1+|\xi|^2)^{k/2}\hat{f}(\xi)$ belongs to L^2, and that $\|(1+|\xi|^2)^{k/2}\hat{f}(\xi)\|_{L^2}$ is a Hilbert space norm equivalent to $\|f\|_{L^2_k(\mathbb{R}^d)}$.

Therefore, if k is any positive number, it is natural to define L^2_k as those functions f in L^2 for which $(1+|\xi|^2)^{k/2}\hat{f}(\xi)$ belongs to L^2, and we can equip L^2_k with the norm $\|f\|_{L^2_k(\mathbb{R}^d)} = \|(1+|\xi|^2)^{k/2}\hat{f}(\xi)\|_{L^2}$.

3.2 Linear functionals and the dual of a Banach space

For the sake of simplicity, we restrict ourselves in this and the following two sections to Banach spaces over $\mathbb{R}$; the reader will find in Section 6 the slight modifications necessary to extend the results to Banach spaces over $\mathbb{C}$.

Suppose that $\mathcal{B}$ is a Banach space over $\mathbb{R}$ equipped with a norm $\|\cdot\|$. A **linear functional** is a linear mapping ℓ from $\mathcal{B}$ to $\mathbb{R}$, that is, $\ell : \mathcal{B} \to \mathbb{R}$, which satisfies

$$\ell(\alpha f + \beta g) = \alpha\ell(f) + \beta\ell(g), \qquad \text{for all } \alpha, \beta \in \mathbb{R}, \text{ and } f, g \in \mathcal{B}.$$

A linear functional ℓ is **continuous** if given $\epsilon > 0$ there exists $\delta > 0$ so that $|\ell(f) - \ell(g)| \leq \epsilon$ whenever $\|f - g\| \leq \delta$. Also we say that a linear functional is **bounded** if there is $M > 0$ with $|\ell(f)| \leq M\|f\|$ for all $f \in \mathcal{B}$. The linearity of ℓ shows that these two notions are in fact equivalent.

Proposition 3.1 *A linear functional on a Banach space is continuous, if and only if it is bounded.*

Proof. The key is to observe that ℓ is continuous if and only if ℓ is continuous at the origin.

Indeed, if ℓ is continuous, we choose $\epsilon = 1$ and $g = 0$ in the above definition so that $|\ell(f)| \leq 1$ whenever $\|f\| \leq \delta$, for some $\delta > 0$. Hence, given any non-zero h, an element of $\mathcal{B}$, we see that $\delta h/\|h\|$ has norm equal to δ, and hence $|\ell(\delta h/\|h\|)| \leq 1$. Thus $|\ell(h)| \leq M\|h\|$ with $M = 1/\delta$.

Conversely, if ℓ is bounded it is clearly continuous at the origin, hence continuous.

The significance of continuous linear functionals in terms of closed hyperplanes in $\mathcal{B}$ is a noteworthy geometric point to which we return later on. Now we take up analytic aspects of linear functionals.

The set of all continuous linear functionals over $\mathcal{B}$ is a vector space since we may add linear functionals and multiply them by scalars:

$$(\ell_1 + \ell_2)(f) = \ell_1(f) + \ell_2(f) \quad \text{and} \quad (\alpha\ell)(f) = \alpha\ell(f).$$

This vector space may be equipped with a norm as follows. The **norm** $\|\ell\|$ of a continuous linear functional ℓ is the infimum of all values M for which $|\ell(f)| \leq M\|f\|$ for all $f \in \mathcal{B}$. From this definition and the linearity of ℓ it is clear that

$$\|\ell\| = \sup_{\|f\| \leq 1} |\ell(f)| = \sup_{\|f\|=1} |\ell(f)| = \sup_{f \neq 0} \frac{|\ell(f)|}{\|f\|}.$$

The vector space of all continuous linear functionals on $\mathcal{B}$ equipped with $\|\cdot\|$ is called the **dual space** of $\mathcal{B}$, and is denoted by $\mathcal{B}^*$.

Theorem 3.2 *The vector space $\mathcal{B}^*$ is a Banach space.*

Proof. It is clear that $\|\cdot\|$ defines a norm, so we only check that $\mathcal{B}^*$ is complete. Suppose that $\{\ell_n\}$ is a Cauchy sequence in $\mathcal{B}^*$. Then, for each $f \in \mathcal{B}$, the sequence $\{\ell_n(f)\}$ is Cauchy, hence converges to a limit, which we denote by $\ell(f)$. Clearly, the mapping $\ell : f \mapsto \ell(f)$ is linear. If M is so that $\|\ell_n\| \leq M$ for all n, we see that

$$|\ell(f)| \leq |(\ell - \ell_n)(f)| + |\ell_n(f)| \leq |(\ell - \ell_n)(f)| + M\|f\|,$$

so that in the limit as $n \to \infty$, we find $|\ell(f)| \leq M\|f\|$ for all $f \in \mathcal{B}$. Thus ℓ is bounded. Finally, we must show that ℓ_n converges to ℓ in $\mathcal{B}^*$. Given $\epsilon > 0$ choose N so that $\|\ell_n - \ell_m\| < \epsilon/2$ for all $n, m > N$. Then, if $n > N$, we see that for all $m > N$ and any f

$$|(\ell - \ell_n)(f)| \leq |(\ell - \ell_m)(f)| + |(\ell_m - \ell_n)(f)| \leq |(\ell - \ell_m)(f)| + \frac{\epsilon}{2}\|f\|.$$

We can also choose m so large (and dependent on f) so that we also have $|(\ell - \ell_m)(f)| \leq \epsilon\|f\|/2$. In the end, we find that for $n > N$,

$$|(\ell - \ell_n)(f)| \leq \epsilon\|f\|.$$

This proves that $\|\ell - \ell_n\| \to 0$, as desired.

In general, given a Banach space $\mathcal{B}$, it is interesting and very useful to be able to describe its dual $\mathcal{B}^*$. This problem has an essentially complete answer in the case of the L^p spaces introduced before.

4 The dual space of L^p when $1 \le p < \infty$

Suppose that $1 \le p \le \infty$ and q is the conjugate exponent of p, that is, $1/p + 1/q = 1$. The key observation to make is the following: Hölder's inequality shows that every function $g \in L^q$ gives rise to a bounded linear functional on L^p by

$$\ell(f) = \int_X f(x)g(x)\,d\mu(x), \tag{5}$$

and that $\|\ell\| \le \|g\|_{L^q}$. Therefore, if we associate g to ℓ above, then we find that $L^q \subset (L^p)^*$ when $1 \le p \le \infty$. The main result in this section is to prove that when $1 \le p < \infty$, every linear functional on L^p is of the form (5) for some $g \in L^q$. This implies that $(L^p)^* = L^q$ whenever $1 \le p < \infty$. We remark that this result is in general not true when $p = \infty$; the dual of L^∞ contains L^1, but it is larger. (See the end of Section 5.3 below.)

Theorem 4.1 *Suppose $1 \le p < \infty$, and $1/p + 1/q = 1$. Then, with $\mathcal{B} = L^p$ we have*

$$\mathcal{B}^* = L^q,$$

in the following sense: For every bounded linear functional ℓ on L^p there is a unique $g \in L^q$ so that

$$\ell(f) = \int_X f(x)g(x)\,d\mu(x), \qquad \textit{for all } f \in L^p.$$

Moreover, $\|\ell\|_{\mathcal{B}^*} = \|g\|_{L^q}$.

This theorem justifies the terminology whereby q is usually called the dual exponent of p.

The proof of the theorem is based on two ideas. The first, as already seen, is Hölder's inequality; to which a converse is also needed. The second is the fact that a linear functional ℓ on L^p, $1 \le p < \infty$, leads naturally to a (signed) measure ν. Because of the continuity of ℓ the measure ν is absolutely continuous with respect to the underlying measure μ, and our desired function g is then the density function of ν in terms of μ.

We begin with:

Lemma 4.2 *Suppose $1 \le p, q \le \infty$, are conjugate exponents.*

(i) *If $g \in L^q$, then* $\|g\|_{L^q} = \sup_{\|f\|_{L^p} \le 1} \left| \int fg \right|$.

(ii) *Suppose g is integrable on all sets of finite measure, and*

$$\sup_{\substack{\|f\|_{L^p} \le 1 \\ f \text{ simple}}} \left| \int fg \right| = M < \infty.$$

Then $g \in L^q$, and $\|g\|_{L^q} = M$.

For the proof of the lemma, we recall the **signum** of a real number defined by

$$\text{sign}(x) = \begin{cases} 1 & \text{if } x > 0 \\ -1 & \text{if } x < 0 \\ 0 & \text{if } x = 0. \end{cases}$$

Proof. We start with (i). If $g = 0$, there is nothing to prove, so we may assume that g is not 0 a.e., and hence $\|g\|_{L^q} \neq 0$. By Hölder's inequality, we have that

$$\|g\|_{L^q} \ge \sup_{\|f\|_{L^p} \le 1} \left| \int fg \right|.$$

To prove the reverse inequality we consider several cases.

- First, if $q = 1$ and $p = \infty$, we may take $f(x) = \text{sign}\, g(x)$. Then, we have $\|f\|_{L^\infty} = 1$, and clearly, $\int fg = \|g\|_{L^1}$.

- If $1 < p, q < \infty$, then we set $f(x) = |g(x)|^{q-1} \text{sign}\, g(x) / \|g\|_{L^q}^{q-1}$. We observe that $\|f\|_{L^p}^p = \int |g(x)|^{p(q-1)}\, d\mu / \|g\|_{L^q}^{p(q-1)} = 1$ since $p(q-1) = q$, and that $\int fg = \|g\|_{L^q}$.

- Finally, if $q = \infty$ and $p = 1$, let $\epsilon > 0$, and E a set of finite positive measure, where $|g(x)| \ge \|g\|_{L^\infty} - \epsilon$. (Such a set exists by the definition of $\|g\|_{L^\infty}$ and the fact that the measure μ is σ-finite.) Then, if we take $f(x) = \chi_E(x)\, \text{sign}\, g(x) / \mu(E)$, where χ_E denotes the characteristic function of the set E, we see that $\|f\|_{L^1} = 1$, and also

$$\left| \int fg \right| = \frac{1}{\mu(E)} \int_E |g| \ge \|g\|_\infty - \epsilon.$$

This completes the proof of part (i).

To prove (ii) we recall[3] that we can find a sequence $\{g_n\}$ of simple functions so that $|g_n(x)| \le |g(x)|$ while $g_n(x) \to g(x)$ for each x. When $p > 1$ (so $q < \infty$), we take $f_n(x) = |g_n(x)|^{q-1} \operatorname{sign} g(x)/\|g_n\|_{L^q}^{q-1}$. As before, $\|f_n\|_{L^p} = 1$. However

$$\int f_n g = \frac{\int |g_n(x)|^q}{\|g_n\|_{L^q}^{q-1}} = \|g_n\|_{L^q},$$

and this does not exceed M. By Fatou's lemma it follows that $\int |g|^q \le M^q$, so $g \in L^q$ with $\|g\|_{L^q} \le M$. The direction $\|g\|_{L^q} \ge M$ is of course implied by Hölder's inequality.

When $p = 1$ the argument is parallel with the above but simpler. Here we take $f_n(x) = (\operatorname{sign} g(x))\chi_{E_n}(x)$, where E_n is an increasing sequence of sets of finite measure whose union is X. The details may be left to the reader.

With the lemma established we turn to the proof of the theorem. It is simpler to consider first the case when the underlying space has finite measure. In this case, with ℓ the given functional on L^p, we can then define a set function ν by

$$\nu(E) = \ell(\chi_E),$$

where E is any measurable set. This definition makes sense because χ_E is now automatically in L^p since the space has finite measure. We observe that

$$|\nu(E)| \le c(\mu(E))^{1/p}, \tag{6}$$

where c is the norm of the linear functional, taking into account the fact that $\|\chi_E\|_{L^p} = (\mu(E))^{1/p}$.

Now the linearity of ℓ clearly implies that ν is finitely-additive. Moreover, if $\{E_n\}$ is a countable collection of disjoint measurable sets, and we put $E = \bigcup_{n=1}^{\infty} E_n$, $E_N^* = \bigcup_{n=N+1}^{\infty} E_n$, then obviously

$$\chi_E = \chi_{E_N^*} + \sum_{n=1}^{N} \chi_{E_n}.$$

Thus $\nu(E) = \nu(E_N^*) + \sum_{n=1}^{N} \nu(E_n)$. However $\nu(E_N^*) \to 0$, as $N \to \infty$, because of (6) and the assumption $p < \infty$. This shows that ν is countably

[3]See for instance Section 2 in Chapter 6 of Book III.

additive and, moreover, (6) also shows us that ν is absolutely continuous with respect to μ.

We can now invoke the key result about absolutely continuous measures, the Lebesgue-Radon-Nykodim theorem. (See for example Theorem 4.3, Chapter 6 in Book III.) It guarantees the existence of an integrable function g so that $\nu(E) = \int_E g\, d\mu$ for every measurable set E. Thus we have $\ell(\chi_E) = \int \chi_E g\, d\mu$. The representation $\ell(f) = \int fg\, d\mu$ then extends immediately to simple functions f, and by a passage to the limit, to all $f \in L^p$ since the simple functions are dense in L^p, $1 \leq p < \infty$. (See Exercise 6.) Also by Lemma 4.2, we see that $\|g\|_{L^q} = \|\ell\|$.

To pass from the situation where the measure of X is finite to the general case, we use an increasing sequence $\{E_n\}$ of sets of finite measure that exhaust X, that is, $X = \bigcup_{n=1}^{\infty} E_n$. According to what we have just proved, for each n there is an integrable function g_n on E_n (which we can set to be zero in E_n^c) so that

$$\ell(f) = \int f g_n\, d\mu \tag{7}$$

whenever f is supported in E_n and $f \in L^p$. Moreover by conclusion (ii) of the lemma $\|g_n\|_{L^q} \leq \|\ell\|$.

Now it is easy to see because of (7) that $g_n = g_m$ a.e. on E_m, whenever $n \geq m$. Thus $\lim_{n\to\infty} g_n(x) = g(x)$ exists for almost every x, and by Fatou's lemma, $\|g\|_{L^q} \leq \|\ell\|$. As a result we have that $\ell(f) = \int fg\, d\mu$ for each $f \in L^p$ supported in E_n, and then by a simple limiting argument, for all $f \in L^p$. The fact that $\|\ell\| \leq \|g\|_{L^q}$, is already contained in Hölder's inequality, and therefore the proof of the theorem is complete.

5 More about linear functionals

First we turn to the study of certain geometric aspects of linear functionals in terms of the hyperplanes that they define. This will also involve understanding some elementary ideas about convexity.

5.1 Separation of convex sets

Although our ultimate focus will be on Banach spaces, we begin by considering an arbitrary vector space V over the reals. In this general setting we can define the following notions.

First, a **proper hyperplane** is a linear subspace of V that arises as the zero set of a (non-zero) linear functional on V. Alternatively, it is a linear subspace of V so that it, together with any vector not in V, spans V. Related to this notion is that of an **affine hyperplane** (which for brevity we will always refer to as a **hyperplane**) defined to be a translate of a proper hyperplane by a vector in V. To put it another way: H is a hyperplane if there is a non-zero linear functional ℓ, and a real number a, so that

$$H = \{v \in V : \quad \ell(v) = a\}.$$

Another relevant notion is that of a convex set. The subset $K \subset V$ is said to be **convex** if whenever v_0 and v_1 are both in K then the straight-line segment joining them

$$v(t) = (1-t)v_0 + tv_1, \qquad 0 \le t \le 1 \tag{8}$$

also lies entirely in K.

A key heuristic idea underlying our considerations can be enunciated as the following general principle:

> *If K is a convex set and $v_0 \notin K$, then K and v_0 can be separated by a hyperplane.*

This principle is illustrated in Figure 1.

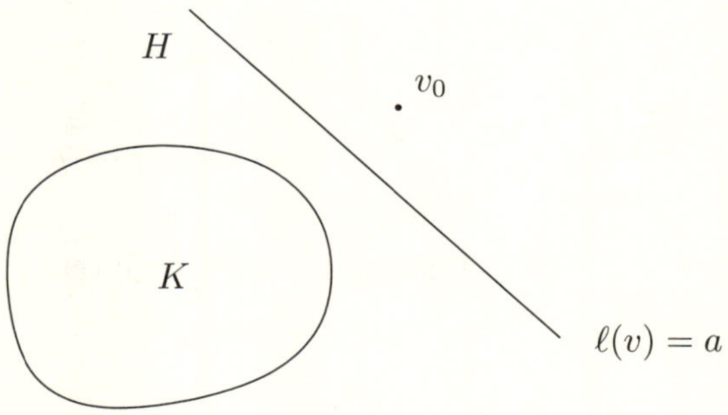

Figure 1. Separation of a convex set and a point by a hyperplane

The sense in which this is meant is that there is a non-zero linear functional ℓ and a real number a, so that

$$\ell(v_0) \ge a, \qquad \text{while} \qquad \ell(v) < a \text{ if } v \in K.$$

To give an idea of what is behind this principle we show why it holds in a nice special case. (See also Section 5.2.)

Proposition 5.1 *The assertion above is valid if $V = \mathbb{R}^d$ and K is convex and open.*

Proof. Since we may assume that K is non-empty, we can also suppose that (after a possible translation of K and v_0) we have $0 \in K$. The key construct used will be that of the Minkowski **gauge function** p associated to K, which measures (the inverse of) how far we need to go, starting from 0 in the direction of a vector v, to reach the exterior of K. The precise definition of p is as follows:

$$p(v) = \inf_{r>0}\{r : \quad v/r \in K\}.$$

Observe that since we have assumed that the origin is an interior point of K, for each $v \in \mathbb{R}^d$ there is an $r > 0$, so that $v/r \in K$. Hence $p(v)$ is well-defined.

Figure 2 below gives an example of a gauge function in the special case where $V = \mathbb{R}$ and $K = (a, b)$, an open interval that contains the origin.

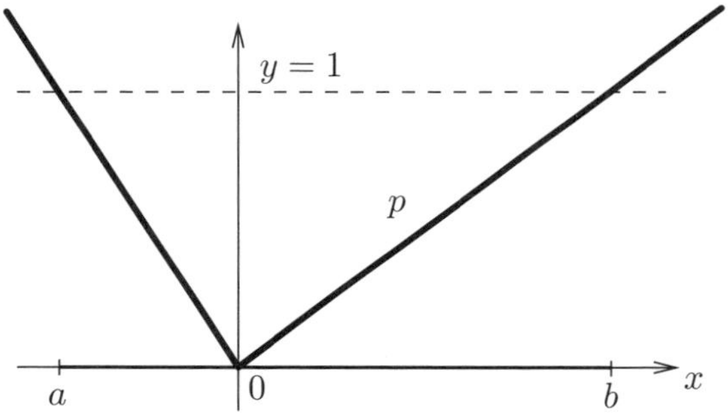

Figure 2. The gauge function of the interval (a, b) in $\mathbb{R}$

We note, for example, that if V is normed and K is the unit ball $\{\|v\| < 1\}$, then $p(v) = \|v\|$.

In general, the non-negative function p completely characterizes K in that

$$p(v) < 1 \quad \text{if and only if } v \in K. \tag{9}$$

Moreover p has an important sub-linear property:

$$\begin{cases} p(av) = ap(v), & \text{if } a \geq 0, \text{ and } v \in V. \\ p(v_1 + v_2) \leq p(v_1) + p(v_2), & \text{if } v_1 \text{ and } v_2 \in V. \end{cases} \tag{10}$$

In fact, if $v \in K$ then $v/(1-\epsilon) \in K$ for some $\epsilon > 0$, since K is open, which gives that $p(v) < 1$. Conversely if $p(v) < 1$, then $v = (1-\epsilon)v'$, for some $0 < \epsilon < 1$, and $v' \in K$. Then since $v = (1-\epsilon)v' + \epsilon \cdot 0$ this shows $v \in K$, because $0 \in K$ and K is convex.

To verify (10) we merely note that $(v_1 + v_2)/(r_1 + r_2)$ belongs to K, if both v_1/r_1 and v_2/r_2 belong to K, in view of property (8) defining the convexity of K with $t = r_2/(r_1 + r_2)$ and $1 - t = r_1/(r_1 + r_2)$.

Now our proposition will be proved once we find a linear functional ℓ, so that

$$\ell(v_0) = 1, \quad \text{and} \quad \ell(v) \leq p(v), \quad v \in \mathbb{R}^d. \tag{11}$$

This is because $\ell(v) < 1$, for all $v \in K$ by (9). We shall construct ℓ in a step-by-step manner.

First, such an ℓ is already determined in the one-dimensional subspace V_0 spanned by v_0, $V_0 = \{\mathbb{R}v_0\}$, since $\ell(bv_0) = b\ell(v_0) = b$, when $b \in \mathbb{R}$, and this is consistent with (11). Indeed, if $b \geq 0$ then $p(bv_0) = bp(v_0) \geq b\ell(v_0) = \ell(bv_0)$ by (10) and (9), while (11) is immediate when $b < 0$.

The next step is to choose any vector v_1 linearly independent from v_0 and extend ℓ to the subspace V_1 spanned by v_0 and v_1. Thus we can make a choice for the value of ℓ on v_1, $\ell(v_1)$, so as to satisfy (11) if

$$a\ell(v_1) + b = \ell(av_1 + bv_0) \leq p(av_1 + bv_0), \quad \text{for all } a, b \in \mathbb{R}.$$

Setting $a = 1$ and $bv_0 = w$ yields

$$\ell(v_1) + \ell(w) \leq p(v_1 + w) \quad \text{for all } w \in V_0,$$

while setting $a = -1$ implies

$$-\ell(v_1) + \ell(w') \leq p(-v_1 + w'), \quad \text{for all } w' \in V_0.$$

Altogether then it is required that for all w, $w' \in V_0$

$$-p(-v_1 + w') + \ell(w') \leq \ell(v_1) \leq p(v_1 + w) - \ell(w). \tag{12}$$

Notice that there is a number that lies between the two extremes of the above inequality. This is a consequence of the fact that $-p(-v_1 + w') + \ell(w')$ never exceeds $p(v_1 + w) - \ell(w)$, which itself follows from the fact that $\ell(w) + \ell(w') \leq p(w + w') \leq p(-v_1 + w') + p(v_1 + w)$, by (11) on V_0 and the sub-linearity of p. So a choice of $\ell(v_1)$ can be made that is consistent with (12) and this allows one to extend ℓ to V_1. In the same way we can proceed inductively to extend ℓ to all of $\mathbb{R}^d$.

The argument just given here in this special context will now be carried over in a general setting to give us an important theorem about constructing linear functionals.

5.2 The Hahn-Banach Theorem

We return to the general situation where we deal with an arbitrary vector space V over the reals. We assume that with V we are given a real-valued function p on V that satisfies the sub-linear property (10). However, as opposed to the example of the gauge function considered above, which by its nature is non-negative, here we do not assume that p has this property. In fact, certain p's which may take on negative values are needed in some of our applications later.

Theorem 5.2 *Suppose V_0 is a linear subspace of V, and that we are given a linear functional ℓ_0 on V_0 that satisfies*

$$\ell_0(v) \leq p(v), \qquad \text{for all } v \in V_0.$$

Then ℓ_0 can be extended to a linear functional ℓ on V that satisfies

$$\ell(v) \leq p(v), \qquad \text{for all } v \in V.$$

Proof. Suppose $V_0 \neq V$, and pick v_1 a vector not in V_0. We will first extend ℓ_0 to the subspace V_1 spanned by V_0 and v_1, as we did before. We can do this by defining a putative extension ℓ_1 of ℓ_0, defined on V_1 by $\ell_1(\alpha v_1 + w) = \alpha\ell_1(v_1) + \ell_0(w)$, whenever $w \in V_0$ and $\alpha \in \mathbb{R}$, if $\ell_1(v_1)$ is chosen so that

$$\ell_1(v) \leq p(v), \qquad \text{for all } v \in V_1.$$

However, exactly as above, this happens when

$$-p(-v_1 + w') + \ell_0(w') \leq \ell_1(v_1) \leq p(v_1 + w) - \ell_0(w)$$

for all w, $w' \in V_0$.

The right-hand side exceeds the left-hand side because of $\ell_0(w') + \ell_0(w) \leq p(w' + w)$ and the sub-linearity of p. Thus an appropriate choice of $\ell_1(v_1)$ is possible, giving the desired extension of ℓ_0 from V_0 to V_1.

We can think of the extension we have constructed as the key step in an inductive procedure. This induction, which in general is necessarily trans-finite, proceeds as follows. We well-order all vectors in V that do not belong to V_0, and denote this ordering by $<$. Among these vectors we call a vector v "extendable" if the linear functional ℓ_0 has an extension of the kind desired to the subspace spanned by V_0, v, and all vectors $< v$. What we want to prove is in effect that all vectors not in V_0 are extendable. Assume the contrary, then because of the well-ordering we can find the smallest v_1 that is not extendable. Now if V_0' is the space spanned by V_0 and all the vectors $< v_1$, then by assumption ℓ_0 extends to V_0'. The previous step, with V_0' in place of V_0 allows us then to extend ℓ_0 to the subspace spanned by V_0' and v_1, reaching a contradiction. This proves the theorem.

5.3 Some consequences

The Hahn-Banach theorem has several direct consequences for Banach spaces. Here $\mathcal{B}^*$ denotes the dual of the Banach space $\mathcal{B}$ as defined in Section 3.2, that is, the space of continuous linear functionals on $\mathcal{B}$.

Proposition 5.3 *Suppose f_0 is a given element of $\mathcal{B}$ with $\|f_0\| = M$. Then there exists a continuous linear functional ℓ on $\mathcal{B}$ so that $\ell(f_0) = M$ and $\|\ell\|_{\mathcal{B}^*} = 1$.*

Proof. Define ℓ_0 on the one-dimensional subspace $\{\alpha f_0\}_{\alpha \in \mathbb{R}}$ by $\ell_0(\alpha f_0) = \alpha M$, for each $\alpha \in \mathbb{R}$. Note that if we set $p(f) = \|f\|$ for every $f \in \mathcal{B}$, the function p satisfies the basic sub-linear property (10). We also observe that

$$|\ell_0(\alpha f_0)| = |\alpha| M = |\alpha| \|f_0\| = p(\alpha f_0),$$

so $\ell_0(f) \leq p(f)$ on this subspace. By the extension theorem ℓ_0 extends to an ℓ defined on $\mathcal{B}$ with $\ell(f) \leq p(f) = \|f\|$, for all $f \in \mathcal{B}$. Since this inequality also holds for $-f$ in place of f we get $|\ell(f)| \leq \|f\|$, and thus $\|\ell\|_{\mathcal{B}^*} \leq 1$. The fact that $\|\ell\|_{\mathcal{B}^*} \geq 1$ is implied by the defining property $\ell(f_0) = \|f_0\|$, thereby proving the proposition.

Another application is to the duality of linear transformations. Suppose $\mathcal{B}_1$ and $\mathcal{B}_2$ are a pair of Banach spaces, and T is a bounded linear transformation from $\mathcal{B}_1$ to $\mathcal{B}_2$. By this we mean that T maps $\mathcal{B}_1$ to $\mathcal{B}_2$; it satisfies $T(\alpha f_1 + \beta f_2) = \alpha T(f_1) + \beta T(f_2)$ whenever $f_1, f_2 \in \mathcal{B}$ and α and β are real numbers; and that it has a bound M so that $\|T(f)\|_{\mathcal{B}_2} \leq M\|f\|_{\mathcal{B}_1}$ for all $f \in \mathcal{B}_1$. The least M for which this inequality holds is called the **norm** of T and is denoted by $\|T\|$.

Often a linear transformation is initially given on a dense subspace. In this connection, the following proposition is very useful.

Proposition 5.4 *Let $\mathcal{B}_1$, $\mathcal{B}_2$ be a pair of Banach spaces and $\mathcal{S} \subset \mathcal{B}_1$ a dense linear subspace of $\mathcal{B}_1$. Suppose T_0 is a linear transformation from $\mathcal{S}$ to $\mathcal{B}_2$ that satisfies $\|T_0(f)\|_{\mathcal{B}_2} \leq M\|f\|_{\mathcal{B}_1}$ for all $f \in \mathcal{S}$. Then T_0 has a unique extension T to all of $\mathcal{B}_1$ so that $\|T(f)\|_{\mathcal{B}_2} \leq M\|f\|_{\mathcal{B}_1}$ for all $f \in \mathcal{B}_1$.*

Proof. If $f \in \mathcal{B}_1$, let $\{f_n\}$ be a sequence in $\mathcal{S}$ which converges to f. Then since $\|T_0(f_n) - T_0(f_m)\|_{\mathcal{B}_2} \leq M\|f_n - f_m\|_{\mathcal{B}_1}$ it follows that $\{T_0(f_n)\}$ is a Cauchy sequence in $\mathcal{B}_2$, and hence converges to a limit, which we define to be $T(f)$. Note that the definition of $T(f)$ is independent of the chosen sequence $\{f_n\}$, and that the resulting transformation T has all the required properties.

We now discuss duality of linear transformations. Whenever we have a linear transformation T from a Banach space $\mathcal{B}_1$ to another Banach space $\mathcal{B}_2$, it induces a **dual transformation**, T^* of $\mathcal{B}_2^*$ to $\mathcal{B}_1^*$, that can be defined as follows.

Suppose $\ell_2 \in \mathcal{B}_2^*$, (a continuous linear functional on $\mathcal{B}_2$), then $\ell_1 = T^*(\ell_2) \in \mathcal{B}_1^*$, is defined by $\ell_1(f_1) = \ell_2(T(f_1))$, whenever $f_1 \in \mathcal{B}_1$. More succinctly

$$T^*(\ell_2)(f_1) = \ell_2(T(f_1)). \tag{13}$$

Theorem 5.5 *The operator T^* defined by* (13) *is a bounded linear transformation from $\mathcal{B}_2^*$ to $\mathcal{B}_1^*$. Its norm $\|T^*\|$ satisfies $\|T\| = \|T^*\|$.*

Proof. First, if $\|f_1\|_{\mathcal{B}_1} \leq 1$, we have that

$$|\ell_1(f_1)| = |\ell_2(T(f_1))| \leq \|\ell_2\| \, \|T(f_1)\|_{\mathcal{B}_2} \leq \|\ell_2\| \, \|T\|.$$

Thus taking the supremum over all $f_1 \in \mathcal{B}_1$ with $\|f_1\|_{\mathcal{B}_1} \leq 1$, we see that the mapping $\ell_2 \mapsto T^*(\ell_2) = \ell_1$ has norm $\leq \|T\|$.

To prove the reverse inequality we can find for any $\epsilon > 0$ an $f_1 \in \mathcal{B}_1$ with $\|f_1\|_{\mathcal{B}_1} = 1$ and $\|T(f_1)\|_{\mathcal{B}_2} \geq \|T\| - \epsilon$. Next, with $f_2 = T(f_1) \in \mathcal{B}_2$, by Proposition 5.3 (with $\mathcal{B} = \mathcal{B}_2$) there is an ℓ_2 in $\mathcal{B}_2^*$ so that $\|\ell_2\|_{\mathcal{B}_2^*} = 1$ but $\ell_2(f_2) \geq \|T\| - \epsilon$. Thus by (13) one has $T^*(\ell_2)(f_1) \geq \|T\| - \epsilon$, and since $\|f_1\|_{\mathcal{B}_1} = 1$, we conclude $\|T^*(\ell_2)\|_{\mathcal{B}_1^*} \geq \|T\| - \epsilon$. This gives $\|T^*\| \geq \|T\| - \epsilon$ for any $\epsilon > 0$, which proves the theorem.

A further quick application of the Hahn-Banach theorem is the observation that in general L^1 is *not* the dual of L^∞ (as opposed to the case $1 \leq p < \infty$ considered in Theorem 4.1).

Let us first recall that whenever $g \in L^1$, the linear functional $f \mapsto \ell(f)$ given by

$$\ell(f) = \int fg\,d\mu \tag{14}$$

is bounded on L^∞, and its norm $\|\ell\|_{(L^\infty)^*}$ is $\|g\|_{L^1}$. In this way L^1 can be viewed as a subspace of $(L^\infty)^*$, with the L^1 norm of g being identical with its norm as a linear functional. One can, however, produce a continuous linear functional of L^∞ not of this form. For simplicity we do this when the underlying space is $\mathbb{R}$ with Lebesgue measure.

We let $\mathcal{C}$ denote the subspace of $L^\infty(\mathbb{R})$ consisting of continuous bounded functions on $\mathbb{R}$. Define the linear function ℓ_0 on $\mathcal{C}$ (the "Dirac delta") by

$$\ell_0(f) = f(0), \qquad f \in \mathcal{C}.$$

Clearly $|\ell_0(f)| \leq \|f\|_{L^\infty}$, $f \in \mathcal{C}$. Thus by the extension theorem, with $p(f) = \|f\|_{L^\infty}$, we see that there is a linear functional ℓ on L^∞, extending ℓ_0, that satisfies $|\ell(f)| \leq \|f\|_{L^\infty}$, for all $f \in L^\infty$.

Suppose for a moment that ℓ were of the form (14) for some $g \in L^1$. Since $\ell(f) = \ell_0(f) = 0$ whenever f is a continuous trapezoidal function that excludes the origin, we would have $\int fg\,dx = 0$ for such functions f; by a simple limiting argument this gives $\int_I g\,dx = 0$ for all intervals excluding the origin, and from there for all intervals I. Hence the indefinite integrals $G(y) = \int_0^y g(x)\,dx$ vanish, and therefore $G' = g = 0$ by the differentiation theorem.[4] This gives a contradiction, hence the linear functional ℓ is not representable as (14).

5.4 The problem of measure

We now consider an application of the Hahn-Banach theorem of a different kind. We present a rather stunning assertion, answering a basic question of the "problem of measure." The result states that there is a finitely-additive[5] measure defined on *all* subsets of $\mathbb{R}^d$ that agrees with Lebesgue measure on the measurable sets, and is translation invariant. We formulate the theorem in one dimension.

Theorem 5.6 *There is an extended-valued non-negative function $\hat{m}$, defined on all subsets of $\mathbb{R}$ with the following properties:*

(i) $\hat{m}(E_1 \cup E_2) = \hat{m}(E_1) + \hat{m}(E_2)$ *whenever E_1 and E_2 are disjoint subsets of $\mathbb{R}$.*

---

[4] See for instance Theorem 3.11, in Chapter 3 of Book III.

[5] The qualifier "finitely-additive" is crucial.

(ii) $\hat{m}(E) = m(E)$ *if* E *is a measurable set and* m *denotes the Lebesgue measure.*

(iii) $\hat{m}(E+h) = \hat{m}(E)$ *for every set* E *and real number* h.

From (i) we see that $\hat{m}$ is finitely additive; however it cannot be countably additive as the proof of the existence of non-measurable sets shows. (See Section 3, Chapter 1 in Book III.)

This theorem is a consequence of another result of this kind, dealing with an extension of the Lebesgue integral. Here the setting is the circle $\mathbb{R}/\mathbb{Z}$, instead of $\mathbb{R}$, with the former realized as $(0,1]$. Thus functions on $\mathbb{R}/\mathbb{Z}$ can be thought of as functions on $(0,1]$, extended to $\mathbb{R}$ by periodicity with period 1. In the same way, translations on $\mathbb{R}$ induce corresponding translations on $\mathbb{R}/\mathbb{Z}$. The assertion now is the existence of a generalized integral (the "Banach integral") defined on all bounded functions on the circle.

Theorem 5.7 *There is a linear functional* $f \mapsto I(f)$ *defined on all bounded functions* f *on* $\mathbb{R}/\mathbb{Z}$ *so that:*

(a) $I(f) \geq 0$, *if* $f(x) \geq 0$ *for all* x.

(b) $I(\alpha f_1 + \beta f_2) = \alpha I(f_1) + \beta I(f_2)$ *for all* α *and* β *real.*

(c) $I(f) = \int_0^1 f(x)\,dx$, *whenever* f *is measurable.*

(d) $I(f_h) = I(f)$, *for all* $h \in \mathbb{R}$ *where* $f_h(x) = f(x-h)$.

The right-hand side of (c) denotes the usual Lebesgue integral.

Proof. The idea is to consider the vector space V of all (real-valued) bounded functions on $\mathbb{R}/\mathbb{Z}$, with V_0 the subspace of those functions that are measurable. We let I_0 denote the linear functional given by the Lebesgue integral, $I_0(f) = \int_0^1 f(x)\,dx$ for $f \in V_0$. The key is to find the appropriate sub-linear p defined on V so that

$$I_0(f) \leq p(f), \qquad \text{for all } f \in V_0.$$

Banach's ingenious definition of p is as follows: We let $A = \{a_1, \ldots, a_N\}$ denote an arbitrary collection of N real numbers, with $\#(A) = N$ denoting its cardinality. Given A, we define $M_A(f)$ to be the real number

$$M_A(f) = \sup_{x \in \mathbb{R}} \left(\frac{1}{N} \sum_{j=1}^{N} f(x + a_j) \right),$$

and set

$$p(f) = \inf_A \{M_A(f)\},$$

where the infimum is taken over all finite collections A.

It is clear that $p(f)$ is well-defined, since f is assumed to be bounded; also $p(cf) = cp(f)$ if $c \geq 0$. To prove $p(f_1 + f_2) \leq p(f_1) + p(f_2)$, we find for each ϵ, finite collections A and B so that

$$M_A(f_1) \leq p(f_1) + \epsilon \quad \text{and} \quad M_B(f_2) \leq p(f_2) + \epsilon.$$

Let C be the collection $\{a_i + b_j\}_{1 \leq i \leq N_1,\ 1 \leq j \leq N_2}$ where $N_1 = \#(A)$, and $N_2 = \#(B)$. Now it is easy to see that

$$M_C(f_1 + f_2) \leq M_C(f_1) + M_C(f_2).$$

Next, we note as a general matter that $M_A(f)$ is the same as $M_{A'}(f')$ where $f' = f_h$ is a translate of f and $A' = A - h$. Also the averages corresponding to C arise as averages of translates of the averages corresponding to A and B, so it is easy to verify that

$$M_C(f_1) \leq M_A(f_1) \quad \text{and also} \quad M_C(f_2) \leq M_B(f_2).$$

Thus

$$p(f_1 + f_2) \leq M_C(f_1 + f_2) \leq M_A(f_1) + M_B(f_2) \leq p(f_1) + p(f_2) + 2\epsilon.$$

Letting $\epsilon \to 0$ proves the sub-linearity of p.

Next if f is Lebesgue measurable (and hence integrable since it is bounded), then for each A

$$I_0(f) = \frac{1}{N} \int_0^1 \left(\sum_{j=1}^{N} f(x + a_j) \right) dx \leq \int_0^1 M_A(f)\, dx = M_A(f),$$

and hence $I_0(f) \leq p(f)$. Let therefore I be the linear functional extending I_0 from V_0 to V, whose existence is guaranteed by Theorem 5.2. It is obvious from its definition that $p(f) \leq 0$ if $f \leq 0$. From this it follows that $I(f) \leq 0$ when $f \leq 0$, and replacing f by $-f$ we see that conclusion (a) holds.

Next we observe that for each real h

$$p(f - f_h) \leq 0. \tag{15}$$

In fact, for h fixed and N given, define the set A_N to be $\{h, 2h, 3h, \ldots, Nh\}$. Then the sum that enters in the definition of $M_{A_N}(f - f_h)$ is

$$\frac{1}{N}\sum_{j=1}^{N}(f(x+jh) - f(x+(j-1)h)),$$

and thus $|M_{A_N}(f - f_h)| \leq 2M/N$, where M is an upper bound for $|f|$. Since $p(f - f_h) \leq M_{A_N}(f - f_h) \to 0$, as $N \to \infty$, we see that (15) is proved. This shows that $I(f - f_h) \leq 0$, for all f and h. However, replacing f by f_h and then h by $-h$, we see that $I(f_h - f) \leq 0$ and thus (d) is also established, finishing the proof of Theorem 5.7.

As a direct consequence we have the following.

Corollary 5.8 *There is a non-negative function $\hat{m}$ defined on all subsets of $\mathbb{R}/\mathbb{Z}$ so that:*

(i) $\hat{m}(E_1 \cup E_2) = \hat{m}(E_1) + \hat{m}(E_2)$ *for all disjoint subsets E_1 and E_2.*

(ii) $\hat{m}(E) = m(E)$ *if E is measurable.*

(iii) $\hat{m}(E + h) = \hat{m}(E)$ *for every h in $\mathbb{R}$.*

We need only take $\hat{m}(E) = I(\chi_E)$, with I as in Theorem 5.7, where χ_E denotes the characteristic function of E.

We now turn to the proof of Theorem 5.6. Let $\mathcal{I}_j$ denote the interval $(j, j+1]$, where $j \in \mathbb{Z}$. Then we have a partition $\bigcup_{j=-\infty}^{\infty} \mathcal{I}_j$ of $\mathbb{R}$ into disjoint sets.

For clarity of exposition, we temporarily relabel the measure $\hat{m}$ on $(0,1] = \mathcal{I}_0$ given by the corollary and call it $\hat{m}_0$. So whenever $E \subset \mathcal{I}_0$ we defined $\hat{m}(E)$ to be $\hat{m}_0(E)$. More generally, if $E \subset \mathcal{I}_j$ we set $\hat{m}(E) = \hat{m}_0(E - j)$.

With these things said, for any set E define $\hat{m}(E)$ by

$$\hat{m}(E) = \sum_{j=-\infty}^{\infty} \hat{m}(E \cap \mathcal{I}_j) = \sum_{j=-\infty}^{\infty} \hat{m}_0((E \cap \mathcal{I}_j) - j). \tag{16}$$

Thus $\hat{m}(E)$ is given as an extended non-negative number. Note that if E_1 and E_2 are disjoint so are $(E_1 \cap \mathcal{I}_j) - j$ and $(E_2 \cap \mathcal{I}_j) - j$. It follows that $\hat{m}(E_1 \cup E_2) = \hat{m}(E_1) + \hat{m}(E_2)$. Moreover if E is measurable then $\hat{m}(E \cap \mathcal{I}_j) = m(E \cap \mathcal{I}_j)$ and so $\hat{m}(E) = m(E)$.

To prove $\hat{m}(E + h) = \hat{m}(E)$, consider first the case $h = k \in \mathbb{Z}$. This is an immediate consequence of the definition (16) once one observes that $((E + k) \cap \mathcal{I}_{j+k}) - (j + k) = (E \cap \mathcal{I}_j) - j$, for all j, $k \in \mathbb{Z}$.

Next suppose $0 < h < 1$. We then decompose $E \cap \mathcal{I}_j$ as $E'_j \cup E''_j$, with $E'_j = E \cap (j, j+1-h]$ and $E''_j = E \cap (j+1-h, j+1]$. The point of this decomposition is that $E'_j + h$ remains in $\mathcal{I}_j$ but $E''_j + h$ is placed in $\mathcal{I}_{j+1}$. In any case, $E = \bigcup_j E'_j \cup \bigcup_j E''_j$, and the union is disjoint.

Thus using the first additivity property proved above and then (16) we see that

$$\hat{m}(E) = \sum_{j=-\infty}^{\infty} \left(\hat{m}(E'_j) + \hat{m}(E''_j)\right).$$

Similarly

$$\hat{m}(E+h) = \sum_{j=-\infty}^{\infty} \left(\hat{m}(E'_j + h) + \hat{m}(E''_j + h)\right).$$

Now both E'_j and $E'_j + h$ are in $\mathcal{I}_j$, hence $\hat{m}(E'_j) = \hat{m}(E'_j + h)$ by the translation invariance of $\hat{m}_0$ and the definition of $\hat{m}$ on subsets of $\mathcal{I}_j$. Also E''_j is in $\mathcal{I}_j$ and $E''_j + h$ is in $\mathcal{I}_{j+1}$, and their measures agree for the same reasons. This establishes that $\hat{m}(E) = \hat{m}(E+h)$, for $0 < h < 1$. Now combining this with the translation invariance with respect to $\mathbb{Z}$ already proved, we obtain conclusion (iii) of Theorem 5.6 for all h, and hence the theorem is completely proved.

For the corresponding extension of Lebesgue measure in $\mathbb{R}^d$ and other related results, see Exercise 36 and Problems 8* and 9*.

6 Complex L^p and Banach spaces

We have supposed in Section 3.2 onwards that our L^p and Banach spaces are taken over the reals. However, the statements and the proofs of the corresponding theorems for those spaces taken with respect to the complex scalars are for the most part routine adaptations of the real case. There are nevertheless several instances that require further comment. First, in the argument concerning the converse of Hölder's inequality (Lemma 4.2), the definition of f should read

$$f(x) = |g(x)|^{q-1} \frac{\overline{\operatorname{sign} g(x)}}{\|g\|_{L^q}^{q-1}},$$

where now "sign" denotes the complex version of the signum function, defined by $\operatorname{sign} z = z/|z|$ if $z \neq 0$, and $\operatorname{sign} 0 = 0$. There are similar occurrences with g replaced by g_n.

Second, while the Hahn-Banach theorem is valid as stated only for real vector spaces, a version of the complex case (sufficient for the applications in Section 5.3 where $p(f) = \|f\|$) can be found in Exercise 33 below.

7 Appendix: The dual of $C(X)$

In this appendix, we describe the bounded linear functionals of the space $C(X)$ of continuous real-valued functions on X. To begin with, we assume that X is a compact metric space. Our main result then states that if $\ell \in C(X)^*$, then there exists a finite signed Borel measure μ (this measure is sometimes referred to as a Radon measure) so that

$$\ell(f) = \int_X f(x)\, d\mu(x) \qquad \text{for all } f \in C(X).$$

Before proceeding with the argument leading to this result, we collect some basic facts and definitions.

Let X be a metric space with metric d, and assume that X is compact; that is, every covering of X by open sets contains a finite sub-covering. The vector space $C(X)$ of real-valued continuous functions on X equipped with the sup-norm

$$\|f\| = \sup_{x\in X} |f(x)|, \qquad f \in C(X)$$

is a Banach space over $\mathbb{R}$. Given a continuous function f on X we define the **support** of f, denoted $\text{supp}(f)$, as the closure of the set $\{x \in X : f(x) \neq 0\}$.[6]

We recall some simple facts about continuous functions and open and closed sets in X that we shall use below.

(i) **Separation.** If A and B are two disjoint closed subsets of X, then there exists a continuous function f with $f = 1$ on A, $f = 0$ on B, and $0 < f < 1$ in the complements of A and B.

Indeed, one can take for instance

$$f(x) = \frac{d(x, B)}{d(x, A) + d(x, B)},$$

where $d(x, B) = \inf_{y\in B} d(x, y)$, with a similar definition for $d(x, A)$.

(ii) **Partition of unity.** If K is a compact set which is covered by finitely many open sets $\{\mathcal{O}_k\}_{k=1}^N$, then there exist continuous functions η_k for $1 \leq k \leq N$ so that $0 \leq \eta_k \leq 1$, $\text{supp}(\eta_k) \subset \mathcal{O}_k$, and $\sum_{k=1}^N \eta_k(x) = 1$ whenever $x \in K$. Moreover, $0 \leq \sum_{k=1}^N \eta_k(x) \leq 1$ for all $x \in X$.

One can argue as follows. For each $x \in K$, there exists a ball $B(x)$ centered at x and of positive radius such that $\overline{B(x)} \subset \mathcal{O}_i$ for some i. Since $\bigcup_{x\in K} B(x)$ covers K, we can select a finite subcovering, say $\bigcup_{j=1}^M B(x_j)$. For each $1 \leq k \leq N$, let U_k be the union of all open balls $B(x_j)$ so that $\overline{B(x_j)} \subset \mathcal{O}_k$; clearly $K \subset \bigcup_{k=1}^N U_k$. By (i) above, there exists a continuous function $0 \leq \varphi_k \leq 1$ so that $\varphi_k = 1$ on $\overline{U_k}$ and $\text{supp}(\varphi_k) \subset \mathcal{O}_k$. If we define

$$\eta_1 = \varphi_1, \quad \eta_2 = \varphi_2(1 - \varphi_1), \quad \ldots, \quad \eta_N = \varphi_N(1 - \varphi_1)\cdots(1 - \varphi_{N-1})$$

[6]This is the common usage of the terminology "support." In Book III, Chapter 2, we used "support of f" to indicate the set where $f(x) \neq 0$, which is convenient when dealing with measurable functions.

then $\text{supp}(\eta_k) \subset \mathcal{O}_k$ and

$$\eta_1 + \cdots + \eta_N = 1 - (1 - \varphi_1) \cdots (1 - \varphi_N),$$

thus guaranteeing the desired properties.

Recall[7] that the **Borel σ-algebra** of X, which is denoted by $\mathcal{B}_X$, is the smallest σ-algebra of X that contains the open sets. Elements of $\mathcal{B}_X$ are called **Borel sets**, and a measure defined on $\mathcal{B}_X$ is called a **Borel measure**. If a Borel measure is finite, that is $\mu(X) < \infty$, then it satisfies the following "regularity property": for any Borel set E and any $\epsilon > 0$, there are an open set $\mathcal{O}$ and a closed set F such that $E \subset \mathcal{O}$ and $\mu(\mathcal{O} - E) < \epsilon$, while $F \subset E$ and $\mu(E - F) < \epsilon$.

In general we shall be interested in finite *signed* Borel measures on X, that is, measures which can take on negative values. If μ is such a measure, and μ^+ and μ^- denote the positive and negative variations of μ, then $\mu = \mu^+ - \mu^-$, and integration with respect to μ is defined by $\int f\, d\mu = \int f\, d\mu^+ - \int f\, d\mu^-$. Conversely, if μ_1 and μ_2 are two finite Borel measures, then $\mu = \mu_1 - \mu_2$ is a finite signed Borel measure, and $\int f\, d\mu = \int f\, d\mu_1 - \int f\, d\mu_2$.

We denote by $M(X)$ the space of finite signed Borel measures on X. Clearly, $M(X)$ is a vector space which can be equipped with the following norm

$$\|\mu\| = |\mu|(X),$$

where $|\mu|$ denotes the total variation of μ. It is a simple fact that $M(X)$ with this norm is a Banach space.

7.1 The case of positive linear functionals

We begin by considering only linear functionals $\ell : C(X) \to \mathbb{R}$ which are **positive**, that is, $\ell(f) \geq 0$ whenever $f(x) \geq 0$ for all $x \in X$. Observe that positive linear functionals are automatically bounded and that $\|\ell\| = \ell(1)$. Indeed, note that $|f(x)| \leq \|f\|$, hence $\|f\| \pm f \geq 0$, and therefore $|\ell(f)| \leq \ell(1)\|f\|$.

Our main result goes as follows.

Theorem 7.1 *Suppose X is a compact metric space and ℓ a positive linear functional on $C(X)$. Then there exists a unique finite (positive) Borel measure μ so that*

$$\ell(f) = \int_X f(x)\, d\mu(x) \qquad \text{for all } f \in C(X). \tag{17}$$

Proof. The existence of the measure μ is proved as follows. Consider the function ρ on the open subsets of X defined by

$$\rho(\mathcal{O}) = \sup \{\ell(f), \text{ where } \text{supp}(f) \subset \mathcal{O}, \text{ and } 0 \leq f \leq 1\},$$

[7] The definitions and results on measure theory needed in this section, in particular the extension of a premeasure used in the proof of Theorem 7.1, can be found in Chapter 6 of Book III.

and let the function μ_* be defined on all subsets of X by

$$\mu_*(E) = \inf\{\rho(\mathcal{O}), \quad \text{where } E \subset \mathcal{O} \text{ and } \mathcal{O} \text{ is open}\}.$$

We contend that μ_* is a metric exterior measure on X.

Indeed, we clearly must have $\mu_*(E_1) \leq \mu_*(E_2)$ whenever $E_1 \subset E_2$. Also, if $\mathcal{O}$ is open, then $\mu_*(\mathcal{O}) = \rho(\mathcal{O})$. To show that μ_* is countably sub-additive on subsets of X, we begin by proving that μ_* is in fact sub-additive on open sets $\{\mathcal{O}_k\}$, that is,

$$\mu_*\left(\bigcup_{k=1}^{\infty} \mathcal{O}_k\right) \leq \sum_{k=1}^{\infty} \mu_*(\mathcal{O}_k). \tag{18}$$

To do so, suppose $\{\mathcal{O}_k\}_{k=1}^{\infty}$ is a collection of open sets in X, and let $\mathcal{O} = \bigcup_{k=1}^{\infty} \mathcal{O}_k$. If f is any continuous function that satisfies $\operatorname{supp}(f) \subset \mathcal{O}$ and $0 \leq f \leq 1$, then by compactness of $K = \operatorname{supp}(f)$ we can pick a sub-cover so that (after relabeling the sets $\mathcal{O}_k$, if necessary) $K \subset \bigcup_{k=1}^{N} \mathcal{O}_k$. Let $\{\eta_k\}_{k=1}^{N}$ be a partition of unity of $\{\mathcal{O}_1, \ldots, \mathcal{O}_N\}$ (as discussed above in (ii)); this means that each η_k is continuous with $0 \leq \eta_k \leq 1$, $\operatorname{supp}(\eta_k) \subset \mathcal{O}_k$ and $\sum_{k=1}^{N} \eta_k(x) = 1$ for all $x \in K$. Hence recalling that $\mu_* = \rho$ on open sets, we get

$$\ell(f) = \sum_{k=1}^{N} \ell(f\eta_k) \leq \sum_{k=1}^{N} \mu_*(\mathcal{O}_k) \leq \sum_{k=1}^{\infty} \mu_*(\mathcal{O}_k),$$

where the first inequality follows because $\operatorname{supp}(f\eta_k) \subset \mathcal{O}_k$ and $0 \leq f\eta_k \leq 1$. Taking the supremum over f we find that $\mu_*\left(\bigcup_{k=1}^{\infty} \mathcal{O}_k\right) \leq \sum_{k=1}^{\infty} \mu_*(\mathcal{O}_k)$.

We now turn to the proof of the sub-additivity of μ_* on all sets. Suppose $\{E_k\}$ is a collection of subsets of X and let $\epsilon > 0$. For each k, pick an open set $\mathcal{O}_k$ so that $E_k \subset \mathcal{O}_k$ and $\mu_*(\mathcal{O}_k) \leq \mu_*(E_k) + \epsilon 2^{-k}$. Since $\mathcal{O} = \bigcup \mathcal{O}_k$ covers $\bigcup E_k$, we must have by (18) that

$$\mu_*(\bigcup E_k) \leq \mu_*(\mathcal{O}) \leq \sum_k \mu_*(\mathcal{O}_k) \leq \sum_k \mu_*(E_k) + \epsilon,$$

and consequently $\mu_*(\bigcup E_k) \leq \sum_k \mu_*(E_k)$ as desired.

The last property we must verify is that μ_* is metric, in the sense that if $d(E_1, E_2) > 0$, then $\mu_*(E_1 \cup E_2) = \mu_*(E_1) + \mu_*(E_2)$. Indeed, the separation condition implies that there exist disjoint open sets $\mathcal{O}_1$ and $\mathcal{O}_2$ so that $E_1 \subset \mathcal{O}_1$ and $E_2 \subset \mathcal{O}_2$. Therefore, if $\mathcal{O}$ is any open subset which contains $E_1 \cup E_2$, then $\mathcal{O} \supset (\mathcal{O} \cap \mathcal{O}_1) \cup (\mathcal{O} \cap \mathcal{O}_2)$, where this union is disjoint. Hence the additivity of μ_* on disjoint open sets, and its monotonicity give

$$\mu_*(\mathcal{O}) \geq \mu_*(\mathcal{O} \cap \mathcal{O}_1) + \mu_*(\mathcal{O} \cap \mathcal{O}_2) \geq \mu_*(E_1) + \mu_*(E_2),$$

since $E_1 \subset (\mathcal{O} \cap \mathcal{O}_1)$ and $E_2 \subset (\mathcal{O} \cap \mathcal{O}_2)$. So $\mu_*(E_1 \cup E_2) \geq \mu_*(E_1) + \mu_*(E_2)$, and since the reverse inequality has already been shown above, this concludes the proof that μ_* is a metric exterior measure.

By Theorems 1.1 and 1.2 in Chapter 6 of Book III, there exists a Borel measure μ on $\mathcal{B}_X$ which extends μ_*. Clearly, μ is finite with $\mu(X) = \ell(1)$.

We now prove that this measure satisfies (17). Let $f \in C(X)$. Since f can be written as the difference of two continuous non-negative functions, we can assume after rescaling, that $0 \leq f(x) \leq 1$ for all $x \in X$. The idea now is to slice f, that is, write $f = \sum f_n$ where each f_n is continuous and relatively small in the sup-norm. More precisely, let N be a fixed positive integer, define $\mathcal{O}_0 = X$, and for every integer $n \geq 1$, let

$$\mathcal{O}_n = \{x \in X : f(x) > (n-1)/N\}.$$

Thus $\mathcal{O}_n \supset \mathcal{O}_{n+1}$ and $\mathcal{O}_{N+1} = \emptyset$. Now if we define

$$f_n(x) = \begin{cases} 1/N & \text{if } x \in \mathcal{O}_{n+1}, \\ f(x) - (n-1)/N & \text{if } x \in \mathcal{O}_n - \mathcal{O}_{n+1}, \\ 0 & \text{if } x \in \mathcal{O}_n^c, \end{cases}$$

then the functions f_n are continuous and they "pile up" to yield f, that is, $f = \sum_{n=1}^{N} f_n$. Since $Nf_n = 1$ on $\mathcal{O}_{n+1}$, $\operatorname{supp}(Nf_n) \subset \overline{\mathcal{O}_n} \subset \mathcal{O}_{n-1}$, and also $0 \leq Nf_n \leq 1$ we have $\mu(\mathcal{O}_{n+1}) \leq \ell(Nf_n) \leq \mu(\mathcal{O}_{n-1})$, and therefore by linearity

$$\frac{1}{N}\sum_{n=1}^{N} \mu(\mathcal{O}_{n+1}) \leq \ell(f) \leq \frac{1}{N}\sum_{n=1}^{N} \mu(\mathcal{O}_{n-1}). \tag{19}$$

The properties of Nf_n also imply $\mu(\mathcal{O}_{n+1}) \leq \int Nf_n \, d\mu \leq \mu(\mathcal{O}_n)$, hence

$$\frac{1}{N}\sum_{n=1}^{N} \mu(\mathcal{O}_{n+1}) \leq \int f \, d\mu \leq \frac{1}{N}\sum_{n=1}^{N} \mu(\mathcal{O}_n). \tag{20}$$

Consequently, combining the inequalities (19) and (20) yields

$$\left| \ell(f) - \int f \, d\mu \right| \leq \frac{2\mu(X)}{N}.$$

In the limit as $N \to \infty$ we conclude that $\ell(f) = \int f \, d\mu$ as desired.

Finally, we prove uniqueness. Suppose μ' is another finite positive Borel measure on X that satisfies $\ell(f) = \int f \, d\mu'$ for all $f \in C(X)$. If $\mathcal{O}$ is an open set, and $0 \leq f \leq 1$ with $\operatorname{supp}(f) \subset \mathcal{O}$, then

$$\ell(f) = \int f \, d\mu' = \int_{\mathcal{O}} f \, d\mu' \leq \int_{\mathcal{O}} 1 \, d\mu' = \mu'(\mathcal{O}).$$

Taking the supremum over f and recalling the definition of μ yields $\mu(\mathcal{O}) \leq \mu'(\mathcal{O})$. For the reverse inequality, recall the inner regularity condition satisfied by a finite Borel measure: given $\epsilon > 0$, there exists a closed set K so that $K \subset \mathcal{O}$, and $\mu'(\mathcal{O} - K) < \epsilon$. By the separation property (i) noted above applied to K and $\mathcal{O}^c$, we can pick a continuous function f so that $0 \le f \le 1$, $\mathrm{supp}(f) \subset \mathcal{O}$ and $f = 1$ on K. Then

$$\mu'(\mathcal{O}) \le \mu'(K) + \epsilon \le \int_K f \, d\mu' + \epsilon \le \ell(f) + \epsilon \le \mu(\mathcal{O}) + \epsilon.$$

Since ϵ was arbitrary, we obtain the desired inequality, and therefore $\mu(\mathcal{O}) = \mu'(\mathcal{O})$ for all open sets $\mathcal{O}$. This implies that $\mu = \mu'$ on all Borel sets, and the proof of the theorem is complete.

7.2 The main result

The main point is to write an arbitrary bounded linear functional on $C(X)$ as the difference of two positive linear functionals.

Proposition 7.2 *Suppose X is a compact metric space and let ℓ be a bounded linear functional on $C(X)$. Then there exist positive linear functionals ℓ^+ and ℓ^- so that $\ell = \ell^+ - \ell^-$. Moreover, $\|\ell\| = \ell^+(1) + \ell^-(1)$.*

Proof. For $f \in C(X)$ with $f \ge 0$, we define

$$\ell^+(f) = \sup\{\ell(\varphi) : 0 \le \varphi \le f\}.$$

Clearly, we have $0 \le \ell^+(f) \le \|\ell\|\|f\|$ and $\ell(f) \le \ell^+(f)$. If $\alpha \ge 0$ and $f \ge 0$, then $\ell^+(\alpha f) = \alpha\ell^+(f)$. Now suppose that $f, g \ge 0$. On the one hand we have $\ell^+(f) + \ell^+(g) \le \ell^+(f+g)$, because if $0 \le \varphi \le f$ and $0 \le \psi \le g$, then $0 \le \varphi + \psi \le f + g$. On the other hand, suppose $0 \le \varphi \le f + g$, and let $\varphi_1 = \min(\varphi, f)$ and $\varphi_2 = \varphi - \varphi_1$. Then $0 \le \varphi_1 \le f$ and $0 \le \varphi_2 \le g$, and $\ell(\varphi) = \ell(\varphi_1) + \ell(\varphi_2) \le \ell^+(f) + \ell^+(g)$. Taking the supremum over φ, we get $\ell^+(f+g) \le \ell^+(f) + \ell^+(g)$. We conclude from the above that $\ell^+(f+g) = \ell^+(f) + \ell^+(g)$ whenever $f, g \ge 0$.

We can now extend ℓ^+ to a positive linear functional on $C(X)$ as follows. Given an arbitrary function f in $C(X)$ we can write $f = f^+ - f^-$, where $f^+, f^- \ge 0$, and define ℓ^+ on f by $\ell^+(f) = \ell^+(f^+) - \ell^+(f^-)$. Using the linearity of ℓ^+ on non-negative functions, one checks easily that the definition of $\ell^+(f)$ is independent of the decomposition of f into the difference of two non-negative functions. From the definition we see that ℓ^+ is positive, and it is easy to check that ℓ^+ is linear on $C(X)$, and that $\|\ell^+\| \le \|\ell\|$.

Finally, we define $\ell^- = \ell^+ - \ell$, and see immediately that ℓ^- is also a positive linear functional on $C(X)$.

Now since ℓ^+ and ℓ^- are positive, we have $\|\ell^+\| = \ell^+(1)$ and $\|\ell^-\| = \ell^-(1)$, therefore $\|\ell\| \le \ell^+(1) + \ell^-(1)$. For the reverse inequality, suppose $0 \le \varphi \le 1$. Then $|2\varphi - 1| \le 1$, hence $\|\ell\| \ge \ell(2\varphi - 1)$. By linearity of ℓ, and taking the supremum over φ we obtain $\|\ell\| \ge 2\ell^+(1) - \ell(1)$. Since $\ell(1) = \ell^+(1) - \ell^-(1)$ we get $\|\ell\| \ge \ell^+(1) + \ell^-(1)$, and the proof is complete.

We are now ready to state and prove the main result.

Theorem 7.3 *Let X be a compact metric space and $C(X)$ the Banach space of continuous real-valued functions on X. Then, given any bounded linear functional ℓ*

on $C(X)$, there exists a unique finite signed Borel measure μ on X so that

$$\ell(f) = \int_X f(x)\, d\mu(x) \qquad \textit{for all } f \in C(X).$$

Moreover, $\|\ell\| = \|\mu\| = |\mu|(X)$. In other words $C(X)^*$ is isometric to $M(X)$.

Proof. By the proposition, there exist two positive linear functionals ℓ^+ and ℓ^- so that $\ell = \ell^+ - \ell^-$. Applying Theorem 7.1 to each of these positive linear functionals yields two finite Borel measures μ_1 and μ_2. If we define $\mu = \mu_1 - \mu_2$, then μ is a finite signed Borel measure and $\ell(f) = \int f\, d\mu$.

Now we have

$$|\ell(f)| \leq \int |f|\, d|\mu| \leq \|f\|\, |\mu|(X),$$

and thus $\|\ell\| \leq |\mu|(X)$. Since we also have $|\mu|(X) \leq \mu_1(X) + \mu_2(X) = \ell^+(1) + \ell^-(1) = \|\ell\|$, we conclude that $\|\ell\| = |\mu|(X)$ as desired.

To prove uniqueness, suppose $\int f\, d\mu = \int f\, d\mu'$ for some finite signed Borel measures μ and μ', and all $f \in C(X)$. Then if $\nu = \mu - \mu'$, one has $\int f d\nu = 0$, and consequently, if ν^+ and ν^- are the positive and negative variations of f, one finds that the two positive linear functionals defined on $C(X)$ by $\ell^+(f) = \int f\, d\nu^+$ and $\ell^-(f) = \int f\, d\nu^-$ are identical. By the uniqueness in Theorem 7.1, we conclude that $\nu^+ = \nu^-$, hence $\nu = 0$ and $\mu = \mu'$, as desired.

7.3 An extension

Because of its later application, it is useful to observe that Theorem 7.1 has an extension when we drop the assumption that the space X is compact. Here we define the space $C_b(X)$ of continuous bounded functions f on X, with norm $\|f\| = \sup_{x\in X} |f(x)|$.

Theorem 7.4 *Suppose X is a metric space and ℓ a positive linear functional on $C_b(X)$. For simplicity assume that ℓ is normalized so that $\ell(1) = 1$. Assume also that for each $\epsilon > 0$, there is a compact set $K_\epsilon \subset X$ so that*

$$|\ell(f)| \leq \sup_{x\in K_\epsilon} |f(x)| + \epsilon\|f\|, \qquad \textit{for all } f \in C_b(X). \tag{21}$$

Then there exists a unique finite (positive) Borel measure μ so that

$$\ell(f) = \int_X f(x)\, d\mu(x), \qquad \textit{for all } f \in C_b(X).$$

The extra hypothesis (21) (which is vacuous when X is compact) is a "tightness" assumption that will be relevant in Chapter 6. Note that as before $|\ell(f)| \leq \|f\|$ since $\ell(1) = 1$, even without the assumption (21).

The proof of this theorem proceeds as that of Theorem 7.1, save for one key aspect. First we define

$$\rho(\mathcal{O}) = \sup \{\ell(f), \text{ where } f \in C_b(X), \operatorname{supp}(f) \subset \mathcal{O}, \text{ and } 0 \leq f \leq 1\}.$$

The change that is required is in the proof of the countable sub-additivity of ρ, in that the support of f's (in the definition of $\rho(\mathcal{O})$) are now not necessarily compact. In fact, suppose $\mathcal{O} = \bigcup_{k=1}^{\infty} \mathcal{O}_k$ is a countable union of open sets. Let C be the support of f, and given a fixed $\epsilon > 0$, set $K = C \cap K_\epsilon$, with K_ϵ the compact set arising in (21). Then K is compact and $\bigcup_{k=1}^{\infty} \mathcal{O}_k$ covers K. Proceeding as before, we obtain a partition of unity $\{\eta_k\}_{k=1}^N$, with η_k supported in $\mathcal{O}_k$ and $\sum_{k=1}^N \eta_k(x) = 1$, for $x \in K$. Now $f - \sum_{k=1}^N f\eta_k$ vanishes on K_ϵ. Thus by (21)

$$\left| \ell(f) - \sum_{k=1}^{N} \ell(f\eta_k) \right| \leq \epsilon,$$

and hence

$$\ell(f) \leq \sum_{k=1}^{\infty} \rho(\mathcal{O}_k) + \epsilon.$$

Since this holds for each ϵ, we obtain the required sub-additivity of ρ and thus of μ_*. The proof of the theorem can then be concluded as before.

Theorem 7.4 did not require that the metric space X be either complete or separable. However if we make these two further assumptions on X, then the condition (21) is actually necessary.

Indeed, suppose $\ell(f) = \int_X f\, d\mu$, where μ is a positive finite Borel measure on X, which we may assume is normalized, $\mu(X) = 1$. Under the assumption that X is complete and separable, then for each fixed $\epsilon > 0$ there is a compact set K_ϵ so that $\mu(K_\epsilon^c) < \epsilon$. Indeed, let $\{c_k\}$ be a dense sequence in X. Since for each m the collection of balls $\{B_{1/m}(c_k)\}_{k=1}^{\infty}$ covers X, there is a finite N_m so that if $\mathcal{O}_m = \bigcup_{k=1}^{N_m} B_{1/m}(c_k)$, then $\mu(\mathcal{O}_m) \geq 1 - \epsilon/2^m$.

Take $K_\epsilon = \bigcap_{m=1}^{\infty} \overline{\mathcal{O}_m}$. Then $\mu(K_\epsilon) \geq 1 - \epsilon$; also, K_ϵ is closed and totally bounded, in the sense that for every $\delta > 0$, the set K_ϵ can be covered by finitely many balls of radius δ. Since X is complete, K_ϵ must be compact. Now (21) follows immediately.

8 Exercises

1. Consider $L^p = L^p(\mathbb{R}^d)$ with Lebesgue measure. Let $f_0(x) = |x|^{-\alpha}$ if $|x| < 1$, $f_0(x) = 0$ for $|x| \geq 1$; also let $f_\infty(x) = |x|^{-\alpha}$ if $|x| \geq 1$, $f_\infty(x) = 0$ when $|x| < 1$. Show that:

(a) $f_0 \in L^p$ if and only if $p\alpha < d$.

(b) $f_\infty \in L^p$ if and only if $d < p\alpha$.

(c) What happens if in the definitions of f_0 and f_∞ we replace $|x|^{-\alpha}$ by $|x|^{-\alpha}/(\log(2/|x|))$ for $|x| < 1$, and $|x|^{-\alpha}$ by $|x|^{-\alpha}/(\log(2|x|))$ for $|x| \geq 1$?

2. Consider the spaces $L^p(\mathbb{R}^d)$, when $0 < p < \infty$.

(a) Show that if $\|f+g\|_{L^p} \leq \|f\|_{L^p} + \|g\|_{L^p}$ for all f and g, then necessarily $p \geq 1$.

(b) Consider $L^p(\mathbb{R})$ where $0 < p < 1$. Show that there are no bounded linear functionals on this space. In other words, if ℓ is a linear function $L^p(\mathbb{R}) \to \mathbb{C}$ that satisfies

$$|\ell(f)| \leq M\|f\|_{L^p(\mathbb{R})} \qquad \text{for all } f \in L^p(\mathbb{R}) \text{ and some } M > 0,$$

then $\ell = 0$.

[Hint: For (a), prove that if $0 < p < 1$ and $x, y > 0$, then $x^p + y^p > (x+y)^p$. For (b), let F be defined by $F(x) = \ell(\chi_x)$, where χ_x is the characteristic function of $[0, x]$, and consider $F(x) - F(y)$.]

3. If $f \in L^p$ and $g \in L^q$, both not identically equal to zero, show that equality holds in Hölder's inequality (Theorem 1.1) if and only if there exist two non-zero constants $a, b \geq 0$ such that $a|f(x)|^p = b|g(x)|^q$ for a.e. x.

4. Suppose X is a measure space and $0 < p < 1$.

(a) Prove that $\|fg\|_{L^1} \geq \|f\|_{L^p}\|g\|_{L^q}$. Note that q, the conjugate exponent of p, is negative.

(b) Suppose f_1 and f_2 are non-negative. Then $\|f_1 + f_2\|_{L^p} \geq \|f_1\|_{L^p} + \|f_2\|_{L^p}$.

(c) The function $d(f,g) = \|f-g\|^p_{L^p}$ for $f, g \in L^p$ defines a metric on $L^p(X)$.

5. Let X be a measure space. Using the argument to prove the completeness of $L^p(X)$, show that if the sequence $\{f_n\}$ converges to f in the L^p norm, then a subsequence of $\{f_n\}$ converges to f almost everywhere.

6. Let $(X, \mathcal{F}, \mu)$ be a measure space. Show that:

(a) The simple functions are dense in $L^\infty(X)$ if $\mu(X) < \infty$, and;

(b) The simple functions are dense in $L^p(X)$ for $1 \leq p < \infty$.

[Hint: For (a), use $E_{\ell,j} = \{x \in X : \frac{M\ell}{j} \leq f(x) < \frac{M(\ell+1)}{j}\}$ where $-j \leq \ell \leq j$, and $M = \|f\|_{L^\infty}$. Then consider the functions f_j that equal $M\ell/j$ on $E_{\ell,j}$. For (b) use a construction similar to that in (a).]

7. Consider the L^p spaces, $1 \leq p < \infty$, on $\mathbb{R}^d$ with Lebesgue measure. Prove that:

(a) The family of continuous functions with compact support is dense in L^p, and in fact:

(b) The family of indefinitely differentiable functions with compact support is dense in L^p.

The cases of L^1 and L^2 are in Theorem 2.4, Chapter 2 of Book III, and Lemma 3.1, Chapter 5 of Book III.

8. Suppose $1 \leq p < \infty$, and that $\mathbb{R}^d$ is equipped with Lebesgue measure. Show that if $f \in L^p(\mathbb{R}^d)$, then

$$\|f(x+h) - f(x)\|_{L^p} \to 0 \qquad \text{as } |h| \to 0.$$

Prove that this fails when $p = \infty$.

[Hint: By the previous exercise, the continuous functions with compact support are dense in $L^p(\mathbb{R}^d)$ for $1 \leq p < \infty$. See also Theorem 2.4 and Proposition 2.5 in Chapter 2 of Book III.]

9. Suppose X is a measure space and $1 \leq p_0 < p_1 \leq \infty$.

(a) Consider $L^{p_0} \cap L^{p_1}$ equipped with

$$\|f\|_{L^{p_0} \cap L^{p_1}} = \|f\|_{L^{p_0}} + \|f\|_{L^{p_1}}.$$

Show that $\|\cdot\|_{L^{p_0} \cap L^{p_1}}$ is a norm, and that $L^{p_0} \cap L^{p_1}$ (with this norm) is a Banach space.

(b) Suppose $L^{p_0} + L^{p_1}$ is defined as the vector space of measurable functions f on X that can be written as a sum $f = f_0 + f_1$ with $f_0 \in L^{p_0}$ and $f_1 \in L^{p_1}$. Consider

$$\|f\|_{L^{p_0} + L^{p_1}} = \inf \{\|f_0\|_{L^{p_0}} + \|f_1\|_{L^{p_1}}\},$$

where the infimum is taken over all decompositions $f = f_0 + f_1$ with $f_0 \in L^{p_0}$ and $f_1 \in L^{p_1}$. Show that $\|\cdot\|_{L^{p_0} + L^{p_1}}$ is a norm, and that $L^{p_0} + L^{p_1}$ (with this norm) is a Banach space.

(c) Show that $L^p \subset L^{p_0} + L^{p_1}$ if $p_0 \leq p \leq p_1$.

10. A measure space (X, μ) is **separable** if there is a countable family of measurable subsets $\{E_k\}_{k=1}^{\infty}$ so that if E is any measurable set of finite measure, then

$$\mu(E \triangle E_{n_k}) \to 0 \qquad \text{as } k \to 0$$

for an appropriate subsequence $\{n_k\}$ which depends on E. Here $A \triangle B$ denotes the symmetric difference of the sets A and B, that is,

$$A \triangle B = (A - B) \cup (B - A).$$

(a) Verify that $\mathbb{R}^d$ with the usual Lebesgue measure is separable.

(b) The space $L^p(X)$ is **separable** if there exists a countable collection of elements $\{f_n\}_{n=1}^{\infty}$ in L^p that is dense. Prove that if the measure space X is separable, then L^p is separable when $1 \leq p < \infty$.

11. In light of the previous exercise, prove the following:

(a) Show that the space $L^\infty(\mathbb{R})$ is not separable by constructing for each $a \in \mathbb{R}$ an $f_a \in L^\infty$, with $\|f_a - f_b\| \geq 1$, if $a \neq b$.

(b) Do the same for the dual space of $L^\infty(\mathbb{R})$.

12. Suppose the measure space (X, μ) is separable as defined in Exercise 10. Let $1 \leq p < \infty$ and $1/p + 1/q = 1$. A sequence $\{f_n\}$ with $f_n \in L^p$ is said to converge to $f \in L^p$ **weakly** if

$$\int f_n g \, d\mu \to \int f g \, d\mu \qquad \text{for every } g \in L^q. \tag{22}$$

(a) Verify that if $\|f - f_n\|_{L^p} \to 0$, then f_n converges to f weakly.

(b) Suppose $\sup_n \|f_n\|_{L^p} < \infty$. Then, to verify weak convergence it suffices to check (22) for a dense subset of functions g in L^q.

(c) Suppose $1 < p < \infty$. Show that if $\sup_n \|f_n\|_{L^p} < \infty$, then there exists $f \in L^p$, and a subsequence $\{n_k\}$ so that f_{n_k} converges weakly to f.

Part (c) is known as the "weak compactness" of L^p for $1 < p < \infty$, which fails when $p = 1$ as is seen in the exercise below.

[Hint: For (b) use Exercise 10 (b).]

13. Below are some examples illustrating weak convergence.

(a) $f_n(x) = \sin(2\pi n x)$ in $L^p([0,1])$. Show that $f_n \to 0$ weakly.

(b) $f_n(x) = n^{1/p}\chi(nx)$ in $L^p(\mathbb{R})$. Then $f_n \to 0$ weakly if $p > 1$, but not when $p = 1$. Here χ denotes the characteristic function of $[0,1]$.

(c) $f_n(x) = 1 + \sin(2\pi n x)$ in $L^1([0,1])$. Then $f_n \to 1$ weakly also in $L^1([0,1])$, $\|f_n\|_{L^1} = 1$, but $\|f_n - 1\|_{L^1}$ does not converge to zero. Compare with Problem 6 part (d).

14. Suppose X is a measure space, $1 < p < \infty$, and suppose $\{f_n\}$ is a sequence of functions with $\|f_n\|_{L^p} \leq M < \infty$.

(a) Prove that if $f_n \to f$ a.e. then $f_n \to f$ weakly.

(b) Show that the above result may fail if $p = 1$.

(c) Show that if $f_n \to f_1$ a.e. and $f_n \to f_2$ weakly, then $f_1 = f_2$ a.e.

15. Minkowski's inequality for integrals. Suppose (X_1, μ_1) and (X_2, μ_2) are two measure spaces, and $1 \leq p \leq \infty$. Show that if $f(x_1, x_2)$ is measurable on $X_1 \times X_2$ and non-negative, then

$$\left\| \int f(x_1, x_2)\, d\mu_2 \right\|_{L^p(X_1)} \leq \int \|f(x_1, x_2)\|_{L^p(X_1)}\, d\mu_2.$$

Extend this statement to the case when f is complex-valued and the right-hand side of the inequality is finite.

[Hint: For $1 < p < \infty$, use a combination of Hölder's inequality, and its converse in Lemma 4.2.]

16. Prove that if $f_j \in L^{p_j}(X)$, where X is a measure space, $j = 1, \ldots, N$, and $\sum_{j=1}^N 1/p_j = 1$ with $p_j \geq 1$, then

$$\| \prod_{j=1}^N f_j \|_{L^1} \leq \prod_{j=1}^N \|f\|_{L^{p_j}}.$$

This is the multiple Hölder inequality.

17. The **convolution** of f and g on $\mathbb{R}^d$ equipped with the Lebesgue measure is defined by

$$(f * g)(x) = \int_{\mathbb{R}^d} f(x-y)g(y)\,dy.$$

(a) If $f \in L^p$, $1 \leq p \leq \infty$, and $g \in L^1$, then show that for almost every x the integrand $f(x-y)g(y)$ is integrable in y, hence $f * g$ is well defined. Moreover, $f * g \in L^p$ with

$$\|f * g\|_{L^p} \leq \|f\|_{L^p} \|g\|_{L^1}.$$

(b) A version of (a) applies when g is replaced by a finite Borel measure μ: if $f \in L^p$, with $1 \leq p \leq \infty$, define

$$(f * \mu)(x) = \int_{\mathbb{R}^d} f(x-y)\,d\mu(y),$$

and show that $\|f * \mu\|_{L^p} \leq \|f\|_{L^p} |\mu|(\mathbb{R}^d)$.

(c) Prove that if $f \in L^p$ and $g \in L^q$, where p and q are conjugate exponents, then $f * g \in L^\infty$ with $\|f * g\|_{L^\infty} \leq \|f\|_{L^p} \|g\|_{L^q}$. Moreover, the convolution $f * g$ is uniformly continuous on $\mathbb{R}$, and if $1 < p < \infty$, then $\lim_{|x| \to \infty} (f * g)(x) = 0$.

[Hint: For (a) and (b) use the Minkowski inequality for integrals in Exercise 15. For part (c), use Exercise 8.]

18. We consider the L^p spaces with **mixed norm**, in a special case that is useful is several contexts.

We take as our underlying space the product space $\{(x,t)\} = \mathbb{R}^d \times \mathbb{R}$, with the product measure $dx\,dt$, where dx and dt are Lebesgue measures on $\mathbb{R}^d$ and $\mathbb{R}$ respectively. We define $L_t^r(L_x^p) = L^{p,r}$, with $1 \leq p \leq \infty$, $1 \leq r \leq \infty$, to be the space of equivalence classes of jointly measurable functions $f(x,t)$ for which the norm

$$\|f\|_{L^{p,r}} = \left(\int_{\mathbb{R}} \left(\int_{\mathbb{R}^d} |f(x,t)|^p \, dx \right)^{\frac{r}{p}} dt \right)^{\frac{1}{r}}$$

is finite (when $p < \infty$ and $r < \infty$), and an obvious variant when $p = \infty$ or $r = \infty$.

(a) Verify that $L^{p,r}$ with this norm is complete, and hence is a Banach space.

(b) Prove the general form of Hölder's inequality in this context

$$\int_{\mathbb{R}^d \times \mathbb{R}} |f(x,t)g(x,t)| \, dx \, dt \leq \|f\|_{L^{p,r}} \|g\|_{L^{p',r'}},$$

with $1/p + 1/p' = 1$ and $1/r + 1/r' = 1$.

(c) Show that if f is integrable over all sets of finite measure, then

$$\|f\|_{L^{p,r}} = \sup \left| \int_{\mathbb{R}^d \times \mathbb{R}} f(x,t)g(x,t) \, dx dt \right|,$$

with the sup taken over all g that are simple and $\|g\|_{L^{p',r'}} \leq 1$.

(d) Conclude that the dual space of $L^{p,r}$ is $L^{p',r'}$, if $1 \leq p < \infty$, and $1 \leq r < \infty$.

19. Young's inequality. Suppose $1 \leq p, q, r \leq \infty$. Prove the following on $\mathbb{R}^d$:

$$\|f * g\|_{L^q} \leq \|f\|_{L^p} \|g\|_{L^r} \qquad \text{whenever } 1/q = 1/p + 1/r - 1.$$

Here, $f * g$ denotes the convolution of f and g as defined in Exercise 17. [Hint: Assume $f, g \geq 0$, and use the decomposition

$$f(y)g(x-y) = f(y)^a g(x-y)^b [f(y)^{1-a} g(x-y)^{1-b}]$$

for appropriate a and b, together with Exercise 16 to find that

$$\left| \int f(y)g(x-y) \, dy \right| \leq \|f\|_{L^p}^{1-p/q} \|g\|_{L^q}^{1-r/q} \left(\int |f(y)|^p |g(x-y)|^r \, dy \right)^{\frac{1}{q}}.]$$

20. Suppose X is a measure space, $0 < p_0 < p < p_1 \leq \infty$, and $f \in L^{p_0}(X) \cap L^{p_1}(X)$. Then $f \in L^p(X)$ and

$$\|f\|_{L^p} \leq \|f\|_{L^{p_0}}^{1-t} \|f\|_{L^{p_1}}^{t}, \qquad \text{if } t \text{ is chosen so that } \tfrac{1}{p} = \tfrac{1-t}{p_0} + \tfrac{t}{p_1}.$$

21. Recall the definition of a convex function. (See Problem 4, Chapter 3, in Book III.) Suppose φ is a non-negative convex function on $\mathbb{R}$ and f is real-valued and integrable on a measure space X, with $\mu(X) = 1$. Then we have **Jensen's inequality**:

$$\varphi\left(\int_X f\,d\mu\right) \le \int_X \varphi(f)\,d\mu.$$

Note that if $\varphi(t) = |t|^p$, $1 \le p$, then φ is convex and the above can be obtained from Hölder's inequality. Another interesting case is $\varphi(t) = e^{at}$.

[Hint: Since φ is convex, one has, $\varphi(\sum_{j=1}^N a_j x_j) \le \sum_{j=1}^N a_j \varphi(x_j)$, whenever a_j, x_j are real, $a_j \ge 0$, and $\sum_{j=1}^N a_j = 1$.]

22. Another inequality of Young. Suppose φ and ψ are both continuous, *strictly* increasing functions on $[0, \infty)$ that are inverses of each other, that is, $(\varphi \circ \psi)(x) = x$ for all $x \ge 0$. Let

$$\Phi(x) = \int_0^x \varphi(u)\,du \quad \text{and} \quad \Psi(x) = \int_0^x \psi(u)\,du.$$

(a) Prove: $ab \le \Phi(a) + \Psi(b)$ for all $a, b \ge 0$.

In particular, if $\varphi(x) = x^{p-1}$ and $\psi(y) = y^{q-1}$ with $1 < p < \infty$ and $1/p + 1/q = 1$, then we get $\Phi(x) = x^p/p$, $\Psi(y) = y^q/q$, and

$$A^\theta B^{1-\theta} \le \theta A + (1-\theta)B \quad \text{for all } A, B \ge 0 \text{ and } 0 \le \theta \le 1.$$

(b) Prove that we have equality in Young's inequality only if $b = \varphi(a)$ (that is, $a = \psi(b)$).

[Hint: Consider the area ab of the rectangle whose vertices are $(0,0)$, $(a,0)$, $(0,b)$ and (a,b), and compare it to areas "under" the curves $y = \Phi(x)$ and $x = \Psi(y)$.]

23. Let (X, μ) be a measure space and suppose $\Phi(t)$ is a continuous, convex, and increasing function on $[0, \infty)$, with $\Phi(0) = 0$. Define

$$L^\Phi = \{f \text{ measurable} : \int_X \Phi(|f(x)|/M)\,d\mu < \infty \text{ for some } M > 0\},$$

and

$$\|f\|_\Phi = \inf\{M > 0 : \int_X \Phi(|f(x)|/M)\,d\mu \le 1\}.$$

Prove that:

(a) L^Φ is a vector space.

(b) $\|\cdot\|_{L^\Phi}$ is a norm.

(c) L^Φ is complete in this norm.

The Banach spaces L^Φ are called **Orlicz spaces**. Note that in the special case $\Phi(t) = t^p$, $1 \le p < \infty$, then $L^\Phi = L^p$.

[Hint: Observe that if $f \in L^\Phi$, then $\lim_{N\to\infty} \int_X \Phi(|f|/N)\, d\mu = 0$. Also, use the fact that there exists $A > 0$ so that $\Phi(t) \ge At$ for all $t \ge 0$.]

24. Let $1 \le p_0 < p_1 < \infty$.

(a) Consider the Banach space $L^{p_0} \cap L^{p_1}$ with norm $\|f\|_{L^{p_0}\cap L^{p_1}} = \|f\|_{L^{p_0}} + \|f\|_{L^{p_1}}$. (See Exercise 9.) Let

$$\Phi(t) = \begin{cases} t^{p_0} & \text{if } 0 \le t \le 1, \\ t^{p_1} & \text{if } 1 \le t < \infty. \end{cases}$$

Show that L^Φ with its norm is equivalent to the space $L^{p_0} \cap L^{p_1}$. In other words, there exist $A, B > 0$, so that

$$A\|f\|_{L^{p_0}\cap L^{p_1}} \le \|f\|_{L^\Phi} \le B\|f\|_{L^{p_0}\cap L^{p_1}}.$$

(b) Similarly, consider the Banach space $L^{p_0} + L^{p_1}$ with its norm as defined in Exercise 9. Let

$$\Psi(t) = \int_0^t \psi(u)\, du \quad \text{where} \quad \psi(u) = \begin{cases} u^{p_1-1} & \text{if } 0 \le u \le 1, \\ u^{p_0-1} & \text{if } 1 \le u < \infty. \end{cases}$$

Show that L^Ψ with its norm is equivalent to the space $L^{p_0} + L^{p_1}$.

25. Show that a Banach space $\mathcal{B}$ is a Hilbert space if and only if the parallelogram law holds

$$\|f+g\|^2 + \|f-g\|^2 = 2(\|f\|^2 + \|g\|^2).$$

As a consequence, prove that if $L^p(\mathbb{R}^d)$ with the Lebesgue measure is a Hilbert space, then necessarily $p = 2$.

[Hint: For the first part, in the real case, let $(f, g) = \frac{1}{4}(\|f+g\|^2 + \|f-g\|^2)$.]

26. Suppose $1 < p_0, p_1 < \infty$ and $1/p_0 + 1/q_0 = 1$ and $1/p_1 + 1/q_1 = 1$. Show that the Banach spaces $L^{p_0} \cap L^{p_1}$ and $L^{q_0} + L^{q_1}$ are duals of each other up to an equivalence of norms. (See Exercise 9 for the relevant definitions of these spaces. Also, Problem 5* gives a generalization of this result.)

27. The purpose of this exercise is to prove that the unit ball in L^p is strictly convex when $1 < p < \infty$, in the following sense. Here L^p is the space of real-valued functions whose p^{th} power are integrable. Suppose $\|f_0\|_{L^p} = \|f_1\|_{L^p} = 1$, and let

$$f_t = (1-t)f_0 + tf_1$$

be the straight-line segment joining the points f_0 and f_1. Then $\|f_t\|_{L^p} < 1$ for all t with $0 < t < 1$, unless $f_0 = f_1$.

(a) Let $f \in L^p$ and $g \in L^q$, $1/p + 1/q = 1$, with $\|f\|_{L^p} = 1$ and $\|g\|_{L^q} = 1$. Then

$$\int fg\, d\mu = 1$$

only when $f(x) = \operatorname{sign} g(x)|g(x)|^{q-1}$.

(b) Suppose $\|f_{t'}\|_{L^p} = 1$ for some $0 < t' < 1$. Find $g \in L^q$, $\|g\|_{L^q} = 1$, so that

$$\int f_{t'} g\, d\mu = 1$$

and let $F(t) = \int f_t g\, d\mu$. Observe as a result that $F(t) = 1$ for all $0 \leq t \leq 1$. Conclude that $f_t = f_0$ for all $0 \leq t \leq 1$.

(c) Show that the strict convexity fails when $p = 1$ or $p = \infty$. What can be said about these cases?

A stronger assertion is given in Problem 6*.
[Hint: To prove (a) show that the case of equality in $A^\theta B^{1-\theta} \leq \theta A + (1-\theta)B$, for $A, B > 0$ and $0 < \theta < 1$ holds only when $A = B$.]

28. Verify the completeness of $\Lambda^\alpha(\mathbb{R}^d)$ and $L^p_k(\mathbb{R}^d)$.

29. Consider further the spaces $\Lambda^\alpha(\mathbb{R}^d)$.

(a) Show that when $\alpha > 1$ the only functions in $\Lambda^\alpha(\mathbb{R}^d)$ are the constants.

(b) Motivated by (a), one defines $C^{k,\alpha}(\mathbb{R}^d)$ to be the class of functions f on $\mathbb{R}^d$ whose partial derivatives of order less than or equal to k belong to $\Lambda^\alpha(\mathbb{R}^d)$. Here k is an integer and $0 < \alpha \leq 1$. Show that this space, endowed with the norm

$$\|f\|_{C^{k,\alpha}} = \sum_{|\beta| \leq k} \left\|\partial_x^\beta f\right\|_{\Lambda^\alpha(\mathbb{R}^d)},$$

is a Banach space.

30. Suppose $\mathcal{B}$ is a Banach space and $\mathcal{S}$ is a closed linear subspace of $\mathcal{B}$. The subspace $\mathcal{S}$ defines an equivalence relation $f \sim g$ to mean $f - g \in \mathcal{S}$. If $\mathcal{B}/\mathcal{S}$ denotes the collection of these equivalence classes, then show that $\mathcal{B}/\mathcal{S}$ is a Banach space with norm $\|f\|_{\mathcal{B}/\mathcal{S}} = \inf(\|f'\|_{\mathcal{B}},\ f' \sim f)$.

31. If Ω is an open subset of $\mathbb{R}^d$ then one definition of $L^p_k(\Omega)$ can be taken to be the quotient Banach space $\mathcal{B}/\mathcal{S}$, as defined in the previous exercise, with $\mathcal{B} = L^p_k(\mathbb{R}^d)$ and $\mathcal{S}$ the subspace of those functions which vanish a.e. on Ω. Another possible space, that we will denote by $L^p_k(\Omega^0)$, consists of the closure in $L^p_k(\mathbb{R}^d)$ of all f that have compact support in Ω. Observe that the natural mapping of $L^p_k(\Omega^0)$ to

$L^p_k(\Omega)$ has norm equal to 1. However, this mapping is in general not surjective. Prove this in the case when Ω is the unit ball and $k \geq 1$.

32. A Banach space is said to be separable if it contains a countable dense subset. In Exercise 11 we saw an example of a Banach space $\mathcal{B}$ that is separable, but where $\mathcal{B}^*$ is not separable. Prove, however, that in general when $\mathcal{B}^*$ is separable, then $\mathcal{B}$ is separable. Note that this gives another proof that in general L^1 is not the dual of L^∞.

33. Let V be a vector space over the complex numbers $\mathbb{C}$, and suppose there exists a real-valued function p on V satisfying:

$$\begin{cases} p(\alpha v) = |\alpha| p(v), & \text{if } \alpha \in \mathbb{C}, \text{ and } v \in V, \\ p(v_1 + v_2) \leq p(v_1) + p(v_2), & \text{if } v_1 \text{ and } v_2 \in V. \end{cases}$$

Prove that if V_0 is a subspace of V and ℓ_0 a linear functional on V_0 which satisfies $|\ell_0(f)| \leq p(f)$ for all $f \in V_0$, then ℓ_0 can be extended to a linear functional ℓ on V that satisfies $|\ell(f)| \leq p(f)$ for all $f \in V$.

[Hint: If $u = \text{Re}(\ell_0)$, then $\ell_0(v) = u(v) - iu(iv)$. Apply Theorem 5.2 to u.]

34. Suppose $\mathcal{B}$ is a Banach space and $\mathcal{S}$ a closed proper subspace, and assume $f_0 \notin \mathcal{S}$. Show that there is a continuous linear functional ℓ on $\mathcal{B}$, so that $\ell(f) = 0$ for $f \in \mathcal{S}$, and $\ell(f_0) = 1$. The linear functional ℓ can be chosen so that $\|\ell\| = 1/d$ where d is the distance from f_0 to $\mathcal{S}$.

35. A linear functional ℓ on a Banach space $\mathcal{B}$ is continuous if and only if $\{f \in \mathcal{B} : \ell(f) = 0\}$ is closed.

[Hint: This is a consequence of Exercise 34.]

36. The results in Section 5.4 can be extended to d-dimensions.

(a) Show that there exists an extended-valued non-negative function $\hat{m}$ defined on all subsets of $\mathbb{R}^d$ so that (i) $\hat{m}$ is finitely additive; (ii) $\hat{m}(E) = m(E)$ whenever E is Lebesgue measurable, where m is Lebesgue measure; and $\hat{m}(E + h) = \hat{m}(E)$ for all sets E and every $h \in \mathbb{R}^d$. Prove this is as a consequence of (b) below.

(b) Show that there is an "integral" I, defined on all bounded functions on $\mathbb{R}^d/\mathbb{Z}^d$, so that $I(f) \geq 0$ whenever $f \geq 0$; the map $f \mapsto I(f)$ is linear; $I(f) = \int_{\mathbb{R}^d/\mathbb{Z}^d} f\, dx$ whenever f is measurable; and $I(f_h) = I(f)$ where $f_h(x) = f(x - h)$, and $h \in \mathbb{R}^d$.

9 Problems

1. The spaces L^∞ and L^1 play universal roles with respect to all Banach spaces in the following sense.

(a) If $\mathcal{B}$ is any separable Banach space, show that it can be realized without change of norm as a linear subspace of $L^\infty(\mathbb{Z})$. Precisely, prove that there is a linear operator i of $\mathcal{B}$ into $L^\infty(\mathbb{Z})$ so that $\|i(f)\|_{L^\infty(\mathbb{Z})} = \|f\|_{\mathcal{B}}$ for all $f \in \mathcal{B}$.

(b) Each such $\mathcal{B}$ can also be realized as a quotient space of $L^1(\mathbb{Z})$. That is, there is a linear surjection P of $L^1(\mathbb{Z})$ onto $\mathcal{B}$, so that if $\mathcal{S} = \{x \in L^1(\mathbb{Z}) : P(x) = 0\}$, then $\|P(x)\|_{\mathcal{B}} = \inf_{y \in \mathcal{S}} \|x + y\|_{L^1(\mathbb{Z})}$, for each $x \in L^1(\mathbb{Z})$. This gives an identification of $\mathcal{B}$ (and its norm) with the quotient space $L^1(\mathbb{Z})/\mathcal{S}$ (and its norm), as defined in Exercise 30.

Note that similar conclusions hold for $L^\infty(X)$ and $L^1(X)$ if X is a measure space that contains a countable disjoint collection of measurable sets of positive and finite measure.

[Hint: For (a), let $\{f_n\}$ be a dense set of non-zero vectors in $\mathcal{B}$, and let $\ell_n \in \mathcal{B}^*$ be such that $\|\ell_n\|_{\mathcal{B}^*} = 1$ and $\ell_n(f_n) = \|f_n\|$. If $f \in \mathcal{B}$, set $i(f) = \{\ell_n(f)\}_{-\infty}^{\infty}$. For (b), if $x = \{x_n\} \in L^1(\mathbb{Z})$, with $\sum_{-\infty}^{\infty} |x_n| = \|x\|_{L^1(\mathbb{Z})} < \infty$, define P by $P(x) = \sum_{-\infty}^{\infty} x_n f_n / \|f_n\|$.]

2. There is a "generalized limit" L defined on the vector space V of all real sequences $\{s_n\}_{n=1}^{\infty}$ that are bounded, so that:

(i) L is a linear functional on V.

(ii) $L(\{s_n\}) \geq 0$ if $s_n \geq 0$, for all n.

(iii) $L(\{s_n\}) = \lim_{n\to\infty} s_n$ if the sequence $\{s_n\}$ has a limit.

(iii) $L(\{s_n\}) = L(\{s_{n+k}\})$ for every $k \geq 1$.

(iii) $L(\{s_n\}) = L(\{s_{n'}\})$ if $s_n - s'_n \neq 0$ for only finitely many n.

[Hint: Let $p(\{s_n\}) = \limsup_{n\to\infty} \left(\frac{s_1 + \cdots + s_n}{n}\right)$, and extend the linear functional L defined by $L(\{s_n\}) = \lim_{n\to\infty} s_n$, defined on the subspace consisting of sequences that have limits.]

3. Show that the closed unit ball in a Banach space $\mathcal{B}$ is compact (that is, if $f_n \in \mathcal{B}$, $\|f_n\| \leq 1$, then there is a subsequence that converges in the norm) if and only if $\mathcal{B}$ is finite dimensional.

[Hint: If $\mathcal{S}$ is a closed subspace of $\mathcal{B}$, then there exists $x \in \mathcal{B}$ with $\|x\| = 1$ and the distance between x and $\mathcal{S}$ is greater than 1/2.]

4. Suppose X is a compact metric space and $C(X)$ denotes the Banach space of continuous functions on X with the sup-norm.

(a) If $\{\mu_n\}_{n=1}^{\infty}$ is a bounded sequence in $M(X)$, then there exists a $\mu \in M(X)$ and a subsequence $\{\mu_{n_j}\}_{j=1}^{\infty}$, so that μ_{n_j} converges to μ in the following (weak*) sense:

$$\int_X g(x)\, d\mu_{n_j}(x) \to \int_X g(x)\, d\mu(x), \qquad \text{for all } g \in C(X).$$

(b) Start with a $\mu_0 \in M(X)$ that is positive, and for each $f \in L^1(\mu_0)$ consider the mapping $f \mapsto f d\mu_0$. This mapping is an isometry of $L^1(\mu_0)$ to the subspace of $M(X)$ consisting of signed measures which are absolutely continuous with respect to μ_0.

(c) Hence if $\{f_n\}$ is a bounded sequence of functions in $L^1(\mu_0)$, then there exist a $\mu \in M(X)$ and a subsequence $\{f_{n_j}\}$ such that the measures $f_{n_j} d\mu_0$ converge to μ in the above sense.

5.* Let X be a measure space. Suppose φ and ψ are both continuous, strictly increasing functions on $[0, \infty)$ which are inverses of each other, that is, $(\varphi \circ \psi)(x) = x$ for all $x \geq 0$. Let

$$\Phi(x) = \int_0^x \varphi(u)\, du \quad \text{and} \quad \Psi(x) = \int_0^x \psi(u)\, du.$$

Consider the Orlicz spaces $L^\Phi(X)$ and $L^\Psi(X)$ introduced in Exercise 23.

(a) In connection with Exercise 22 the following Hölder-like inequality holds:

$$\int |fg| \leq C\|f\|_{L^\Phi}\|g\|_{L^\Psi} \qquad \text{for some } C > 0, \text{ and all } f \in L^\Phi \text{ and } g \in L^\Psi.$$

(b) Suppose there exists $c > 0$ so that $\Phi(2t) \leq c\Phi(t)$ for all $t \geq 0$. Then the dual of L^Φ is equivalent to L^Ψ.

6.* There are generalizations of the parallelogram law for L^2 (see Exercise 25) that hold for L^p. These are the Clarkson inequalities:

(a) For $2 \leq p \leq \infty$ the statement is that

$$\left\|\frac{f+g}{2}\right\|_{L^p}^p + \left\|\frac{f-g}{2}\right\|_{L^p}^p \leq \frac{1}{2}\left(\|f\|_{L^p}^p + \|g\|_{L^p}^p\right).$$

(b) For $1 < p \leq 2$ the statement is that

$$\left\|\frac{f+g}{2}\right\|_{L^p}^q + \left\|\frac{f-g}{2}\right\|_{L^p}^q \leq \left(\frac{\|f\|_{L^p}^p + \|g\|_{L^p}^p}{2}\right)^{q/p},$$

where $1/p + 1/q = 1$.

(c) As a result, L^p is **uniformly convex** when $1 < p < \infty$. This means that there is a function $\delta = \delta(\epsilon) = \delta_p(\epsilon)$, with $0 < \delta < 1$, (and $\delta(\epsilon) \to 0$ as $\epsilon \to 0$), so that whenever $\|f\|_{L^p} = \|g\|_{L^p} = 1$, then $\|f - g\|_{L^p} \geq \epsilon$ implies that $\|\frac{f+g}{2}\| \leq 1 - \delta$.

This is stronger than the conclusion of strict convexity in Exercise 27.

(d) Using the result in (c), prove the following: suppose $1 < p < \infty$, and the sequence $\{f_n\}$, $f_n \in L^p$, converges weakly to f. If $\|f_n\|_{L^p} \to \|f\|_{L^p}$, then f_n converges to f strongly, that is, $\|f_n - f\|_{L^p} \to 0$ as $n \to \infty$.

7.* An important notion is that of the equivalence of Banach spaces. Suppose $\mathcal{B}_1$ and $\mathcal{B}_2$ are a pair of Banach spaces. We say that $\mathcal{B}_1$ and $\mathcal{B}_2$ are **equivalent** (also said to be "isomorphic") if there is a linear bijection T between $\mathcal{B}_1$ and $\mathcal{B}_2$ that is bounded and whose inverse is also bounded. Note that any pair of finite-dimensional Banach spaces are equivalent if and only if their dimensions are the same.

Suppose now we consider $L^p(X)$ for a general class of X (which contains for instance, $X = \mathbb{R}^d$ with Lebesgue measure). Then:

(a) L^p and L^q are equivalent if and only if $p = q$.

(b) However, for any p with $1 \leq p \leq \infty$, L^2 is equivalent with a closed infinite-dimensional subspace of L^p.

8.* There is no finitely-additive rotationally-invariant measure extending Lebesgue measure to all subsets of the sphere S^d when $d \geq 2$, in distinction to what happens on the torus $\mathbb{R}^d/\mathbb{Z}^d$ when $d \geq 2$. (See Exercise 36). This is due to a remarkable construction of Hausdorff that uses the fact that the corresponding rotation group of S^d is non-commutative. In fact, one can decompose S^2 into four disjoint sets A, B, C and Z so that (i) Z is denumerable, (ii) $A \sim B \sim C$, but $A \sim (B \cup C)$.

Here the notation $A_1 \sim A_2$ means that A_1 can be transformed into A_2 via a rotation.

9.* As a consequence of the previous problem one can show that it is not possible to extend Lebesgue measure on $\mathbb{R}^d$, $d \geq 3$, as a finitely-additive measure on all subsets of $\mathbb{R}^d$ so that it is both translation and rotation invariant (that is, invariant under Euclidean motions). This is graphically shown by the "Banach-Tarski paradox": There is a finite decomposition of the unit ball $B_1 = \bigcup_{j=1}^N E_j$, with the sets E_j disjoint, and there are corresponding sets $\tilde{E}_j$ that are each obtained from E_j by a Euclidean motion, with the $\tilde{E}_j$ also disjoint, so that $\bigcup_{j=1}^N \tilde{E}_j = B_2$ the ball of radius 2.

2 L^p Spaces in Harmonic Analysis

> The important part played in Hilbert's treatment of Fredholm theory of integral equations by functions whose squares are summable is well-known, and it was inevitable that members of the Göttingen school of mathematics should be led to set themselves the task of proving the converse of Parseval's theorem.... On the other hand, efforts made to extend these isolated results to embrace cases in which the known or unknown index of summability is other than 2, appear to have failed...
>
> *W. H. Young,* 1912

> ...I have proved that two conjugate trigonometric series are at the same time the Fourier series of L^p functions, $p > 1$. That is, if one is, so is the other. My proof is unrelated to the theorem of Young-Hausdorff...
>
> *M. Riesz, letter to G. H. Hardy,* 1923

> Some months ago you wrote "... I have proved that two conjugate... L^p functions, $p > 1$". *I want the proof.* Both I and my pupil Titchmarsh have tried in vain to prove it..."
>
> *G. H. Hardy, letter to M. Riesz* 1923

The fact that L^p spaces were bound to play a significant role in harmonic analysis was understood not long after their introduction. Viewed from that early perspective, these spaces stood at the nexus between Fourier series and complex analysis, this connection having been given by the Cauchy integral and the related conjugate function. For this reason methods of complex function theory predominated in the beginning stages of the subject, but they had to give way to "real" methods so as to allow the extension of much of the theory to higher dimensions.

It is the aim of this chapter to show the reader something about both of these methods. In fact, the real-variable ideas that will be introduced here will also be further exploited in the next chapter, when studying singular integral operators in $\mathbb{R}^d$.

The present chapter is organized as follows. We begin with an initial view of the role of L^p in the context of Fourier series, together with a related convexity theorem for operators acting on these spaces. Then we pass to M. Riesz's proof of the L^p boundedness of the Hilbert transform, an iconic example of the use of complex analysis in this setting.

From this we turn to the real-variable methods, starting with the maximal function and its attendant "weak-type" estimate. The importance of the weak-type space is that it provides a useful substitute for L^1 when, as in many instances, L^1 estimates fail. We also study another significant substitute for L^1, the "real" Hardy space $\mathbf{H}^1_r$. It has the advantage that it is a Banach space and that its dual space (a substitute for L^∞) is the space of functions of bounded mean oscillation. This last function space is itself of wide interest in analysis.

1 Early Motivations

An initial problem considered was that of formulating an L^p analog of the basic L^2 Parseval relation for functions on $[0, 2\pi]$. This theorem states that if $a_n = \frac{1}{2\pi}\int_0^{2\pi} f(\theta)e^{-in\theta}\,d\theta$ denotes the Fourier coefficients of a function f in $L^2([0, 2\pi])$, usually written as

$$f(\theta) \sim \sum_{n=-\infty}^{\infty} a_n e^{in\theta}, \tag{1}$$

then the following fundamental identity holds:

$$\sum_{n=-\infty}^{\infty} |a_n|^2 = \frac{1}{2\pi}\int_0^{2\pi} |f(\theta)|^2\,d\theta. \tag{2}$$

Conversely, if $\{a_n\}$ is a sequence for which the left-hand side of (2) is finite, then there exists a unique f in $L^2([0, 2\pi])$ so that both (1) and (2) hold. Notice, in particular, if $f \in L^2([0, 2\pi])$, then its Fourier coefficients $\{a_n\}$ belong to $L^2(\mathbb{Z}) = \ell^2(\mathbb{Z})$.[1] The question that arose was: is there an analog of this result for L^p when $p \neq 2$?

Here an important dichotomy between the case $p > 2$ and $p < 2$ occurs. In the first case, when $f \in L^p([0, 2\pi])$, although f is automatically in $L^2([0, 2\pi])$, examples show that no better conclusion than $\sum |a_n|^2 < \infty$

[1] See for instance Section 3 in Chapter 4 of Book III.

is possible. On the other hand, when $p < 2$ one can see that essentially there can be no better conclusion than $\sum |a_n|^q < \infty$, with q the dual exponent of p. Analogous restrictions must be envisaged when the roles of f and $\{a_n\}$ are reversed.

In fact, what does hold is the Hausdorff-Young inequality:

$$\left(\sum |a_n|^q\right)^{1/q} \leq \left(\frac{1}{2\pi}\int_0^{2\pi} |f(\theta)|^p \, d\theta\right)^{1/p}, \tag{3}$$

and its "dual"

$$\left(\frac{1}{2\pi}\int_0^{2\pi} |f(\theta)|^q \, d\theta\right)^{1/q} \leq \left(\sum |a_n|^p\right)^{1/p}, \tag{4}$$

both valid when $1 \leq p \leq 2$ and $1/p + 1/q = 1$. (The case $q = \infty$ corresponds to the usual L^∞ norm.) These may be viewed as intermediate results, between the case $p = 2$ corresponding to Parseval's theorem, and its "trivial" case $p = 1$ and $q = \infty$.

A few words about how the inequalities (3) and (4) were first attacked are in order, because they contain a useful insight about L^p spaces: often, the simplest case arises when p (or its dual) is an *even integer*. Indeed, when, for example $q = 4$, a function belonging to L^4 is the same as its square belonging to L^2, and this sometimes allows reduction to the easier situation when $p = 2$. To see how this works in the present situation, let us take $q = 4$ (and $p = 4/3$) in (3). With f given in L^p, we denote by $\mathcal{F}$ the convolution of f with itself,

$$\mathcal{F}(\theta) = \frac{1}{2\pi}\int_0^{2\pi} f(\theta - \varphi) f(\varphi) \, d\varphi.$$

By the multiplicative property of Fourier coefficients of convolutions we have

$$\mathcal{F}(\theta) \sim \sum_{n=-\infty}^{\infty} a_n^2 e^{in\theta},$$

with $\{a_n\}$ the Fourier coefficients of f. Parseval's identity applied to $\mathcal{F}$ then yields

$$\sum |a_n|^4 = \frac{1}{2\pi}\int_0^{2\pi} |\mathcal{F}(\theta)|^2 \, d\theta,$$

and Young's inequality for convolutions (the periodic analog of Exercise 19, Chapter 1) gives

$$\|\mathcal{F}\|_{L^2} \leq \|f\|^2_{L^{4/3}},$$

proving (3) when $p = 4/3$ and $q = 4$.

Once the case $q = 4$ has been established, the cases corresponding to $q = 2k$, where k is a positive integer, can be handled in a similar way. However the general situation, $2 \leq q \leq \infty$, corresponding to $1 \leq p \leq 2$, involves further ideas.

In contrast to the above ingenious but special argument, in turns out that there is a general principle of great interest that underlies such inequalities, which in fact leads to direct and abstract proofs of both (3) and (4). This is the M. Riesz interpolation theorem. Stated succinctly, it asserts that whenever a linear operator satisfies a pair of inequalities (like (3) for $p = 2$ and $p = 1$), then automatically the operator satisfies the corresponding inequalities for the intermediate exponents: here all p for $1 \leq p \leq 2$, and q with $1/p + 1/q = 1$. The formulation and proof of this general theorem will be our first task in the next section.

Before we turn to that, we will describe briefly another initial source for the role of L^p in harmonic analysis, one which highlights its connection with complex analysis.

Together with the Fourier series (1) for f in L^2, one considers its "conjugate function" or "allied series", defined by

$$\tilde{f}(\theta) \sim \sum_{n=-\infty}^{\infty} \frac{\text{sign}(n)}{i} a_n e^{in\theta}, \tag{5}$$

where $\text{sign}(n) = 1$ if $n > 0$, $\text{sign}(n) = -1$ if $n < 0$, and $\text{sign}(n) = 0$ when $n = 0$.[2]

The significance of this definition is that

$$\frac{1}{2}(f(\theta) + i\tilde{f}(\theta) + a_0) \sim \sum_{n=0}^{\infty} a_n e^{in\theta} = F(e^{i\theta}),$$

where $F(z) = \sum_{n=0}^{\infty} a_n z^n$ is the analytic function in the unit disc $|z| < 1$ given as the Cauchy integral (projection) of f, namely:

$$F(z) = \frac{1}{2\pi i} \int_0^{2\pi} \frac{f(\theta)}{e^{i\theta} - z} ie^{i\theta}\, d\theta.$$

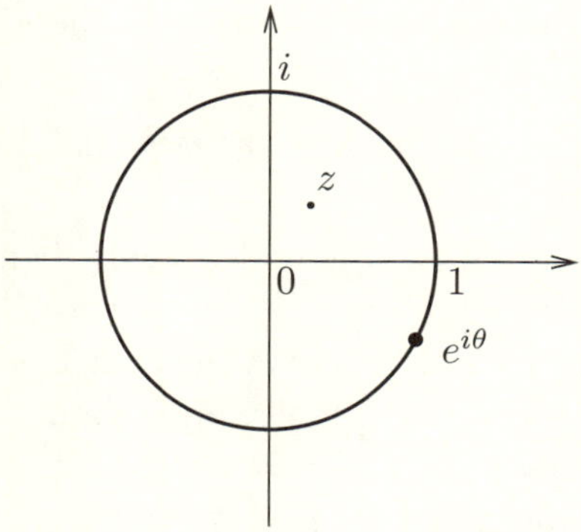

Figure 1. The Cauchy integral $F(z)$ is defined for $|z| < 1$, while $f(\theta)$ is defined for $z = e^{i\theta}$.

Moreover, if f is real-valued (that is $a_n = \overline{a_{-n}}$), so is $\tilde{f}$ and thus $f + a_0$ and $\tilde{f}$ represent respectively the real and imaginary parts of the boundary values of the analytic function $2F$ in the unit disc.

The key L^2 identity linking f and $\tilde{f}$ is a simple consequence of Parseval's relation:

$$\frac{1}{2\pi}\int_0^{2\pi} |\tilde{f}(\theta)|^2\, d\theta + |a_0|^2 = \frac{1}{2\pi}\int_0^{2\pi} |f(\theta)|^2\, d\theta. \tag{6}$$

An early goal of the subject was the extension of this theory to L^p, and it was also achieved by M. Riesz.

As he tells it, he was led to the discovery of his result when preparing to administer a "licenciat" exam to a rather mediocre student. One of the problems on the exam was to prove (6). To quote Riesz: " ... However it was quite obvious that my candidate did not know Parseval's theorem. Before giving him the problem, I had therefore to think if there was another way for him to arrive at the required conclusion. I immediately realized that it was Cauchy's theorem that was at the source of the result, and this observation led me quite directly to the solution of the general problem, a question that had longtime occupied me."

What Riesz had in mind was the following argument. If we assume for simplicity that $a_0 = 0$, then under the (technical) assumption that the analytic function F is actually continuous in the closure of the unit disc, one has by the mean-value theorem (as a simple consequence of Cauchy's

---

[2]Incidentally the conjugate function is the "symmetry-breaking" operator relevant to the divergence of Fourier series considered in Book I.

theorem) applied to the analytic function F^2, the identity

$$\frac{1}{2\pi}\int_0^{2\pi}(F(e^{i\theta}))^2\,d\theta = 0. \tag{7}$$

If we suppose, as above, that f is real-valued, then by considering the real part of $4(F(e^{i\theta}))^2$, which is $(f(e^{i\theta}))^2 - (\tilde{f}(e^{i\theta}))^2$, we immediately get (6). What became clear to Riesz is that when we replace F^2 by F^{2k} in the above, with k a positive integer, and again consider its real part, the boundedness of $f \mapsto \tilde{f}$ in L^p, where $p = 2k$ follows. Similar but more involved arguments worked for all p, $1 < p < \infty$.

Here, once again, the Riesz interpolation theorem can play a crucial role. We will present these ideas below in the context where the unit disc is replaced by the upper half-plane.

2 The Riesz interpolation theorem

Suppose (p_0, q_0) and (p_1, q_1) are two pairs of indices with $1 \leq p_j, q_j \leq \infty$, and assume that

$$\|T(f)\|_{L^{q_0}} \leq M_0\|f\|_{L^{p_0}} \quad \text{and} \quad \|T(f)\|_{L^{q_1}} \leq M_1\|f\|_{L^{p_1}}$$

where T is a linear operator. Does it follow that

$$\|T(f)\|_{L^q} \leq M\|f\|_{L^p}, \qquad \text{for other pairs } (p,q)?$$

We shall see that this inequality will hold with values of p and q determined by a linear expression involving the reciprocals of the indexes p_0, p_1, q_0 and q_1. (Linearity in the reciprocals of the exponents already arises in the relation $1/p + 1/p' = 1$ of dual exponents.)

The precise statement of the theorem requires that we fix some notation. Let (X, μ) and (Y, ν) be a pair of measure spaces. We shall abbreviate the L^p norm on (X, μ) by writing $\|f\|_{L^p} = \|f\|_{L^p(X,\mu)}$, and similarly for the L^q norm for functions on $(Y, d\nu)$. We will also consider the space $L^{p_0} + L^{p_1}$ that consists of functions on (X, μ) that can be written as $f_0 + f_1$, with $f_j \in L^{p_j}(X, \mu)$, with a similar definition for $L^{q_0} + L^{q_1}$.

Theorem 2.1 *Suppose T is a linear mapping from $L^{p_0} + L^{p_1}$ to $L^{q_0} + L^{q_1}$. Assume that T is bounded from L^{p_0} to L^{q_0} and from L^{p_1} to L^{q_1}*

$$\begin{cases} \|T(f)\|_{L^{q_0}} \leq M_0\|f\|_{L^{p_0}}, \\ \|T(f)\|_{L^{q_1}} \leq M_1\|f\|_{L^{p_1}}. \end{cases}$$

Then T is bounded from L^p to L^q,

$$\|T(f)\|_{L^q} \leq M\|f\|_{L^p},$$

whenever the pair (p, q) can be written as

$$\frac{1}{p} = \frac{1-t}{p_0} + \frac{t}{p_1} \quad \text{and} \quad \frac{1}{q} = \frac{1-t}{q_0} + \frac{t}{q_1}$$

for some t with $0 \leq t \leq 1$. Moreover, the bound M satisfies $M \leq M_0^{1-t}M_1^t$.

We should emphasize that the theorem holds for L^p spaces of *complex-valued* functions because the proof of it depends on complex analysis. Starting with the strip $0 \leq \text{Re}(z) \leq 1$ in the complex plane, our operator T will lead us to an analytic function Φ, so that the hypotheses $\|T(f)\|_{L^{q_0}} \leq M_0\|f\|_{L^{p_0}}$ and $\|T(f)\|_{L^{q_1}} \leq M_1\|f\|_{L^{p_1}}$ are encoded in the boundedness of Φ on the boundary lines $\text{Re}(z) = 0$ and $\text{Re}(z) = 1$, respectively. Moreover, the conclusion will follow from the boundedness of Φ at the point t on the real axis. (See Figure 2.)

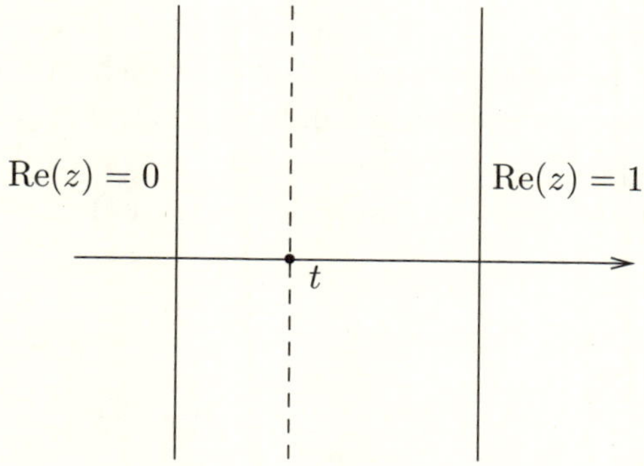

Figure 2. The domain of the function Φ

The analysis of the function Φ will depend on the following lemma.

Lemma 2.2 (Three-lines lemma) *Suppose $\Phi(z)$ is a holomorphic function in the strip $S = \{z \in \mathbb{C} : 0 < \text{Re}(z) < 1\}$, that is also continuous and bounded on the closure of S. If*

$$M_0 = \sup_{y \in \mathbb{R}} |\Phi(iy)| \quad \text{and} \quad M_1 = \sup_{y \in \mathbb{R}} |\Phi(1 + iy)|,$$

then

$$\sup_{y \in \mathbb{R}} |\Phi(t + iy)| \leq M_0^{1-t}M_1^t, \quad \text{for all } 0 \leq t \leq 1.$$

The term "three-lines" describes the fact that the size of Φ on the line $\text{Re}(z) = t$ is controlled by its size on the two boundary lines $\text{Re}(z) = 0$ and $\text{Re}(z) = 1$. The reader may note that this lemma belongs to the family of results of the Phragmén-Lindelöf type that were discussed in Chapter 4, Book II. As with other assertions of this kind, it is deduced from the more familiar maximum modulus principle, and it is here that the global assumption that Φ is bounded throughout the strip is used. Notice, however, that the size of the assumed global bound of Φ does not occur in the conclusion. (That some condition on the growth of Φ is necessary is shown in Exercise 5.)

Proof. We begin by proving the lemma under the assumption that $M_0 = M_1 = 1$ and $\sup_{0\leq x\leq 1} |\Phi(x+iy)| \to 0$ as $|y| \to \infty$. In this case, let $M = \sup|\Phi(z)|$ where the sup is taken over all z in the closure of the strip S. We may clearly assume that $M > 0$, and let $z_1, z_2, \dots$ be a sequence of points in the strip with $|\Phi(z_n)| \to M$ as $n \to \infty$. By the decay condition imposed on Φ, the points z_n cannot go to infinity, hence there exists z_0 in the closure of the strip, so that a subsequence of $\{z_n\}$ converges to z_0. By the maximum modulus principle, z_0 cannot be in the interior of the strip, (unless Φ is constant, in which case the conclusion is trivial) hence z_0 must be on its boundary, where $|\Phi| \leq 1$. Thus $M \leq 1$, and the result is proved in this special case.

If we only assume now that $M_0 = M_1 = 1$, we define

$$\Phi_\epsilon(z) = \Phi(z)e^{\epsilon(z^2-1)}, \qquad \text{for each } \epsilon > 0.$$

Since $e^{\epsilon[(x+iy)^2-1]} = e^{\epsilon(x^2-1-y^2+2ixy)}$, we find that $|\Phi_\epsilon(z)| \leq 1$ on the lines $\text{Re}(z) = 0$ and $\text{Re}(z) = 1$. Moreover,

$$\sup_{0\leq x\leq 1} |\Phi_\epsilon(x+iy)| \to 0 \qquad \text{as } |y| \to \infty,$$

since Φ is bounded. Therefore, by the first case, we know that $|\Phi_\epsilon(z)| \leq 1$ in the closure of the strip. Letting $\epsilon \to 0$, we see that $|\Phi| \leq 1$ as desired.

Finally, for arbitrary positive values of M_0 and M_1, we let $\tilde{\Phi}(z) = M_0^{z-1}M_1^{-z}\Phi(z)$, and note that $\tilde{\Phi}$ satisfies the condition of the previous case, that is, $\tilde{\Phi}$ is bounded by 1 on the lines $\text{Re}(z) = 0$ and $\text{Re}(z) = 1$. Thus $|\tilde{\Phi}| \leq 1$ in the strip, which completes the proof of the lemma.

To prove the interpolation theorem, we begin by establishing the inequality when f is a simple function, and it clearly suffices to do so with $\|f\|_{L^p} = 1$. Also, we recall that to show $\|Tf\|_{L^q} \leq M\|f\|_{L^p}$ it suffices to prove, by Lemma 4.2 in Chapter 1, that

$$\left|\int (Tf)g\,d\nu\right| \le M\|f\|_{L^p}\|g\|_{L^{q'}},$$

where $1/q + 1/q' = 1$, and g simple with $\|g\|_{L^{q'}} = 1$.

For now, we also assume that $p < \infty$ and $q > 1$. Suppose $f \in L^p$ is simple with $\|f\|_{L^p} = 1$, and define

$$f_z = |f|^{\gamma(z)}\frac{f}{|f|} \quad \text{where} \quad \gamma(z) = p\left(\frac{1-z}{p_0} + \frac{z}{p_1}\right),$$

and

$$g_z = |g|^{\delta(z)}\frac{g}{|g|} \quad \text{where} \quad \delta(z) = q'\left(\frac{1-z}{q_0'} + \frac{z}{q_1'}\right),$$

with q', q_0' and q_1' denoting the duals of q, q_0, and q_1 respectively. Then, we note that $f_t = f$, while

$$\begin{cases} \|f_z\|_{L^{p_0}} = 1 & \text{if } \mathrm{Re}(z) = 0 \\ \|f_z\|_{L^{p_1}} = 1 & \text{if } \mathrm{Re}(z) = 1. \end{cases}$$

Similarly $\|g_z\|_{L^{q_0'}} = 1$ if $\mathrm{Re}(z) = 0$ and $\|g_z\|_{L^{q_1'}} = 1$ if $\mathrm{Re}(z) = 1$, and also $g_t = g$. The trick now is to consider

$$\Phi(z) = \int (Tf_z)g_z\,d\nu.$$

Since f is a finite sum, $f = \sum a_k\chi_{E_k}$ where the sets E_k are disjoint and of finite measure, then f_z is also simple with

$$f_z = \sum |a_k|^{\gamma(z)}\frac{a_k}{|a_k|}\chi_{E_k}.$$

Since $g = \sum b_j\chi_{F_j}$ is also simple, then

$$g_z = \sum |b_j|^{\delta(z)}\frac{b_j}{|b_j|}\chi_{F_j}.$$

With the above notation, we find

$$\Phi(z) = \sum_{j,k} |a_k|^{\gamma(z)}|b_j|^{\delta(z)}\frac{a_k}{|a_k|}\frac{b_j}{|b_j|}\left(\int T(\chi_{E_k})\chi_{F_j}\,d\nu\right),$$

so that the function Φ is a holomorphic function in the strip $0 < \mathrm{Re}(z) < 1$ that is bounded and continuous in its closure. After an application of Hölder's inequality and using the fact that T is bounded on L^{p_0} with bound M_0, we find that if $\mathrm{Re}(z) = 0$, then

$$|\Phi(z)| \leq \|Tf_z\|_{L^{q_0}} \|g_z\|_{L^{q_0'}} \leq M_0 \|f_z\|_{L^{p_0}} = M_0.$$

Similarly we find $|\Phi(z)| \leq M_1$ on the line $\mathrm{Re}(z) = 1$. Therefore, by the three-lines lemma, we conclude that Φ is bounded by $M_0^{1-t} M_1^t$ on the line $\mathrm{Re}(z) = t$. Since $\Phi(t) = \int (Tf) g \, d\nu$, this gives the desired result, at least when f is simple.

In general, when $f \in L^p$ with $1 \leq p < \infty$, we may choose a sequence $\{f_n\}$ of simple functions in L^p so that $\|f_n - f\|_{L^p} \to 0$ (as in Exercise 6, Chapter 1). Since $\|T(f_n)\|_{L^q} \leq M \|f_n\|_{L^p}$, we find that $T(f_n)$ is a Cauchy sequence in L^q and if we can show that $\lim_{n \to \infty} T(f_n) = T(f)$ almost everywhere, it would follow that we also have $\|T(f)\|_{L^q} \leq M \|f\|_{L^p}$.

To do this, write $f = f^U + f^L$, where $f^U(x) = f(x)$ if $|f(x)| \geq 1$ and 0 elsewhere, while $f^L(x) = f(x)$ if $|f(x)| < 1$ and 0 elsewhere. Similarly, set $f_n = f_n^U + f_n^L$. Now assume that $p_0 \leq p_1$ (the case $p_0 \geq p_1$ is parallel). Then $p_0 \leq p \leq p_1$, and since $f \in L^p$, it follows that $f^U \in L^{p_0}$ and $f^L \in L^{p_1}$. Moreover, since $f_n \to f$ in the L^p norm, then $f_n^U \to f^U$ in the L^{p_0} norm and $f_n^L \to f^L$ in the L^{p_1} norm. By hypothesis, then $T(f_n^U) \to T(f^U)$ in L^{q_0} and $T(f_n^L) \to T(f^L)$ in L^{q_1}, and selecting appropriate subsequences we see that $T(f_n) = T(f_n^U) + T(f_n^L)$ converges to $T(f)$ almost everywhere, which establishes the claim.

It remains to consider the cases $q = 1$ and $p = \infty$. In the latter case then necessarily $p_0 = p_1 = \infty$, and the hypotheses $\|T(f)\|_{L^{q_0}} \leq M_0 \|f\|_{L^\infty}$ and $\|T(f)\|_{L^{q_1}} \leq M_1 \|f\|_{L^\infty}$ imply the conclusion

$$\|T(f)\|_{L^q} \leq M_0^{1-t} M_1^t \|f\|_{L^\infty}$$

by Hölder's inequality (as in Exercise 20 in Chapter 1).

Finally if $p < \infty$ and $q = 1$, then $q_0 = q_1 = 1$, then we may take $g_z = g$ for all z, and argue as in the case when $q > 1$. This completes the proof of the theorem.

We shall now describe a slightly different but useful way of stating the essence of the theorem. Here we assume that our linear operator T is initially defined on simple functions of X, mapping these to functions on Y that are integrable on sets of finite measure. We then ask: for which (p, q) is the operator of **type** (p, q), in the sense that there is a bound M so that

$$\|T(f)\|_{L^q} \leq M \|f\|_{L^p}, \quad \text{whenever } f \text{ is simple?} \tag{8}$$

In this formulation of the question, the useful role of simple functions is that they are at once common to all the L^p spaces. Moreover, if (8) holds then T has a unique extension to all of L^p, with the same bound M in (8), as long as either $p < \infty$; or $p = \infty$ in the case X has finite measure. This is a consequence of the density of the simple functions in L^p, and the extension argument in Proposition 5.4 of Chapter 1.

With these remarks in mind, we define the **Riesz diagram** of T to consist of all all points in the unit square $\{(x, y) : \ 0 \leq x \leq 1, \ 0 \leq y \leq 1\}$ that arise when we set $x = 1/p$ and $y = 1/q$ whenever T is of type (p, q). We then also define $M_{x,y}$ as the least M for which (8) holds when $x = 1/p$ and $y = 1/q$.

Corollary 2.3 *With T as before:*

(a) *The Riesz diagram of T is a convex set.*

(b) $\log M_{x,y}$ *is a convex function on this set.*

Conclusion (a) means that if $(x_0, y_0) = (1/p_0, 1/q_0)$ and $(x_1, y_1) = (1/p_1, 1/q_1)$ are points in the Riesz diagram of T, then so is the line segment joining them. This is an immediate consequence of Theorem 2.1. Similarly the convexity of the function $\log M_{x,y}$ is its convexity on each line segment, and this follows from the conclusion $M \leq M_0^{1-t} M_1^t$ guaranteed also by Theorem 2.1.

In view of this corollary, the theorem is often referred to as the "Riesz convexity theorem."

2.1 Some examples

EXAMPLE 1. The first application of Theorem 2.1 is the Hausdorff-Young inequality (3). Here X is $[0, 2\pi]$ with the normalized Lebesgue measure $d\theta/(2\pi)$, and $Y = \mathbb{Z}$ with its usual counting measure. The mapping T is defined by $T(f) = \{a_n\}$, with

$$a_n = \frac{1}{2\pi} \int_0^{2\pi} f(\theta) e^{-in\theta} \, d\theta.$$

Corollary 2.4 *If $1 \leq p \leq 2$ and $1/p + 1/q = 1$, then*

$$\|T(f)\|_{L^q(\mathbb{Z})} \leq \|f\|_{L^p([0,2\pi])}.$$

Note that since $L^2([0, 2\pi]) \subset L^1([0, 2\pi])$ and $L^2(\mathbb{Z}) \subset L^\infty(\mathbb{Z})$ we have $L^2([0, 2\pi]) + L^1([0, 2\pi]) = L^1([0, 2\pi])$, and also $L^2(\mathbb{Z}) + L^\infty(\mathbb{Z}) = L^\infty(\mathbb{Z})$.

The inequality for $p_0 = q_0 = 2$ is a consequence of Parseval's identity, while the one for $p_1 = 1$, $q_1 = \infty$ follows from the observation that for all n,

$$|a_n| \leq \frac{1}{2\pi} \int_0^{2\pi} |f(\theta)| \, d\theta.$$

Thus Riesz's theorem guarantees the conclusion when $1/p = \frac{(1-t)}{2} + t$, $1/q = \frac{(1-t)}{2}$ for any t with $0 \leq t \leq 1$. This gives all p with $1 \leq p \leq 2$, and q related to p by $1/p + 1/q = 1$.

EXAMPLE 2. We next come to the dual Hausdorff-Young inequality (4). Here we define the operator T' mapping functions on $\mathbb{Z}$ to functions on $[0, 2\pi]$ by

$$T'(\{a_n\}) = \sum_{n=-\infty}^{\infty} a_n e^{in\theta}.$$

Notice that since $L^p(\mathbb{Z}) \subset L^2(\mathbb{Z})$ when $p \leq 2$, then the above is a well-defined function on $L^2([0, 2\pi])$ when $\{a_n\} \in L^p(\mathbb{Z})$, by the unitary character of Parseval's identity.

Corollary 2.5 *If* $1 \leq p \leq 2$ *and* $1/p + 1/q = 1$, *then*

$$\|T'(\{a_n\})\|_{L^q([0,2\pi])} \leq \|\{a_n\}\|_{L^p(\mathbb{Z})}.$$

The proof is parallel to that of the previous corollary. The case $p_0 = q_0 = 2$ is, as has already been mentioned, a consequence of Parseval's identity, while the case $p_1 = 1$ and $q_1 = \infty$ follows directly from the fact that

$$\left| \sum_{n=-\infty}^{\infty} a_n e^{in\theta} \right| \leq \sum_{n=-\infty}^{\infty} |a_n|.$$

An alternative proof of this corollary uses Corollary 2.4 as well as Theorem 4.1 and Theorem 5.5 in the previous chapter.

EXAMPLE 3. We consider the analog for the Fourier transform. Here the setting is $\mathbb{R}^d$ and the L^p spaces are taken with respect to the usual Lebesgue measure. We initially define the Fourier transform (denoted here by T) on simple functions by

$$T(f)(\xi) = \int_{\mathbb{R}^d} f(x) e^{-2\pi i x \cdot \xi} \, dx.$$

Then clearly, $\|T(f)\|_{L^\infty} \leq \|f\|_{L^1}$, and T has an extension (by Proposition 5.4 in Chapter 1 for instance) to $L^1(\mathbb{R}^d)$ for which this inequality continues to hold. Also, T has an extension to $L^2(\mathbb{R}^d)$ as a unitary mapping. (This is essentially the content of Plancherel's theorem. See Section 1, Chapter 5 in Book III.) Thus in particular $\|T(f)\|_{L^2} \leq \|f\|_{L^2}$, for f simple.

The same arguments as before then prove:

Corollary 2.6 *If $1 \leq p \leq 2$ and $1/p + 1/q = 1$, then the Fourier transform T has a unique extension to a bounded map from L^p to L^q, with $\|T(f)\|_{L^q} \leq \|f\|_{L^p}$.*

We summarize these results by describing in Figure 3 the Riesz diagrams for each of the above versions of the Hausdorff-Young theorem. The three variants are as follows:

(i) The operator T in Corollary 2.4: the closed triangle I.

(ii) The operator T' in Corollary 2.5: the closed triangle II.

(iii) The operator T in Corollary 2.6: the line segment joining $(1, 0)$ to $(1/2, 1/2)$, that is, the common boundary of these two triangles.

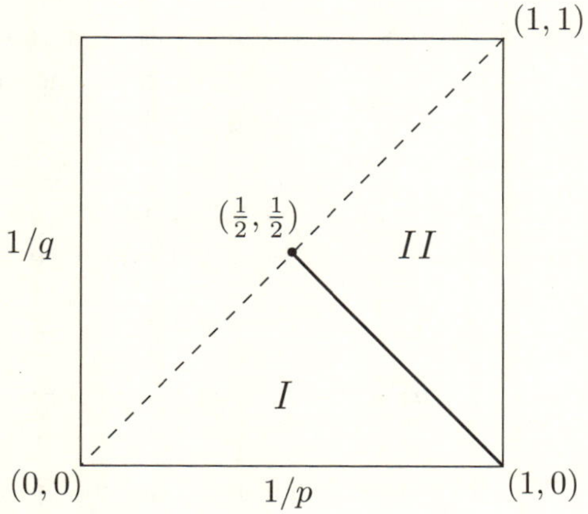

Figure 3. Riesz diagrams for the Hausdorff-Young theorem

More precisely, the results above guarantee the inequality for the segment joining $(1, 0)$ to $(1/2, 1/2)$ in each of the three cases. If we use the trivial inequality $\|f\|_{L^1} \leq \|f\|_{L^\infty}$ in Example 1 above, we get that the point $(0,0)$ also belongs to the Riesz diagram of T, yielding the closed triangle I. Similarly, because $\|T'(\{a_n\})\|_{L^\infty} \leq \|\{a_n\}\|_{L^1}$, we obtain the triangle II for Example 2. Finally, we should note that in Example 3, the Fourier transform, the Riesz diagram consists of no more than the segment joining $(1,0)$ to $(1/2,1/2)$. (See Exercises 2 and 3.)

EXAMPLE 4. Our last illustration is Young's inequality for convolutions in $\mathbb{R}^d$. It states that whenever f and g are a pair of functions in L^p and L^r respectively, then the convolution

$$(f * g)(x) = \int_{\mathbb{R}^d} f(x-y)g(y)\,dy$$

is well-defined (that is, the function $f(x-y)g(y)$ is integrable for almost every x), and moreover

$$\|f * g\|_{L^q} \leq \|f\|_{L^p}\|g\|_{L^r}, \tag{9}$$

under the assumption that $1/q = 1/p + 1/r - 1$, (with $1 \leq q \leq \infty$). One proof of this has been outlined in Exercise 19 of the previous chapter. Here we point out that it is also a consequence of the similar special cases corresponding to $p = 1$, and p the dual exponent of r. In fact it suffices to prove (9) for simple functions f and g, and then pass to the general case by an easy limiting argument. With this in mind, fix g, and consider the map T defined by $T(f) = f * g$. We know (see Exercise 17 (a) in Chapter 1, where the role of f are g are interchanged) that $\|T(f)\|_{L^r} \leq M\|f\|_{L^1}$, with $M = \|g\|_{L^r}$. Also by Hölder's inequality, $\|T(f)\|_{L^\infty} \leq M\|f\|_{L^{r'}}$, where $1/r' + 1/r = 1$. Now applying the Riesz interpolation theorem gives the desired result.

There is of course the parallel situation of the periodic case. For example, in one dimension, taking the functions with period 2π, the **convolution** of f and g is defined by

$$(f * g)(\theta) = \frac{1}{2\pi}\int_0^{2\pi} f(\theta-\varphi)g(\varphi)\,d\varphi.$$

If we set $L^p = L^p([0,2\pi])$ with the underlying measure $d\theta/(2\pi)$, then one has again $\|f * g\|_{L^q} \leq \|f\|_{L^p}\|g\|_{L^r}$, but automatically in a larger range because $\|g\|_{L^{\overline{r}}} \leq \|g\|_{L^r}$, whenever $\overline{r} \leq r$.

The Riesz diagrams are described as follows (Figure 4):

The solid line segment joining $(1-1/r,0)$ to $(1,1/r)$ represents Young's inequality for $\mathbb{R}^d$. The closed (shaded) trapezoid represents the inequality in the periodic case.

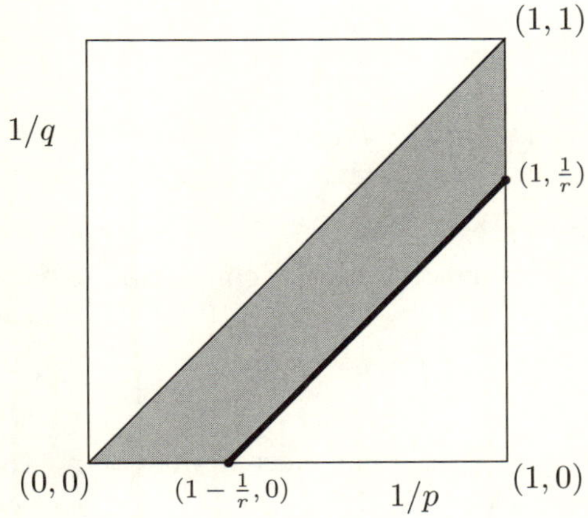

Figure 4. Riesz diagrams for $T(f) = f * g$, with $g \in L^r$

3 The L^p theory of the Hilbert transform

We carry out the theory of the "conjugate function," alluded to earlier in Section 1, but we do it in the parallel framework where the unit circle and the unit disc are replaced by $\mathbb{R}$ and the upper half-plane $\mathbb{R}^2_+ = \{z = x + iy,\ x \in \mathbb{R},\ y > 0\}$, respectively. While the technical details of the proofs are a little more involved in the latter context, the resulting formulas are more elegant and their form leads more directly to important generalizations in higher dimensions.

3.1 The L^2 formalism

We begin by setting down the basic formalism connecting the Hilbert transform and the projection operator arising from the Cauchy integral. Starting with an appropriate function f on $\mathbb{R}$ we define its Cauchy integral by

$$F(z) = C(f)(z) = \frac{1}{2\pi i}\int_{-\infty}^{\infty} \frac{f(t)}{t - z}\, dt, \qquad \mathrm{Im}(z) > 0. \tag{10}$$

For the moment we restrict ourselves to f in $L^2(\mathbb{R})$. Then of course the integral converges for all $z = x + iy$ with $y > 0$, (because $1/(t - z)$ is in $L^2(\mathbb{R})$ as a function of t) and $F(z)$ is holomorphic in the upper half-plane. We can also represent the Cauchy integral F in terms of the L^2 Fourier transform $\hat{f}$ of f as[3]

$$F(z) = \int_0^\infty \hat{f}(\xi)e^{2\pi i z\xi}\,d\xi, \qquad \text{Im}(z) > 0. \tag{11}$$

This integral converges because $e^{-2\pi y\xi}$ as a function of ξ is in $L^2(0,\infty)$, for $y > 0$. The above representation comes about because of the formula

$$\int_0^\infty e^{2\pi i z\xi}\,d\xi = -\frac{1}{2\pi i z}, \tag{12}$$

which holds for $\text{Im}(z) > 0$. (For more details about these assertions, and their connection to the Hardy space H^2, see Section 2, Chapter 5 in Book III.)

As is clear from (11) and Plancherel's theorem, one has $F(x+iy) \to P(f)(x)$, as $y \to 0$, in the $L^2(\mathbb{R})$ norm with

$$P(f)(x) = \int_{-\infty}^\infty \hat{f}(\xi)\chi(\xi)e^{2\pi i x\xi}\,d\xi$$

and χ the characteristic function of $(0,\infty)$. Thus P is the orthogonal projection of $L^2(\mathbb{R})$ to the subspace of those f for which $\hat{f}(\xi) = 0$ for almost every $\xi < 0$. So as in (5) of Section 1, one is led to define the **Hilbert transform** H by

$$H(f)(x) = \int_{-\infty}^\infty \hat{f}(\xi)\frac{\text{sign}(\xi)}{i}e^{2\pi i x\xi}\,d\xi. \tag{13}$$

Some elementary facts, following directly from the definitions of P and H, are worth noting:

- $P = \frac{1}{2}(I + iH)$, where I is the identity operator.
- H is unitary on L^2, and $H \circ H = H^2 = -I$.

In other words, $\|H(f)\|_{L^2} = \|f\|_{L^2}$, and H is invertible with $H^{-1} = -H$.

We now come to the important realization of the Hilbert transform as a "singular integral." It can be stated as follows.

Proposition 3.1 *If $f \in L^2(\mathbb{R})$ then*

$$H(f)(x) = \lim_{\epsilon\to 0}\frac{1}{\pi}\int_{|t|\geq\epsilon} f(x-t)\frac{dt}{t}. \tag{14}$$

[3]The Fourier transforms in the definitions below are taken in the L^2 sense, via Plancherel's theorem.

That is, with $H_\epsilon(f)$ denoting the integral on the right-hand side above, we have $H_\epsilon(f) \in L^2(\mathbb{R})$ for every $\epsilon > 0$, and the convergence asserted in (14) is in the $L^2(\mathbb{R})$ norm.

First, we make a few observations. Note that with $z = x + iy$, then

$$-\frac{1}{i\pi z} = \mathcal{P}_y(x) + i\mathcal{Q}_y(x) \tag{15}$$

where

$$\mathcal{P}_y(x) = \frac{y}{\pi(x^2+y^2)} \quad \text{and} \quad \mathcal{Q}_y(x) = \frac{x}{\pi(x^2+y^2)}$$

are called the **Poisson kernel** and **conjugate Poisson kernel**, respectively. Then because of (10), (11) and (15)

$$\int_0^\infty \hat{f}(\xi)e^{2\pi i z\xi}\,d\xi = \frac{1}{2}\left[(f*\mathcal{P}_y)(x) + i(f*\mathcal{Q}_y)(x)\right], \tag{16}$$

where $(f*\mathcal{P}_y)(x) = \int f(x-t)\mathcal{P}_y(t)\,dt = \int f(t)\mathcal{P}_y(x-t)\,dt$, with similar formulas for $f*\mathcal{Q}_y$.

Next define the reflection $\varphi \mapsto \varphi^\sim$ by $\varphi^\sim(x) = \varphi(-x)$, and observe that $(f*\mathcal{P}_y)^\sim = f^\sim * \mathcal{P}_y$, while $(f*\mathcal{Q}_y)^\sim = -(f^\sim * \mathcal{Q}_y)$, since $\mathcal{P}_y$ and $\mathcal{Q}_y$ are respectively even and odd functions of x. Also $\widehat{(f^\sim)} = (\hat{f})^\sim$. Therefore using (16) with f and $f^\sim$ we then obtain

$$\begin{aligned} (f*\mathcal{P}_y)(x) &= \int_{-\infty}^{\infty} \hat{f}(\xi)e^{2\pi i x\xi}e^{-2\pi y|\xi|}\,d\xi \\ (f*\mathcal{Q}_y)(x) &= \int_{-\infty}^{\infty} \hat{f}(\xi)e^{2\pi i x\xi}e^{-2\pi y|\xi|}\frac{\operatorname{sign}(\xi)}{i}\,d\xi. \end{aligned} \tag{17}$$

As a result, we obtain that the Fourier transforms of $\mathcal{P}_y$ and $\mathcal{Q}_y$ (taken in L^2) are given by

$$\begin{aligned} \widehat{\mathcal{P}_y}(\xi) &= e^{-2\pi y|\xi|} \\ \widehat{\mathcal{Q}_y}(\xi) &= e^{-2\pi y|\xi|}\frac{\operatorname{sign}(\xi)}{i}. \end{aligned} \tag{18}$$

With this we turn to the proof of the proposition. We note, by (13), (17), (18), and Plancherel's theorem, that $f*\mathcal{Q}_\epsilon \to H(f)$ in the L^2 norm, as $\epsilon \to 0$. Now consider

$$\frac{1}{\pi}\int_{|t|\geq\epsilon} f(x-t)\frac{dt}{t} - (f*\mathcal{Q}_\epsilon)(x) = H_\epsilon(f)(x) - (f*\mathcal{Q}_\epsilon)(x).$$

This difference equals $f*\Delta_\epsilon$, where

$$\begin{aligned} \Delta_\epsilon(x) &= \tfrac{1}{\pi x} - \mathcal{Q}_\epsilon(x), &\quad \text{for } |x| \geq \epsilon \\ &= -\mathcal{Q}_\epsilon(x), &\quad \text{for } |x| < \epsilon. \end{aligned}$$

It is important to observe that $\Delta_\epsilon(x) = \epsilon^{-1}\Delta_1(\epsilon^{-1}x)$, while $|\Delta_1(x)| \leq A/(1+x^2)$, since $1/x - x/(x^2+1) = O(1/x^3)$, if $|x| \geq 1$.[4] In particular Δ_1 is integrable over $\mathbb{R}$ and the family of kernels $\Delta_\epsilon(x)$ satisfies the usual size conditions for an approximation to the identity,[5] but not the condition $\int \Delta_\epsilon(x)\,dx = 1$. Instead $\int \Delta_\epsilon(x)\,dx = 0$, for all $\epsilon \neq 0$, because $\Delta_\epsilon(x)$ is an odd function of x. As a consequence

$$f * \Delta_\epsilon \to 0 \quad \text{in the } L^2 \text{ norm, as } \epsilon \to 0, \tag{19}$$

and this gives that $H_\epsilon(f) \to H(f)$ in the L^2 norm, as $\epsilon \to 0$.

We recall briefly how (19) can be proved. First

$$\begin{aligned}(f * \Delta_\epsilon)(x) &= \int f(x-t)\Delta_\epsilon(t)\,dt = \int (f(x-t) - f(x))\Delta_\epsilon(t)\,dt \\ &= \int (f(x-\epsilon t) - f(x))\Delta_1(t)\,dt.\end{aligned}$$

Then by Minkowski's inequality

$$\|f * \Delta_\epsilon\|_{L^2} \leq \int \|f(x-\epsilon t) - f(x)\|_{L^2}|\Delta_1(t)|\,dt.$$

Now, the integral tends to zero with ϵ by the dominated convergence theorem. This is because $\|f(x-\epsilon t) - f(x)\|_{L^2} \leq 2\|f\|_{L^2}$, and $\|f(x-\epsilon t) - f(x)\|_{L^2} \to 0$ as $\epsilon \to 0$ for each t. (For the continuity of the L^2 norm used here, see Exercise 8 in Chapter 1.)

Remark. The above argument shows also that $\|H_\epsilon(f)\|_{L^2} \leq A\|f\|_{L^2}$ with A independent of ϵ and f.

3.2 The L^p theorem

With the elementary properties of the Hilbert transform established we can now turn to our goal: the theorem of M. Riesz. It states that the Hilbert transform is bounded on L^p, $1 < p < \infty$. One way to formulate this is as follows.

Theorem 3.2 *Suppose $1 < p < \infty$. Then the Hilbert transform H, initially defined on $L^2 \cap L^p$ by (13) or (14), satisfies the inequality*

$$\|H(f)\|_{L^p} \leq A_p\|f\|_{L^p}, \quad \textit{whenever } f \in L^2 \cap L^p, \tag{20}$$

[4] We remind the reader of the notation $f(x) = O(g(x))$, which means that $|f(x)| \leq C|g(x)|$ for some constant C and all x in a given range.

[5] A discussion of approximations to the identity can be found, for instance, in Book III, Section 2 and Exercise 2 of Chapter 3.

with a bound A_p independent of f. The Hilbert transform then has a unique extension to all of L^p satisfying the same bound.[6]

To have a better appreciation of the nature of this theorem it may help to see why the conclusions fail for $p = 1$ or $p = \infty$. For this, an explicit calculation does the job. Let I denote the interval $(-1, 1)$, and $f = \chi_I$ be the characteristic function of that interval. Now f is an even function, so its Hilbert transform is odd, and in fact a simple calculation gives $H(f)(x) = \lim_{\epsilon\to 0} H_\epsilon(f)(x) = \frac{1}{\pi}\log\left|\frac{x+1}{x-1}\right|$. Hence $H(f)$ is unbounded near $x = -1$ and $x = 1$, with mild (logarithmic) singularities there. However $H(f)(x) \sim \frac{2}{\pi x}$ as $|x| \to \infty$, so it is obvious that $H(f)$ does not belong to L^1.

It is also instructive to consider instead of $f = \chi_I$, the odd function $g(x) = \chi_J(x) - \chi_J(-x)$, where $J = (0, 1)$. Then the Hilbert transform of g equals $H(g)(x) = \frac{1}{\pi}\log\left|\frac{x^2}{x^2-1}\right|$, and is an even function. While $H(g)$ is still unbounded (with mild logarithmic singularities at -1, 0 and 1), it is integrable on $\mathbb{R}$, since $H(g)(x) \sim \frac{1}{\pi x^2}$, as $|x| \to \infty$. (See Figure 5.)

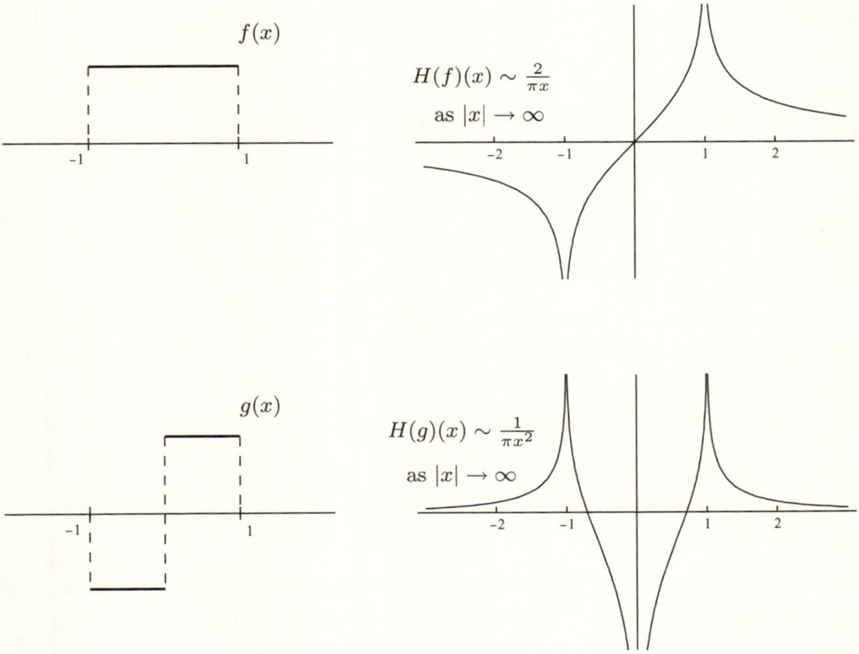

Figure 5. Two examples of Hilbert transforms

[6]For the general extension principle used, see Proposition 5.4 in Chapter 1.

There is a nice lesson here whose significance will be clear at several stages later on: namely, if f is (say) a bounded function with compact support on $\mathbb{R}$, then $H(f)$ is in $L^1(\mathbb{R})$ if and only if $\int f(x)\,dx = 0$. (See Exercise 7.)

3.3 Proof of Theorem 3.2

The main idea of the proof was already outlined at the end of Section 1 in the context of Fourier series and the corresponding theorem for the conjugate function. While this proof, which depends on complex analysis, is elegant, its approach is essentially limited to this operator and cannot deal with the generalizations of the Hilbert transform in the setting of $\mathbb{R}^d$. The "real-variable" theory of those operators will be described in Section 3 of the next chapter.

We turn to the proof of the Theorem 3.2, and in preparation we invoke two technical devices. The first is very simple and is the realization that it suffices to prove the theorem for real-valued functions, from which its extension to complex-valued functions is immediate (with a resulting bound which is not more than twice the bound A_p for real-valued functions).

The second device depends on the use of the space $C_0^\infty(\mathbb{R})$ of indefinitely differentiable functions of compact support. There are two useful facts concerning this space. First, it is dense in $L^p(\mathbb{R})$, and more particularly, if $f \in L^2 \cap L^p$, with $p < \infty$, there is a sequence $\{f_n\}$ with $f_n \in C_0^\infty$, and $f_n \to f$ both in the L^2 and L^p norms. (This follows from the argument to solve Exercise 7 in Chapter 1 as well as the references therein.)

For our purposes, a particularly helpful observation is that whenever $f \in C_0^\infty(\mathbb{R})$ then its Cauchy integral $F(z) = \frac{1}{2\pi i}\int_{-\infty}^{\infty} \frac{f(t)}{t-z}\,dt$ extends as a continuous function on the closure of the upper half-plane, is bounded there, and moreover satisfies the decay inequality

$$|F(z)| \leq \frac{M}{1+|z|}, \qquad z = x+iy,\ y \geq 0, \tag{21}$$

for an appropriate constant M. The simplest way to prove this is to use the Fourier transform representation (11). Then the rapid decrease at infinity of $\hat{f}$ shows that F is continuous and bounded in the closed half-plane $\overline{\mathbb{R}^2_+}$. Moreover the smoothness of $\hat{f}$ lets us integrate by parts, giving

$$F(z) = \frac{1}{2\pi i z}\int_0^\infty \frac{d(e^{2\pi i z\xi})}{d\xi}\hat{f}(\xi)\,d\xi = \frac{1}{2\pi i z}\left[-\int_0^\infty e^{2\pi i z\xi}\hat{f}'(\xi)\,d\xi - \hat{f}(0)\right].$$

As a result, $|F(z)| \leq M_0/|z|$, so together with the boundedness of F the estimate (21) is established. Notice also that the continuity of F with (11), (16) and (17) yields

$$2F(x) = 2\lim_{y\to 0} F(x+iy) = f(x) + iH(f)(x). \tag{22}$$

It is also important to remark here that if f is real-valued (as we have assumed), then by (14) the Hilbert transform $H(f)$ is also real-valued.

With these matters out of the way, the main conclusions can be obtained in a few strokes.

Step 1: *Cauchy's theorem.* We see first that

$$\int_{-\infty}^{\infty} (F(x))^k\, dx = 0, \qquad \text{whenever } k \text{ is an integer, } k \geq 2. \tag{23}$$

Indeed, if we integrate the analytic function $(F(z))^k$ over the contour γ in the upper half-plane consisting of the rectangle (see Figure 6) whose vertices are $R+i\epsilon$, $R+iR$, $-R+iR$, and $-R+i\epsilon$, then by Cauchy's theorem $\int_\gamma (F(z))^k\, dz = 0$. Letting $\epsilon \to 0$ and $R \to \infty$, also taking into account the continuity of F and the decay (21) then gives (23). (Note also that by (21), we have $H(f) \in L^p$ for all $p > 1$.)

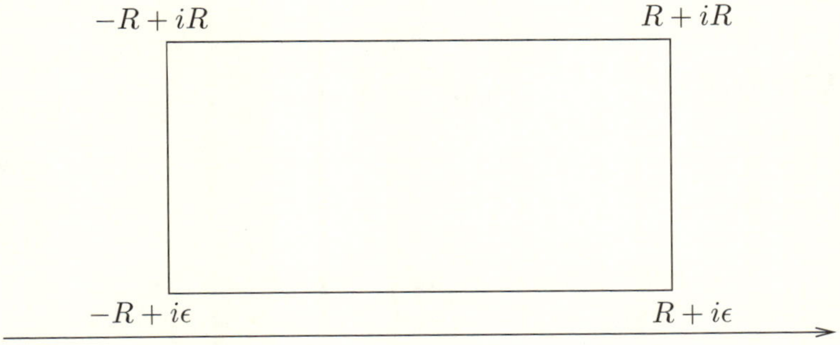

Figure 6. The rectangle of integration γ

We now exploit (23). Observe that when $k = 2$, if we take the real parts of this identity (using that f and $H(f)$ are real-valued), we have $\int_{-\infty}^{\infty}(f^2 - (Hf)^2)\, dx = 0$. This is essentially the unitarity of H on L^2 that we mentioned previously.

Next we consider other values of $k \geq 2$, those when k is even $k = 2\ell$. (When k is odd, the identity (23) does not have an immediately useful consequence.) Suppose, for example, that $k = 4$. Then the real part of (23) gives us

$$\int f^4\,dx - 6\int f^2(Hf)^2\,dx + \int (Hf)^4\,dx = 0.$$

As a result,

$$\int (Hf)^4\,dx \le 6\int f^2(Hf)^2\,dx \le 6\left(\int f^4\,dx\right)^{1/2}\left(\int (Hf)^4\,dx\right)^{1/2},$$

the last majorization following by Schwarz's inequality. Hence

$$\left(\int (Hf)^4\,dx\right)^{1/2} \le 6\left(\int f^4\,dx\right)^{1/2},$$

which means

$$\|H(f)\|_{L^4} \le 6^{1/2}\|f\|_{L^4}.$$

In the same way, if we take $p = 2\ell$, with ℓ an integer ≥ 1, we obtain

$$\|H(f)\|_{L^p} \le A_p\|f\|_{L^p}, \qquad p = 2\ell. \tag{24}$$

Indeed, the real part of $(f + iH(f))^{2\ell}$ is

$$\sum_{r=0}^{\ell} f^{2r}(Hf)^{2\ell-2r}c_r, \qquad \text{where } c_r = (-1)^{\ell-r}\binom{2\ell}{2r},\ r = 0, 1, \ldots, \ell.$$

Hence

$$\int (Hf)^{2\ell}\,dx \le \sum_{r=1}^{\ell} a_r \int f^{2r}(Hf)^{2\ell-2r}\,dx,$$

with $a_r = \binom{2\ell}{2r}$. Now Hölder's inequality (with dual exponents $\frac{2\ell}{2r}$, $\frac{2\ell}{2\ell-2r}$) shows that

$$\int f^{2r}(Hf)^{2\ell-2r}\,dx \le \|f\|_{L^p}^{2r}\|H(f)\|_{L^p}^{2\ell-2r},$$

with $p = 2\ell$. Thus

$$\|H(f)\|_{L^p}^{2\ell} \le \sum_{r=1}^{\ell} a_r\|f\|_{L^p}^{2r}\|H(f)\|_{L^p}^{2\ell-2r}.$$

Note that this inequality is jointly homogeneous of degree 2ℓ in $\|f\|_{L^p}$ and $\|H(f)\|_{L^p}$. Moreover the right-hand side is of degree at most $2\ell - 2$ in $\|H(f)\|_{L^p}$. Upon normalizing f so that $\|f\|_{L^p} = 1$, and setting $X = \|H(f)\|_{L^p}$ we have $X^{2\ell} \leq \sum_{r=1}^{\ell} a_r X^{2\ell-2r}$. Now either $X < 1$ or $X \geq 1$. In the second case, then $X^{2\ell} \leq (\sum_{r=1}^{\ell} a_r) X^{2\ell-2}$. As a result $X^2 \leq \sum_{r=1}^{\ell} a_r \leq 2^{2\ell}$. In either case $X \leq 2^{\ell}$, and therefore (24) is proved with $A_p = 2^{p/2}$.

To carry out the next step we extend the basic inequality (24), proved for $f \in C_0^\infty$, to f that are simple functions. Recall that we have already defined $H(f)$ whenever f is in L^2, and in particular if f is simple. Next, since such f belongs to $L^2 \cap L^p$, we can find a sequence $\{f_n\}$, with $f_n \in C_0^\infty$, so that $f_n \to f$ both in the L^2 and L^p norms. As a result, $\{H(f_n)\}$ are Cauchy sequences in both the L^2 and L^p norms, while $H(f_n) \to H(f)$ in the L^2 norm. Thus (24) is established when f is simple.

Step 2: *Interpolation.* Having proved (24) for simple functions and p even, we can apply the Riesz interpolation theorem once we have extended H to complex-valued functions. But this is easily done by setting $H(f_1 + if_2) = H(f_1) + iH(f_2)$, for f_1 and f_2 real-valued. Note that as a result, the inequality (24) extends to this case, but with A_p replaced by $2A_p$. (By a further argument we can show that the original bound A_p holds in this case also. See Exercise 8.)

With this in mind Riesz interpolation yields the inequality

$$\|H(f)\|_{L^p} \leq A_p \|f\|_{L^p}$$

for all p such that $2 \leq p \leq 2\ell$, where ℓ is any positive integer. This follows by taking $p_0 = q_0 = 2$, $p_1 = q_1 = 2\ell$ and noting that if $1/p = (1-t)/2 + t/(2\ell)$, then p ranges over the interval $2 \leq p \leq 2\ell$, when t ranges over $0 \leq t \leq 1$. Since ℓ may be taken to be arbitrarily large, we get (20) for all $2 \leq p < \infty$ and f simple.

Step 3: *Duality.* We pass from the case $2 \leq p < \infty$ to the case $1 < p \leq 2$ by duality. This passage is based on the simple identity

$$\int_{-\infty}^{\infty} (Hf)\overline{g}\, dx = -\int_{-\infty}^{\infty} f(\overline{Hg})\, dx \tag{25}$$

whenever f and g belong to $L^2(\mathbb{R})$ and are now allowed to be complex-valued. In fact this follows immediately from Plancherel's identity $(f, g) = (\hat{f}, \hat{g})$, and the definition (13), which can be restated as

$$\widehat{H(f)}(\xi) = \frac{\text{sign}(\xi)}{i} \hat{f}(\xi).$$

One can invoke the abstract duality principle in Theorem 5.5 of Chapter 1 or proceed directly as follows. Restricting attention to f and g simple, one has by Lemma 4.2 in the previous chapter, with $1 < p \leq 2$,

$$\|H(f)\|_{L^p} = \sup_g \left| \int H(f)\overline{g}\, dx \right|,$$

where the supremum is taken over g simple, with $\|g\|_{L^q} \leq 1$, $1/p + 1/q = 1$. However, by (25) and Hölder's inequality, this is equal to

$$\sup_g \left| \int f\overline{H(g)}\, dx \right| \leq \sup_g \|f\|_{L^p} \|\overline{H(g)}\|_{L^q} \leq \|f\|_{L^p} A_q,$$

using (20) for q in place of p, and noting that $2 \leq q < \infty$.

Therefore (20) holds for all p, $1 < p < \infty$, for all simple functions f. The passage to all $f \in L^2 \cap L^p$, and thus to the general result, is by now a familiar limiting argument.

4 The maximal function and weak-type estimates

Another important illustration of the occurrence of L^p spaces is in connection with the maximal function f^*. For appropriate functions f given on $\mathbb{R}^d$, the **maximal function** f^* is defined by

$$f^*(x) = \sup_{x \in B} \frac{1}{m(B)} \int_B |f(y)|\, dy,$$

where the supremum is taken over all balls B containing x, and m (as well as dy) denote the Lebesgue measure.[7]

It is a fact that f^* plays a role in a wide variety of questions in analysis, and it is there that its L^p inequality

$$\text{(26)} \qquad \|f^*\|_{L^p} \leq A_p \|f\|_{L^p}, \qquad 1 < p \leq \infty,$$

is of crucial interest.

Before we come to the proof of (26) a few observations are in order. First, the mapping $f \mapsto f^*$ is not linear, but does satisfy the sub-additive property that $f^* \leq f_1^* + f_2^*$, whenever $f = f_1 + f_2$.

Next, while (26) obviously holds for $p = \infty$ (with $A_\infty = 1$), the inequality for $p = 1$ fails. This can be seen directly by taking f to be

[7] An introduction to f^*, and a complete proof of (27) below can be found for instance in Chapter 3 of Book III.

the characteristic function of the unit ball B, and noticing that then $f^*(x) \geq 1/(1+|x|)^d$. This function clearly fails to be integrable at infinity. The asserted inequality follows immediately from the fact that for each $x \in \mathbb{R}^d$ the ball of radius $1+|x|$ centered at x contains B. There are also simple examples where the integrability of f^* fails locally. (See Exercise 12.)

There is nevertheless a very useful substitute for L^1 boundedness for f^*. It is the **weak-type** inequality: there is a bound A (independent of f), so that

$$(27) \qquad m(\{x:\ f^*(x) > \alpha\}) \leq \frac{A}{\alpha}\|f\|_{L^1(\mathbb{R}^d)}, \qquad \text{for all } \alpha > 0.$$

We briefly recall the main steps in the proof of (27). If we denote by $E_\alpha = \{x:\ f^*(x) > \alpha\}$, then to obtain the above majorization for $m(E_\alpha)$ it suffices to have the same for $m(K)$, where K is any compact subset of E_α. Now, using the definition of f^* we can cover K by a finite collection of balls $B_1, B_2, \dots, B_N$ with $\int_{B_i} |f(x)|\,dx \geq \alpha\, m(B_i)$, for each i. If we then apply a Vitali covering lemma, we can select a disjoint sub-collection of these balls $B_{i_1}, B_{i_2}, \dots, B_{i_n}$ with $\sum_{j=1}^n m(B_{i_j}) \geq 3^{-d} m(K)$. Adding the above inequalities over the disjoint balls then gives $m(K) \leq \frac{3^d}{\alpha}\|f\|_{L^1}$, which leads to (27).

4.1 The L^p inequality

We turn to the proof of the L^p inequality for the maximal function. It is formulated as follows.

Theorem 4.1 *Suppose $f \in L^p(\mathbb{R}^d)$ with $1 < p \leq \infty$. Then $f^* \in L^p(\mathbb{R}^d)$, and (26) holds, namely*

$$\|f^*\|_{L^p} \leq A_p \|f\|_{L^p}.$$

The bound A_p depends on p but is independent of f.

Let us first see why $f^*(x) < \infty$, for a.e. x, whenever $f \in L^p$. Observe that we can decompose $f = f_1 + f_\infty$, where $f_1(x) = f(x)$ if $|f(x)| > 1$, and $f_1(x) = 0$ elsewhere; also $f_\infty(x) = f(x)$ if $|f(x)| \leq 1$ and $f_\infty(x) = 0$ elsewhere. Then $f_1 \in L^1$ and $f_\infty \in L^\infty$. But clearly $f^* \leq f_1^* + f_\infty^* \leq f_1^* + 1$, since $|f_\infty(x)| \leq 1$ everywhere. Now from (27) (with f_1 in place of f), we see that f_1^* is finite almost everywhere. Thus the same is true for f^*.

The proof that $f^* \in L^p$ relies on a more quantitative version of the argument just given. We strengthen the weak-type inequality (27) by incorporating in it the L^∞ boundedness of the mapping $f \mapsto f^*$. The stronger version states

$$m(\{x:\ f^*(x) > \alpha\}) \leq \frac{A'}{\alpha} \int_{|f|>\alpha/2} |f|\, dx, \qquad \text{for all } \alpha > 0. \tag{28}$$

Here A' is a different constant; it can be taken to be $2A$. The improvement of (27), (except for a different constant, which is inessential), is that here we only integrate over the set where $|f(x)| > \alpha/2$, instead of the whole of $\mathbb{R}^d$.

To prove (28) we write $f = f_1 + f_\infty$, where now $f_1(x) = f(x)$, if $|f(x)| > \alpha/2$, and $f_\infty(x) = f(x)$ if $|f(x)| \leq \alpha/2$. Then $f^* \leq f_1^* + f_\infty^* \leq f_1^* + \alpha/2$, since $|f_\infty(x)| \leq \alpha/2$ for all x. Therefore $\{x:\ f^*(x) > \alpha\} \subset \{x:\ f_1^* > \alpha/2\}$, and applying the weak-type inequality (27) to f_1 in place of f (and $\alpha/2$ in place of α) then immediately yields (28), with $A' = 2A$.

Distribution function

We will next need an observation concerning the quantity occurring on the left-hand side of the inequalities (27) and (28), which we formulate more generally as follows. Suppose F is any non-negative measurable function. Then its **distribution function**, $\lambda(\alpha) = \lambda_F(\alpha)$ is defined for positive α by

$$\lambda(\alpha) = m(\{x:\ F(x) > \alpha\}).$$

The key point here is that for any $0 < p < \infty$,

$$\int_{\mathbb{R}^d} (F(x))^p\, dx = \int_0^\infty \lambda(\alpha^{1/p})\, d\alpha, \tag{29}$$

and this holds in the extended sense (that is, both sides are simultaneously finite and equal, or both sides are infinite).

To see this, consider first the case $p = 1$. Then the identity is an immediate consequence of Fubini's theorem, in the setting $\mathbb{R}^d \times \mathbb{R}^+$, applied to the characteristic function of the set $\{(x, \alpha):\ F(x) > \alpha > 0\}$. Indeed, integrating the characteristic function first in α then in x gives $\int_{\mathbb{R}^d} \left(\int_0^{F(x)} d\alpha\right) dx$, while integrating in the reverse order yields $\int_0^\infty m(\{x:\ F(x) > \alpha\})\, d\alpha$, and this shows (29) for $p = 1$. Finally, let $G(x) = (F(x))^p$, so $\{x:\ G(x) > \alpha\} = \{x : F(x) > \alpha^{1/p}\}$. Using (29) for $p = 1$ (and G instead of F) then gives the conclusion for general p.

We also note that

$$\lambda(\alpha) \leq \frac{1}{\alpha} \int_{\mathbb{R}^d} F(x)\, dx,$$

which is **Tchebychev's inequality**. In fact,

$$\int_{\mathbb{R}^d} F(x)\, dx \geq \int_{F(x)>\alpha} F(x)\, dx \leq \alpha\, m(\{x:\ F(x) > \alpha\}),$$

and this proves the assertion. One also sees, more generally, $\lambda(\alpha) \leq \frac{1}{\alpha^p} \int (F(x))^p\, dx$ for $p > 0$.

We now apply (29) to $F(x) = f^*(x)$, utilizing (28). Then

$$\begin{aligned}\int_{\mathbb{R}^d} (f^*(x))^p\, dx &= \int_0^\infty \lambda(\alpha^{1/p})\, d\alpha \\ &\leq A' \int_0^\infty \alpha^{-1/p} \left(\int_{|f|>\alpha^{1/p}/2} |f|\, dx \right) d\alpha.\end{aligned}$$

We evaluate the integral on the right-hand side by interchanging the order of integration. It then becomes

$$A' \int_{\mathbb{R}^d} |f(x)| \left(\int_0^{|2f(x)|^p} \alpha^{-1/p}\, d\alpha \right) dx.$$

However, if $p > 1$, $\int_0^t \alpha^{-1/p}\, d\alpha = a_p t^{1-1/p}$, for all $t \geq 0$, (with $a_p = p/(p-1)$). So the double integral equals $A' a_p 2^{p-1} \int_{\mathbb{R}^d} |f(x)|\, |f(x)|^{p-1}\, dx$, which is $A_p^p \|f\|_{L^p}^p$, with ($A_p^p = A' a_p 2^{p-1}$), and this gives (26), proving the theorem.

Note, as a result of the above proof, that the constant A_p in (26) satisfies $A_p = O(1/(p-1))$ as $p \to 1$.

Remark. The Hilbert transform $H(f)$, like the maximal function f^*, also satisfies a weak-type L^1 inequality, a result we will prove in a more general setting in the next chapter. In fact, this weak-type inequality will then be used to prove L^p inequalities for the generalizations of the Hilbert transform to $\mathbb{R}^d$, in much the same way as they are used above for the maximal function.

5 The Hardy space $\mathbf{H}^1_r$

We now come to the real Hardy space $\mathbf{H}^1_r(\mathbb{R}^d)$, which plays a significant role as another substitute for $L^1(\mathbb{R}^d)$, in the context where important

L^p inequalities for $p > 1$ break down at $p = 1$. This space is a Banach space that is "near" L^1, and whose dual space also occurs naturally in many applications. Moreover, $\mathbf{H}_r^1$ stands in sharp contrast to the space of weak-type functions considered above: the latter space cannot be made into a Banach space, nor does it have any bounded linear functionals. (See Exercise 15.)

The space $\mathbf{H}_r^1(\mathbb{R}^d)$ arose first for $d = 1$ in the setting of complex analysis as the "real parts" of the boundary values of functions of the complex Hardy space H^p, when $p = 1$. The Hardy space H^p, in the version of the upper half-plane, consists of holomorphic functions F on $\mathbb{R}_+^2$ for which

$$\sup_{y>0} \int_{-\infty}^{\infty} |F(x+iy)|^p \, dx < \infty,$$

and whose norm $\|F\|_{H^p}$, is defined as the pth-root of the quantity on the left-hand side of the above inequality.[8]

Now, it can be shown that whenever $F \in H^p$, $p < \infty$, then the limit $F_0(x) = \lim_{y \to 0} F(x+iy)$ exists in the $L^p(\mathbb{R})$ norm and in fact $\|F\|_{H^p} = \|F_0\|_{L^p(\mathbb{R})}$. Moreover, when $1 < p < \infty$, Riesz's theorem can be reinterpreted to say that $2F_0 = f + iH(f)$ where f is a real-valued function in $L^p(\mathbb{R})$. Conversely, every element $F \in H^p$ arises in this way. Thus, when $1 < p < \infty$ we see that the Banach space H^p is the same, up to equivalence of norms as (real) $L^p(\mathbb{R})$. The equivalence breaks down at $p = 1$, since the Hilbert transform H is not bounded on L^1. This situation led to the original definition of $\mathbf{H}_r^1(\mathbb{R})$: the space of real-valued functions f that arise as $2F_0 = f + iH(f)$ where $F \in H^1$. Equivalently, $f \in \mathbf{H}_r^1(\mathbb{R})$ if and only if $f \in L^1(\mathbb{R})$ and $H(f)$, defined in an appropriate "weak" sense, also belongs to $L^1(\mathbb{R})$. (An outline of the proof of these assertions can be found in Problems 2, 7*, and 8*.)

The notion of $\mathbf{H}_r^1$ was later extended to $\mathbb{R}^d$, $d > 1$, and various equivalent defining properties were ultimately found. It turns out that the simplest of these to state, and the most useful in applications, is the definition in terms of decompositions into "atoms." To this we now turn.

5.1 Atomic decomposition of $\mathbf{H}_r^1$

A bounded measurable function $\mathfrak{a}$ on $\mathbb{R}^d$ is an **atom** associated to a ball $B \subset \mathbb{R}^d$, if:

(i) $\mathfrak{a}$ is supported in B, with $|\mathfrak{a}(x)| \leq 1/m(B)$, for all x; and,

[8] The case $p = 2$ is treated in Section 2, Chapter 5 of Book III.

(ii) $\int_{\mathbb{R}^d} \mathfrak{a}(x)\,dx = 0$.

Note that (i) guarantees that for each atom $\mathfrak{a}$ we have $\|\mathfrak{a}\|_{L^1(\mathbb{R}^d)} \leq 1$.

The **space $\mathbf{H}_r^1(\mathbb{R}^d)$** consists of all L^1 functions f that can be written as

$$f = \sum_{k=1}^{\infty} \lambda_k \mathfrak{a}_k, \tag{30}$$

where the $\mathfrak{a}_k$ are atoms and the λ_k are scalars with

$$\sum_{k=1}^{\infty} |\lambda_k| < \infty. \tag{31}$$

Observe that (31) insures that the sum (30) converges in the L^1 norm. The infimum of the values $\sum |\lambda_k|$, taken over all possible decompositions of f of the form (30) is, by definition, the **$\mathbf{H}_r^1$ norm** of f, written as $\|f\|_{\mathbf{H}_r^1}$.

One can then observe the following properties of $\mathbf{H}_r^1$:

- With the above norm the space $\mathbf{H}_r^1$ is complete, hence is a Banach space. If f belongs to $\mathbf{H}_r^1$ then f belongs to L^1 and $\|f\|_{L^1(\mathbb{R}^d)} \leq \|f\|_{\mathbf{H}_r^1}$; also obviously $\int f(x)\,dx = 0$.

- However, the above necessary conditions are far from sufficient to imply $f \in \mathbf{H}_r^1$.

- The significance of the cancelation condition (ii) was already indicated at the end of Section 3.2. Moreover, if one drops this cancelation property for atoms, then sums of the kind (30) represent arbitrary functions in $L^1(\mathbb{R}^d)$.

- However, in the opposite direction if f is any $L^p(\mathbb{R}^d)$ function, $1 < p$, (say) of bounded support that satisfies the cancelation condition $\int f(x)\,dx = 0$, then f belongs to $\mathbf{H}_r^1$.

Proofs of the first three assertions are outlined in Exercises 16, 17, and 18. The fourth assertion is the deepest of these. Its proof, which follows below, provides us with valuable insight into the nature of $\mathbf{H}_r^1$, and its ideas will be exploited in several circumstances later.

We state the result mentioned above.

Proposition 5.1 *Suppose $f \in L^p(\mathbb{R}^d)$, $p > 1$, and f has bounded support. Then f belongs to $\mathbf{H}_r^1(\mathbb{R}^d)$ if and only if $\int_{\mathbb{R}^d} f(x)\,dx = 0$.*

Note that f is automatically in L^1, by Hölder's inequality (see Proposition 1.4 in Chapter 1), and the cancelation condition is necessary as has been pointed out.

To prove the sufficiency we assume that f is supported in a ball B_1 of unit radius, and that $\int_{B_1} |f(x)|\,dx \leq 1$. These normalizations can be achieved by a simple change of scale and multiplication of f by an appropriate constant. We next consider a truncated version of the maximal function f^*. We define $f^\dagger$ by

$$f^\dagger(x) = \sup \frac{1}{m(B)} \int_B |f(y)|\,dy,$$

where the supremum is taken over all balls B of radius ≤ 1 that contain x. We note that under our assumptions we have

$$\int_{\mathbb{R}^d} f^\dagger(x)\,dx < \infty. \tag{32}$$

Indeed, $f^\dagger(x) = 0$ if $x \notin B_3$, where B_3 is the ball with same center as B_1, but with radius 3. This is because $x \notin B_3$ and if $x \in B$ with the radius of B less than or equal to 1, then B must be disjoint from B_1, the support of f. Thus

$$\int_{\mathbb{R}^d} f^\dagger(x)\,dx = \int_{B_3} f^\dagger(x)\,dx \leq c \left(\int_{B_3} (f^\dagger(x))^p\,dx \right)^{1/p}$$

by Hölder's inequality. However the last integral is finite by Theorem 4.1, since clearly $f^\dagger(x) \leq f^*(x)$.

Now for each $\alpha \geq 1$, we consider a basic decomposition of f at "height" α, carried out with respect to the set $E_\alpha = \{x : \ f^\dagger(x) > \alpha\}$. This is a variant of the important "Calderón-Zygmund decomposition." It will be a little simpler to carry out the steps when $d = 1$, and this we do first; we return to the general case $d \geq 2$ immediately afterwards. The reader who is impatient with the technicalities of the next few pages may want to glance ahead to the lemma in Section 3.2 of the next chapter, where a more streamlined version of the decomposition appears.

This decomposition allows us to write $f = g + b$ where

$$|g| \leq c\alpha, \qquad \text{for an appropriate constant } c,^{9} \tag{33}$$

[9] Here we continue the practice of using c, c_1, etc. to denote constants that may not be the same in different places.

and where b is supported in E_α. In fact, since, as is easily seen, the set E_α is open, we can write $E_\alpha = \bigcup I_j$, where I_j are disjoint open intervals, and we will be able to construct b so that $b = \sum b_j$, with b_j supported on I_j and satisfying

$$\int b_j(x)\, dx = 0, \qquad \text{for all } j. \tag{34}$$

The key observation used in this construction is

$$\frac{1}{m(I_j)} \int_{I_j} |f(x)|\, dx \leq \alpha, \qquad \text{for all } j. \tag{35}$$

When $m(I_j) \geq 1$, the inequality (35) is automatic in view of our assumptions that $\int |f(x)|\, dx \leq 1$ and $\alpha \geq 1$. Otherwise, writing $I_j = (x_1, x_2)$ we note that (35) follows because $x_1 \in E_\alpha^c$, and hence $f^\dagger(x_1) \leq \alpha$ while $f^\dagger(x_1) \geq \frac{1}{m(I_j)} \int_{I_j} |f(x)|\, dx$.

As a result, if

$$m_j = \frac{1}{m(I_j)} \int_{I_j} f(x)\, dx$$

denotes the mean of f on I_j, then $|m_j| \leq \alpha$. Since $1 = \chi_{E_\alpha^c} + \sum_j \chi_{I_j}$, we can write $f = g + b$ with

$$g = f\chi_{E_\alpha^c} + \sum_j m_j \chi_{I_j},$$

and

$$b = \sum_j (f - m_j)\chi_{I_j} = \sum_j b_j,$$

where the b_j's are defined by $b_j = (f - m_j)\chi_{I_j}$, and the χ's designate the characteristic functions of the indicated sets. Note that on E_α^c we have $f^\dagger(x) \leq \alpha$, so that $|f(x)| \leq \alpha$ for a.e. x on this set by the differentiation theorem.[10] Since the I_j are disjoint, (35) then guarantees that (33) holds, with $c = 1$. The cancelation property (34) is also clear because

$$\int b_j(x)\, dx = \int_{I_j} (f(x) - m_j)\, dx = m(I_j)(m_j - m_j) = 0.$$

With the decomposition $f = g + b$ given for each α, we now consider simultaneously all decompositions of this form for $\alpha = 2^k$, $k = 0, 1, 2, \ldots$.

[10] See for instance Theorem 1.3 in Chapter 3 of Book III.

Thus for each k we can write $f = g^k + b^k$, with $|g^k| \leq c2^k$, $b^k = \sum_j b_j^k$, where b_j^k is supported on open intervals I_j^k, which for fix k are disjoint, and moreover $E_{2^k} = \{x : f^\dagger(x) > 2^k\} = \bigcup_j I_j^k$, while $\int b_j^k(x)\,dx = 0$.

Now since b^k is supported in the set E_{2^k}, and the sets E_{2^k} are decreasing with $m(E_{2^k}) \to 0$, as $k \to \infty$, we have that $b^k \to 0$ almost everywhere, as $k \to \infty$. Thus $f = \lim_{k\to\infty} g^k$ a.e., and

$$f = g^0 + \sum_{k=0}^{\infty}(g^{k+1} - g^k).$$

However,

$$g^{k+1} - g^k = b^k - b^{k+1} = \sum_j b_j^k - \sum_i b_i^{k+1} = \sum_j A_j^k,$$

where $A_j^k = b_j^k - \sum_{I_i^{k+1} \subset I_j^k} b_i^{k+1}$; the last identity holds because each I_i^{k+1} is contained in exactly one I_j^k. The A_j^k are supported in the intervals I_j^k, and by the cancelation properties of b_j^k and b_i^{k+1}, we have that $\int A_j^k(x)\,dx = 0$. Also since $|g^{k+1} - g^k| \leq c2^{k+1} + c2^k = 3c2^k$, and $g^{k+1} - g^k = b^k - b^{k+1}$, the disjointness of the intervals $\{I_j^k\}_j$ shows that $|A_j^k| \leq 3c2^k$. As a result we will see that the sum

$$f = g^0 + \sum_{k,j} A_j^k \tag{36}$$

will give us an atomic decomposition of f. In fact we set $\mathfrak{a}_j^k = \frac{1}{m(I_j^k)3c2^k} A_j^k$, $\lambda_j^k = m(I_j^k)3c2^k$, and $f = g^0 + \sum_{k,j} \lambda_j^k \mathfrak{a}_j^k$. Now the $\mathfrak{a}_j^k$ are atoms (associated to the intervals I_j^k) while

$$\sum_{k,j} \lambda_j^k = \sum_k \left(\sum_j \lambda_j^k\right) = 3c \sum_k 2^k \left(\sum_j m(I_j^k)\right)$$
$$= 3c \sum_{k=0}^{\infty} 2^k m(\{f^\dagger(x) > 2^k\}).$$

However, because $m(\{f^\dagger(x) > \alpha\})$ is decreasing in α,

$$2^k m(\{f^\dagger(x) > 2^k\}) \leq 2 \int_{2^{k-1}}^{2^k} m(\{f^\dagger(x) > \alpha\})\,d\alpha,$$

and hence summing in k we find that $\sum_{k,j} \lambda_j^k < \infty$, because

$$\int_0^\infty m(\{f^\dagger(x) > \alpha\})\, d\alpha = \int_{\mathbb{R}} f^\dagger(x)\, dx < \infty$$

as we saw by (29) and (32). Finally, g^0 is bounded and supported in B_3, while $\int g^0(x)\, dx = 0$ because of the cancelation properties of f and A_j^k. Hence g^0 is a multiple of an atom, and this yields that (36) is an atomic decomposition of f.

To extend the result to general d we need to modify the argument just given in one point: the appropriate analog of the decomposition of the open set $E_\alpha = \{x : \ f^\dagger(x) > \alpha\}$ into a disjoint union of open intervals is its decomposition into a union of (closed) cubes whose interiors are disjoint and so that the distance from each cube to E_α^c is comparable to the diameter of the cube.[11] It is also helpful to take the cubes entering in this union to be **dyadic** cubes. These cubes are defined as follows.

The dyadic cubes of the 0th-generation are the closed cubes of side-length 1, whose vertices are points with integral coordinates. The dyadic cubes of the kth-generation are the cubes of the form $2^{-k}Q$, where Q is a cube of the 0th-generation. Notice that bisecting the edges of any dyadic cube of the kth-generation decomposes it into 2^d cubes of the $(k+1)^{\text{th}}$-generation whose interiors are disjoint. Observe also that if Q_1 and Q_2 are dyadic cubes (of possibly different generations), and their interiors intersect, then either $Q_1 \subset Q_2$, or $Q_2 \subset Q_1$.

The decomposition we need of an open set into a union of such cubes is as follows.

Lemma 5.2 *Suppose $\Omega \subset \mathbb{R}^d$ is a non-trivial open set. Then there is a collection $\{Q_j\}$ of dyadic cubes with disjoint interiors so that $\Omega = \bigcup_{j=1}^\infty Q_j$, and*

$$\text{diam}(Q_j) \leq d(Q_j, \Omega^c) \leq 4\ \text{diam}(Q_j). \tag{37}$$

Proof. We claim first that every point $\overline{x} \in \Omega$ belongs to some dyadic cube $Q_{\overline{x}}$ for which (37) holds (with $Q_{\overline{x}}$ in place of Q_j).

Let $\delta = d(\overline{x}, \Omega^c) > 0$. Now the dyadic cubes containing $\overline{x}$ have diameters varying over $\{\sqrt{d}2^{-k}\}$, $k \in \mathbb{Z}$. Hence we can find a dyadic cube $Q_{\overline{x}}$ which contains $\overline{x}$, with $\delta/4 \leq \text{diam}(Q_{\overline{x}}) \leq \delta/2$. Now $d(Q_{\overline{x}}, \Omega^c) \leq \delta \leq 4\ \text{diam}(Q_{\overline{x}})$, since $\overline{x} \in Q_{\overline{x}}$. Also

$$d(Q_{\overline{x}}, \Omega^c) \geq \delta - \text{diam}(Q_{\overline{x}}) \geq \delta/2 \geq \text{diam}(Q_{\overline{x}}),$$

[11] This kind of decomposition already arose in Chapter 1 of Book III.

thus (37) is proved for $Q_{\overline{x}}$. Now let $\tilde{\mathcal{Q}}$ be the collection of all cubes $Q_{\overline{x}}$ obtained as $\overline{x}$ ranges over Ω. Their union clearly covers Ω but their interiors are far from disjoint. To achieve the desired disjointness select from $\tilde{\mathcal{Q}}$ the *maximal* cubes, that is, those cubes in $\tilde{\mathcal{Q}}$ not contained in larger cubes of $\tilde{\mathcal{Q}}$. Clearly, by what has been said above, each Q is contained in a maximal cube and these maximal cubes necessarily have disjoint interiors. The lemma is therefore proved.

With the above lemma, we can redo the decomposition of f in the setting $d \geq 2$. The argument is essentially the same as before except for some small changes. For $\alpha \geq 1$, we apply the lemma to the open set $E_\alpha = \{x : f^\dagger(x) > \alpha\}$; therefore we have a decomposition $f = g + b$, with $g = f\chi_{E_\alpha^c} + \sum_{j=1}^\infty m_j \chi_{Q_j}$, and $b = \sum_{j=1}^\infty b_j$, with $b_j = (f - m_j)\chi_{Q_j}$. Now as in the case $d = 1$ we see that $|m_j| \leq c\alpha$. In fact, $\int_{Q_j} |f|\, dx \leq \int_B |f|\, dx$ for any ball $B \supset Q_j$. We choose B so that it contains a point $\overline{x}$ of E_α^c. We can do this with a ball whose radius is 5 diam(Q_j), since $d(Q, E_\alpha^c) \leq$ 4 diam(Q_j). If we choose such a ball and it has radius ≤ 1 (that is, diam$(Q_j) \leq 1/5$), then

$$\frac{1}{m(B)} \int_B |f(x)|\, dx \leq f^\dagger(\overline{x}) \leq \alpha,$$

and hence $|m_j| \leq c_1\alpha$ where $m(B)/m(Q_j) = c_1$. (The ratio c_1 is independent of j). Otherwise, if diam$(Q_j) \geq 1/5$, the inequality $|m_j| \leq c_2\alpha$ is automatic (with c_2 independent of j), since $\int |f(x)|\, dx \leq 1$ by assumption, and $\alpha \geq 1$. In either case, therefore $|m_j| \leq c\alpha$. Next, since each dyadic cube arising in the decomposition of $\{x : f^\dagger(x) > 2^{k+1}\}$ must be a sub-cube of a dyadic cube arising for $\{x : f^\dagger(x) > 2^k\}$ we can proceed as before to obtain

$$f = g^0 + \sum_{k,j} A_j^k$$

with A_j^k supported in the cube Q_j^k, and $\{x : f^\dagger(x) > 2^k\} = \bigcup_j Q_j^k$.

As a result we can write $A_j^k = \lambda_j^k \mathfrak{a}_j^k$ where $\lambda_j^k = c' 2^k m(Q_j^k)$ and $\mathfrak{a}_j^k$ are atoms associated to the balls B_j^k where the ball B_j^k is defined to be, for each k and j, the smallest ball containing the cube Q_j^k. Note that $m(B_j^k)/m(Q_j^k)$ is independent of k and j. (See Figure 7.)

Finally, since

$$\sum_{k,j} 2^k m(Q_j^k) = \sum_k 2^k m(\{x : f^\dagger(x) > 2^k\}) < \infty,$$

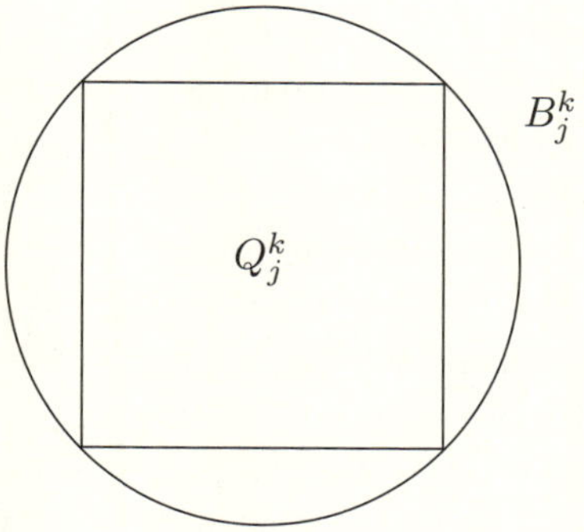

Figure 7. The cube Q^k_j and the ball B^k_j

as above, we have established the atomic decomposition of f, concluding the proof of the proposition.

5.2 An alternative definition of $\mathbf{H}^1_r$

A nearly immediate consequence of Proposition 5.1 allows us to recast the atomic decomposition of $\mathbf{H}^1_r$ in a more general form. For any p with $p > 1$ we define a **p-atom** (associated to a ball B) to be a measurable function $\mathfrak{a}$ which satisfies:

(i′) $\mathfrak{a}$ is supported in B, and $\|\mathfrak{a}\|_{L^p} \leq m(B)^{-1+1/p}$.

(ii′) $\int_{\mathbb{R}^d} \mathfrak{a}(x)\, dx = 0$.

We reserve the terminology of "atom" for the atoms defined previously in Section 5.1, which correspond to p-atoms for $p = \infty$. Note that any atom is automatically a p-atom.

Corollary 5.3 *Fix $p > 1$. Then any p-atom $\mathfrak{a}$ is in $\mathbf{H}^1_r$. Moreover there is a bound c_p, independent of the atom $\mathfrak{a}$, so that*

$$\|\mathfrak{a}\|_{\mathbf{H}^1_r} \leq c_p. \tag{38}$$

Note that the proof below yields that $c_p = O(1/(p-1))$ as $p \to 1$. Also, the requirement $p > 1$ for the conclusion of the corollary is necessary, as can be seen by using the reasoning in Exercise 17.

Proof. One can rescale a p-atom $\mathfrak{a}$, associated to a ball B of radius r, by replacing $\mathfrak{a}$ by $\mathfrak{a}_r$, with $\mathfrak{a}_r(x) = r^d \mathfrak{a}(rx)$. Then clearly $\mathfrak{a}_r(x)$ is supported where $rx \in B$, that is, $x \in \frac{1}{r}B = B^r$ and the latter ball has radius one. Also since $m(B^r) = r^{-d}m(B)$ and $\|\mathfrak{a}_r\|_{L^p} = r^{d-d/p}\|\mathfrak{a}\|_{L^p}$, we have $\|\mathfrak{a}_r\|_{L^p} \leq m(B^r)^{-1+1/p}$. Thus $\mathfrak{a}_r$ is a p-atom for the (unit) ball B^r. Moreover, as has already been observed $\|r^d f(rx)\|_{\mathbf{H}^1_r} = \|f\|_{\mathbf{H}^1_r}$, for every $r > 0$. Thus (38) has been reduced to the case of p-atoms associated to balls of unit radius. Observe that automatically for such p-atoms one has $\int |\mathfrak{a}(x)|\,dx \leq 1$, therefore we see that we find ourselves exactly in the setting of the proof of Proposition 5.1 with $f(x) = \mathfrak{a}(x)$. In fact one notes that what is proved there amounts to (38), with the constant c_p incorporating the bound A_p in (26) for the maximal function, since the calculation for $\int_{\mathbb{R}^d} f^\dagger(x)\,dx$ used to establish (32) shows that this quantity is bounded by $cA_p\|f\|_{L^p}$. We have already noted that $A_p = O(1/(p-1))$ as $p \to 1$. Because $f = \mathfrak{a}$, the proof of (38) is complete.

As a result, if $f = \sum_{k=1}^\infty \lambda_k \mathfrak{a}_k$ with p-atoms $\mathfrak{a}_k$, and $\sum |\lambda_k| < \infty$, then f is in $\mathbf{H}^1_r$ and

$$\|f\|_{\mathbf{H}^1_r} \leq c_p \sum_{k=1}^\infty |\lambda_k|.$$

Conversely, whenever $f \in \mathbf{H}^1_r$, it has a decomposition with respect to ($p = \infty$) atoms and therefore has such a decomposition with respect to p-atoms. We may summarize this as follows.

In defining $\mathbf{H}^1_r$ via (30) and (31), we may replace atoms by p-atoms, $p > 1$, and obtain an equivalent norm.

5.3 Application to the Hilbert transform

The result below exemplifies the role of the Hardy space $\mathbf{H}^1_r$ as an improvement over the space L^1. In contrast with the failure of the boundedness of the Hilbert transform on L^1, we have that it is bounded from $\mathbf{H}^1_r$ to L^1.

Theorem 5.4 *If f belongs to the Hardy space $\mathbf{H}^1_r(\mathbb{R})$, then $H_\epsilon(f) \in L^1(\mathbb{R})$, for every $\epsilon > 0$. Moreover $H_\epsilon(f)$ (see (14)) converges in the L^1 norm, as $\epsilon \to 0$. Its limit, defined as $H(f)$, satisfies*

$$\|H(f)\|_{L^1(\mathbb{R})} \leq A\|f\|_{\mathbf{H}^1_r(\mathbb{R})}.$$

Proof. The argument below illustrates a nice feature of $\mathbf{H}^1_r(\mathbb{R})$: to show the boundedness of an operator on $\mathbf{H}^1_r$ it often suffices merely to verify it for atoms, and this is usually a simple task.

Let us first see that for all atoms $\mathfrak{a}$, we have

$$\|H_\epsilon(\mathfrak{a})\|_{L^1(\mathbb{R})} \leq A, \tag{39}$$

with A independent of the atom $\mathfrak{a}$ and ϵ. Indeed, we can avail ourselves of the translation-invariance and scale-invariance of the Hilbert transform to simplify matters even further by restricting ourselves in proving (39) for the case of atoms associated to the (unit) interval $I = [-1/2, 1/2]$. This reduction proceeds, on the one hand, by recalling that if $\mathfrak{a}_r(x) = r\mathfrak{a}(rx)$, then $H(\mathfrak{a}_r)(x) = rH(\mathfrak{a})(rx)$; that $\mathfrak{a}_r$ is an atom associated to the interval $I_r = \frac{1}{r}I$ whenever $\mathfrak{a}$ is supported in I; and that $\|rF(rx)\|_{L^1(\mathbb{R})} = \|F(x)\|_{L^1(\mathbb{R})}$, whenever $F \in L^1$. On the other hand, the translations $f(x) \mapsto f(x+h)$, $h \in \mathbb{R}$, commute with the operator H, as is evident from (14); also translation clearly preserves atoms and the radii of their associated balls.

Thus in proving (39) we may assume that $\mathfrak{a}$ is an atom associated to the interval $|x| \leq 1/2$. We will estimate $H_\epsilon(\mathfrak{a})(x)$ differently, according to whether $|x| \leq 1$, (x belongs to the "double" of the support of $\mathfrak{a}$), or $|x| > 1$. In the first case, we have

$$\int_{|x|\leq 1} |H_\epsilon(\mathfrak{a})(x)|\,dx \leq 2^{1/2}\left(\int_{|x|\leq 1} |H_\epsilon(\mathfrak{a})(x)|^2\,dx\right)^{1/2} \leq 2^{1/2}\|H_\epsilon(\mathfrak{a})\|_{L^2}$$
$$\leq c\|\mathfrak{a}\|_{L^2} \leq c,$$

using the Cauchy-Schwarz inequality and the L^2 theory studied earlier.

Next when $|x| > 1$ we write (for small ϵ)

$$H_\epsilon(\mathfrak{a})(x) = \frac{1}{\pi}\int_{|t|\geq\epsilon} \mathfrak{a}(x-t)\frac{dt}{t} = \frac{1}{\pi}\int_{|x-t|\geq\epsilon} \mathfrak{a}(t)\frac{dt}{x-t}$$
$$= \frac{1}{\pi}\int_{|x-t|\geq\epsilon} \mathfrak{a}(t)\left[\frac{1}{x-t} - \frac{1}{x}\right]dt,$$

since $\int \mathfrak{a}(t)\,dt = 0$. Hence if $|x| > 1$, then $|H_\epsilon(\mathfrak{a})(x)| \leq c/x^2$ because $\left|\frac{1}{x-t} - \frac{1}{x}\right| \leq \frac{1}{x^2}$ when $|x| \geq 1$ and $|t| \leq 1/2$, and $|\mathfrak{a}(t)| \leq 1$. Therefore $\int_{|x|\geq 1} |H_\epsilon(\mathfrak{a})(x)|\,dx \leq 2c$, and this proves (39) for atoms associated to the interval $[-1/2, 1/2]$, and thus for all atoms.

At the same time, the inequality $|H_\epsilon(\mathfrak{a})(x)| \leq c/x^2$ when $|x| > 1$, and the convergence in the L^2 norm, guaranteed by Proposition 3.1, shows that $H_\epsilon(\mathfrak{a})$ converges in the L^1 norm to $H(\mathfrak{a})$, as $\epsilon \to 0$, for every atom $\mathfrak{a}$.

Now if $f = \sum_{k=1}^\infty \lambda_k \mathfrak{a}_k$ is an $\mathbf{H}_r^1$ function with the indicated atomic decomposition, then by (39)

$$\|H_\epsilon(f)\|_{L^1} \leq A\sum_{k=1}^\infty |\lambda_k|,$$

and if we take the infimum over atomic decompositions, we obtain

$$\|H_\epsilon(f)\|_{L^1} \leq A\|f\|_{\mathbf{H}^1_r}, \qquad \text{for every } f \in \mathbf{H}^1_r. \tag{40}$$

Next, let $f_N = \sum_{k=1}^N \lambda_k \mathfrak{a}_k$, so that $f = f_N + (f - f_N)$. Now since f_N is a finite linear combination of atoms, it is itself a constant multiple of an atom. So we know that $H_\epsilon(f_N)$ converges in the L^1 norm as $\epsilon \to 0$. Also,

$$\|H_{\epsilon_1}(f) - H_{\epsilon_2}(f)\|_{L^1} \leq \|H_{\epsilon_1}(f_N) - H_{\epsilon_2}(f_N)\|_{L^1} + 2A\|f - f_N\|_{\mathbf{H}^1_r}.$$

However, $\|f - f_N\|_{\mathbf{H}^1_r} \to 0$, as $N \to \infty$. Thus given $\delta > 0$ and choosing first N sufficiently large, then with both ϵ_1 and ϵ_2 sufficiently small, we get that $\|H_{\epsilon_1}(f) - H_{\epsilon_2}(f)\|_{L^1} < \delta$, which shows that $H_\epsilon(f)$ converges in the L^1 norm. The conclusion asserted by the theorem then follows from (40), and the proof is complete.

Remark. A more elaborate form of the argument given above shows that in fact the Hilbert transform maps the Hardy space $\mathbf{H}^1_r$ to itself. This is outlined in a more general setting in Problem 2 of the next chapter.

6 The space $\mathbf{H}^1_r$ and maximal functions

The real Hardy space $\mathbf{H}^1_r$ also leads to interesting insights regarding maximal functions. The fact that this might be the case was already suggested by the use of f^* (more precisely, its truncated version $f^\dagger$) in the proof of Proposition 5.1. In parallel with what we saw for the Hilbert transform, our goal will be to find a suitable maximal function that maps $\mathbf{H}^1_r$ to L^1. In doing this we must keep in mind the following points.

First, neither f^* nor $f^\dagger$ can be used as such because by their definitions both f^* and $f^\dagger$ involve f only through its absolute value, and therefore cannot take into account the cancelation properties of f that enter to exploit the fact that $f \in \mathbf{H}^1_r$.

Second, even if one removed the absolute values in the definitions of these maximal functions this would not be enough, because the cut-off functions involved (the characteristic functions of balls) are not smooth.

It is the notion of nice "approximations to the identity," and the resulting family of convolution operators that lead us to the version of the maximal function relevant for $\mathbf{H}^1_r$. Recall that if we fix a suitable function Φ that, for example, is bounded and has compact support, then for any $f \in L^1$, if $\Phi_\epsilon = \epsilon^{-d}\Phi(x/\epsilon)$, then

$$(f * \Phi_\epsilon)(x) \to f(x), \qquad \text{as } \epsilon \to 0, \text{ for a.e. } x,$$

under the assumption $\int \Phi(x)\,dx = 1$.

Given this Φ we define the **maximal function** M corresponding to the limit above by

$$M(f)(x) = \sup_{\epsilon>0} |(f * \Phi_\epsilon)(x)|. \tag{41}$$

Note that by what we have said it is easy to observe that for every $f \in L^1(\mathbb{R}^d)$

$$|f(x)| \leq M(f)(x) \leq cf^*(x), \qquad \text{for a.e. } x,$$

where c is a suitable constant.

We shall also want to assume that Φ has some smoothness, as indicated above. With this in mind we can state our result as follows.

Theorem 6.1 *Suppose Φ is a C^1 function with compact support on $\mathbb{R}^d$. With M defined by* (41) *we have that $M(f) \in L^1(\mathbb{R}^d)$, whenever $f \in \mathbf{H}_r^1(\mathbb{R}^d)$. Moreover*

$$\|M(f)\|_{L^1(\mathbb{R})} \leq A\|f\|_{\mathbf{H}_r^1(\mathbb{R}^d)}. \tag{42}$$

Before coming to the proof, which is very similar to that of the Hilbert transform, we make some additional remarks.

- In the definition of M we have assumed that the function Φ that enters has one degree of smoothness. Less could be assumed with the same result (for example a Hölder condition of exponent α with $0 < \alpha < 1$), but some degree of smoothness is necessary. (See Exercise 22.)

- In fact, the inequality (42) can be reversed. Thus there is a converse theorem that gives the maximal characterization of $\mathbf{H}_r^1$. This is formulated in Problem 6*.

Proof. Suppose f is in $\mathbf{H}_r^1(\mathbb{R}^d)$ and $f = \sum \lambda_k \mathfrak{a}_k$ is an atomic decomposition. Then clearly $M(f) \leq \sum |\lambda_k| M(\mathfrak{a}_k)$, and thus it suffices to prove (42) when f is an atom $\mathfrak{a}$.

In fact, note that with $\mathfrak{a}_r$ defined as $\mathfrak{a}_r(x) = r^d\mathfrak{a}(rx)$, $r > 0$, we have $(\mathfrak{a}_r * \Phi_\epsilon)(x) = r^d(\mathfrak{a} * \Phi_{\epsilon r})(rx)$, and hence $M(\mathfrak{a}_r)(x) = r^d M(\mathfrak{a})(rx)$. Also the mapping $\mathfrak{a} \mapsto M(\mathfrak{a})$ clearly commutes with translations. Therefore in proving (42) we may assume that the atom $\mathfrak{a}$ is associated to the unit ball (centered at the origin).

Now we consider two cases; when $|x| \leq 2$ and when $|x| > 2$. In the first case clearly $M(\mathfrak{a})(x) \leq c$ and hence $\int_{|x|\leq 2} M(\mathfrak{a})(x)\,dx \leq c'$. In the second case, we write

$$\begin{aligned}(\mathfrak{a} * \Phi_\epsilon)(x) &= \epsilon^{-d} \int_{\mathbb{R}^d} \mathfrak{a}(y)\Phi\left(\frac{x-y}{\epsilon}\right) dy \\ &= \epsilon^{-d} \int_{\mathbb{R}^d} \mathfrak{a}(y) \left[\Phi\left(\frac{x-y}{\epsilon}\right) - \Phi\left(\frac{x}{\epsilon}\right)\right] dy,\end{aligned}$$

since $\int \mathfrak{a}(y)\,dy = 0$. However since $|x| \geq 2$ and $|y| \leq 1$, we have that $|x - y| \geq |x|/2$. Moreover since $\Phi \in C^1$ we have that $\left|\Phi\left(\frac{x-y}{\epsilon}\right) - \Phi\left(\frac{x}{\epsilon}\right)\right| \leq c|y|/\epsilon \leq c/\epsilon$. In addition the fact that Φ has compact support implies that $(\mathfrak{a} * \Phi_\epsilon)(x)$ vanishes unless $\left|\frac{x-y}{\epsilon}\right| \leq A$ for some bound A, which in turn means that $\epsilon > |x|/(2A)$. Altogether then

$$\epsilon^{-d}\left|\Phi\left(\frac{x-y}{\epsilon}\right) - \Phi\left(\frac{x}{\epsilon}\right)\right| \leq c\epsilon^{-d-1} \leq c'|x|^{-d-1}$$

for those x. As a result $\int_{|x|>2} M(\mathfrak{a})(x)\,dx \leq c$. Therefore (42) is established and the theorem is proved.

6.1 The space BMO

In the same sense that the real Hardy space $\mathbf{H}^1_r(\mathbb{R}^d)$ is a substitute for $L^1(\mathbb{R}^d)$, the space $\mathrm{BMO}(\mathbb{R}^d)$ is the corresponding natural substitute for the space $L^\infty(\mathbb{R}^d)$.

A locally integrable function f on $\mathbb{R}^d$ is said to be of **bounded mean oscillation** (abbreviated by BMO) if

$$\sup \frac{1}{m(B)} \int_B |f(x) - f_B|\,dx < \infty, \tag{43}$$

where the supremum is taken over all balls B. Here f_B denotes the mean-value of f over B, namely

$$f_B = \frac{1}{m(B)} \int_B f(x)\,dx.$$

The quantity (43) is taken as the norm in the space BMO, and is denoted by $\|f\|_{\mathrm{BMO}}$.

We first make some observations about the space of BMO functions.

- The null elements of the norm are the constant functions. Thus, strictly speaking, elements of BMO should be thought of as equivalence classes of functions, modulo constants.

- Note that if (43) holds with possibly different constants c_B instead of f_B, then f would still be in BMO. Indeed, if for all B

$$\frac{1}{m(B)}\int_B |f(x) - c_B|\,dx \leq A$$

then necessarily $|f_B - c_B| \leq A$ and hence $\|f\|_{\mathrm{BMO}} \leq 2A$. It is also easy to verify that one would obtain the same space, (with an equivalence of norms), if the balls appearing in (43), were replaced by, say, the family of all cubes.

- If $f \in L^\infty$ then it is obvious that f is in BMO. A more typical example of a BMO function is $f(x) = \log|x|$. Like the general BMO function it has the property that it belongs (locally) to every L^q space, with $q < \infty$. It also exemplifies a property shared by BMO and the L^∞ space: whenever $f(x)$ belongs to one of these spaces, then so does the scaled function $f(rx)$, $r > 0$, with the norm remaining unchanged. (For more about the above remarks, see Exercise 23, and Problems 3 and 4.)

- The space of real-valued BMO functions forms a lattice, that is, if f and g belong to BMO then so do $\min(f,g)$ and $\max(f,g)$. This is because $|f|$ is in BMO whenever f is, which in turn follows from the fact $||f| - |f|_B| \leq |f - f_B| + \|f\|_{\mathrm{BMO}}$. However, if $f \in \mathrm{BMO}$ and $|g| \leq |f|$, it is not necessarily true that g belongs to BMO.

- From the above, we also deduce that if $f \in \mathrm{BMO}$ is real-valued, and $f^{(k)}$ is the truncation of f defined by $f^{(k)}(x) = f(x)$, if $|f(x)| \leq k$; $f^{(k)}(x) = k$ if $f(x) > k$; and $f^{(k)}(x) = -k$, if $f(x) < -k$, then $\{f^{(k)}\}$ is a sequence of bounded BMO functions so that $|f^{(k)}| \leq |f|$ for all k, $f^{(k)} \to f$ for a.e x as $k \to \infty$, and hence $\|f^{(k)}\|_{\mathrm{BMO}} \to \|f\|_{\mathrm{BMO}}$ as $k \to \infty$.

  If f is complex-valued, one may apply this to both the real and imaginary parts of f.

Our focus now will be on the key fact that BMO is the dual space of the Hardy space $\mathbf{H}^1_r$. This assertion means that every continuous linear functional ℓ on $\mathbf{H}^1_r$ can be realized as

$$\ell(f) = \int_{\mathbb{R}^d} f(x)g(x)\,dx, \qquad f \in \mathbf{H}^1_r, \tag{44}$$

for some element g in BMO, when (44) is suitably defined. In fact, a little care must be exercised when dealing with the pairing (44): for general

$f \in \mathbf{H}^1_r$, and $g \in \mathrm{BMO}$, the integral need not converge. See Exercise 24. Thus we proceed indirectly, defining ℓ first on a dense subspace of $\mathbf{H}^1_r$. This will be H^1_0, the subspace of finite linear combinations of atoms. Note that every element of H^1_0 is itself a multiple of an atom. Also, if $f \in H^1_0$ the integral converges, and the ambiguity of the BMO element g (that is, the additive constant) disappears because $\int f\,dx = 0$.

Our basic result is Fefferman's duality theorem:

Theorem 6.2 *Suppose $g \in \mathrm{BMO}$. Then the linear functional ℓ defined by* (44), *initially considered for $f \in H^1_0$, has a unique extension to $\mathbf{H}^1_r$ that satisfies*

$$\|\ell\| \le c\|g\|_{\mathrm{BMO}}.$$

Conversely, every bounded linear functional ℓ on $\mathbf{H}^1_r$ can be written as (44) *with $g \in \mathrm{BMO}$ and*

$$\|g\|_{\mathrm{BMO}} \le c'\|\ell\|.$$

Here $\|\ell\|$ stands for $\|\ell\|_{(\mathbf{H}^1_r)^*}$, the norm of ℓ as a linear functional on $\mathbf{H}^1_r$.

Proof. Let us first assume that $g \in \mathrm{BMO}$ is bounded. Start with a general $f \in \mathbf{H}^1_r$, and let $f = \sum_{k=1}^{\infty} \lambda_k \mathfrak{a}_k$ be an atomic decomposition. Then by the convergence of the sum in the L^1 norm we get $\ell(f) = \sum \lambda_k \int \mathfrak{a}_k g$. But

$$\int \mathfrak{a}_k(x) g(x)\,dx = \int \mathfrak{a}_k(x)[g(x) - g_{B_k}]\,dx,$$

where $\mathfrak{a}_k$ is supported in the ball B_k. However $|\mathfrak{a}_k(x)| \le \frac{1}{m(B_k)}$ and thus

$$|\ell(f)| \le \sum_k |\lambda_k| \frac{1}{m(B_k)} \int_{B_k} |g(x) - g_{B_k}|\,dx.$$

Therefore considering all possible decompositions of f then gives

$$\left| \int f(x) g(x)\,dx \right| \le \|f\|_{\mathbf{H}^1_r} \|g\|_{\mathrm{BMO}},$$

under the assumption that g is bounded. Next, if we restrict ourselves to $f \in H^1_0$ (in particular to f that are bounded) with a general g in BMO, and let $g^{(k)}$ be the truncation of g (as defined above), then the fact that

$$\left| \int f(x) g^{(k)}(x)\,dx \right| \le \|f\|_{\mathbf{H}^1_r} \|g^{(k)}\|_{\mathrm{BMO}}$$

just proved, together with a passage to the limit as $k \to \infty$ using the dominated convergence theorem, shows that

$$|\ell(f)| = \left| \int f(x)g(x)\,dx \right| \leq c\|f\|_{\mathbf{H}_r^1}\|g\|_{\mathrm{BMO}},$$

whenever $f \in H_0^1$ and $g \in \mathrm{BMO}$. Thus the direct conclusion of the theorem is established.

To prove the converse we will test the given linear functional ℓ on atoms, and here it will be convenient to test ℓ on p-atoms, with $p = 2$.

For this purpose fix a ball B and consider the L^2 space on B with norm

$$\|f\|_{L_B^2} = \left(\int_B |f(x)|^2\,dx \right)^{1/2},$$

and let $L_{B,0}^2$ denote the subspace of those $f \in L_B^2$ for which $\int f(x)\,dx = 0$. Note that the ball $\|f\|_{L_{B,0}^2} \leq m(B)^{-1/2}$ of $L_{B,0}^2$ consists of exactly the 2-atoms associated to B.

Let us assume our linear functional ℓ has been normalized so that its norm is less than or equal to 1. Then restricting ourselves to $f \in L_{B,0}^2$ we see that $|\ell(f)| \leq \|f\|_{\mathbf{H}_r^1} \leq cm(B)^{1/2}\|f\|_{L_{B,0}^2}$, the last inequality being a consequence of (38) in Corollary 5.3. Thus by the Riesz representation theorem for $L_{B,0}^2$ (or as a simple consequence of the self-duality of L^2 spaces) there is a $g^B \in L_{B,0}^2$, so that $\ell(f) = \int fg^B\,dx$, for $f \in L_{B,0}^2$. We also have that $\|g^B\|_{L_{B,0}^2} \leq cm(B)^{1/2}$, because $\|\ell\|_{L_{B,0}^2} \leq cm(B)^{1/2}$, as we have seen above. Hence for each ball B we have a function g^B defined on B. What we want is a single function g so that for each B, g and g^B differ by a constant on B. To construct this g note that if $B_1 \subset B_2$ then $g^{B_1} - g^{B_2}$ is a constant on B_1, since both g^{B_1} and g^{B_2} give the same linear functional on $L_{B_1,0}^2$. Now replace each g^B by $\tilde{g}^B = g^B + c_B$, where the constant c_B is so chosen that $\int_{|x|\leq 1} \tilde{g}^B\,dx = 0$. As a result $\tilde{g}^{B_1} = \tilde{g}^{B_2}$, on B_1 if $B_1 \subset B_2$. Therefore we can unambiguously define g on $\mathbb{R}^d$ by taking $g(x) = \tilde{g}^B(x)$, for $x \in B$ and B any ball. Now observe that $\frac{1}{m(B)}\int_B |g(x) - c_B|\,dx \leq m(B)^{-1/2}\|\tilde{g}^B - c_B\|_{L_B^2} \leq m(B)^{-1/2}\|g^B\|_{L_{B,0}^2} \leq c$. Therefore $g \in \mathrm{BMO}$, with $\|g\|_{\mathrm{BMO}} \leq c$. Since the representation has been established for $f \in L_{B,0}^2$ and all B, it holds for the dense subspace H_0^1. The proof of the theorem is therefore concluded.

7 Exercises

1. Show that an inequality

$$\|\{a_n\}\|_{L^q} \le A\|f\|_{L^p}, \qquad \text{for all } f \in L^p,$$

with $a_n = \frac{1}{2\pi}\int_0^{2\pi} f(\theta)e^{-in\theta}\,d\theta$, is possible only if $1/p + 1/q \le 1$.
[Hint: Let $D_N(\theta) = \sum_{|n|\le N} e^{in\theta}$ be the Dirichlet kernel. Then $\|D_N\|_{L^p} \approx N^{1-1/p}$ as $N \to \infty$, if $p > 1$ and $\|D_N\|_{L^1} \approx \log N$.]

2. The following are simple generalizations of the Hausdorff-Young inequalities.

(a) Suppose $\{\varphi_n\}$ is an orthonormal sequence on $L^2(X, \mu)$. Assume also that $|\varphi_n(x)| \le M$ for all n. If $a_n = \int f\overline{\varphi_n}\,d\mu$, then $\|a_n\|_{L^q} \le M^{(2/p)-1}\|f\|_{L^p(X)}$, $1 \le p \le 2$, $1/p + 1/q = 1$.

(b) Suppose $f \in L^p$ on the torus $\mathbb{T}^d$, $1 \le p \le 2$, and $a_n = \int_{\mathbb{T}^d} f(x)e^{-2\pi i n\cdot x}\,dx$, $n \in \mathbb{Z}^d$. Then $\|\{a_n\}\|_{L^q(\mathbb{Z}^d)} \le \|f\|_{L^p(\mathbb{T}^d)}$, where $1/q \le 1 - 1/p$.

3. Check that an inequality of the form $\|\hat{f}\|_{L^q(\mathbb{R}^d)} \le A\|f\|_{L^p(\mathbb{R}^d)}$ (holding for all simple functions f) is possible if and only if $1/p + 1/q = 1$.
[Hint: Let $f_r(x) = f(rx)$, $r > 0$. Then $\hat{f_r}(\xi) = \hat{f}(\xi/r)r^{-d}$.]

4. Prove that another necessary condition for the inequality in the previous exercise is that $p \le 2$. In fact the estimate

$$\int_{|\xi|\le 1} |\hat{f}(\xi)|\,d\xi \le A\|f\|_{L^p}$$

can hold only if $p \le 2$.
[Hint: Let $f^s(x) = s^{-d/2}e^{-\pi|x|^2/s}$, $s = \sigma + it$, $\sigma > 0$. Then $(\widehat{f^s})(\xi) = e^{-\pi s|\xi|^2}$. Note that $\|f^s\|_{L^p} \le ct^{d(1/p-1/2)}$ when $\sigma = 1$, and let $t \to \infty$.]

5. Let ψ be the conformal map of the strip $0 < \mathrm{Re}(z) < 1$ to the upper half-plane defined by $\psi(z) = e^{i\pi z}$. Check that $\Phi(z) = e^{-i\psi(z)}$ is continuous on the closure of the strip, $|\Phi(z)| = 1$ on the boundary lines, but $\Phi(z)$ is unbounded in the strip.

6. Extend the Riesz convexity theorem (in Section 2) to the $L^{p,r}$ spaces discussed in Exercise 18 of Chapter 1. We assume T is a linear transformation from simple functions to locally integrable functions. Suppose

$$\|T(f)\|_{L^{q_0,s_0}} \le M_0\|f\|_{L^{p_0,r_0}}, \qquad \text{and} \qquad \|T(f)\|_{L^{q_1,s_1}} \le M_1\|f\|_{L^{p_1,r_1}}$$

for all simple f. Prove that as a consequence $\|T(f)\|_{L^{q,s}} \le M_\theta\|f\|_{L^{p,r}}$ where $\frac{1}{p} = \frac{1-\theta}{p_0} + \frac{\theta}{p_1}$, $\frac{1}{r} = \frac{1-\theta}{r_0} + \frac{\theta}{r_1}$, $\frac{1}{q} = \frac{1-\theta}{q_0} + \frac{\theta}{q_1}$, $\frac{1}{s} = \frac{1-\theta}{s_0} + \frac{\theta}{s_1}$, and $0 \le \theta \le 1$.

[Hint: Suppose f and g are a pair of simple functions with $\|f\|_{L^{p,r}} \leq 1$ and $\|f\|_{L^{q',s'}} \leq 1$. Define

$$f_z = |f(x,t)|^{p\alpha(z)} \frac{f(x,t)}{|f(x,t)|} \|f(\cdot,t)\|_{L^p(dx)}^{r\beta(z)-p\alpha(z)},$$

where $\alpha(z) = \frac{1-z}{p_0} + \frac{z}{p_1}$, $\beta(z) = \frac{1-z}{r_0} + \frac{z}{r_1}$. Note that when $z = \theta$, then $f_z = f$. Also

$$\|f_{1+it}\|_{L^{p_1,r_1}} \leq 1 \quad \text{and} \quad \|f_{0+it}\|_{L^{p_0,r_0}} \leq 1.$$

Make an analogous definition for g_z and consider $\int T(f_z)g_z\,dxdt$.]

7. Suppose f is a bounded function on $\mathbb{R}$ with compact support. Then $H(f) \in L^1(\mathbb{R})$ if and only if $\int f\,dx = 0$.
[Hint: If $a = \int f\,dx$, then $H(f)(x) = \frac{a}{\pi x} + O(1/x^2)$ as $|x| \to \infty$.]

8. Suppose T is a bounded linear transformation mapping the space of real-valued L^p functions into itself with

$$\|T(f)\|_{L^p} \leq M\|f\|_{L^p}.$$

(a) Let T' be the extension of T to complex-valued functions: $T'(f_1 + if_2) = T(f_1) + iT(f_2)$. Then T' has the same bound: $\|T'(f)\|_{L^p} \leq M\|f\|_{L^p}$.

(b) More generally, fix any N, then

$$\|(\sum_{j=1}^{N} |T(f_j)|^2)^{1/2}\|_{L^p} \leq M\|(\sum_{j=1}^{N} |f_j|^2)^{1/2}\|_{L^p}.$$

[Hint: For part (b), let ξ denote a unit vector in $\mathbb{R}^N$, and let $F_\xi = \sum_{j=1}^N \xi_j f_j$, $\xi = (\xi_1, \ldots, \xi_N)$. Then $\int |(TF_\xi)(x)|^p \leq M^p \int |F_\xi(x)|^p$. Integrate this inequality for ξ on the unit sphere.]

9. Show the identity of the following two classes of harmonic functions u in the upper half-plane $\mathbb{R}^2_+ = \{z : \ x + iy, \ y > 0\}$.

(a) The harmonic functions u that are continuous in the closure $\overline{\mathbb{R}^2_+}$, and that vanish at infinity (that is, $u(x,y) \to 0$, if $|x| + y \to \infty$).

(b) The functions representable as $u(x,y) = (f * \mathcal{P}_y)(x)$, where $\mathcal{P}_y(x)$ is the Poisson kernel $\frac{1}{\pi}\frac{y}{x^2+y^2}$, and f is a continuous function on $\mathbb{R}$ that vanishes at infinity (that is, $f(x) \to 0$, as $|x| \to \infty$).

[Hint: To show that (a) implies (b), let $f(x) = u(x,0)$. Then $\mathcal{D}(x,y) = u(x,y) - (f * \mathcal{P}_y)(x)$ is harmonic in $\mathbb{R}^2_+$, continuous on $\overline{\mathbb{R}^2_+}$, vanishes at infinity, and moreover $\mathcal{D}(x,0) = 0$. Thus by the maximum principle, $\mathcal{D}(x,y) = 0$.]

10. Suppose $f \in L^p(\mathbb{R})$. Verify that:

(a) $\|f * \mathcal{P}_y\|_{L^p(\mathbb{R})} \leq \|f\|_{L^p}$, $1 \leq p \leq \infty$.

(b) $f * \mathcal{P}_y \to f$, as $y \to 0$, in the L^p norm, when $1 \leq p < \infty$.

11. Assume $f \in L^p(\mathbb{R})$, $1 < p < \infty$. Prove that:

(a) $f * \mathcal{Q}_y = H(f) * \mathcal{P}_y$, where H, $\mathcal{P}_y$ and $\mathcal{Q}_y$ are respectively the Hilbert transform, the Poisson kernel and the conjugate Poisson kernel.

(b) $f * \mathcal{Q}_y \to H(f)$ in the L^p norm, as $y \to 0$.

(c) $H_\epsilon(f) \to H(f)$ in the L^p norm, as $\epsilon \to 0$.

[Hint: Verify (a) first for $f \in L^2$ by noting that the Fourier transform of both sides equals $\hat{f}(\xi)\frac{\text{sign}(\xi)}{\xi}e^{-2\pi|\xi|y}$.]

12. In $\mathbb{R}^d$, suppose $f(x) = |x|^{-d}(\log 1/|x|)^{-1-\delta}$ if $|x| \leq 1/2$, $f(x) = 0$ otherwise. Then observe that $f^*(x) \geq c|x|^{-d}(\log 1/|x|)^{-\delta}$, if $|x| \leq 1/2$. Hence if $0 < \delta \leq 1$, we have $f \in L^1(\mathbb{R}^d)$ but $f^*(x)$ is not integrable over the unit ball.

13. Prove that the basic distribution function inequality (28) for the maximal function can essentially be reversed, that is, there is a constant A so that

$$m(\{x:\ f^*(x) > \alpha\}) \geq (A/\alpha)\int_{|f(x)|>\alpha} |f(x)|\,dx.$$

[Hint: Write $E_\alpha = \{x:\ f^*(x) > \alpha\}$ as $\bigcup_{j=1}^{\infty} Q_j$, with Q_j closed cubes satisfying (37), with $\Omega = E_\alpha$. For each Q_j let B_j be the smallest ball so that $Q_j \subset \overline{B_j}$, and $\overline{B_j}$ intersects E_α^c. First $m(B_j) \leq cm(Q_j)$, then $\frac{1}{m(B_j)}\int_{B_j}|f|\,dx \leq \alpha$. Thus $m(Q_j) \geq \frac{c^{-1}}{\alpha}\int_{B_j}|f(x)|\,dx \geq \frac{c^{-1}}{\alpha}\int_{Q_j}|f(x)|\,dx$. Now add in j, and use the fact that $\{x:\ |f(x)| > \alpha\} \subset \{x:\ f^*(x) > \alpha\}$.]

14. Deduce the following important consequence from (28) and the previous exercise. Suppose f is an integrable function on $\mathbb{R}^d$, and B_1, B_2 are a pair of balls with $\overline{B_1} \subset B_2$.

(a) f^* is integrable on B_1 if $|f|\log(1+|f|)$ is integrable on B_2.

(b) In the converse direction, whenever f^* is integrable on B_1 then $|f|\log(1+|f|)$ is also integrable there.

[Hint: Integrate the inequalities in α, for $\alpha \geq 1$.]

15. Consider the weak-type space, consisting of all functions f for which $m(\{x:\ |f(x)| > \alpha\}) \leq \frac{A}{\alpha}$ for some A and all $\alpha > 0$. One might hope to define a norm on this space by taking the "norm" of f to be the least A for which the above inequality holds. Denote this quantity by $\mathcal{N}(f)$.

(a) Show, however, that $\mathcal{N}$ is not a genuine norm; moreover there is no norm $\|\cdot\|$ on this space so that $\|f\|$ is equivalent with $\mathcal{N}(f)$.

(b) Prove that any bounded linear functional on this space must vanish on the sub-space of functions that are integrable.

[Hint: Consider $\mathbb{R}$. The function $f(x) = 1/|x|$ has $\mathcal{N}(f) = 2$. But if $f_N(x) = \frac{1}{N}[f(x+1) + f(x+2) + \cdots + f(x+N)]$, then $\mathcal{N}(f_N) \geq c \log N$.]

16. Prove that the space $\mathbf{H}_r^1$ is complete as follows. Let $\{f_n\}$ be a Cauchy sequence in $\mathbf{H}_r^1$. Then since $\{f_n\}$ is also Cauchy in L^1, there is an L^1 function f so that $f = \lim_{n\to\infty} f_n$ in the L^1 norm. Now for an appropriate sub-sequence $\{n_k\}$, write $f = f_{n_1} + \sum_{k=1}^{\infty}(f_{n_{k+1}} - f_{n_k})$.

17. Consider the function f defined by $f(x) = 1/(x(\log x)^2)$ for $0 < x \leq 1/2$ and $f(x) = 0$ if $x > 1/2$, and extended to $x < 0$ by $f(x) = -f(-x)$. Then f is integrable on $\mathbb{R}$, with $\int f = 0$, hence f is a multiple of a 1-atom in the terminology of Section 5.2.

Verify that $M(f) \geq c/(|x| \log |x|)$ for $|x| \leq 1/2$, hence $M(f) \notin L^1$, thus by Theorem 6.1 we know that $f \notin \mathbf{H}_r^1$.

18. Show that there exists a $c > 1$ so that every $f \in L^1(\mathbb{R}^d)$ can be written as $f(x) = \sum_{k=1}^{\infty} \lambda_k \mathfrak{a}_k(x)$, with $\sum |\lambda_k| \leq c\|f\|_{L^1}$, where the $\mathfrak{a}_k$ are "faux" atoms: each $\mathfrak{a}_k$ is supported in a ball B_k; $|\mathfrak{a}_k(x)| \leq 1/m(B_k)$ for all x; but $\mathfrak{a}_k$ does not necessarily satisfy the cancelation condition $\int \mathfrak{a}_k(x)\, dx = 0$.

[Hint: Let $f_n = \mathbb{E}_n(f)$, where $\mathbb{E}_n$ replaces f by its average over each dyadic cube of the nth-generation. Then $\|f_n - f\|_{L^1} \to 0$. Pick $\{n_k\}$ so that $\|f_{n_{k+1}} - f_{n_k}\|_{L^1} < 1/2^k$, and write $f = f_{n_1} + \sum_{k=1}^{\infty}(f_{n_{k+1}} - f_{n_k})$.]

19. The following illustrates two senses in which $\mathbf{H}_r^1$ is near L^1, but yet different.

(a) Suppose $f_0(x)$ is a positive decreasing function on $(0, \infty)$ that is integrable on $(0, \infty)$. Then show that there is a function $f \in \mathbf{H}_r^1(\mathbb{R})$ so that $|f(x)| \geq f_0(|x|)$.

(b) However if $f \in \mathbf{H}_r^1(\mathbb{R}^d)$, and f is positive on an open set, then its size must be "smaller" on that open set than a general integrable function. In fact, prove that if $f \in \mathbf{H}_r^1$, and $f \geq 0$ in a ball B_1, then $f \log(1+f)$ must be integrable over any proper sub-ball $B_0 \subset B_1$.

[Hint: For (a) take $f(x) = \text{sign}(x) f_0(|x|)$, and find an atomic decomposition for f. For (b) use Exercise 14, together with the maximal theorem in Section 6, with Φ positive.]

20. When $f \in L^1(\mathbb{R}^d)$ we know that its Fourier transform $\hat{f}$ is bounded and $\hat{f}(\xi)$ tends to 0 as $|\xi| \to \infty$ (the Riemann-Lebesgue lemma), but no better assertion about the "smallness" of $\hat{f}$ can be made. (For the analogous result for Fourier series, see Chapter 3 in Book III.) Show, however, that for $f \in \mathbf{H}_r^1$ we have

$$\int_{\mathbb{R}^d} |\hat{f}(\xi)| \frac{d\xi}{|\xi|^d} \leq A\|f\|_{\mathbf{H}_r^1}.$$

[Hint: Verify this for atoms.]

21. Prove that if $|f(x)| \le A(1+|x|)^{-d-1}$, and $\int_{\mathbb{R}^d} f(x)\,dx = 0$, then $f \in \mathbf{H}_r^1(\mathbb{R}^d)$.
[Hint: While this is elementary, it is a little tricky. Write $f = \sum_{k=0}^{\infty} f_k$, where $f_0(x) = f(x)$ if $|x| \le 1$, 0 elsewhere, and $f_k(x) = f(x)$ if $2^{k-1} < |x| \le 2^k$ and 0 elsewhere, and $k \ge 1$. Let $c_k = \int f_k\,dx$, $s_k = \sum_{j \ge k} c_j$, then $s_0 = 0$. Fix a bounded function η supported in $|x| \le 1$, with $\int \eta(x)\,dx = 1$. Now write $f(x) = \sum_{k=0}^{\infty}(f_k - c_k\eta_k) + \sum_{k=0}^{\infty} c_k\eta_k$, where $\eta_k(x) = 2^{-kd}\eta(2^{-k}x)$ and $\int \eta_k = 1$. The first sum is clearly a sum of multiples of atoms (which are $O(2^{-k})$) supported on the balls $|x| \le 2^k$. That the second sum is similar can be seen by rewriting it as $\sum_{k=1}^{\infty} s_k(\eta_k - \eta_{k-1})$.]

22. Let f be the atom on $\mathbb{R}$ supported in $|x| \le 1/2$ given by $f(x) = \text{sign}(x)$. Apply to f the maximal function f_0^* defined by

$$f_0^*(x) = \sup_{\epsilon > 0} |(f * \chi_\epsilon)(x)|,$$

where χ is the characteristic function of $|x| \le 1/2$ and $\chi_\epsilon(x) = \epsilon^{-1}\chi(x/\epsilon)$.

Verify that $|f_0^*(x)| \ge 1/(2|x|)$ if $|x| \ge 1/2$ hence $f_0^* \notin L^1$. Thus the maximal function f_0^*, defined in terms of χ, cannot be used to characterize the real Hardy space $\mathbf{H}_r^1$.

23. Verify the following examples related to BMO:

(a) $\log|x| \in \text{BMO}(\mathbb{R}^d)$.

(b) If $f(x) = \log x$, when $x > 0$, and $= 0$ when $x \le 0$, then $f \notin \text{BMO}(\mathbb{R})$.

(c) If $\delta \ge 0$, $(\log|x|)^\delta \in \text{BMO}(\mathbb{R}^d)$ if and only if $\delta \le 1$.

[Hint: With $f(x) = \log|x|$, note that $f(rx) = f(x) + c_r$ and so we may assume the ball B has radius 1 in testing the condition (43). For (b), test f on small intervals centered at the origin.]

24. Using Exercises 19 (a) and 23, give examples of $f \in \mathbf{H}_r^1$ and $g \in \text{BMO}$ so that $|f(x)g(x)|$ is not integrable over $\mathbb{R}^d$.

8 Problems

1. Another way $\mathbf{H}_r^1$ is an improvement over L^1 is in its weak compactness of the unit ball. The following can be proved. Suppose $\{f_n\}$ is a sequence in $\mathbf{H}_r^1$ with $\|f_n\|_{\mathbf{H}_r^1} \le A$. Then we can select a subsequence $\{f_{n_k}\}$ and find an $f \in \mathbf{H}_r^1$ so that $\int f_{n_k}(x)\varphi(x)\,dx \to \int f(x)\varphi(x)\,dx$, as $k \to \infty$, for every φ that is a continuous function of compact support.

This is to be compared with L^1, and the failure there of weak compactness as described in Exercises 12 and 13 in the previous chapter.

[Hint: Apply the result in Problem 4 (c) of the previous chapter to obtain a subsequence $\{f_{n_k}\}$ and a finite measure μ so that $f_{n_k} \to \mu$ in the weak* sense. Next use the fact that if $\sup_{\epsilon>0} |\mu * \varphi_\epsilon| \in L^1$, for an appropriate φ, then μ is absolutely continuous.]

2. Suppose H^p is the complex Hardy space defined in Section 5. For $1 < p < \infty$, prove the following:

(a) If $F \in H^p$, then $\lim_{y\to 0} F(x+iy) = F_0(x)$ exists in the $L^p(\mathbb{R})$ norm.

(b) $\|F\|_{H^p} = \|F_0\|_{L^p}$.

(c) One has $2F_0 = f + iH(f)$, with f real-valued in $L^p(\mathbb{R})$, and $\|F_0\|_{L^p} \approx \|f\|_{L^p}$. Moreover, every F_0 (and thus F) arises this way. This gives a linear isomorphism (over the reals) of H^p, with L^p with an equivalence of norms.

[Hint: Here is an outline of the proof. For each $y_1 > 0$, write $F_{y_1}(z) = F(z + iy_1)$ and $F^\epsilon_{y_1}(z) = F_{y_1}(z)/(1 - i\epsilon z)$, $\epsilon > 0$. One has that F_{y_1} is bounded $\overline{\mathbb{R}^2_+}$ (see Section 2, in Chapter 5, Book III). Thus by Exercise 9, $F^\epsilon_{y_1}(z) = (F^\epsilon_{y_1} * \mathcal{P}_y)(x)$. Now using the weak compactness of the unit ball in L^p, (Exercise 12 in Chapter 1), we can find $F_0 \in L^p$ so that $F^\epsilon_{y_1}(x) \to F_0(x)$ weakly as ϵ and $y_1 \to 0$. Observe that this breaks down for $p = 1$. Conclusion (c) is then essentially a restatement of the boundedness of the Hilbert transform for $1 < p < \infty$.]

3. Let P be any non-zero polynomial of degree k in $\mathbb{R}^d$. Then $f = \log |P(x)|$ is in BMO and $\|f\|_{\text{BMO}} \leq c_k$, where c_k depends only on the degree k of the polynomial.

[Hint: Verify the result first when $d = 1$. Then use induction in the dimension and the following assertion, stated for $\mathbb{R}^2$. Suppose $f(x, y)$, $(x, y) \in \mathbb{R}^2$ is for each y a BMO($\mathbb{R}$) function in x, uniformly in y. Assume also that this holds when the roles of x and y are interchanged. Then $f \in \text{BMO}(\mathbb{R}^2)$.]

4. Prove the following John-Nirenberg inequalities for every $f \in \text{BMO}(\mathbb{R}^d)$:

(a) For every $q < \infty$ there is a bound b_q so that

$$\sup_B \frac{1}{m(B)} \int_B |f - f_B|^q \, dx \leq b_q^q \|f\|^q_{\text{BMO}}.$$

(b) There are positive constants μ and A, so that

$$\sup_B \frac{1}{m(B)} \int_B e^{\mu|f - f_B|} \, dx \leq A, \qquad \text{whenever } \|f\|_{\text{BMO}} \leq 1.$$

[Hint: For (a) test f against p-atoms, where p is dual to q. For (b) use the bound $c_p = O(1/(p-1))$ as $p \to 1$ (in (38)) to obtain $b_q = O(q)$, as $q \to \infty$. Then write $e^u = \sum_{q=0}^\infty u^q/q!$.]

5. The Hilbert transform of a bounded function is in BMO. Show this in two different ways.

(a) Directly: Suppose f is bounded (and belongs to some L^p, $1 \leq p < \infty$). Then $H(f) \in \text{BMO}$ with

$$\|H(f)\|_{\text{BMO}} \leq A\|f\|_{L^\infty},$$

with A not depending on the L^p norm of f.

(b) By duality, using Theorem 5.4.

[Hint: For (a), fix any ball B, and let B_1 be its double. Consider separately, $f\chi_{B_1}$ and $f\chi_{B_1^c}$.]

6.* The following is the maximal characterization of $\mathbf{H}_r^1(\mathbb{R}^d)$. Suppose Φ belongs to the Schwartz space $\mathcal{S}$ and $\int \Phi(x)\,dx \neq 0$. Let $M(f)(x) = \sup_{\epsilon>0} |(f * \Phi_\epsilon)(x)|$, for $f \in L^1$. Then

(a) f is in $\mathbf{H}_r^1$ if and only if $M(f)$ belongs to L^1.

(b) The condition $\Phi \in \mathcal{S}$ can be relaxed to require only

$$|\partial_x^\alpha \Phi(x)| \leq c_\alpha(1 + |x|)^{-d-1-|\alpha|}.$$

(c) Note two interesting examples, first $\Phi_{t^{1/2}}(x) = (4\pi t)^{-d/2} e^{-|x|^2/(4t)}$: then $u(x,t) = (f * \Phi_{t^{1/2}})(x)$ is the solution of the Heat equation $\triangle_x u = \partial_t u$, with initial data $u(x,0) = f(x)$. Also, $\Phi_t(x) = \frac{c_d t}{(t^2+|x|^2)^{\frac{d+1}{2}}}$ with $c_d = \Gamma(\frac{d+1}{2})/\pi^{\frac{d+1}{2}}$ so that $u(x,t) = (f * \Phi_t)(x)$ is the solution to Laplace's equation $\triangle_x u + \partial_t^2 u = 0$, with initial data $u(x,0) = f(x)$. (Here Γ denotes the gamma function.)

7.* H^p, when $p = 1$. The results in (a) and (b) of Problem 2 also hold for $p = 1$, but require a different proof. The analog of (c) is as follows. One has $2F_0 = f + iH(f)$, where f belongs to the real Hardy space $\mathbf{H}_r^1$. Also $\|F_0\|_{L^1} \approx \|f\|_{\mathbf{H}_r^1}$. As a consequence, a necessary and sufficient condition that $f \in \mathbf{H}_r^1$ is that both f and $H(f)$ are in L^1.

The conclusions (a) and (b) may be proved by showing that any $F \in H^1$ can be written as $F = F_1 \cdot F_2$ with $F_j \in H^2$ and $\|F_j\|_{H^2}^2 = \|F\|_{H^1}$, and then using the corresponding results in H^2.

8.* Suppose $f \in L^1(\mathbb{R})$. Then we can define $H(f) \in L^1(\mathbb{R})$ in the weak sense to mean that there exists $g \in L^1(\mathbb{R})$ so that

$$\int_{\mathbb{R}} g\varphi\,dx = \int_{\mathbb{R}} fH(\varphi)\,dx, \qquad \text{for all functions } \varphi \text{ in the Schwartz space.}$$

Then we say $g = H(f)$ in the weak sense.

As a consequence of Problem 7*, one has that $f \in \mathbf{H}_r^1(\mathbb{R})$ if and only if $f \in L^1(\mathbb{R})$ and $H(f)$, taken in the weak sense, also belongs to $L^1(\mathbb{R})$.

9.* Let $\{f_n\}$ be a sequence of elements in $\mathbf{H}_r^1$ so that $\|f_n\|_{\mathbf{H}_r^1} \leq M < \infty$ for all n. Assume that f_n converges to f almost everywhere. Then:

(a) $f \in \mathbf{H}^1_r$.

(b) $\int f_n g \to \int fg$, as $n \to \infty$, for all g continuous with compact support.

A corresponding result holds for L^p, $p > 1$, but fails for $p = 1$. See Exercise 14 in Chapter 1.

10.* The following result illustrates the application of $\mathbf{H}^1_r$ to the theory of compensated compactness.

Suppose $A = (A_1, \ldots, A_d)$ and $B = (B_1, \ldots, B_d)$ are vector fields in $\mathbb{R}^d$ with $A_i, B_i \in L^2(\mathbb{R}^d)$ for all i. The divergence of A is defined by

$$\operatorname{div}(A) = \sum_{k=1}^{d} \frac{\partial A_k}{\partial_{x_k}},$$

and the curl of B is the $d \times d$ matrix whose ij-entry is

$$(\operatorname{curl}(B))_{ij} = \frac{\partial B_i}{\partial x_j} - \frac{\partial B_j}{\partial x_i}.$$

(The derivatives here are taken in the sense of distributions, as in the next chapter.) If $\operatorname{div}(A) = 0$ and $\operatorname{curl}(B) = 0$ then $\sum_{k=1}^d A_k B_k \in \mathbf{H}^1_r$.
This is in contrast with the result that in general, if $f, g \in L^2$, then one only has $fg \in L^1$.

3 Distributions: Generalized Functions

> The heart of analysis is the concept of function, and functions "belong" to analysis, even if, nowadays, they occur everywhere and anywhere, in and out of mathematics, in thought, cognition, even perception.
>
> Functions came into being in "modern" mathematics, that is, in mathematics since the Renaissance. By a rough division into centuries, the 17th and 18th centuries made preparations, the 19th century created functions of one variable, real and complex, and the 20th century has turned to functions in several variables, real and complex.
>
> *S. Bochner,* 1969

> ... It was not accidental that the notion of function generally accepted now was first formulated in the celebrated memoir of Dirichlet (1837) dealing with the convergence of Fourier series; or that the definition of Riemann's integral in its general form appeared in Riemann's *Habilitationsschrift* devoted to trigonometric series; or that the theory of sets, one of the most important developments of nineteenth-century mathematics, was created by Cantor in his attempts to solve the problem of the sets of uniqueness for trigonometric series. In more recent times, the integral of Lebesgue was developed in close connection with the theory of Fourier series, and the theory of generalized functions (distributions) with that of Fourier integrals.
>
> *A. Zygmund,* 1959

The growth of analysis can be traced by the evolution of the idea of what a function is. The formulation of the notion of "generalized functions" (or "distributions" as they are commonly called) represents a significant stage in that development with ramifications in many different areas. Looking back, one can see that this concept had many antecedents.

Among these were: Riemann's formal integration and differentiation of trigonometric series in his study of uniqueness; the necessity of using weak solutions in the theory of partial differential equations; and the possibility of realizing a function (say in L^p) as a linear functional on an appropriate dual space. The importance of distributions derives from the ease with which this tool permits us to carry out formal manipulations, finessing numerous technical issues. While as such it is not a panacea, it allows us, in many instances, to arrive more quickly at the heart of the matter.

We divide our treatment of distributions in two parts. First, we set down the basic properties of general distributions and the rules of their manipulation. Thus we see that an ordinary function has derivatives of all orders in the sense of distributions. Also in that sense, any function that does not increase too fast at infinity has a Fourier transform.

Next, we study specific distributions of particular importance, beginning with the principal-value distribution defining the Hilbert transform, and more general homogeneous distributions. We also consider distributions that arise as fundamental solutions of partial differential equations. Finally, we take up the Calderón-Zygmund distributions that occur as kernels of singular integrals generalizing the Hilbert transform, and for these we obtain basic L^p estimates.

1 Elementary properties

Classically a function f (defined on $\mathbb{R}^d$) assigns a definite value $f(x)$ for each $x \in \mathbb{R}^d$. For many purposes, it is often convenient to relax this requirement by allowing f to remain undefined at certain "exceptional" points x. This is particularly so when dealing with integration and measure theory. Thus in that context a function can be unspecified on a set of measure zero.[1]

In contrast to this, a **distribution** or **generalized function** F will not be given by assigning values of F at "most" points, but will instead be determined by its averages taken with respect to (smooth) functions. Thus if we are to think of a function f as a distribution F, we determine F by the quantities

$$F(\varphi) = \int_{\mathbb{R}^d} f(x)\varphi(x)\, dx, \tag{1}$$

where the φ's range over an appropriate space of "test" functions. Therefore, in keeping with (1), our starting point in defining a distribution F

[1]More precisely, a function is then really an equivalence class of functions that agree almost everywhere.

will be to think of F as a linear functional on a suitable space of these test functions.

Actually, there will be two classes of distributions (each with its space of test functions) that we will consider: the broader class, which we deal with first, and which can be defined on any open set Ω of $\mathbb{R}^d$; also later, a narrower class of distributions defined on $\mathbb{R}^d$, those which are suitably "tempered" at infinity, and that arise naturally in the context of Fourier transforms.

1.1 Definitions

We fix an open set Ω in $\mathbb{R}^d$. The **test functions** for the larger class of distributions will be the functions that belong to $C_0^\infty(\Omega)$, the complex-valued indefinitely differentiable functions of compact support in Ω. In keeping with a common notation used in this context we denote this space of test functions as $\mathcal{D}$ (or more explicitly as $\mathcal{D}(\Omega)$). Now if $\{\varphi_n\}$ is a sequence of elements in $\mathcal{D}$, and also $\varphi \in \mathcal{D}$, we say that $\{\varphi_n\}$ converges to φ in $\mathcal{D}$, and write $\varphi_n \to \varphi$ in $\mathcal{D}$, if the supports of the φ_n are contained in a common compact set and for each multi-index α, one has $\partial_x^\alpha \varphi_n \to \partial_x^\alpha \varphi$ uniformly in x as $n \to \infty$.[2] With this in mind we come to our basic definition. A **distribution** F on Ω is a complex-valued linear functional $\varphi \mapsto F(\varphi)$, defined for $\varphi \in \mathcal{D}(\Omega)$, that is continuous in the sense that $F(\varphi_n) \to F(\varphi)$ whenever $\varphi_n \to \varphi$ in $\mathcal{D}$. The vector space of distributions on Ω is denoted by $\mathcal{D}^*(\Omega)$.

In what follows we shall tend to reserve the upper case letters F, $G, \ldots$ for distributions, and the lower case letters f, $g, \ldots$ for ordinary functions. First, we look at a few quick examples of distributions.

EXAMPLE 1. *Ordinary functions.* Let f be any locally integrable function on Ω.[3] Then f defines a distribution $F = F_f$, according to the formula (1). Distributions arising this way are of course referred to as "functions."

EXAMPLE 2. Let μ be a (signed) Borel measure on Ω which is finite on compact subsets of Ω (sometimes called a Radon measure). Then

$$F(\varphi) = \int \varphi(x)\, d\mu(x)$$

---

[2] We recall the notation: $\partial_x^\alpha = (\partial/\partial x)^\alpha = (\partial/\partial x_1)^{\alpha_1} \cdots (\partial/\partial x_d)^{\alpha_d}$, $|\alpha| = \alpha_1 + \cdots + \alpha_d$, and $\alpha! = \alpha_1 \cdots \alpha_d$, where $\alpha = (\alpha_1, \ldots, \alpha_d)$.

[3] By this we mean that f is measurable and Lebesgue integrable over any compact subset of Ω. (Compare this definition with the one in Chapter 3 of Book III, where it has a slightly different meaning.)

is a distribution which, in general, is not a function as above. The special case, when μ is the point-mass which assigns total mass of 1 to the origin, gives the **Dirac delta function** δ, that is, $\delta(\varphi) = \varphi(0)$. (Note, however, that δ is not a function!)

Further examples arise from the above by differentiation. In fact, a key feature of distributions is that, as opposed to ordinary functions, these can be differentiated any number of times. The **derivative** $\partial_x^\alpha F$ of a distribution generalizes that of a differentiable function. Indeed, whenever f is a smooth function on Ω and (say) $\varphi \in \mathcal{D}(\Omega)$, then an integration by parts yields

$$\int (\partial_x^\alpha f)\, \varphi\, dx = (-1)^{|\alpha|} \int f\, (\partial_x^\alpha \varphi)\, dx.$$

Hence in keeping with (1) we define $\partial_x^\alpha F$ as the distribution given by

$$(\partial_x^\alpha F)(\varphi) = (-1)^{|\alpha|} F\,(\partial_x^\alpha \varphi), \qquad \text{whenever } \varphi \in \mathcal{D}(\Omega).$$

Thus in particular, if f is a locally integrable function, we can define its partial derivatives as distributions. A few examples may be useful here.

- Suppose h is the Heaviside function on $\mathbb{R}$, that is, $h(x) = 1$ for $x > 0$, and $h(x) = 0$, for $x < 0$. Then dh/dx, taken in the sense of distributions equals the Dirac delta δ. This is because $-\int_0^\infty \varphi'(x)\, dx = \varphi(0)$, whenever $\varphi \in \mathcal{D}(\mathbb{R})$. Note however that the usual derivative of h is zero when $x \neq 0$, and is undefined at $x = 0$. So we must be careful to distinguish the distribution derivative of a function, from its usual derivative (when it exists), if the function is not smooth. (See also Exercises 1 and 2.)

  A higher dimensional variant of the Heaviside function is given in Exercise 15.

- Suppose the function f is of **class** C^k on Ω, that is, all the partial derivatives $\partial_x^\alpha f$ with $|\alpha| \leq k$, taken in the usual sense, are continuous on Ω. Then these derivatives of f agree with the corresponding derivatives taken in the sense of distributions.

- More generally, suppose f and g are a pair of functions in $L^2(\Omega)$ and $\partial_x^\alpha f = g$ in the "weak sense" as discussed in Section 3.1 of Chapter 1, or in Section 3.1, Chapter 5 of Book III. If F and G are the distributions determined by f and g respectively, according to (1), then $\partial_x^\alpha F = G$.

1.2 Operations on distributions

As in the case of differentiation, one can carry over various operations on distributions by transforming the corresponding actions on test functions. We first give some simple examples.

- Whenever F belongs to $\mathcal{D}^*$ and ψ is a C^∞ function, then we can define the product $\psi \cdot F$ by $(\psi \cdot F)(\varphi) = F(\psi\varphi)$, for every $\varphi \in \mathcal{D}$. This agrees with the usual pointwise definition of the product when F is a function.

- For a distribution on $\mathbb{R}^d$ the actions of translations, dilations and more generally non-singular linear transformations can be defined by the corresponding actions on test functions via "duality." Thus for the translation operator τ_h, defined for functions by $\tau_h(f)(x) = f(x-h)$, $h \in \mathbb{R}^d$, the corresponding definition on distributions is:

$$\tau_h(F)(\varphi) = F(\tau_{-h}(\varphi)), \qquad \text{for every test function } \varphi.$$

  Similarly, for dilations given on functions f by the simple relation $f_a(x) = f(ax)$, $a > 0$, one defines F_a by $F_a(\varphi) = a^{-d}F(\varphi_{a^{-1}})$. More generally, if L is a non-singular linear transformation then the extension of $f_L(x) = f(L(x))$ to distributions is given by the rule $F_L(\varphi) = |\det L|^{-1}F(\varphi_{L^{-1}})$ for every $\varphi \in \mathcal{D}$.

It is important that one can also extend the notion of **convolution**, defined for appropriate functions on $\mathbb{R}^d$ by

$$(f * g)(x) = \int_{\mathbb{R}^d} f(x-y)g(y)\,dy$$

to large classes of distributions.

To begin with, suppose that F is a distribution on $\mathbb{R}^d$ and ψ a test function. Then there are two ways that we might define $F * \psi$ (in keeping with (1) when F is a function). The first is as a *function* (of x) given by $F(\psi_x^\sim)$, with $\psi_x^\sim(y) = \psi(x-y)$.

The second is that $F * \psi$ is the *distribution* determined by

$$(F * \psi)(\varphi) = F(\psi^\sim * \varphi), \qquad \text{with } \psi^\sim = \psi_0^\sim.$$

Proposition 1.1 *Suppose F is a distribution and $\psi \in \mathcal{D}$. Then*

(a) *The two definitions of $F * \psi$ given above coincide.*

(b) *The distribution $F * \psi$ is a C^∞ function.*

Proof. Let us observe first that $F(\psi_x^\sim)$ is continuous in x and in fact indefinitely differentiable. Note that if $x_n \to x_0$ as $n \to \infty$, then $\psi_{x_n}^\sim(y) = \psi(x_n - y) \to \psi(x_0 - y) = \psi_{x_0}^\sim(y)$ uniformly in y, and the same is true for all partial derivatives. Therefore $\psi_{x_n}^\sim \to \psi_{x_0}^\sim$ in $\mathcal{D}$ (as functions of y) as $n \to \infty$, and thus by the assumed continuity of F on $\mathcal{D}$ we have that $F(\psi_x^\sim)$ is continuous in x. Similarly, all corresponding difference quotients converge and the result is that $F(\psi_x^\sim)$ is indefinitely differentiable, with $\partial_x^\alpha F(\psi_x^\sim) = F(\partial_x^\alpha \psi_x^\sim)$.

It remains to prove conclusion (a), and for this it suffices to show that

$$\int F(\psi_x^\sim)\varphi(x)\,dx = F(\psi^\sim * \varphi), \qquad \text{for each } \varphi \in \mathcal{D}. \tag{2}$$

However since $\psi \in \mathcal{D}$, and of course φ is continuous with compact support, then it is easily seen that

$$(\psi^\sim * \varphi)(x) = \int \psi^\sim(x-y)\varphi(y)\,dy = \lim_{\epsilon \to 0} S(\epsilon)$$

where $S(\epsilon) = \epsilon^d \sum_{n \in \mathbb{Z}^d} \psi^\sim(x - n\epsilon)\varphi(n\epsilon)$. Here the convergence of the Riemann sums $S(\epsilon)$ to $\psi^\sim * \varphi$ is in $\mathcal{D}$. Clearly, $S(\epsilon)$ is finite for each $\epsilon > 0$, and thus $F(S_\epsilon) = \epsilon^d \sum_{n \in \mathbb{Z}^d} F(\psi_{n\epsilon}^\sim)\varphi(n\epsilon)$. Hence by the continuity of $x \mapsto F(\psi_x^\sim)$, a passage to the limit $\epsilon \to 0$ yields (2), proving the proposition.

A simple application of the proposition is the observation that every distribution F in $\mathbb{R}^d$ is the limit of C^∞ functions. We say that a sequence of distributions $\{F_n\}$ converges to a distribution F in the **weak sense** (or in the **sense of distributions**), if $F_n(\varphi) \to F(\varphi)$, for every $\varphi \in \mathcal{D}$.

Corollary 1.2 *Suppose F is a distribution on $\mathbb{R}^d$. Then there exists a sequence $\{f_n\}$, with $f_n \in C^\infty$, and $f_n \to F$ in the weak sense.*

Proof. Let $\{\psi_n\}$ be an approximation to the identity constructed as follows. Fix a $\psi \in \mathcal{D}$ with $\int_{\mathbb{R}^d} \psi(x)\,dx = 1$ and set $\psi_n(x) = n^d \psi(nx)$.

Form $F_n = F * \psi_n$. Then by the second conclusion of the proposition, each F_n is a C^∞ function. However by the first conclusion

$$F_n(\varphi) = F(\psi_n^\sim * \varphi) \qquad \text{for every } \varphi \in \mathcal{D}.$$

Moreover, as is easily verified, $\psi_n^\sim * \varphi \to \varphi$ in $\mathcal{D}$. Thus $F_n(\varphi) \to F(\varphi)$, for each $\varphi \in \mathcal{D}$, and the corollary is established.

1.3 Supports of distributions

We come next to the notion of the support of a distribution. If f is a *continuous* function its **support** is defined as the closure of the set where $f(x) \neq 0$. Or put another way, it is the complement of the largest open set on which f vanishes. For a distribution F we say that F vanishes in an open set if $F(\varphi) = 0$, for all test functions $\varphi \in \mathcal{D}$ which have their supports in that open set. Thus we define the **support of a distribution** F as the complement of the largest open set on which F vanishes.

This definition is unambiguous because if F vanishes on any collection of open sets $\{\mathcal{O}_i\}_{i \in \mathcal{I}}$, then F vanishes on the union $\mathcal{O} = \bigcup_{i \in \mathcal{I}} \mathcal{O}_i$. Indeed suppose φ is a test function supported in the compact set $K \subset \mathcal{O}$. Since $\mathcal{O}$ covers the compact set K, we may select a sub-cover which (after possibly relabeling the sets $\mathcal{O}_i$) we can write as $K \subset \bigcup_{k=1}^{N} \mathcal{O}_k$. A regularization applied to the partition of unity obtained in Section 7 in Chapter 1 yields *smooth* functions η_k for $1 \leq k \leq N$ so that $0 \leq \eta_k \leq 1$, $\text{supp}(\eta_k) \subset \mathcal{O}_k$, and $\sum_{k=1}^{N} \eta_k(x) = 1$ whenever $x \in K$. Then $F(\varphi) = F(\sum_{k=1}^{N} \varphi\eta_k) = \sum_{k=1}^{N} F(\varphi\eta_k) = 0$, since F vanishes on each $\mathcal{O}_k$. Thus F vanishes on $\mathcal{O}$ as claimed.[4]

Note the following simple facts about the supports of distributions. The supports of $\partial_x^{\alpha} F$ and $\psi \cdot F$ (with $\psi \in C^{\infty}$) are contained in the support of F. The support of the Dirac delta function (as well as its derivatives) is the origin. Finally, $F(\varphi) = 0$ whenever the supports of F and φ are disjoint

We observe next the additivity of the supports under convolution.

Proposition 1.3 *Suppose F is a distribution whose support is C_1, and ψ is in $\mathcal{D}$ and has support C_2. Then the support of $F * \psi$ is contained in $C_1 + C_2$.*

Indeed for each x for which $F(\psi_x^{\sim}) \neq 0$, we must have that the support of F intersects the support of $\psi_x^{\sim}$. Since the support of $\psi_x^{\sim}$ is the set $x - C_2$ this means that the set C_1 and $x - C_2$ have a point, say y, in common. Because $x = y + x - y$, while $y \in C_1$ and $x - y \in C_2$ (since $y \in x - C_2$) we have that $x \in C_1 + C_2$, and thus our assertion is established. Note that the set $C_1 + C_2$ is closed because C_1 is closed and C_2 is compact.

We can now extend the definition of convolution to a pair of distributions if one of them has compact support. Indeed, if F and F_1 are given distributions with F_1 having compact support, then we define $F * F_1$ as

[4]One must take care that this notion of support does not coincide with the "support" defined in Chapter 2 of Book III for an integrable function, when such function is considered as a distribution. A further clarification is in Exercise 5.

the distribution $(F * F_1)(\varphi) = F(F_1^{\sim} * \varphi)$, where $F_1^{\sim}$ is the reflected distribution given by $F_1^{\sim}(\varphi) = F_1(\varphi^{\sim})$. This extends the definition given above when $F_1 = \psi \in \mathcal{D}$. Notice that if C is the support of F_1, then $-C$ is the support of $F_1^{\sim}$. Therefore by the previous proposition $F_1^{\sim} * \varphi$ has compact support and is C^∞, hence it belongs to $\mathcal{D}$. The fact that the mapping $\varphi \mapsto (F * F_1)(\varphi)$ has the required continuity in $\mathcal{D}$ is then straightforward and is left to the reader to verify.

Other properties of convolutions that are direct consequences of the above reasoning are as follows:

- If F_1 and F_2 have compact support, then $F_1 * F_2 = F_2 * F_1$. (For this reason we shall sometimes also write $F_1 * F$ for $F * F_1$, when only F_1 has compact support.)
- With δ the Dirac delta function

$$F * \delta = \delta * F = F.$$

- If F_1 has compact support, then for every multi-index α

$$\partial_x^\alpha (F * F_1) = (\partial_x^\alpha F) * F_1 = F * (\partial_x^\alpha F_1).$$

- If F and F_1 have supports C and C_1 respectively, and C is compact, then the support of $F * F_1$ is contained in $C + C_1$. (This follows from the previous proposition and the approximation stated in part (b) of Exercise 4.)

1.4 Tempered distributions

There are distributions on $\mathbb{R}^d$ that, roughly speaking, are of at most polynomial growth at infinity. The restricted growth of these distributions is reflected in the space $\mathcal{S}$ of its test functions. This space $\mathcal{S} = \mathcal{S}(\mathbb{R}^d)$ of **test functions** (the Schwartz space[5]) consists of indefinitely differentiable functions on $\mathbb{R}^d$ that are rapidly decreasing at infinity with all their derivatives. More precisely, we consider the increasing sequence of norms $\|\cdot\|_N$, with N ranging over the positive integers, defined by[6]

$$\|\varphi\|_N = \sup_{x \in \mathbb{R}^d, |\alpha|, |\beta| \leq N} \left| x^\beta (\partial_x^\alpha \varphi)(x) \right|.$$

[5]The space $\mathcal{S}$ occurred already in Chapters 5 and 6 of Book I.

[6]We shall use the notation $\|\cdot\|_N$ throughout this chapter. This is not to be confused with the L^p norm, $\|\cdot\|_{L^p}$.

We define $\mathcal{S}$ to consist of all smooth functions φ such that $\|\varphi\|_N < \infty$ for every N. Moreover, one says that $\varphi_k \to \varphi$ in $\mathcal{S}$, whenever $\|\varphi_k - \varphi\|_N \to 0$, as $k \to \infty$, for every N.

With this in mind we say that F is a **tempered distribution** if it is a linear functional on $\mathcal{S}$ which is continuous in the sense that $F(\varphi_k) \to F(\varphi)$ whenever $\varphi_k \to \varphi$ in $\mathcal{S}$. We shall write $\mathcal{S}^*$ for the vector space of tempered distributions. Since the test space $\mathcal{D} = \mathcal{D}(\mathbb{R}^d)$ is contained in $\mathcal{S}$, and convergence in $\mathcal{D}$ implies convergence in $\mathcal{S}$, we see that any tempered distribution is automatically a distribution on $\mathbb{R}^d$ in the previous sense. However the converse is not true. (See Exercise 9). It is worthwhile to note that $\mathcal{D}$ is dense in $\mathcal{S}$ in that for every function $\varphi \in \mathcal{S}$, there exists a sequence of functions $\varphi_k \in \mathcal{D}$ such that $\varphi_k \to \varphi$ in $\mathcal{S}$ as $k \to \infty$. (See Exercise 10.)

It is also useful to observe that any tempered distribution is already controlled by finitely many of the norms $\|\cdot\|_N$.

Proposition 1.4 *Suppose F is a tempered distribution. Then there is a positive integer N and a constant $c > 0$, so that*

$$|F(\varphi)| \leq c\|\varphi\|_N, \qquad \textit{for all } \varphi \in \mathcal{S}.$$

Proof. Assume otherwise. Then the conclusion fails and for each positive integer n there is a $\psi_n \in \mathcal{S}$ with $\|\psi_n\|_n = 1$, while $|F(\psi_n)| \geq n$. Take $\varphi_n = \psi_n/n^{1/2}$. Then $\|\varphi_n\|_N \leq \|\varphi_n\|_n$ as soon as $n \geq N$, and thus $\|\varphi_n\|_N \leq n^{-1/2} \to 0$ as $n \to \infty$, while $|F(\varphi_n)| \geq n^{1/2} \to \infty$, contradicting the continuity of F.

The following are some simple examples of tempered distributions.

- A distribution F of compact support is also tempered. This follows from the fact that if C is the support of F, there is an $\eta \in \mathcal{D}$, with $\eta(x) = 1$ for all x in a neighborhood of C, hence $F(\varphi) = F(\eta\varphi)$ if $\varphi \in \mathcal{D}$. Thus the linear functional F defined on $\mathcal{D}$ has an obvious extension to $\mathcal{S}$ given by $\varphi \mapsto F(\eta\varphi)$, and this gives the corresponding distribution.

- Suppose f is locally integrable on $\mathbb{R}^d$ and for some $N \geq 0$,

$$\int_{|x|\leq R} |f(x)|\, dx = O(R^N), \qquad \text{as } R \to \infty.$$

  Then the distribution corresponding to f is tempered. Hence in particular this holds if $f \in L^p(\mathbb{R}^d)$ for some p with $1 \leq p \leq \infty$.

- Whenever F is tempered so is $\partial_x^\alpha F$ for all α; also $x^\beta F(x)$ is tempered for all multi-index $\beta \geq 0$.

The last assertion can be generalized as follows: let ψ be any C^∞ function on $\mathbb{R}^d$ which is **slowly increasing**: this means that for each α, $\partial_x^\alpha \psi(x) = O(|x|^{N_\alpha})$ as $|x| \to \infty$, for some $N_\alpha \geq 0$. Then ψF defined by $(\psi F)(\varphi) = F(\psi\varphi)$ is also a tempered distribution, whenever F is tempered.

The properties of convolutions of distributions discussed in Sections 1.2 and 1.3 have modifications for tempered distributions. The proofs of the assertions below are routine adaptations of previous arguments.

(a) If F is tempered and $\psi \in \mathcal{S}$, then $F * \psi$, defined as the function $F(\psi_x^\sim)$ is C^∞ and slowly increasing. Moreover the alternate definition $(F * \psi)(\varphi) = F(\psi^\sim * \varphi)$, for $\varphi \in \mathcal{S}$, continues to be valid here. To verify this we need the fact that $\psi^\sim * \varphi \in \mathcal{S}$, whenever ψ and φ are in $\mathcal{S}$. (See Exercise 11.)

(b) If F is a tempered distribution and F_1 is a distribution of compact support, then $F * F_1$ is also tempered. Note that $(F * F_1)(\varphi) = F(F_1^\sim * \varphi)$, and to establish the claim we need the implication that $F_1^\sim * \varphi \in \mathcal{S}$, if F_1 has compact support and $\varphi \in \mathcal{S}$. (See Exercise 12.)

1.5 Fourier transform

The main interest of tempered distributions is that this class is mapped into itself by the Fourier transform, and this is a reflection of the fact that the space $\mathcal{S}$ is also closed under the Fourier transform.

Recall that whenever $\varphi \in \mathcal{S}$, its Fourier transform $\varphi^\wedge$ (also sometimes denoted by $\hat{\varphi}$) is defined as the convergent integral[7]

$$\varphi^\wedge(\xi) = \int_{\mathbb{R}^d} \varphi(x) e^{-2\pi i x\cdot\xi}\, dx.$$

The mapping $\varphi \mapsto \varphi^\wedge$ is a continuous bijection of $\mathcal{S}$ to $\mathcal{S}$ whose inverse is given by the mapping $\psi \mapsto \psi^\vee$, where

$$\psi^\vee(x) = \int_{\mathbb{R}^d} \psi(\xi) e^{2\pi i x\cdot\xi}\, d\xi.$$

In this connection it is useful to keep in mind the simple norm estimates

$$\|\hat{\varphi}\|_N \leq C_N \|\varphi\|_{N+d+1},$$

[7]For the elementary facts about the Fourier transform on $\mathcal{S}$ that are used here, see for example Chapters 5 and 6 of Book I.

which holds for every $\varphi \in \mathcal{S}$ and every $N \geq 0$. (This estimate is itself immediate from the observation $\sup_\xi |\hat{\varphi}(\xi)| \leq \int_{\mathbb{R}^d} |\varphi(x)|\, dx \leq A\|\varphi\|_{d+1}$.)

The multiplication identity

$$\int_{\mathbb{R}^d} \hat{\psi}(x)\varphi(x)\, dx = \int_{\mathbb{R}^d} \psi(x)\hat{\varphi}(x)\, dx$$

(which holds for all $\varphi, \psi \in \mathcal{S}$) suggests the definition of the **Fourier transform** $F^\wedge$ (sometimes denoted by $\hat{F}$) for a tempered distribution F. It is

$$F^\wedge(\varphi) = F(\varphi^\wedge), \qquad \text{for all } \varphi \in \mathcal{S}.$$

From this it follows that the mapping $F \mapsto F^\wedge$ is a bijection of the space of tempered distributions, with inverse the mapping $F \mapsto F^\vee$, where $F^\vee$ is defined by $F^\vee(\varphi) = F(\varphi^\vee)$. Indeed

$$(F^\wedge)^\vee(\varphi) = F^\wedge(\varphi^\vee) = F((\varphi^\vee)^\wedge) = F(\varphi).$$

Moreover the mappings $F \mapsto F^\wedge$ and $F \mapsto F^\vee$ are continuous with convergence of distributions taken in the weak sense, that is, $F_n \to F$ if $F_n(\varphi) \to F(\varphi)$, as $n \to \infty$ for all $\varphi \in \mathcal{S}$. (This convergence is also said to be in the sense of tempered distributions.)

Next it is worthwhile to point out that the definition of the Fourier transform in the general context of tempered distributions is consistent with (and generalizes) previous definitions given in various particular settings. Let us take for example the L^2 definition via Plancherel's theorem.[8] Starting with an $f \in L^2(\mathbb{R}^d)$, we write $F = F_f$ for the corresponding tempered distribution. Now f can be approximated (in the L^2 norm) by a sequence $\{f_n\}$, with $f_n \in \mathcal{S}$. Thus taken as distributions, $f_n \to F$ in the weak sense above. Hence $\hat{f}_n \to \hat{F}$ also in the weak sense, but since $\hat{f}_n$ converges in the L^2 norm to $\hat{f}$, we see that $\hat{F}$ is the function $\hat{f}$. Similar arguments hold for $f \in L^p(\mathbb{R}^d)$, with $1 \leq p \leq 2$, and $\hat{f}$ defined in $L^q(\mathbb{R}^d)$, $1/p + 1/q = 1$, in accordance with the Hausdorff-Young theorem in Section 2 of the previous chapter.

Let us next remark that the usual formal rules involving differentiation and multiplication by monomials apply to the Fourier transform in this general context. Thus, if $F \in \mathcal{S}^*$, we have

$$(\partial_x^\alpha F)^\wedge = (2\pi i x)^\alpha F^\wedge,$$

[8]See Section 1, Chapter 5 in Book III.

because

$$\begin{aligned}(\partial_x^\alpha F)^\wedge(\varphi) &= \partial_x^\alpha F(\varphi^\wedge)\\ &= (-1)^{|\alpha|}F\left(\partial_x^\alpha(\varphi^\wedge)\right)\\ &= F\left(((2\pi ix)^\alpha\varphi)^\wedge\right)\\ &= (2\pi ix)^\alpha F^\wedge(\varphi).\end{aligned}$$

Similarly $((-2\pi ix)^\alpha F)^\wedge = \partial_x^\alpha(F^\wedge)$. One should also observe that if $\mathbf{1}$ is the function that is identically equal to 1, then as tempered distributions

$$\hat{\mathbf{1}} = \delta \quad \text{and} \quad \hat{\delta} = \mathbf{1},$$

and by the above

$$((-2\pi ix)^\alpha)^\wedge = \partial_x^\alpha\delta \quad \text{while} \quad (\partial_x^\alpha\delta)^\wedge = (2\pi ix)^\alpha.$$

The following additional properties elucidate the nature of the Fourier transform in the context of tempered distributions.

Proposition 1.5 *Suppose F is a tempered distribution and $\psi \in \mathcal{S}$. Then $F * \psi$ is a slowly increasing C^∞ function, which when considered as a tempered distribution satisfies $(F*\psi)^\wedge = \psi^\wedge F^\wedge$.*

Proof. The fact that $F(\psi_x^\sim)$ is slowly increasing follows from the proposition in Section 1.4 together with the observation that for any function $\psi \in \mathcal{D}$ and N, $\|\psi_x^\sim\|_N \le c(1+|x|)^N\|\psi\|_N$, and more generally,

$$\|\partial_x^\alpha\psi_x^\sim\|_N \le c(1+|x|)^N\|\psi\|_{N+|\alpha|}.$$

Since $(F*\psi)(\varphi) = F(\psi^\sim * \varphi)$, it follows that $(F*\psi)^\wedge(\varphi) = F(\psi^\sim * \varphi^\wedge)$. On the other hand, $\psi^\wedge F^\wedge(\varphi) = F^\wedge(\psi^\wedge\varphi) = F((\psi^\wedge\varphi)^\wedge)$. Thus the desired identity, $(F*\psi)^\wedge(\varphi) = (\psi^\wedge F^\wedge)(\varphi)$ is proved because, as is easily verified, $(\psi^\wedge\varphi)^\wedge = \psi^\sim * \varphi^\wedge$.

Proposition 1.6 *If F is a distribution of compact support then its Fourier transform $F^\wedge$ is a slowly increasing C^∞ function. In fact, as a function of ξ, one has $F^\wedge(\xi) = F(e_\xi)$ where e_ξ is the element of $\mathcal{D}$ given by $e_\xi(x) = \eta(x)e^{-2\pi ix\xi}$, with η a function in $\mathcal{D}$ that equals 1 in a neighborhood of the support of F.*

Proof. If we invoke Proposition 1.4, we see immediately that $|F(e_\xi)| \le C\|e_\xi\|_N \le c'(1+|\xi|)^N$. By the same estimate, every difference quotient of $F(e_\xi)$ converges and $\left|\partial_\xi^\alpha F(e_\xi)\right| \leq c_\alpha(1+|\xi|)^{N+|\alpha|}$. Therefore $F(e_\xi)$ is C^∞ and slowly increasing. To prove that the function $F(e_\xi)$ is the Fourier transform of F it suffices to see that

$$(3) \qquad \int_{\mathbb{R}^d} F(e_\xi)\varphi(\xi)\,d\xi = F(\hat{\varphi}) \quad \text{for every } \varphi \in \mathcal{S}.$$

We prove this first when $\varphi \in \mathcal{D}$.

Now by what we have already seen, the function $g(\xi) = F(e_\xi)\varphi(\xi)$ is continuous and certainly has compact support. Thus

$$\int_{\mathbb{R}^d} F(e_\xi)\varphi(\xi)\,d\xi = \int_{\mathbb{R}^d} g(\xi)\,d\xi = \lim_{\epsilon\to 0} S_\epsilon,$$

where for each $\epsilon > 0$, S_ϵ is the (finite) sum $\epsilon^d \sum_{n\in\mathbb{Z}^d} g(n\epsilon)$. However $S_\epsilon = F(s_\epsilon)$, with $s_\epsilon = \epsilon^d \sum_{n\in\mathbb{Z}^d} e_{n\epsilon}(x)\varphi(n\epsilon)$. Clearly as $\epsilon \to 0$, we have

$$s_\epsilon(x) \to \eta(x)\int_{\mathbb{R}^d} e^{-2\pi i x\cdot\xi}\varphi(\xi)\,d\xi = \eta(x)\hat{\varphi}(x)$$

in the $\|\cdot\|_N$ norm. Thus, using Proposition 1.4 again, we get that $S_\epsilon \to F(\eta\hat{\varphi})$. Now since $\eta = 1$ in a neighborhood of the support of F, then $F(\eta\hat{\varphi}) = F(\hat{\varphi})$. Altogether we have (3) when $\varphi \in \mathcal{D}$, and to extend this result to $\varphi \in \mathcal{S}$ it suffices to recall that $\mathcal{D}$ is dense in $\mathcal{S}$.

1.6 Distributions with point supports

Unlike continuous functions, distributions can have isolated points as their support. This is the case of the Dirac delta function and each of its derivatives. That these examples represent essentially the general case of this phenomenon, is contained in the following theorem.

Theorem 1.7 *Suppose F is a distribution supported at the origin. Then F is a finite sum*

$$F = \sum_{|\alpha|\leq N} a_\alpha \partial_x^\alpha \delta.$$

That is,

$$F(\varphi) = \sum_{|\alpha|\leq N} (-1)^{|\alpha|} a_\alpha (\partial_x^\alpha \varphi)(0), \qquad \textit{for } \varphi \in \mathcal{D}.$$

The argument is based on the following.

Lemma 1.8 *Suppose F_1 is a distribution supported at the origin that satisfies for some N the following two conditions:*

(a) $|F_1(\varphi)| \leq c\|\varphi\|_N$, *for all* $\varphi \in \mathcal{D}$.

(b) $F_1(x^\alpha) = 0$, *for all* $|\alpha| \leq N$.

Then $F_1 = 0$.

In fact, let $\eta \in \mathcal{D}$, with $\eta(x) = 0$ for $|x| \geq 1$, and $\eta(x) = 1$ when $|x| \leq 1/2$, and write $\eta_\epsilon(x) = \eta(x/\epsilon)$. Then since F_1 is supported at the origin, $F_1(\eta_\epsilon \varphi) = F_1(\varphi)$. Moreover, by the same token $F_1(\eta_\epsilon x^\alpha) = F_1(x^\alpha) = 0$ for all $|\alpha| \leq N$, and hence

$$F_1(\varphi) = F_1\left(\eta_\epsilon(\varphi(x) - \sum_{|\alpha| \leq N} \frac{\varphi^{(\alpha)}(0)}{\alpha!} x^\alpha)\right),$$

with $\varphi^{(\alpha)} = \partial_x^\alpha \varphi(0)$. If $R(x) = \varphi(x) - \sum_{|\alpha| \leq N} \frac{\varphi^{(\alpha)}(0)}{\alpha!} x^\alpha$ is the remainder, then $|R(x)| \leq c|x|^{N+1}$ and $|\partial_x^\beta R(x)| \leq c_\beta |x|^{N+1-|\beta|}$, when $|\beta| \leq N$. However $|\partial_x^\beta \eta_\epsilon(x)| \leq c_\beta \epsilon^{-|\beta|}$ and $\partial_x^\beta \eta_\epsilon(x) = 0$ if $|x| \geq \epsilon$. Thus by Leibnitz's rule, $\|\eta_\epsilon R\|_N \leq c\epsilon$, and our assumption (a) gives $|F_1(\varphi)| \leq c'\epsilon$, which yields the desired conclusion upon letting $\epsilon \to 0$.

Proceeding with the proof of the theorem, we now apply the above lemma to $F_1 = F - \sum_{|\alpha| \leq N} a_\alpha \partial_x^\alpha \delta$ where N is the index that guarantees the conclusion of Proposition 1.4, while the a_α are chosen so that $a_\alpha = \frac{(-1)^{|\alpha|}}{\alpha!} F(x^\alpha)$. Then since $\partial_x^\alpha(\delta)(x^\beta) = (-1)^{|\alpha|}\alpha!$, if $\alpha = \beta$, and zero otherwise, we see that $F_1 = 0$, which proves the theorem.

2 Important examples of distributions

Having described the elementary properties of distributions, we now intend to illustrate their occurrence in several areas of analysis.

2.1 The Hilbert transform and $\text{pv}(\frac{1}{x})$

We consider the function $1/x$, defined for real x with $x \neq 0$. As it stands, this function is not a distribution on $\mathbb{R}$ because it is not integrable near the origin. However, there is a distribution that can be naturally associated to the function $1/x$. It is defined as the **principal value**

$$\varphi \mapsto \lim_{\epsilon \to 0} \int_{|x| \geq \epsilon} \varphi(x) \frac{dx}{x}.$$

We observe first that the limit exists for every function $\varphi \in \mathcal{S}$. Assuming $\epsilon \leq 1$, we write

$$(4) \qquad \int_{|x|\geq\epsilon} \varphi(x)\,\frac{dx}{x} = \int_{1\geq|x|\geq\epsilon} \varphi(x)\,\frac{dx}{x} + \int_{|x|>1} \varphi(x)\,\frac{dx}{x}.$$

The right most integral clearly converges because of the (rapid) decay of φ at infinity. As to the other integral on the right-hand side, we can write it as

$$\int_{1\geq|x|\geq\epsilon} \frac{\varphi(x)-\varphi(0)}{x}\,dx$$

because $\int_{1\geq|x|\geq\epsilon} \frac{dx}{x} = 0$ due to the fact that $1/x$ is an odd function. However $|\varphi(x)-\varphi(0)| \leq c|x|$ (with $c = \sup|\varphi'(x)|$), thus the limit as $\epsilon \to 0$ of the left-hand side of (4) clearly exists. We denote this limit as

$$\text{pv}\int_{\mathbb{R}} \varphi(x)\,\frac{dx}{x}.$$

It is also evident from the above that

$$\left|\text{pv}\int_{\mathbb{R}} \varphi(x)\,\frac{dx}{x}\right| \leq c'\|\varphi\|_1$$

(where the norm $\|\cdot\|_1$ is defined in Section 1.4), and thus

$$\varphi \mapsto \text{pv}\int_{\mathbb{R}} \varphi(x)\,\frac{dx}{x}$$

is a tempered distribution. We denote this distribution by $\text{pv}(\frac{1}{x})$.

As the reader may have guessed, the distribution $\text{pv}(\frac{1}{x})$ is intimately connected with the Hilbert transform H studied in the previous chapter. We observe first that

$$(5) \qquad H(f) = \frac{1}{\pi}\,\text{pv}(\frac{1}{x}) * f, \qquad \text{for } f \in \mathcal{S}.$$

Indeed, according to the definition of $\text{pv}(\frac{1}{x})$ and the definition of the convolution, we have

$$\frac{1}{\pi}\,\text{pv}(\frac{1}{x}) * f = \lim_{\epsilon\to 0} \frac{1}{\pi}\int_{|y|\geq\epsilon} f(x-y)\,\frac{dy}{y},$$

and this limit exists for every x. However Proposition 3.1 of the previous chapter asserts that the right-hand side also converges in the $L^2(\mathbb{R})$ norm to $H(f)$ as $\epsilon \to 0$, whenever $f \in L^2(\mathbb{R})$. Thus the convolution $\frac{1}{\pi}\mathrm{pv}\left(\frac{1}{x}\right) * f$ equals the L^2 function $H(f)$.

We now give several alternate formulations of $\mathrm{pv}(\frac{1}{x})$. The meanings of the abbreviations used will be explained in the proof of the theorem below.

Theorem 2.1 *The distribution* $\mathrm{pv}(\frac{1}{x})$ *equals:*

(a) $\frac{d}{dx}(\log|x|)$.

(b) $\frac{1}{2}\left(\frac{1}{x-i0} + \frac{1}{x+i0}\right)$.

Also, its Fourier transform equals $\frac{\pi}{i}\ \mathrm{sign}(x)$.

Regarding (a), note that $\log|x|$ is a locally integrable function. Here $\frac{d}{dx}(\log|x|)$ is its derivative taken as a distribution. Now in that sense

$$\left(\frac{d}{dx}\log|x|\right)(\varphi) = -\int_{-\infty}^{\infty}(\log|x|)\frac{d\varphi}{dx}\,dx, \qquad \text{for every } \varphi \in \mathcal{S}.$$

However the integral is the limit as $\epsilon \to 0$ of $-\int_{|x|\geq\epsilon}(\log|x|)\frac{d\varphi}{dx}\,dx$, and an integration by parts shows that this equals

$$\int_{|x|\geq\epsilon}\frac{\varphi(x)}{x}\,dx + \log(\epsilon)[\varphi(\epsilon) - \varphi(-\epsilon)].$$

Moreover, $\varphi(\epsilon) - \varphi(-\epsilon) = O(\epsilon)$ since in particular φ is of class C^1. Therefore $\log(\epsilon)[\varphi(\epsilon) - \varphi(-\epsilon)] \to 0$ as $\epsilon \to 0$, and we have established (a).

We turn to conclusion (b) and consider for $\epsilon > 0$ the bounded function $1/(x - i\epsilon)$. We will see that as $\epsilon \to 0$, the function $1/(x - i\epsilon)$ converges to a limit in the sense of distributions, which we denote by $1/(x - i0)$. We will also see that $1/(x - i0) = \mathrm{pv}(\frac{1}{x}) + i\pi\delta$. Similarly, $\lim_{\epsilon\to 0} 1/(x + i\epsilon) = 1/(x + i0)$ will exist and equals $\mathrm{pv}(\frac{1}{x}) - i\pi\delta$. To prove this, we are thus lead to the function

$$\frac{1}{2}\left(\frac{1}{x - i\epsilon} + \frac{1}{x + i\epsilon}\right) = \frac{x}{x^2 + \epsilon^2}.$$

We claim first that

$$\text{(6)} \qquad \frac{x}{x^2 + \epsilon^2} \to \mathrm{pv}\left(\frac{1}{x}\right), \qquad \text{as } \epsilon \to 0$$

in the sense of distributions.

We are dealing in effect with the conjugate Poisson kernel $\mathcal{Q}_\epsilon(x) = \frac{1}{\pi}\frac{x}{x^2+\epsilon^2}$, defined in Section 3.1 of the previous chapter. The argument there, after the identities (18), shows that

$$\frac{1}{\pi}\int_{|x|\geq\epsilon}\varphi(x)\,\frac{dx}{x} - \int_{\mathbb{R}}\varphi(x)\mathcal{Q}_\epsilon(x)\,dx = \int_{\mathbb{R}}\varphi(x)\Delta_\epsilon(x)\,dx$$
$$= \int_{|x|\leq 1}[\varphi(x)-\varphi(0)]\Delta_\epsilon(x)\,dx + \int_{|x|>1}\varphi(x)\Delta_\epsilon(x)\,dx,$$

since $\Delta_\epsilon(x)$ is an odd function of x. This function satisfies the estimate $|\Delta_\epsilon(x)| \leq A/\epsilon$, and $|\Delta_\epsilon(x)| \leq A\epsilon/x^2$. Moreover if $\varphi \in \mathcal{D}$, then $|\varphi(x) - \varphi(0)| \leq c|x|$ and φ is bounded on $\mathbb{R}$. Therefore

$$\left|\int_{\mathbb{R}}\varphi(x)\Delta_\epsilon(x)\,dx\right| \leq O\left\{\epsilon^{-1}\int_{|x|\leq\epsilon}|x|\,dx + \epsilon\int_{\epsilon<|x|\leq 1}\frac{dx}{|x|} + \epsilon\int_{1<|x|}\frac{dx}{x^2}\right\}.$$

The expression on the right is clearly $O(\epsilon|\log\epsilon|)$ as $\epsilon \to 0$, and hence tends to zero. Therefore we have established (6). Next, recall the identity (15) in the previous chapter

$$-\frac{1}{i\pi z} = \mathcal{P}_y(x) + i\mathcal{Q}_y(x), \qquad z = x+iy,$$

where $\mathcal{P}_y(x)$ is the Poisson kernel $\frac{1}{\pi}\frac{y}{x^2+y^2}$. By letting $y = \epsilon > 0$, and taking complex conjugates we see

$$\frac{1}{x-i\epsilon} = \pi\mathcal{Q}_\epsilon(x) + i\pi\mathcal{P}_\epsilon(x).$$

Since the $\mathcal{P}_y$ form an approximation to the identity (see Chapter 3 in Book III) or by an argument very similar as the one just given for $\mathcal{Q}_y$, we have that $\mathcal{P}_\epsilon \to \delta$ as $\epsilon \to 0$. Thus

$$\frac{1}{x-i0} = \text{pv}(\frac{1}{x}) + i\pi\delta.$$

We may take complex conjugates of the above identity and also obtain, as a limit in the sense of distributions,

$$\frac{1}{x+i0} = \text{pv}(\frac{1}{x}) - i\pi\delta.$$

Adding these two gives conclusion (b). Notice that incidentally, we have obtained the identity

$$i\pi\delta = \frac{1}{2}\left(\frac{1}{x-i0} - \frac{1}{x+i0}\right).$$

To prove the last statement of the theorem we consider the Fourier transform of $x/(x^2+\epsilon^2)$ taken in the sense of distributions. By (17) in Section 3.1 of the previous chapter we have that

$$\int_{\mathbb{R}} f(-x)\frac{x\,dx}{x^2+\epsilon^2} = \pi\int_{\mathbb{R}} \hat{f}(\xi)e^{-2\pi\epsilon|\xi|}\frac{\operatorname{sign}(\xi)}{i}\,d\xi$$

for all $f \in L^2(\mathbb{R})$, and this holds in particular for $f \in \mathcal{S}$. Substituting for f, $f = \hat{\varphi}$, (and noting that $(\varphi^\wedge)^\wedge = \varphi(-x)$) we get

$$\left(\frac{x}{x^2+\epsilon^2}\right)^\wedge(\varphi) = \left(\frac{x}{x^2+\epsilon^2}\right)(\hat{\varphi}) = \pi\int_{\mathbb{R}} \varphi(\xi)e^{-2\pi\epsilon|\xi|}\frac{\operatorname{sign}(\xi)}{i}\,d\xi.$$

Letting $\epsilon \to 0$, this yields

$$\left(\mathrm{pv}\frac{1}{x}\right)^\wedge(\varphi) = \pi\int_{\mathbb{R}} \varphi(\xi)\frac{\operatorname{sign}(\xi)}{i}\,d\xi,$$

which shows that $\left(\mathrm{pv}\frac{1}{x}\right)^\wedge$ is the function $\frac{\pi}{i}\operatorname{sign}(\xi)$, and the proof of the theorem is concluded.

Let us remark that we have seen from the above that the distributions $1/(x-i0)$, $1/(x+i0)$, and $\mathrm{pv}(\frac{1}{x})$, while different, all agree with the function $1/x$ away from the origin.

2.2 Homogeneous distributions

We pass to the next topic by observing that $\mathrm{pv}(\frac{1}{x})$ is a homogeneous distribution. To define this notion, recall that a function f defined on $\mathbb{R}^d - \{0\}$ is said to be **homogeneous of degree** λ, if $f_a = a^\lambda f$, for every $a > 0$, where $f_a(x) = f(ax)$. Now the dilation F_a of a distribution F has been defined by duality:

$$F_a(\varphi) = F(\varphi^a),$$

where φ^a is the dual dilation of φ, that is, $\varphi^a = a^{-d}\varphi_{a^{-1}}$. We can incidentally define the dual dilation F^a by $F^a(\varphi) = F(\varphi_a)$, and note that $F^a = a^{-d}F_{a^{-1}}$.

In view of the above, a distribution F is said to be **homogeneous of degree** λ, if $F_a = a^\lambda F$ for all $a > 0$.

Now the function $1/x$ is clearly homogeneous of degree -1, but what is significant for us is that the distribution $\mathrm{pv}(\frac{1}{x})$ is homogenous of degree -1. In fact

$$\begin{aligned}\mathrm{pv}(\frac{1}{x})_a(\varphi) = \mathrm{pv}(\frac{1}{x})(\varphi^a) &= a^{-1} \lim_{\epsilon\to 0} \int_{|x|\geq\epsilon} \varphi(x/a)\,\frac{dx}{x}\\ &= a^{-1} \lim_{\epsilon\to 0} \int_{|x|\geq\epsilon/a} \varphi(x)\,\frac{dx}{x} = a^{-1}\mathrm{pv}(\frac{1}{x})(\varphi).\end{aligned}$$

The next to the last identity follows from making the change of variables $x \to ax$ and noting that dx/x remains unchanged. The reader may also verify that the distributions $1/(x - i0)$, $1/(x + i0)$, and δ are also homogenous of degree -1.

There is an important interplay between homogeneous distributions and the Fourier transform. A hint that this may be so is the elementary identity $(\varphi^a)^\wedge = (\varphi^\wedge)_a$, that holds for all $\varphi \in \mathcal{S}$, where φ_a and φ^a are the dilations of φ defined earlier. The simplest proposition containing this idea is the following.

Proposition 2.2 *Suppose F is a tempered distribution on $\mathbb{R}^d$ that is homogeneous of degree λ. Then its Fourier transform $F^\wedge$ is homogeneous of degree $-d - \lambda$.*

Remark. The restriction that F be tempered is unnecessary. It can be shown that any homogeneous distribution is automatically tempered. See Exercise 8 for this result.

To deal with $(F^\wedge)_a$ we write successively,

$$\begin{aligned}(F^\wedge)_a(\varphi) &= F^\wedge(\varphi^a) = F((\varphi^a)^\wedge) = F((\varphi^\wedge)_a)\\ &= F^a(\varphi^\wedge) = a^{-d}F_{a^{-1}}(\varphi^\wedge) = a^{-d-\lambda}F(\varphi^\wedge) = a^{-d-\lambda}F^\wedge(\varphi).\end{aligned}$$

Thus $(F^\wedge)_a = a^{-d-\lambda}F^\wedge$, as was to be proved.

A particularly interesting example arises if we consider the function $|x|^\lambda$ which is homogeneous of degree λ and locally integrable if $\lambda > -d$. Let H_λ denote the corresponding distribution (for $\lambda > -d$); this is clearly tempered.

The following identity holds.

Theorem 2.3 *If $-d < \lambda < 0$, then*

$$(H_\lambda)^\wedge = c_\lambda H_{-d-\lambda}, \qquad \textit{with } c_\lambda = \frac{\Gamma(\frac{d+\lambda}{2})}{\Gamma(\frac{-\lambda}{2})}\pi^{-d/2-\lambda}.$$

Note that the assumption $\lambda < 0$ guarantees that $-d-\lambda > -d$ so that $|x|^{-d-\lambda}$, which defines $H_{-d-\lambda}$, is again locally integrable.

To prove the theorem we start with the fact that $\psi(x) = e^{-\pi|x|^2}$ is its own Fourier transform. Then since $(\psi_a)^\wedge = (\psi^\wedge)^a$ we get (with $a = t^{1/2}$)

$$\int_{\mathbb{R}^d} e^{-\pi t|x|^2} \hat{\varphi}(x)\, dx = t^{-d/2} \int_{\mathbb{R}^d} e^{-\pi|x|^2/t} \varphi(x)\, dx.$$

We now multiply both sides by $t^{-\lambda/2-1}$ and integrate over $(0,\infty)$, and then interchange the order of integration. We note that

$$\int_0^\infty e^{-tA} t^{-\lambda/2-1}\, dt = A^{\lambda/2}\Gamma(-\lambda/2),$$

if $A > 0$ and $\lambda > 0$, by making the indicated change of variables that reduces the identity to the case $A = 1$. Thus using the above identity with $A = \pi|x|^2$, we get

$$\int_{\mathbb{R}^d}\int_0^\infty e^{-\pi t|x|^2} \hat{\varphi}(x) t^{-\lambda/2-1}\, dt\, dx = \pi^{\lambda/2}\Gamma(-\lambda/2) \int_{\mathbb{R}^d} |x|^\lambda \hat{\varphi}(x)\, dx.$$

Similarly, we deal with $\int_0^\infty t^{-d/2} t^{-\lambda/2-1} e^{-A/t}\, dt$ by making the change of variables $t \to 1/t$ which shows that this integral equals

$$\int_0^\infty t^{d/2+\lambda/2-1} e^{-At}\, dt = A^{-d/2-\lambda/2}\Gamma\left(\frac{d}{2}+\frac{\lambda}{2}\right).$$

Inserting this in $\int_{\mathbb{R}^d}\int_0^\infty t^{-d/2} e^{-\pi|x|^2/t}\varphi(x)\, dt\, dx$ yields

$$\pi^{\lambda/2}\Gamma(-\lambda/2) \int_{\mathbb{R}^d} |x|^\lambda \hat{\varphi}(x)\, dx = \pi^{-d/2-\lambda/2}\Gamma(d/2+\lambda/2) \int_{\mathbb{R}^d} |x|^{-d-\lambda}\varphi(x)\, dx,$$

and this is our theorem.

The principal value distribution $\mathrm{pv}(\frac{1}{x})$ and the H_λ just considered have in common the property that these distributions agree with C^∞ functions when tested away from the origin. We formulate this notion in the following definition. We say that a distribution K is **regular** if there exists a function k that is C^∞ in $\mathbb{R}^d - \{0\}$, so that $K(\varphi) = \int_{\mathbb{R}^d} k(x)\varphi(x)\, dx$ for all $\varphi \in \mathcal{D}$ whose supports are disjoint from the origin.

We also refer to this by saying that K is C^∞ away from the origin, and calling k the function **associated** to K. (Note that k is uniquely determined by K.) One should remark that the function associated to $\mathrm{pv}(\frac{1}{x})$ is $1/x$.

Returning to the general case one may observe that the function k is automatically homogeneous of degree λ if the distribution K is homogeneous of degree λ. In fact, if $\varphi \in \mathcal{D}$ is supported away from the origin, $K(\varphi) = \int k(x)\varphi(x)\,dx$, while

$$K_a(\varphi) = K(\varphi^a) = a^{-d}\int_{\mathbb{R}^d} k(x)\varphi(x/a)\,dx = \int_{\mathbb{R}^d} k_a(x)\varphi(x)\,dx.$$

Hence

$$\int_{\mathbb{R}^d} (a^\lambda k(x) - k_a(x))\varphi(x)\,dx = 0$$

for all such φ, which means that $k_a(x) = a^\lambda k(x)$.

The above considerations and examples raise the following two questions.

Question 1. Given a function k, homogeneous of degree λ, and C^∞ away from the origin, when does there exist a regular homogeneous distribution K of degree λ such that k is its associated function? If such a distribution exists, to what extent is it uniquely determined by k?

Question 2. How do we characterize the Fourier transform of such K?

We answer first the second question.

Theorem 2.4 *The Fourier transform of a regular homogeneous distribution K of degree λ is a regular homogeneous distribution of degree $-d-\lambda$, and conversely.*

Proof. We already know from Proposition 2.2 that $K^\wedge$ is homogeneous of degree $-d-\lambda$. To prove that $K^\wedge$ agrees with a C^∞ function away from the origin, we decompose $K = K_0 + K_1$, with K_0 supported near the origin and K_1 supported away from the origin. To do this, fix a cut-off function η that is C^∞, is supported in $|x| \le 1$, and that equals 1 on $|x| \le 1/2$. Write $K_0 = \eta K$, $K_1 = (1-\eta)K$. In particular K_1 is the function $(1-\eta)k$, since $1-\eta$ vanishes near the origin. Also $K^\wedge = K_0^\wedge + K_1^\wedge$.

Now by Proposition 1.6, $K_0^\wedge$ is an (everywhere) C^∞ function. To prove that $K_1^\wedge$ is C^∞ away from the origin we observe that by the usual manipulations of the Fourier transform valid for tempered distributions,

$$(-4\pi^2|\xi|^2)^N \partial_\xi^\alpha(K_1^\wedge) = \left(\triangle^N[(-2\pi i x)^\alpha K_1]\right)^\wedge. \tag{7}$$

Recall that $\triangle$ denotes the Laplacian, $\triangle = \partial^2/\partial x_1^2 + \cdots + \partial^2/\partial x_d^2$.

Now when $|x| \geq 1$, $K_1 = k$, so there $\partial_x^\beta(K_1)$ is a bounded homogeneous function of degree $\lambda - |\beta|$ and thus is $O(|x|^{\lambda-|\beta|})$, for $|x| \geq 1$. Therefore $\triangle^N[x^\alpha K_1]$ is $O(|x|^{\lambda+|\alpha|-2N})$ for $|x| \geq 1$ while it is certainly a bounded function for $|x| \leq 1$. Hence for N sufficiently large ($2N > \lambda + |\alpha| + d$) this function belongs to $L^1(\mathbb{R}^d)$. As a result its Fourier transform is continuous. (See Chapter 2 in Book III.) This shows by (7) that $\partial_x^\alpha(K_1^\wedge)$ agrees with a continuous function away from the origin. Since this holds for every α, it follows from Exercise 2 that $K_1^\wedge$ is a C^∞ function away from the origin, as desired.

Note that since the inverse Fourier transform is the Fourier transform followed by reflection, that is, $K^\vee = (K^\wedge)^\sim$, the converse is a consequence of the direction we have just proved.

We now turn to the first question raised above.

Theorem 2.5 *Suppose k is a given C^∞ function on $\mathbb{R}^d - \{0\}$ that is homogeneous of degree λ.*

(a) *If λ is not of the form $-d - m$, with m a non-negative integer, then there exists a unique distribution K homogeneous of degree λ that agrees with k away from the origin.*

(b) *If $\lambda = -d - m$, where m is a non-negative integer, then there exists a distribution K as in* (a) *if and only if k satisfies the cancelation condition*

$$\int_{|x|=1} x^\alpha k(x)\, d\sigma(x) = 0, \qquad \textit{for all } |\alpha| = m.$$

(c) *Every distribution arising in* (b) *is of the form*

$$K + \sum_{|\alpha|=m} c_\alpha \partial_x^\alpha \delta.$$

Proof. We deal first with the question of constructing the distribution K given by k. Note that the function k automatically satisfies the bound $|k(x)| \leq c|x|^\lambda$. Indeed, $k(x)/|x|^\lambda$ is homogeneous of degree 0 and is bounded on the unit sphere (by continuity of k there), thus it is bounded throughout $\mathbb{R}^d - \{0\}$.

So if $\lambda > -d$, the function k is locally integrable on $\mathbb{R}^d$, and thus we can take K to be the distribution defined by k. This local integrability fails when $\lambda \leq -d$.

In the general case we shall proceed by analytic continuation. Our starting point is the integral

$$(8)\qquad I(s) = I(s)(\varphi) = \int_{\mathbb{R}^d} k(x)|x|^{-\lambda+s}\varphi(x)\,dx, \qquad \text{with } \varphi \in \mathcal{S},$$

initially defined for complex s with $\mathrm{Re}(s) > -d$, which we will see continues to a meromorphic function in the entire complex plane. We will then ultimately set

$$K(\varphi) = I(s)|_{s=\lambda}\,.$$

In fact, for our given homogeneous function k, and φ any test function in $\mathcal{S}$, we note by the above bound on k, that the integral (8) converges when $\mathrm{Re}(s) > -d$, thus I is analytic in that half-plane. Moreover I continues to the whole complex plane, with at most simple poles at $s = -d, -d-1, \ldots, -d-m, \ldots$.

To prove this, write $I(s) = \int_{|x|\leq 1} + \int_{|x|>1}$. Given the rapid decrease of φ at infinity, the integral over $|x| > 1$ gives an entire function of s. However, for every $N \geq 0$,

(9)

$$\int_{|x|\leq 1} k(x)|x|^{-\lambda+s}\varphi(x)\,dx = \sum_{|\alpha|<N} \frac{\varphi^{(\alpha)}(0)}{\alpha!}\int_{|x|\leq 1} k(x)|x|^{-\lambda+s}x^\alpha\,dx + \\ + \int_{|x|\leq 1} k(x)|x|^{-\lambda+s}R(x)\,dx,$$

where $R(x) = \varphi(x) - \sum_{|\alpha|<N} \frac{\varphi^{(\alpha)}(0)}{\alpha!}x^\alpha$, with $\varphi^{(\alpha)}(0) = \partial_x^\alpha\varphi(0)$.

Now by the homogeneity of k and the use of polar coordinates, we see that

$$\int_{|x|\leq 1} k(x)|x|^{-\lambda+s}x^\alpha\,dx = \left(\int_{|x|=1} k(x)x^\alpha\,d\sigma(x)\right)\int_0^1 r^{s+|\alpha|+d-1}\,dr,$$

with the last integral equalling $1/(s+|\alpha|+d)$. Moreover the remainder $R(x)$ satisfies $|R(x)| \leq c|x|^N$, and this together with $|k(x)| \leq c|x|^\lambda$ implies that $\int_{|x|\leq 1} k(x)|x|^{-\lambda+s}R(x)\,dx$ is analytic in the half-plane $\mathrm{Re}(s) > -d-N$.

As a result, for each non-negative integer N, we have that $I(s)$ can be continued in the half-plane $\mathrm{Re}(s) > -d-N$ and can be represented as

$$I(s) = \sum_{|\alpha|<N} \frac{C_\alpha}{s+|\alpha|+d} + E_N(s)$$

in that half-plane, with $E_N(s)$ analytic there, and

$$C_\alpha = \frac{\varphi^{(\alpha)}(0)}{\alpha!}\left(\int_{|x|=1} k(x)x^\alpha\, d\sigma(x)\right).$$

Now for our given λ with $\lambda \neq -d, -d-1, \dots$ we need only to take N so large that $\lambda > -d - N$, and define the distribution K by setting $K(\varphi) = I(\lambda)$. (See Figure 1.) Moreover, by keeping track of the bounds that arise, one sees that $|K(\varphi)| \leq c\|\varphi\|_M$, with $M \geq \max(N+1, \lambda+d+1)$, with the norm $\|\cdot\|_M$ defined earlier. Thus K is a tempered distribution.

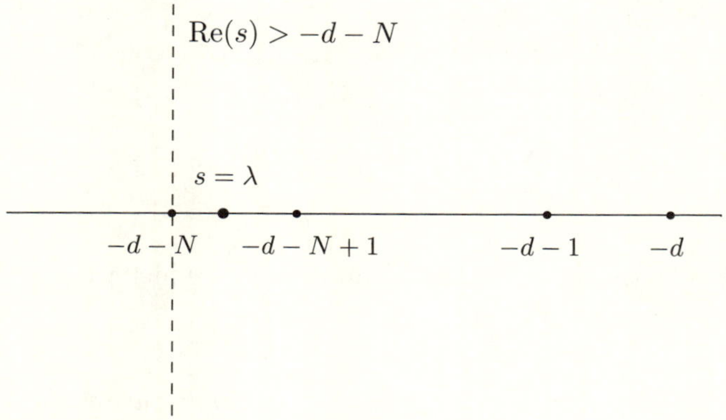

Figure 1. The half-plane $\mathrm{Re}(s) > -d - N$, and the definition of $I(\lambda)$

To verify that K agrees with the function k away from the origin, we note that whenever φ vanishes near the origin, the integral $I(s)$ converges for every complex s and is an entire function. Therefore by (8)

$$K(\varphi) = I(\lambda) = \int_{\mathbb{R}^d} k(x)\varphi(x)\, dx.$$

This proves the claim.

Next notice that for any $a > 0$, whenever $\mathrm{Re}(s) > -d$,

$$\begin{aligned} I(s)(\varphi^a) &= \int_{\mathbb{R}^d} k(x)|x|^{-\lambda+s}a^{-d}\varphi(x/a)\, dx \\ &= a^s \int_{\mathbb{R}^d} k(x)|x|^{-\lambda+s}\varphi(x)\, dx = a^s I(s)(\varphi). \end{aligned}$$

This follows by the homogeneity of k, and the change of variables $x \mapsto ax$. As a result, $I(s)(\varphi^a) = a^s I(s)$ when $\mathrm{Re}(s) > -d$, and thus by analytic continuation this continues to hold at all s at which $I(s)$ is analytic, and hence at $s = \lambda$. Therefore the distribution $K = I(\lambda)$ has the asserted homogeneity, and this proves the existence stated in part (a) of the theorem. If we also note that under the cancelation conditions of part (b) of the theorem one has $C_\alpha = 0$ whenever $|\alpha| = m$, our argument also proves the existence in that case.

We next come to the question of the uniqueness of the distribution K when $\lambda \neq -d, -d-1, \ldots$. Suppose K and K_1 are a pair of regular distributions of degree λ, each of which agrees with k away from the origin. Then $D = K - K_1$ is supported at the origin and hence, by Theorem 1.7, $D = \sum_{|\alpha| \leq M} c_\alpha \partial_x^\alpha \delta$ for some constants c_α. Now on the one hand $D(\varphi^a) = a^\lambda D(\varphi)$, because K and K_1 are homogeneous of degree λ. On the other hand $\partial_x^\alpha \delta(\varphi^a) = a^{-d-|\alpha|} \partial_x^\alpha \delta(\varphi)$, and as a result,

$$a^\lambda D(\varphi) = \sum_{|\alpha| \leq M} c_\alpha \partial_x^\alpha \delta(\varphi) a^{-d-|\alpha|} \qquad \text{for all } a > 0.$$

We now invoke the following simple observation, which we state in a form that will also be useful later.

Lemma 2.6 *Suppose $\lambda_1, \lambda_2, \ldots, \lambda_n$, are distinct real numbers and that for constants a_j and b_j, $1 \leq j \leq n$, we have*

$$\sum_{j=1}^{n} (a_j x^{\lambda_j} + b_j x^{\lambda_j} \log x) = 0 \qquad \textit{for all } x > 0.$$

Then $a_j = b_j = 0$ for all $1 \leq j \leq n$.

For $\lambda \neq -d, -d-1, \ldots$, we apply the lemma to $\lambda_1 = \lambda$, $\lambda_2 = -d$, $\lambda_3 = -d-1$, and so on, and $x = a$, to obtain $D(\varphi) = 0$ as desired. If $\lambda = -d-m$, we get $D(\varphi) = \sum_{|\alpha|=m} c_\alpha \partial_x^\alpha \delta(\varphi)$, proving the relative uniqueness asserted in conclusion (c) of the theorem.

To prove the lemma we assume, as one may, that λ_n is the largest of the λ_j's. Then multiplying the identity by $x^{-\lambda_n}$ and letting x tend to infinity we see that b_n as well as a_n must vanish. Thus we are reduced to the case when n is replaced by $n-1$, and this induction gives the lemma.

Finally we show that when $\lambda = -d-m$ and $\int_{|x|=1} k(x) x^\alpha \, d\sigma(x) \neq 0$, for some α, with $|\alpha| = m$, then there does not exist a homogeneous distribution of degree $-d-m$ that agrees with $k(x)$ away from the origin.

We consider first the case $m = 0$, and examine $I(s)$, given by (8), near $s = -d$, in the special case $k(x) = |x|^{-d}$. In this case we use (9) with $N = 1$, which is valid for $\mathrm{Re}(s) > -d - 1$. With $R(x) = \varphi(x) - \varphi(0)$ this yields

(10)
$$I(s)(\varphi) = A_d \frac{\varphi(0)}{s+d} + \int_{|x|\leq 1} [\varphi(x) - \varphi(0)]|x|^s \, dx + \int_{|x|>1} \varphi(x)|x|^s \, dx.$$

(Here $A_d = 2\pi^{d/2}/\Gamma(d/2)$ denotes the area of the unit sphere in $\mathbb{R}^d$). Since the two integrals are analytic when $\mathrm{Re}(s) > -d - 1$, the factor $A_d\varphi(0)$ represents the residue of the pole of $I(s)(\varphi)$ as $s = -d$, and in particular, as distributions

$$(s + d)I(s) \to A_d\delta, \qquad \text{as } s \to -d.$$

We will temporarily call J the distribution that arises as the next term in the expression of $I(s)$ as $s \to -d$, $I(s) = \frac{A_d\delta}{s+d} + J + O(s + d)$, that is

$$J = ((s + d)I(s))'_{s=-d}.$$

This distribution J, which we shall now write as $\left[\frac{1}{|x|^d}\right]$, is given, because of (10), by

$$\text{(11)} \qquad \left[\frac{1}{|x|^d}\right](\varphi) = \int_{|x|\leq 1} \frac{\varphi(x) - \varphi(0)}{|x|^d} \, dx + \int_{|x|>1} \frac{\varphi(x)}{|x|^d} \, dx.$$

We observe the following facts about $\left[\frac{1}{|x|^d}\right]$:

(i) It is a tempered distribution. Indeed, it is easily verified that $\left|\left[\frac{1}{|x|^d}\right](\varphi)\right| \leq c\|\varphi\|_1$.

(ii) $\left[\frac{1}{|x|^d}\right]$ agrees with the function $1/|x|^d$ away from the origin; this is because when tested with φ that vanishes near the origin, the term $\varphi(0)$ disappears from (11).

(iii) However, $\left[\frac{1}{|x|^d}\right]$ is *not* homogeneous.

What holds is the identity

(12)
$$\left[\frac{1}{|x|^d}\right](\varphi^a) = a^{-d}\left[\frac{1}{|x|^d}\right](\varphi) + a^{-d}\log(a)A_d\varphi(0), \qquad \text{for all } a > 0.$$

To prove this note that

$$\left[\frac{1}{|x|^d}\right](\varphi^a) = a^{-d}\int_{|x|\leq 1/a}[\varphi(x)-\varphi(0)]\frac{dx}{|x|^d} + a^{-d}\int_{|x|>1/a}\varphi(x)\frac{dx}{|x|^d}$$

as a change of variable shows. A comparison of this with the case $a = 1$ immediately yields (12). A consequence of this identity is contained in the following.

Corollary 2.7 *There is no distribution K_0 that is homogeneous of degree $-d$ and that agrees with the function $1/|x|^d$ away from the origin.*

If such a K_0 existed, then $K_0 - \left[\frac{1}{|x|^d}\right]$ would be supported at the origin, and hence equal to $\sum_{|\alpha|\leq M} c_\alpha \partial_x^\alpha \delta$. Applying this difference to φ^a would yield that

$$a^{-d}K_0(\varphi) - a^{-d}\left[\frac{1}{|x|^d}\right](\varphi) - a^{-d}\log(a)A_d\varphi(0) - \\ - \sum_{|\alpha|\leq M} c_\alpha a^{-d-|\alpha|}\partial_x^\alpha\delta(\varphi) = 0,$$

for all $a > 0$. This leads to a contradiction with Lemma 2.6 if we take φ so that $\varphi(0) \neq 0$.

The result of Corollary 2.7 can be restated as follows. If k is homogeneous of degree $-d$, and $\int_{|x|=1} k(x)\,d\sigma(x) \neq 0$, then there is no distribution K homogeneous of degree $-d$, that agrees with k away from the origin.

Indeed, write $k(x) = \frac{c}{|x|^d} + k_1(x)$, where

$$c\int_{|x|=1} d\sigma(x) = \int_{|x|=1} k(x)\,d\sigma(x),$$

and $c \neq 0$, while $\int_{|x|=1} k_1(x)\,d\sigma(x) = 0$. Now if K_1 is the distribution whose associated function is k_1, and whose existence is guaranteed by conclusion (b), then $\frac{1}{c}(K - K_1)$ would be a homogeneous distribution of degree $-d$ agreeing with $1/|x|^d$ away from the origin. This we have seen is precluded by Corollary 2.7.

Finally, turning to the general case, suppose K is a homogeneous distribution of degree $-d-m$, whose associated function is $k(x)$. Let $K' = x^\alpha K$ for some α with $|\alpha| = m$ and $\int_{|x|=1} k(x)x^\alpha\,d\sigma(x) \neq 0$. Then clearly

K' is homogeneous of degree $-d - m + |\alpha| = -d$, while $k'(x) = x^\alpha k(x)$ is its associated function. However now

$$\int_{|x|=1} k'(x)\, d\sigma(x) \neq 0,$$

which contradicts the special case $\lambda = -d$ considered above. The theorem is therefore completely proved.

Remark 1. The results of the theorems continue to hold with minor modifications if λ, which was assumed to be real, is allowed to be complex. In this situation the proof of Lemma 2.6 needs a slight additional argument, which is indicated in Exercise 20.

Remark 2. When $\lambda = -d$ with k satisfying the cancelation condition $\int_{|x|=1} k(x)\, d\sigma(x) = 0$, the resulting distribution K is then a natural generalization of $\text{pv}(\frac{1}{x})$ in $\mathbb{R}$ considered earlier. Indeed, as we have seen

$$K(\varphi) = \int_{|x|\leq 1} k(x)[\varphi(x) - \varphi(0)]\, dx + \int_{|x|>1} k(x)\varphi(x)\, dx$$

and this equals the "principal value"

$$\lim_{\epsilon \to 0} \int_{\epsilon \leq |x|} k(x)\varphi(x)\, dx$$

because $\int_{\epsilon \leq |x| \leq 1} k(x)\, dx = \log(1/\epsilon) \int_{|x|=1} k(x)\, d\sigma(x) = 0$. Distributions of this kind, first studied by Mihlin, Calderón and Zygmund, are often denoted by $\text{pv}(k)$.

2.3 Fundamental solutions

Among the most significant examples of distributions are fundamental solutions of partial differential equations and derivatives of these fundamental solutions. Suppose L is a partial differential operator

$$L = \sum_{|\alpha|\leq m} a_\alpha \partial_x^\alpha \quad \text{on } \mathbb{R}^d,$$

with a_α complex constants. A **fundamental solution** of L is a distribution F so that

$$L(F) = \delta,$$

where δ is the Dirac delta function. The importance of a fundamental solution[9] is that it implies that the operator $f \mapsto T(f) = F * f$, mapping

[9] Note that a fundamental solution is not unique since we can always add to it a solution of the homogeneous equation $L(u) = 0$.

$\mathcal{D}$ to C^∞, is an "inverse" to L. One way to interpret this is the statement that

$$LT = TL = I$$

when acting on $\mathcal{D}$. This holds because as we have seen earlier in this chapter, $\partial_x^\alpha(F * f) = (\partial_x^\alpha F) * f = F * (\partial_x^\alpha f)$ for all α, hence $L(F * f) = (LF) * f = F * (Lf)$, while of course $\delta * f = f$.

Now let

$$P(\xi) = \sum_{|\alpha| \leq m} a_\alpha (2\pi i \xi)^\alpha,$$

be the **characteristic polynomial** of the operator L. Since, for example when f belongs to $\mathcal{S}$, one has $(L(f))^\wedge = P \cdot f^\wedge$, we might hope to find such an F by defining it via $\hat{F}(\xi) = 1/P(\xi)$ or as

$$F = \int_{\mathbb{R}^d} \frac{1}{P(\xi)} e^{2\pi i x \cdot \xi} \, d\xi, \tag{13}$$

taken in an appropriate sense.

The main problem with this approach in the general case is due to the zeros of P and the resulting difficulty of defining $1/P(\xi)$ as a distribution. However in a number of interesting cases this can de done quite directly.

We consider first the **Laplacian**

$$\triangle = \sum_{j=1}^{d} \frac{\partial^2}{\partial x_j^2} \qquad \text{in } \mathbb{R}^d.$$

Here $1/P(\xi) = 1/(-4\pi^2|\xi|^2)$, and when $d \geq 3$ this function is locally integrable, and the required calculation of a fundamental solution is given by Theorem 2.3. This results in the following.

Theorem 2.8 *For $d \geq 3$, the locally integrable function F defined by $F(x) = C_d |x|^{-d+2}$ is a fundamental solution for the operator $\triangle$, with $C_d = -\frac{\Gamma(\frac{d}{2}-1)}{4\pi^{\frac{d}{2}}}$.*

This follows by taking $\lambda = -d + 2$ (in Theorem 2.3), then $\Gamma\left(\frac{d+\lambda}{2}\right) = \Gamma(1) = 1$, while $\Gamma(d/2) = (d/2 - 1)\Gamma(d/2 - 1)$. Therefore $\hat{F}(\xi)$ equals $1/(-4\pi^2|\xi|^2)$, and hence

$$(\triangle F)^\wedge = 1, \qquad \text{which means} \qquad \triangle F = \delta.$$

The case of two dimensions leads to the following variant.

Theorem 2.9 *When $d = 2$, the locally integrable function $\frac{1}{2\pi}\log|x|$ is a fundamental solution of $\triangle$.*

This fundamental solution arises when considering the limiting case $\lambda \to -d+2=0$ in Theorem 2.3. It can be given formally as

$$\frac{-1}{4\pi^2}\int_{\mathbb{R}^2}\frac{1}{|\xi|^2}e^{2\pi i x\xi}\,d\xi,$$

but we need to assign a meaning to this non-convergent integral. In fact, we shall be led to the distribution $\left[\frac{1}{|x|^d}\right]$ considered in (11). We start with the identity

$$\int_{\mathbb{R}^2}\hat{\varphi}(x)|x|^{\lambda}\,dx = c_\lambda\int_{\mathbb{R}^2}\varphi(\xi)|\xi|^{-\lambda-2}\,d\xi, \tag{14}$$

with $-2<\lambda<0$, and $c_\lambda = \frac{\Gamma(1+\lambda/2)}{\Gamma(-\lambda/2)}\pi^{-1-\lambda}$. We examine (14) near $\lambda = 0$ and use the fact that $c_\lambda \sim -\lambda/(2\pi)+c'\lambda^2$ as $\lambda\to 0$, for some constant c'. This follows from the fact that $\Gamma(1)=1$, the function $\Gamma(s)$ is smooth near $s=1$, and the identity $\Gamma(s+1)=s\Gamma(s)$ with $s=-\lambda/2$. Looking back at (10) (with $s=-\lambda-2$), we differentiate both sides of (14) with respect to λ, which is justified by the rapid decay of φ and $\hat{\varphi}$. After a multiplication of $1/2\pi$ the result is, upon letting $\lambda\to 0$,

$$\frac{1}{2\pi}\int_{\mathbb{R}^2}\hat{\varphi}(x)\log|x|\,dx$$
$$= \frac{-1}{4\pi^2}\left\{\int_{|x|\le 1}\frac{\varphi(x)-\varphi(0)}{|x|^2}\,dx + \int_{|x|>1}\varphi(x)\frac{dx}{|x|^2}\right\} - c'\varphi(0).$$

That is, if we take $F=\frac{1}{2\pi}\log|x|$, then

$$\hat{F} = -\frac{1}{4\pi^2}\left[\frac{1}{|x|^2}\right] - c'\delta.$$

Now it is clear that $|x|^2\delta = 0$, because $|x|^2\delta(\varphi) = |x|^2\varphi(x)|_{x=0}=0$. Also, for all $\varphi\in\mathcal{S}$, $|x|^2\left[\frac{1}{|x|^2}\right](\varphi) = \int_{\mathbb{R}^2}\varphi(x)\,dx$, which means $|x|^2\left[\frac{1}{|x|^2}\right]$ equals the function 1. Thus $(\triangle F)^\wedge = -4\pi^2|x|^2\hat{F} = 1$, and so $\triangle F = \delta$, proving that F is a fundamental solution for $\triangle$ on $\mathbb{R}^2$.

We shall next give an explicit fundamental solution for the heat operator

$$L = \frac{\partial}{\partial t} - \triangle_x,$$

taken over $\mathbb{R}^{d+1}$, with $(x,t) \in \mathbb{R}^{d+1} = \mathbb{R}^d \times \mathbb{R}$, and $\triangle_x$ the Laplacian in the x-variables, $x \in \mathbb{R}^d$. We do this by linking the inhomogeneous equation $L(u) = g$ with the homogeneous initial-value problem, $L(u) = 0$ for $t > 0$ with $u(x,t)|_{t=0} = f(x)$ given on $\mathbb{R}^d$.

Recall from Chapters 5 and 6 in Book I that the latter problem is solved by the heat kernel

$$\mathcal{H}_t^{\wedge}(\xi) = e^{-4\pi^2|\xi|^2 t},$$

where the Fourier transform is taken only in the x-variables. This shows that if $f \in \mathcal{S}$, then $u(x,t) = (\mathcal{H}_t * f)(x)$ solves the equation $L(u) = 0$, while $u(x,t) \to f(x)$ in $\mathcal{S}$ as $t \to 0$. Notice also that

$$\frac{\partial \mathcal{H}_t}{\partial t} = \triangle_x \mathcal{H}_t(x), \qquad \text{and} \qquad \int_{\mathbb{R}^d} \mathcal{H}_t(x)\, dx = 1,$$

and $\mathcal{H}_t$ is an "approximation to the identity." (For these properties of $\mathcal{H}_t$, see Chapter 5, Book I and Chapter 3 in Book III.)

Now on $\mathbb{R}^{d+1}$ define F by

$$F(x,t) = \begin{cases} \mathcal{H}_t(x), & \text{if } t > 0, \\ 0, & \text{if } t \leq 0. \end{cases}$$

It follows that F is locally integrable on $\mathbb{R}^{d+1}$ (and in fact one has $\int_{|t|\leq R} \int_{\mathbb{R}^d} F(x,t)\, dx\, dt \leq R$), and so F defines a tempered distribution on $\mathbb{R}^{d+1}$.

Theorem 2.10 *F is a fundamental solution of $L = \frac{\partial}{\partial t} - \triangle_x$.*

Proof. Since $LF(\varphi) = F(L'\varphi)$ with $L' = -\frac{\partial}{\partial t} - \triangle_x$, it suffices to see that $F(L'\varphi)$, which equals

$$\lim_{\epsilon \to 0} \int_{t \geq \epsilon} \int_{\mathbb{R}^d} F(x,t) \left(-\frac{\partial}{\partial t} - \triangle_x \right) \varphi(x,t)\, dx\, dt,$$

is $\delta(\varphi) = \varphi(0,0)$.

Now $F(x,t) = \mathcal{H}_t(x)$ when $t > 0$, so an integration by parts in the

x-variables gives

$$\int_{t\geq\epsilon}\int_{\mathbb{R}^d} F(x,t)\left(-\frac{\partial}{\partial t}-\triangle_x\right)\varphi(x,t)\,dx\,dt =$$

$$= -\int_{\mathbb{R}^d}\left(\int_{t\geq\epsilon}\mathcal{H}_t\frac{\partial\varphi}{\partial t}+(\triangle_x\mathcal{H}_t)\varphi\,dt\right)\,dx$$

$$= -\int_{\mathbb{R}^d}\left(\int_{t\geq\epsilon}\mathcal{H}_t\frac{\partial\varphi}{\partial t}+\frac{\partial\mathcal{H}_t}{\partial t}\varphi\,dt\right)\,dx$$

$$= \int_{\mathbb{R}^d}\mathcal{H}_\epsilon(x)\varphi(x,\epsilon)\,dx.$$

However, because $\varphi\in\mathcal{S}$, one has $|\varphi(x,\epsilon)-\varphi(x,0)|\leq O(\epsilon)$ uniformly in x. Therefore

$$\int_{\mathbb{R}^d}\mathcal{H}_\epsilon(x)\varphi(x,\epsilon)\,dx = \int_{\mathbb{R}^d}\mathcal{H}_\epsilon(x)(\varphi(x,0)+O(\epsilon))\,dx,$$

and this tends to $\varphi(0,0)$, since $\mathcal{H}_t$ is an approximation to the identity.

An alternate proof can be given by computing the Fourier transform of F, as in Exercise 21.

2.4 Fundamental solution to general partial differential equations with constant coefficients

We now tackle the general case of any constant coefficient partial differential operator L on $\mathbb{R}^d$ by addressing the convergence issues raised by (13), where a candidate for a fundamental solution F was written as

$$F = \int_{\mathbb{R}^d}\frac{1}{P(\xi)}e^{2\pi i x\cdot\xi}\,d\xi,$$

with P the characteristic polynomial of the operator L. Ignoring for a moment the problem of convergence, we note that if $\varphi\in\mathcal{D}$, then we would have

$$F(\varphi) = \int_{\mathbb{R}^d}\varphi(x)\int_{\mathbb{R}^d}\frac{1}{P(\xi)}e^{2\pi i x\cdot\xi}\,d\xi\,dx,$$

and hence after interchanging the order of integration,

$$F(\varphi) = \int_{\mathbb{R}^d}\frac{\hat{\varphi}(-\xi)}{P(\xi)}\,d\xi. \tag{15}$$

To circumvent the obstacle that arises in (15) because of possible zeroes of P, we shift the line of integration in the ξ_1-variable to avoid any zeroes of the polynomial $p(z) = P(z, \xi')$, where $\xi' = (\xi_2, \ldots, \xi_d)$ is fixed. The result we obtain is as follows.

Theorem 2.11 *Every constant coefficient (linear) partial differential equation L on $\mathbb{R}^d$ has a fundamental solution.*

Proof. After a possible change of coordinates consisting of a rotation and multiplication by a constant, we may assume that the characteristic polynomial of L will be of the form

$$P(\xi) = P(\xi_1, \xi') = \xi_1^m + \sum_{j=0}^{m-1} \xi_1^j Q_j(\xi'),$$

where each Q_j' is a polynomial of degree at most $m - j$. A proof that a general polynomial P can be written in the above form, can be found for instance in Section 3, Chapter 5, Book III, where an earlier version of the "invertibility" of L appears.

For each ξ', the polynomial $p(z) = P(z, \xi')$ has m roots in $\mathbb{C}$, which can be ordered lexicographically, say $\alpha_1(\xi'), \ldots, \alpha_m(\xi')$. We claim that we can pick an integer $n(\xi')$ so that:

(i) $|n(\xi')| \leq m + 1$ for all ξ'.

(ii) If $\mathrm{Im}(\xi_1) = n(\xi')$, then $|\xi_1 - \alpha_j(\xi')| \geq 1$ for all $j = 1, \ldots, m$.

(iii) The function $\xi' \mapsto n(\xi')$ is measurable.

Indeed, for each ξ' the polynomial p has m zeroes, so at least one of the $m + 1$ intervals $I_\ell = [-m - 1 + 2\ell, -m - 1 + 2(\ell + 1))$ (for $\ell = 0, \ldots, m$) has the property that it does not contain any of the imaginary parts of the zeroes of p. We can then set $n(\xi')$ to be the mid-point of such interval I_ℓ with the smallest ℓ having the above property. Condition (ii) is then automatically satisfied. Finally, Rouché's theorem[10] applied to small circles around the zeroes of p shows that $\alpha_1(\xi'), \ldots, \alpha_m(\xi')$ are continuous functions of ξ', and this implies (iii).

So, instead of (15) we now define

$$(16) \qquad F(\varphi) = \int_{\mathbb{R}^{d-1}} \int_{\mathrm{Im}(\xi_1) = n(\xi')} \frac{\hat{\varphi}(-\xi)}{P(\xi)} \, d\xi_1 \, d\xi', \qquad \text{whenever } \varphi \in \mathcal{D}.$$

---

[10]See for example Chapter 3 in Book II.

In the above, the inner integral is taken over the line $\{\mathrm{Im}(\xi_1) = n(\xi')\}$ in the complex ξ_1-space.

To see that F is well defined as a distribution, recall first that since φ has compact support then $\hat{\varphi}$ is analytic with rapid decay on each line parallel to the real axis, so it suffices to show that P is uniformly bounded from below on the line of integration. To this end, fix ξ on such line, and consider the polynomial in one variable $q(z) = P(\xi_1 + z, \xi')$. Then q is a polynomial of degree m with leading coefficient 1, so if $\lambda_1, \ldots, \lambda_m$ denote the roots of q, then $q(z) = (z - \lambda_1) \cdots (z - \lambda_m)$. By (ii) above we have that $|\lambda_j| \geq 1$ for all j, hence $|P(\xi)| = |q(0)| = |\lambda_1 \cdots \lambda_m| \geq 1$, as desired. Therefore F defines a distribution.

Finally, the rapid decrease also allows us to differentiate under the integral sign, so if $L' = \sum_{|\alpha| \leq m} a_\alpha (-1)^{|\alpha|} \partial_x^\alpha$, then the characteristic polynomial of L' is $P(-\xi)$ therefore $(L'(\varphi))^\wedge = P(-\xi)\hat{\varphi}(\xi)$. Hence

$$(LF)(\varphi) = F(L'(\varphi)) = \int_{\mathbb{R}^{d-1}} \int_{\mathrm{Im}(\xi_1)=n(\xi')} \hat{\varphi}(-\xi)\, d\xi_1\, d\xi'.$$

We can now deform the contour of integration back to the real line, so that

$$(LF)(\varphi) = \int_{\mathbb{R}^d} \hat{\varphi}(-\xi)\, d\xi = \varphi(0) = \delta(\varphi),$$

which completes the proof of the theorem.

Remark. We obtain from this the following existence theorem: whenever $f \in C_0^\infty(\mathbb{R}^d)$, there exists a $u \in C^\infty(\mathbb{R}^d)$ so that $L(u) = f$. This is clear if we take $u = F * f$, with F the fundamental solution above.[11] It should also be pointed out that an analogous solvability fails if L is not constant-coefficient, as is seen in Section 8.3 of Chapter 7.

2.5 Parametrices and regularity for elliptic equations

In many instances it is convenient to replace the notion of a fundamental solution by a more flexible variant, that of an "approximate fundamental solution" or parametrix. Given a differential operator L with constant coefficients, a **parametrix** for L is a distribution Q, so that

$$LQ = \delta + r$$

where the "error" r is in (say) $\mathcal{S}$. In this sense, the difference $LQ - \delta$ is small.

[11] This result may be compared with Section 3 in Chapter 5 of Book III, where not-necessarily smooth solutions are found by a different method.

Of particular interest are parametrices that are smooth away from the origin. Adopting the terminology used earlier, we say that Q is **regular** if this distribution agrees with a C^∞ function away from the origin.

An important class of partial differential operators that have regular parametrices are the elliptic operators. A given partial differential operator $L = \sum_{|\alpha|\leq m} a_\alpha \partial_x^\alpha$, of order m, is said to be **elliptic** if its characteristic polynomial P satisfies the inequality $|P(\xi)| \geq c|\xi|^m$, for some $c > 0$, and all sufficiently large ξ. Note that this is the same as assuming that P_m, the principal part of P (the part of P which is homogeneous of degree m), has the property that $P_m(\xi) = 0$ only when $\xi = 0$.

Note, for example, that the Laplacian $\triangle$ is elliptic.

Theorem 2.12 *Every elliptic operator has a regular parametrix.*

Proof. Observe first by a straightforward inductive argument in k, that whenever $|\alpha| = k$ and P is any polynomial

$$\left(\frac{\partial}{\partial\xi}\right)^\alpha \left(\frac{1}{P(\xi)}\right) = \sum_{0\leq\ell\leq k} \frac{q_\ell(\xi)}{P(\xi)^{\ell+1}},$$

where each q_ℓ is a polynomial of degree $\leq \ell m - k$.

Now suppose $|P(\xi)| \geq c|\xi|^m$, whenever $|\xi| \geq c_1$, and let γ be a C^∞ function which is equal to 1 for all large values of ξ and is supported in $|\xi| \geq c_1$. Then observe from the above identity that

$$\left|\partial_\xi^\alpha \left(\frac{\gamma(\xi)}{P(\xi)}\right)\right| \leq A_\alpha |\xi|^{-m-|\alpha|}. \tag{17}$$

Now let Q be the tempered distribution whose Fourier transform is the (bounded) function $\gamma(\xi)/P(\xi)$. Taking up the same argument as in the proof of Theorem 2.4, we have

$$\left((-4\pi^2|x|^2)^N \partial_x^\beta Q\right)^\wedge = \triangle_\xi^N[(2\pi i\xi)^\beta(\gamma/P)].$$

Because of (17) and Leibnitz's rule, the right-hand side above is clearly dominated by $A'_\alpha|\xi|^{-m-2N+|\beta|}$ for $|\xi| \geq 1$; it is also bounded when $|\xi| \leq 1$. Thus as soon as $2N + m - |\beta| > d$, this function is integrable, and therefore $|x|^{2N}\partial_x^\beta Q$, being its inverse Fourier transform up to a multiplicative constant, is continuous. Since this is true for each β, we see that Q agrees with a C^∞ function away from the origin.

Note moreover that $(LQ)^\wedge = P(\xi)[\gamma(\xi))/P(\xi)] = \gamma(\xi) = 1 + (\gamma(\xi) - 1)$. By its definition, $\gamma(\xi) - 1$ is in $\mathcal{D}$, and hence $\gamma(\xi) - 1 = \hat{r}$, for some

$r \in \mathcal{S}$. Finally, $(LQ)^\wedge = 1 + \hat{r}$, which means $LQ = \delta + r$, as was to be shown.

The following variant is useful.

Corollary 2.13 *Given any $\epsilon > 0$, the elliptic operator L has a regular parametrix Q_ϵ that is supported in the ball $\{x : |x| \leq \epsilon\}$.*

In fact, let η_ϵ be a cut-off function in $\mathcal{D}$, that is 1 when $|x| \leq \epsilon/2$, and that is supported where $|x| \leq \epsilon$. Set $Q_\epsilon = \eta_\epsilon Q$, and observe that $L(\eta_\epsilon Q) - \eta_\epsilon L(Q)$ involves only terms that are derivatives of η_ϵ of positive order, and these vanish when $|x| < \epsilon/2$. The difference is therefore a C^∞ function. However, $\eta_\epsilon L(Q) = \eta_\epsilon(\delta + r) = \delta + \eta_\epsilon r$. Altogether, this gives $L(Q_\epsilon) = \delta + r_\epsilon$, where r_ϵ is a C^∞ function. Notice that r_ϵ is automatically also supported in $|x| \leq \epsilon$.

Elliptic operators satisfy the following basic regularity property.

Theorem 2.14 *Suppose the partial differential operator L has a regular parametrix. Assume U is a distribution given in an open set $\Omega \subset \mathbb{R}^d$ and $L(U) = f$, with f a C^∞ function in Ω. Then U is also a C^∞ function on Ω. In particular, this holds whenever L is elliptic.*

Remark. The terminology **hypo-elliptic** is used to denote operators for which the above regularity holds. The prefix "hypo" reflects the fact that there are non-elliptic operators (for example the heat operator $\frac{\partial}{\partial t} - \triangle_x$) that also have this property as a result of the fact that they have a regular fundamental solution. However, it should be noted that for general partial differential operators, hypo-ellipticity fails; a good example is the wave operator. (See Exercise 22 and Problem 7*.)

Proof of the theorem. It suffices to show that U agrees with a C^∞ function on any ball B with $\overline{B} \subset \Omega$. Fix such a ball (say of radius ρ), and let B_1 be the concentric ball having radius $\rho + \epsilon$, with $\epsilon > 0$ so small that $\overline{B_1} \subset \Omega$. Next, choose a cut-off function η in $\mathcal{D}$, supported in Ω, with $\eta(x) = 1$ in a neighborhood of $\overline{B_1}$. Define $U_1 = \eta U$. Then U_1 and $L(U_1) = F_1$ are distributions of compact support in $\mathbb{R}^d$ and moreover F_1 agrees with a C^∞ function (that is, f) in a neighborhood of $\overline{B_1}$. Thus F_1 agrees in a smaller neighborhood of $\overline{B_1}$ with a C^∞ function f_1 that has compact support.

We now apply the parametrix Q_ϵ supported in $\{|x| \leq \epsilon\}$ whose existence is guaranteed by Corollary 2.13. On the one hand,

$$Q_\epsilon * L(U_1) = L(Q_\epsilon) * U_1 = (\delta + r_\epsilon) * U_1 = U_1 + r_\epsilon * U_1,$$

and since $r_\epsilon * U_1 = U_1 * r_\epsilon$ by Proposition 1.1, we have that $r_\epsilon * U_1$ is a C^∞ function in $\mathbb{R}^d$. On the other hand,

$$Q_\epsilon * L(U_1) = Q_\epsilon * F_1 = Q_\epsilon * f_1 + Q_\epsilon * (F_1 - f_1).$$

Now again, $Q_\epsilon * f_1$ is a C^∞ function, while by Proposition 1.3, $Q_\epsilon * (F_1 - f_1)$ is supported in the closure of the ϵ-neighborhood of the support of $F_1 - f_1$. Since $F_1 - f_1$ vanishes in a neighborhood of $\overline{B_1}$ it follows that $Q_\epsilon * (F_1 - f_1)$ vanishes in B. Altogether then U_1 is a C^∞ function on B. Since $U_1 = \eta U$ and η equals 1 in B, then U is a C^∞ function in B, and the theorem is therefore proved.

3 Calderón-Zygmund distributions and L^p estimates

We will now consider an important class of operators that generalize the Hilbert transform and that have a corresponding L^p theory. These arise as "singular integrals," that is, as convolution operators T given by

$$T(f) = f * K, \tag{18}$$

with K that are appropriate distributions. Among kernels K of this kind the first considered were homogeneous distributions of critical degree $-d$, similar to those described in Remark 2 at the end of Section 2.2.[12] Over time, various generalizations and extensions of these operators have arisen. Here we want to restrict our attention to a narrow but particularly simple and useful class of such operators, which have the added feature that they can be defined either in terms of (18) or in terms of the Fourier transform via

$$(Tf)^\wedge(\xi) = m(\xi)\hat{f}(\xi). \tag{19}$$

The reciprocity of the resulting conditions on the kernel K and the **multiplier** m, with $m = K^\wedge$, can then be seen as a generalization of Theorem 2.4 when $\lambda = -d$.

3.1 Defining properties

We consider a distribution K that is "regular" in the terminology used in Sections 2.2 and 2.5. This means that for such K there is a function k that is C^∞ away from the origin so that K agrees with k away from the origin. Given a K of this kind, we consider the following **differential inequalities** for its associated function k,

$$|\partial_x^\alpha k(x)| \leq c_\alpha |x|^{-d-|\alpha|}, \qquad \text{for all } \alpha. \tag{20}$$

[12] Without however requiring a high degree of smoothness of k.

Notice that the above for $\alpha = 0$, implies that the distribution K is tempered.

In addition to (20) we formulate a **cancelation condition** as follows. Given an integer n, we say that φ is a $C^{(n)}$**-normalized bump function** if φ is a C^∞ function supported in the unit ball and

$$\sup_x |\partial_x^\alpha \varphi(x)| \leq 1, \qquad \text{all } |\alpha| \leq n.$$

We define φ_r by $\varphi_r(x) = \varphi(rx)$, for $r > 0$. Our condition is then that for some fixed $n \geq 1$, there is an A so that

$$\text{(21)} \quad \sup_{0<r} |K(\varphi_r)| \leq A, \qquad \text{for all } C^{(n)}\text{-normalized bump functions } \varphi.$$

Proposition 3.1 *The following three properties of a distribution K are equivalent.*

(i) *K is regular and satisfies the differential inequalities* (20) *together with the cancelation property* (21).

(ii) *K is tempered, and $m = K^\wedge$ is a function that is C^∞ away from the origin that satisfies*

$$\text{(22)} \qquad |\partial_\xi^\alpha m(\xi)| \leq c'_\alpha |\xi|^{-\alpha}, \qquad \textit{for all } \alpha.$$

(iii) *K is a regular distribution that satisfies the differential inequalities* (20) *and $K^\wedge$ is a bounded function.*

We refer to kernels K that satisfy these equivalent properties as **Calderón-Zygmund distributions.**[13]

The proof will be facilitated by noting the dilation-invariance of the *set* of all distributions that satisfy the above conditions. Recall the scaling of a distribution K as defined in Section 2.1. For each $a > 0$, the scaled distribution K^a is given by $K^a(\varphi) = K(\varphi_a)$, with $\varphi_a(x) = \varphi(ax)$. With this we claim that whenever K satisfies (20) and (21), K^a satisfies (20) and (21) with the *same bounds*. In fact, the function associated to K^a is $a^{-d}k(x/a)$, while $K^a(\varphi_r) = K(\varphi_{ar})$, as the reader may easily verify. Moreover, if $m = K^\wedge$, then $m_a = (K^a)^\wedge$, and $m_a(\xi) = m(a\xi)$, so m_a satisfies (22) with the same bounds.

Once this is observed, the proof of the proposition is in the same spirit as that of Theorem 2.4, and so we will be correspondingly brief. Let us

[13]We should note that phrases like "Calderón-Zygmund operators" or "Calderón-Zygmund kernels" have been used in many contexts to denote different but related objects in the theory.

begin by assuming condition (i). We first observe that $m = K^\wedge$ is a C^∞ function away from the origin. This is done by splitting K as $K_0 + K_1$, where $K_0 = \eta K$ and $K_1 = (1-\eta)K$ (with η a C^∞ cut-off function that is supported in the unit ball and is equal to 1 when $|x| \leq 1/2$), and proceeding as in the proof of Theorem 2.4.

To show that the inequalities (22) are satisfied for $m(\xi) = K^\wedge$, $\xi \neq 0$, we can reduce matters to the case $|\xi| = 1$ by the dilation-invariance pointed out above. Now by Proposition 1.6, $K_0^\wedge(\xi) = K(\eta e^{-2\pi i x\cdot\xi})$, and the latter is $K(\varphi)$ with $\varphi(x) = \eta(x)e^{-2\pi i x\cdot\xi}$. Now φ is a multiple (independent of ξ, for $|\xi| = 1$) of a $C^{(n)}$-normalized bump function, so (20) implies $|K_0^\wedge(\xi)| \leq c'$. The same argument gives $|\partial_\xi^\alpha K_0^\wedge(\xi)| \leq c'_\alpha$.

Next, since $K_1 = (1-\eta)K = (1-\eta)k$ is supported where $|x| \geq 1/2$, we have by (7)

$$
\begin{aligned}
|\xi|^{2N}|\partial_\xi^\alpha K_1^\wedge(\xi)| &= c|(\triangle^N(x^\alpha K_1))^\wedge| \\
&\leq c_{\alpha,N}\int_{|x|\geq 1/2} |x|^{-d+|\alpha|-2N}\,dx < \infty
\end{aligned}
$$

if $2N > |\alpha|$. Thus $|\partial_\xi^\alpha K_1^\wedge(\xi)| \leq c'_\alpha$ when $|\xi| = 1$, and therefore combining estimates for $K_0^\wedge$ and $K_1^\wedge$ implies (ii) in the proposition.

To prove that (ii) implies (i), we first assume that m satisfies (22) and, in addition, has bounded support, but we will make our estimates independent of the size of the support of m.

Define $K(x) = \int_{\mathbb{R}^d} m(\xi)e^{2\pi i\xi\cdot x}\,d\xi$. Then clearly K is a bounded C^∞ function on $\mathbb{R}^d$, and $K^\wedge = m$ in the sense of distributions. In proving the differential inequalities (20), it will be sufficient to do this for $|x| = 1$, because of the dilation-invariance used earlier. Now write $K = K_0 + K_1$, with K_j defined like K with m replaced by m_j, where $m_0(\xi) = m(\xi)\eta(\xi)$ and $m_1(\xi) = m(\xi)(1-\eta(\xi))$. Now obviously $|\partial_x^\alpha K_0(x)| \leq c_\alpha$, since m_0 is bounded and is supported in the unit ball. Also in analogy with (7) and the previous argument,

$$
\begin{aligned}
|x|^{2N}|\partial_x^\alpha K_1(x)| &= c\left|\int_{\mathbb{R}^d} \triangle_\xi^N(\xi^\alpha m_1(\xi))\,d\xi\right| \\
&\leq c_{\alpha,N}\int_{|\xi|\geq 1/2} |\xi|^\alpha|\xi|^{-2N}d\xi < \infty
\end{aligned}
$$

if $2N - |\alpha| > d$. Since $|x| = 1$, these estimates for K_0 and K_1 yield (20) for $|x| = 1$, and thus for all $x \neq 0$.

To prove the cancelation condition, take $n = d+1$. Note first that $(2\pi i\xi)^\alpha\hat{\varphi}(\xi) = (\partial_x^\alpha\varphi)^\wedge(\xi)$, so this implies that $\sup_\xi(1+|\xi|)^{d+1}|\hat{\varphi}(\xi)| \leq c$, whenever φ is a $C^{(n)}$-normalized bump function, and as a result

$$\int_{\mathbb{R}^d} |\hat{\varphi}(\xi)|\, d\xi \leq c \int_{\mathbb{R}^d} \frac{d\xi}{(1+|\xi|)^d} \leq c'$$

for such a normalized bump functions.

However, $K(\varphi_r) = K^r(\varphi) = \int m_r(-\xi)\hat{\varphi}(\xi)\, d\xi$. Therefore $|K(\varphi_r)| \leq \sup_\xi |m(\xi)| \int |\hat{\varphi(\xi)}|\, d\xi \leq A$, and the condition (21) is established.

To dispense with the hypothesis that m has compact support, consider the family $m_\epsilon(\xi) = m(\xi)\eta_\epsilon(\xi)$, with $\epsilon > 0$. Observe that each m_ϵ has compact support and (22) is satisfied uniformly in ϵ. Set

$$K_\epsilon(x) = \int_{\mathbb{R}^d} m_\epsilon(\xi) e^{2\pi i x\cdot\xi}\, d\xi.$$

Then since $m_\epsilon \to m$ pointwise and boundedly as $\epsilon \to 0$, the convergence is also in the sense of tempered distributions, and this implies the convergence of K_ϵ to K in the sense of tempered distributions, with $K^\wedge = m$. Now the differential inequalities (20) hold for $x \neq 0$, and K_ϵ, uniformly in ϵ. Thus these estimates hold for K, (more precisely for its associated function k). Similarly, since the cancelation conditions (21) hold for K_ϵ, uniformly in ϵ, these conditions hold for K, and thus altogether we see that (ii) implies (i). We observe that the argument just given shows that (iii) implies (i). Since (iii) is clearly a consequence of (i) and (ii) together, all three conditions are equivalent, finishing the proof of the proposition.

The following points may help clarify the nature of the hypotheses concerning Calderón-Zygmund distributions.

- It is clear that if the cancelation condition holds for $C^{(n)}$-normalized bump functions for a given n, then it also holds with $n' > n$. In the other direction, it can be shown that in the presence of (20), the fact that (21) holds for some n implies that it holds for $n = 1$, and thus for all $n' \geq 1$. This is sketched in Exercise 32.

- Given a function k that satisfies the differential inequalities (20), we may ask if there is a Calderón-Zygmund distribution K that has k as its associated function. The necessary and sufficient condition on k is that

$$\sup_{0<a<b} \left| \int_{a<|x|<b} k(x)\, dx \right| < \infty.$$

  The proof of this fact is outlined in Exercise 33. Note however that K is not uniquely determined by k.

- We make a last remark about the significance of Calderón-Zygmund distributions in the theory of partial differential equations. It is that, whenever Q is a parametrix for an elliptic operator L of order m as in Section 2.5, then $\partial_x^\alpha Q$ is a Calderón-Zygmund distribution, whenever $|\alpha| \leq m$. This follows immediately from the estimate (17) and the characterization of such distributions by the Fourier transform given by assertion (ii) of the proposition.

3.2 The L^p theory

The L^p estimates for operators of the form (18) are given by the following theorem.

Theorem 3.2 *Let T be the operator $T(f) = f * K$, with K as in Proposition 3.1. Then T initially defined for f in $\mathcal{S}$ extends to a bounded operator on $L^p(\mathbb{R}^d)$, for $1 < p < \infty$.*

This means that for each p, $1 < p < \infty$, there is a bound A_p so that

$$\|Tf\|_{L^p(\mathbb{R}^d)} \leq A_p \|f\|_{L^p(\mathbb{R}^d)} \tag{23}$$

for $f \in \mathcal{S}$. Thus by Proposition 5.4 in Chapter 1 we see that T has a (unique) extension to all of L^p that satisfies the bound (23) for $f \in L^p$. We break the proof into five steps.

Step 1: *L^2 estimate.* The case $p = 2$ follows directly from the fact that $(Tf)^\wedge = f^\wedge K^\wedge$, (see Proposition 1.5) and that

$$\|Tf\|_{L^2} = \|(Tf)^\wedge\|_{L^2} \leq (\sup_\xi |K^\wedge(\xi)|)\, \|\hat{f}\|_{L^2} \leq A\|f\|_{L^2},$$

by Plancherel's theorem. The inequality $\sup_\xi |K^\wedge(\xi)| \leq A$ is of course a consequence of Proposition 3.1.

Step 2: *A variant of atoms.* While our operator T does not in general map L^1 to itself (as the example in Section 3.2 of the previous chapter already shows), its L^p theory for $1 < p < \infty$ is bound up with a "weak-type" L^1 estimate, as was the case for the maximal function treated in Section 4 of Chapter 2. Here we arrive at this kind of estimate by studying the action of T on variants of the atoms that are relevant for the Hardy space theory. In the present situation we deal with "1-atoms," the case $p = 1$ of the p-atoms (specifically excluded from Corollary 5.3 in the previous chapter!).

A **1-atom** $\mathfrak{a}$ associated to a ball B is an L^2 function with:

(i) $\mathfrak{a}$ is supported in B, and $\int |\mathfrak{a}(x)|\,dx \le 1$.

(ii) $\int_B \mathfrak{a}(x)\,dx = 0$.

Notice that the L^2 norm of $\mathfrak{a}$ does not enter into the conditions (i) and (ii) above; the requirement that $\mathfrak{a} \in L^2$ is made only for technical convenience.

For each ball B we will denote by B^* its double, that is, the ball with the same center as B but with twice its radius. The key estimate involving our operator T and 1-atoms is that there is a bound A so that

$$\int_{(B^*)^c} |T(\mathfrak{a})(x)|\,dx \le A, \qquad \text{for all 1-atoms } \mathfrak{a}. \tag{24}$$

Now (24) will be a consequence of an inequality satisfied by the function k associated to the distribution kernel K of the operator, namely that for all $r > 0$

$$\int_{|x|\ge 2r} |k(x-y) - k(x)|\,dx \le A, \qquad \text{whenever } |y| \le r. \tag{25}$$

To see (25), note that by the mean-value theorem,

$$|k(x-y) - k(x)| \le |y| \sup_{z\in L} |\nabla k(z)|,$$

where L is the line segment joining x to $x - y$. Since $|x| \ge 2r$ and $|y| \le r$, it follows that $|z| \ge |x|/2$, whenever $z \in L$. Thus the differential inequalities (20) for $|x| = 1$ show that $|k(x-y) - k(x)| \le c|x|^{-d-1}$, and (25) follows because $r \int_{|x|\ge 2r} |x|^{-d-1}\,dx$ is independent of r (and is finite).

To deduce (24) from this, observe first that whenever f is in $\mathcal{S}$ and is supported in the ball B, then for $x \notin B^*$ we have

$$T(f)(x) = \int_B k(x-y) f(x)\,dy.$$

This is so because the distribution K agrees with the function k away from the origin and here $|x - y| \ge r$. Since $k(x-y)$ is bounded there, a passage to the limit shows that the same identity holds if f is supported in B and is assumed merely to be in L^2. So if $\mathfrak{a}$ is a 1-atom associated to B and $x \notin B^*$, we have

$$T(\mathfrak{a})(x) = \int_B k(x-y)\mathfrak{a}(y)\,dy = \int_B (k(x-y) - k(x))\mathfrak{a}(y)\,dy,$$

because $\int_B \mathfrak{a}(y)\,dy = 0$. Therefore,

$$\int_{x\notin B^*} |T(\mathfrak{a})(x)|\,dx \leq \int_B \left\{ \int_{x\notin B^*} |k(x-y)-k(x)|\,dx \right\} |\mathfrak{a}(y)|\,dy,$$

and (24) is established if we invoke (25) with r the radius of the ball B.

Step 3: *The decomposition.* We exploit (24) by decomposing any integrable function f as a sum of a "good" function g, for which the L^2 theory applies, and an infinite sum of multiples of atoms, for which the estimate (24) is used.

Lemma 3.3 *For each f in $L^1(\mathbb{R}^d)$ and $\alpha > 0$, we can find an open set E_α and a decomposition $f = g + b$ so that:*

(a) $m(E_\alpha) \leq \frac{c}{\alpha}\|f\|_{L^1(\mathbb{R}^d)}$.

(b) $|g(x)| \leq c\alpha$, *for all* x.

(c) *E_α is a union $\bigcup Q_k$ of cubes Q_k whose interiors are disjoint. Moreover $b = \sum_k b_k$, with each function b_k supported in Q_k and*

$$\int |b_k(x)|\,dx \leq c\alpha m(Q_k), \qquad \textit{while } \int_{Q_k} b_k(x)\,dx = 0.$$

Note that (c) implies that b is supported in E_α hence $g(x) = f(x)$ if $x \notin E_\alpha$. Observe also that each b_k is of the form $c\alpha m(Q_k)\mathfrak{a}_k$, where $\mathfrak{a}_k$ is a 1-atom.

The proof of the lemma is a simplified version of the argument used to prove Proposition 5.1 in the previous chapter; in particular, here we use the full maximal function f^* instead of the truncated version $f^\dagger$. The guiding idea is to try to cut the domain of f into the set when $|f(x)| > \alpha$ and its complement. However, as before, we must be more subtle and in the present situation cut f according to where $f^*(x) > \alpha$. Thus we take $E_\alpha = \{x :\ f^*(x) > \alpha\}$. The conclusion (a) is therefore the weak-type estimate for f^* given in (27) of the previous chapter.

Next, since E_α is open we can write it as $\bigcup_k Q_k$, where the Q_k are closed cubes with disjoint interiors, with the distance of Q_k from E_α^c comparable to the diameter of Q_k. (This is Lemma 5.2 of the previous chapter.) Now set

$$m_k = \frac{1}{m(Q_k)} \int_{Q_k} f\,dx.$$

Thus if $\overline{x}_k$ is a point of E_α^c closest to Q_k, one has $|m_k| \leq cf^*(\overline{x}_k) \leq c\alpha$. We define $g(x) = f(x)$ for $x \notin E_\alpha^c$ and $g(x) = m_k$ for $x \in Q_k$. As a result $|f(x)| \leq \alpha$ for $x \in E_\alpha^c$, because $f^*(x) \leq \alpha$ there. Altogether then $|g(x)| \leq c\alpha$, proving conclusion (b).

Finally, $b(x) = f(x) - g(x)$ is supported in $E_\alpha = \bigcup_k Q_k$ and hence $b = \sum_k b_k$, where each b_k is supported in Q_k and equals $f(x) - m_k$ there. Thus

$$\int |b_k(x)|\, dx = \int_{Q_k} |f(x) - m_k|\, dx \leq \int_{Q_k} |f(x)|\, dx + |m_k| m(Q_k).$$

Also as before

$$\int_{Q_k} |f(x)|\, dx \leq cm(Q_k) f^*(\overline{x}_k) \leq c\alpha m(Q_k),$$

hence

$$\int |b_k(x)|\, dx \leq c\alpha m(Q_k),$$

since $|m_k| \leq c\alpha$. Clearly, $\int b_k(x)\, dx = \int_{Q_k} (f(x) - m_k)\, dx = 0$, and so the decomposition lemma is proved.

One observes that if we were also given that f was in $L^2(\mathbb{R}^d)$, then it would follow that g, b, and each b_k would also be in $L^2(\mathbb{R}^d)$. Since the supports of the b_k are disjoint, the sum $b = \sum_k b_k$ would converge not only in the obvious pointwise sense, but also in the L^2 norm.

Step 4: Weak-type estimate. Here we show that

$$(26) \qquad m(\{x : |T(f)(x)| > \alpha\}) \leq \frac{A}{\alpha} \|f\|_{L^1}, \qquad \text{for each } \alpha > 0$$

whenever $f \in L^1 \cap L^2$, with the bound A independent of f and α. To do this we decompose $f = g + b$ according to the lemma and note that

$$m(\{x : |T(f)(x)| > \alpha\}) \leq m(\{x : |T(g)(x)| > \alpha/2\}) + m(\{x : |T(b)(x)| > \alpha/2\}),$$

because $T(f) = T(g) + T(b)$. Now by Tchebychev's inequality and the L^2 estimate for T,

$$m(\{x : |T(g)(x)| > \alpha/2\}) \leq \left(\frac{2}{\alpha}\right)^2 \|Tg\|_{L^2}^2 \leq \frac{c}{\alpha^2} \|g\|_{L^2}^2.$$

However $\int |g(x)|^2\,dx = \int_{E_\alpha^c} |g(x)|^2\,dx + \int_{E_\alpha} |g(x)|^2\,dx$. Now on E_α^c we have $g(x) = f(x)$ and $|g(x)| \le c\alpha$, so the first integral on the right is majorized by $c\alpha\|f\|_{L^1}$. Also

$$\int_{E_\alpha} |g(x)|^2\,dx \le c\alpha^2 m(E_\alpha) \le c\alpha\|f\|_{L^1},$$

by conclusion (a) of the lemma. As a result

$$m(\{x: \ |T(g)(x)| > \alpha/2\}) \le \frac{c}{\alpha}\|f\|_{L^1}.$$

To deal with $T(b) = \sum_k T(b_k)$, we let B_k denote the smallest ball whose closure contains Q_k, and B_k^* the double of B_k. We define $E_\alpha^* = \bigcup B_k^*$. Now, again by Tchebychev's inequality, for a bounded set S,

$$\begin{aligned} m(\{x \in S: \ |T(b)(x)| > \alpha/2\}) &\le \frac{2}{\alpha}\int_S |T(b)(x)|\,dx \\ &\le \frac{2}{\alpha}\sum_k \int_S |T(b_k)(x)|\,dx, \end{aligned}$$

since $T(b) = \sum_k T(b_k)$, with convergence in the L^2 norm.

Now set $S = (E_\alpha^*)^c \cap B$, where B is a large ball. Letting the radius of B tend to infinity then yields

$$m(\{x \notin E_\alpha^*: \ |T(b)(x)| > \alpha/2\}) \le \frac{2}{\alpha}\sum_k \int_{(B_k^*)^c} |T(b_k)(x)|\,dx,$$

because $E_\alpha^* = \bigcup B_k^*$ implies that $(E_\alpha^*)^c \subset (B_k^*)^c$ for each k. However as we have noted, b_k is of the form $c\alpha m(Q_k)\mathfrak{a}_k$, where $\mathfrak{a}_k$ is a 1-atom associated to the ball B_k. Hence the estimate (24) gives

$$m(\{x \in (E_\alpha^*)^c: \ |T(b)(x)| > \alpha/2\}) \le c\sum_k m(Q_k) = cm(E_\alpha) \le \frac{c}{\alpha}\|f\|_{L^1}.$$

Finally,

$$m(E_\alpha^*) \le \sum_k m(B_k^*) = c\sum m(Q_k) = cm(E_\alpha) \le \frac{c'}{\alpha}\|f\|_{L^1},$$

because $m(B_k^*) = cm(Q_k)$ for every k.

Gathering the inequalities for $T(g)$ and $T(b)$ together then shows that the weak-type estimate (26) is established.

Step 5: *The L^p inequalities.* We now borrow the idea used in Chapter 2 in the proof of the L^p estimates for the maximal function f^* in which the weak-type inequality is transformed to its more elaborate form, given in equation (28) of that chapter. In our case the stronger version is

(27)
$$m(\{x:\ |T(f)(x)|>\alpha\})\le A\left(\frac{1}{\alpha}\int_{|f|>\alpha}|f|\,dx+\frac{1}{\alpha^2}\int_{|f|\le\alpha}|f|^2\,dx\right),$$

whenever f belongs to both L^1 and L^2. To prove this, we cut f (this time, more simply) into two parts for each $\alpha>0$, according to the size of f. Namely, we set $f=f_1+f_2$ where $f_1(x)=f(x)$ if $|f(x)|>\alpha$, and $f_1(x)=0$ otherwise; also $f_2(x)=f(x)$ if $|f(x)|\le\alpha$, and $f_2(x)=0$ otherwise. Then again

$$m(\{|T(f)(x)|>\alpha\})\le m(\{|T(f_1)(x)|>\alpha/2\})+m(\{|T(f_2)(x)|>\alpha/2\}).$$

By the weak-type estimate just proved,

$$m(\{|T(f_1)(x)|>\alpha/2\})\le\frac{A}{\alpha}\|f_1\|_{L^1}=\frac{A}{\alpha}\int_{|f|>\alpha}|f|\,dx.$$

By the L^2-boundedness of T and Tchebychev's inequality

$$m(\{|T(f_2)(x)|>\alpha/2\})\le\left(\frac{2}{\alpha}\right)^2\|T(f_2)\|_{L^2}^2=\frac{A}{\alpha^2}\int_{|f|\le\alpha}|f|^2\,dx,$$

proving (27).

Now (see (29) in Chapter 2)

$$\int|T(f)(x)|^p\,dx=\int_0^\infty\lambda(\alpha^{1/p})\,d\alpha,$$

where $\lambda(\alpha)=m(\{x:\ |T(f)(x)|>\alpha\})$. Therefore, because of (27), the above integrals are majorized by

$$A\left(\int_0^\infty\alpha^{-1/p}\left(\int_{|f|>\alpha^{1/p}}|f|\,dx\right)d\alpha+\int_0^\infty\alpha^{-2/p}\left(\int_{|f|\le\alpha^{1/p}}|f|^2\,dx\right)d\alpha\right).$$

We have

$$\int_0^\infty\alpha^{-1/p}\left(\int_{|f|>\alpha^{1/p}}|f|\,dx\right)d\alpha=\int|f|\left(\int_0^{|f|^p}\alpha^{-1/p}\,d\alpha\right)dx$$
$$=a_p\int|f|^p\,dx$$

if $p > 1$ where $a_p = p/(p-1)$. Also,

$$\int_0^\infty \alpha^{-2/p} \left(\int_{|f| \leq \alpha^{1/p}} |f|^2 \, dx \right) d\alpha = b_p \int |f|^p \, dx$$

if $p < 2$, with $b_p = p/(2-p)$. Thus we get

$$\|T(f)\|_{L^p} \leq A_p \|f\|_{L^p},$$

with $A_p = A \cdot p \cdot (\frac{1}{p-1} + \frac{1}{2-p})$. This takes care of the case $1 < p < 2$ (the case $p = 2$ having been settled before).

To pass to the case $2 \leq p < \infty$, we use the duality of L^p spaces set forth in Section 4 of the first chapter.

We note that whenever f and g are in $\mathcal{S}$ then by Plancherel's theorem

$$\int_{\mathbb{R}^d} T(f)\overline{g} \, dx = \int_{\mathbb{R}^d} m(\xi)\hat{f}(\xi)\overline{\hat{g}(\xi)} \, d\xi = \int_{\mathbb{R}^d} f\overline{T^*(g)} \, dx.$$

Here $T^*(g) = g * K^*$, where $(K^*)^\wedge = \overline{m}$, with $m = K^\wedge$. Now $\overline{m}$ satisfies the same characterization (22) that m does, and hence the results above apply to T^*. In particular the identity

$$\int_{\mathbb{R}^d} (Tf)\overline{g} \, dx = \int_{\mathbb{R}^d} f\overline{(T^*g)} \, dx \tag{28}$$

extends to f and g in L^2.

Next with $2 \leq p < \infty$, let q be its dual exponent ($1/p + 1/q = 1$), where now $1 < q \leq 2$. Then, by Lemma 4.2 in Chapter 1,

$$\|T(f)\|_{L^p} = \sup_g \left| \int T(f)\overline{g} \, dx \right|,$$

where the supremum is taken over all g that are simple with $\|g\|_{L^q} \leq 1$. However

$$\left| \int T(f)\overline{g} \, dx \right| = \left| \int f\overline{(T^*(g))} \, dx \right| \leq \|f\|_{L^p} \|T^*(g)\|_{L^q} \leq A_q \|f\|_{L^p},$$

by Hölder's inequality and the boundedness of T^* on L^q ($1 < q \leq 2$). The result is now (23) for all $f \in \mathcal{S}$, for $1 < p < \infty$, concluding the proof of the theorem.

We make two closing comments about the theorem just proved.

- The result leads to "interior" estimates for solutions of elliptic equations in terms of L^p based Sobolev spaces. As such, these may be viewed as a quantitative version of Theorem 2.14. This is outlined in Problem 3.

- The essential properties of K that enter in the proof of the L^p theorem are, first, the L^2 boundedness via the Fourier transform, and second, the use of inequality (25). This inequality has natural extensions to a variety of contexts that arise in applications, in particular where the underlying structure of $\mathbb{R}^d$ is replaced by another suitable "geometry." However, obtaining L^2 boundedness in other settings is more problematic, since in general the Fourier transform may be unavailing. For this, further ideas have been developed that use the almost-orthogonality principle in Proposition 7.4 of Chapter 8, but these will not be pursued here.

4 Exercises

1. Suppose F is a distribution on Ω and $F = f$, with f a C^k function in Ω. Show that $\partial_x^\alpha F$, taken in the sense of distributions, agrees with $\partial_x^\alpha f$ for each $|\alpha| \leq k$.

2. The following represent converses to the previous exercise.

(a) Suppose f and g are continuous functions on $(a, b) \subset \mathbb{R}$ and $\frac{df}{dx}$ (taken in the sense of distributions) agrees with g. Show that for every $x \in (a, b)$, $(f(x+h) - f(x))/h \to g(x)$ as $h \to 0$.

(b) If f and g are merely assumed to be in $L^1(a, b)$ with $\frac{df}{dx} = g$ in the sense of distributions, then f is absolutely continuous and $(f(x+h) - f(x))/h \to g(x)$ as $h \to 0$ for a.e. x.

As a result, if f is a continuous but nowhere differentiable function on $\mathbb{R}$, then the distribution derivative of f is not a locally integrable function on any sub-interval.

(c) Generalize (a) as follows: Suppose $k \geq 1$ is an integer, and that f is a continuous function on an open set Ω. If for each multi-index α with $|\alpha| \leq k$, the distribution $\partial_x^\alpha f$ equals a continuous function g_α, then f is of class C^k and $\partial_x^\alpha f = g_\alpha$ as functions, for all $|\alpha| \leq k$.

[Hint: To see (a), let $x_0 \in (a, b)$, $h > 0$, and let η be a test function on (a, b) so that $\int \eta = 1$. With $\delta > 0$, define $\eta^\delta(x) = \delta^{-1}\eta(x/\delta)$ and

$$\varphi(x) = \int_{-\infty}^{x} \eta^\delta(x_0 + h - y) - \eta^\delta(x_0 - y)\, dy.$$

Then $\int f(x) \frac{d}{dx}\varphi(x)\, dx = -\int g(x)\varphi(x)\, dx$ and let $\delta, h \to 0$.

For (b), show as a first step that, up to a constant, f equals the indefinite integral of g, almost everywhere. Then use Theorem 3.8 in Chapter 3, Book III, about the differentiability almost everywhere of an absolutely continuous function.]

3. Show that a bounded function f on $\mathbb{R}^d$ satisfies a Lipschitz condition (also known as a Hölder condition of exponent 1)

$$|f(x) - f(y)| \leq C|x - y|, \qquad \text{for all } x, y \in \mathbb{R}^d,$$

if and only if $f \in L^\infty$ and all the first order partial derivatives $\partial f/\partial x^j$, $1 \leq j \leq d$, belong to L^∞ in the sense of distributions.

[Hint: Let $f_n = f * \psi_n$, where ψ_n is an approximation to the identity as in Corollary 1.2. Then $\partial f_n/\partial x_j \in L^\infty$ uniformly in n.]

4. Suppose F is a distribution on Ω.

(a) There exist $f_n \in C^\infty$, each of compact support in Ω, so that $f_n \to F$ in the sense of distributions.

(b) If F is supported in the compact set C, then for every $\epsilon > 0$ we can choose the f_n so that their supports are in the ϵ-neighborhood of C.

5. Let f be locally integrable on $\mathbb{R}^d$. Then the "support" of f in the measure-theoretic sense is the set $E = \{x : f(x) \neq 0\}$. Note that E is essentially determined only modulo sets of measure zero.

Show that the support of f, as a distribution, is equal to the intersection of all closed sets C such that $E - C$ has measure zero.

6. Assume that Ω is a region in $\mathbb{R}^d$ defined by $\Omega = \{x \in \mathbb{R}^d : x_d > \varphi(x')\}$, with $x = (x', x_d) \in \mathbb{R}^{d-1} \times \mathbb{R}$, and φ a C^1 function. Suppose f is a function that is continuous in $\overline{\Omega}$ and whose first derivatives are also continuous in $\overline{\Omega}$, with $f|_{\partial\Omega} = 0$.

Let $\tilde{f}$ be the extension of f to $\mathbb{R}^d$ defined by $\tilde{f}(x) = f(x)$ if $x \in \overline{\Omega}$, and $\tilde{f}(x) = 0$ if $x \notin \overline{\Omega}$. Then $\frac{\partial \tilde{f}}{\partial x_j}$, taken in the sense of distributions, is the function which is $\frac{\partial f}{\partial x_j}$ in $\overline{\Omega}$, and zero in $\overline{\Omega}^c$. (Note that it is not necessarily true that $\frac{\partial \tilde{f}}{\partial x_j}$ is continuous.)

[Hint: Show that $-\int_\Omega f(x) \frac{\partial \psi}{\partial x_j}\, dx = \int_\Omega \frac{\partial f}{\partial x_j} \psi \, dx$ for all C^∞ functions ψ of compact support in $\mathbb{R}^d$.]

7. Show that the distribution F is tempered if and only if there is an integer N, and a constant A, so that for all $R \geq 1$,

$$|F(\varphi)| \leq A R^N \sup_{|x| \leq R,\ 0 \leq |\alpha| \leq N} |\partial_x^\alpha \varphi(x)|,$$

for all $\varphi \in \mathcal{D}$ supported in $|x| \leq R$.

8. Suppose F is a homogeneous distribution of degree λ. Show that F is tempered.

[Hint: Fix $\eta \in \mathcal{D}$, $\eta(x) = 1$, for $|x| \leq 1$, η supported in $|x| \leq 2$. Let $\eta_R(x) = \eta(x/R)$. Find N so that $|\eta_1 F(\varphi)| \leq c\|\varphi\|_N$. Then deduce that $|(\eta_R F)(\varphi)| \leq cR^{N+|\lambda|}\|\varphi\|_N$.]

9. Check that on the real line, $f(x) = e^x$, considered as a distribution, is not tempered.
[Hint: Show that the criterion in Exercise 7 fails for every N.]

10. Verify that $\mathcal{D}$ is dense in $\mathcal{S}$.
[Hint: Fix $\eta \in \mathcal{D}$ so that $\eta = 1$ in a neighborhood of the origin. Let $\eta_k(x) = \eta(x/k)$ and consider $\varphi_k = \eta_k \varphi$.]

11. Suppose that $\varphi_1, \varphi_2 \in \mathcal{S}$.

(a) Verify that $\varphi_1 \cdot \varphi_2$ belongs to $\mathcal{S}$.

(b) Using the Fourier transform, prove that $\varphi_1 * \varphi_2 \in \mathcal{S}$.

(c) Show directly from the definition of convolution that $\varphi_1 * \varphi_2 \in \mathcal{S}$.

12. Prove that if F_1 is a distribution of compact support and $\varphi \in \mathcal{S}$, then $F_1 * \varphi \in \mathcal{S}$.
[Hint: For each N, there exists a constant c_N so that

$$\|\psi_y^{\sim}\|_N \leq c_N (1 + |y|)^N \|\psi\|_N.]$$

13. Use the previous exercise to prove that if F_1 and F are distributions with F_1 having compact support and F being tempered then:

(a) $F * F_1$ is tempered, and;

(b) $(F * F_1)^\wedge = F_1^\wedge F^\wedge$, ($F_1^\wedge$ is C^∞ and slowly increasing.)

14. Check that $f(x) = \frac{1}{2}|x|$ is a fundamental solution for $\frac{d^2}{dx^2}$ on $\mathbb{R}$.

15. A d-dimensional generalization of the identity for the Heaviside function is the identity

$$\delta = \sum_{j=1}^{d} \left(\frac{\partial}{\partial x_j}\right) h_j,$$

with $h_j(x) = \frac{1}{A_d}\frac{x_j}{|x|^d}$, and $A_d = 2\pi^{d/2}/\Gamma(d/2)$ denotes the area of the unit sphere in $\mathbb{R}^d$.
[Hint: When $d > 2$, write $\delta = \sum_{j=1}^d \frac{\partial}{\partial x_j}\left(\frac{\partial}{\partial x_j} C_d |x|^{-d+2}\right)$.]

16. Consider the complex plane $\mathbb{C} = \mathbb{R}^2$, with $z = x + iy$.

(a) Note that the Cauchy-Riemann operator

$$\partial_{\overline{z}} = \frac{1}{2}\left(\frac{\partial}{\partial x} + i\frac{\partial}{\partial y}\right)$$

is elliptic.

(b) Show that the locally integrable function $1/(\pi z)$ is a fundamental solution for $\partial_{\overline{z}}$.

(c) Suppose f is continuous in Ω, and $\partial_{\overline{z}} f = 0$ in the sense of distributions. Then f is analytic.

[Hint: For (b), use Theorem 2.9, and note that $\triangle = 4\partial_{\overline{z}}\partial_z$, where $\partial_z = \frac{1}{2}\left(\frac{\partial}{\partial x} - i\frac{\partial}{\partial y}\right)$.]

17. Suppose $f(z)$ is a meromorphic function on $\Omega \subset \mathbb{C}$. Prove:

(a) $\log|f(z)|$ is locally integrable.

(b) $\triangle(\log|f(z)|)$ taken in the sense of distributions is equal to $2\pi\sum_j m_j\delta_j - 2\pi\sum_k m'_k\delta_k$. Here the δ_j are the delta functions placed at the distinct zeroes of f, namely $\delta_j(\varphi) = \varphi(z_j)$, and the δ_k are placed at the poles z'_k of f; also m_j, and m'_k are the respective multiplicities.

[Hint: $\frac{1}{2\pi}\log|z|$ is a fundamental solution of $\triangle$.]

18. Prove that a distribution F is homogeneous of degree λ if and only if

$$\sum_{j=1}^{d} x_j \frac{\partial F}{\partial x_j} = \lambda F.$$

[Hint: For the converse, consider $\Phi(a) = F(\varphi^a)$ for $a > 0$, $\varphi \in \mathcal{D}$. Then $\Phi(a)$ is C^∞ for $a > 0$, and $\frac{d\Phi(a)}{da} = \frac{\lambda}{a}\Phi(a)$.]

19. Prove the following facts about distributions in $\mathbb{R}$.

(a) Given a distribution F, there exists a distribution F_1 so that

$$\frac{d}{dx}F_1 = F.$$

(b) Show that F_1 is unique modulo an additive constant.

[Hint: For (a) fix $\varphi_0 \in \mathcal{D}$, with $\int \varphi_0 = 1$, and note that each $\varphi \in \mathcal{D}$ can be written uniquely as $\varphi = \frac{d\psi}{dx} + a\varphi_0$ for some $\psi \in \mathcal{D}$ and a constant a. Then define $F_1(\varphi) = F(\psi)$. For (b), use the fact that d/dx is elliptic.]

20. Show that if $\lambda_1, \ldots, \lambda_d$ are distinct complex exponents and $\sum_{j=1}^{n}(a_j x^{\lambda_j} + b_j x^{\lambda_j}\log x) = 0$ for all $x > 0$, then $a_j = b_j = 0$ for all $1 \leq j \leq n$.

[Hint: Proceed as in the proof of Lemma 2.6, and use the fact that $\int_1^R x^{-1+i\mu_j}\,dx$ is equal to $\log R$ if $\mu_j = 0$ and that this integral is $O(1)$ if μ_j is real and $\neq 0$.]

21. Let $F(x,t) = \mathcal{H}_t(x)$, for $t > 0$, and $F(x,t) = 0$, when $t \leq 0$, as in Theorem 2.10. Prove directly that

$$\hat{F}(\xi,\tau) = \frac{1}{4\pi^2|\xi|^2 + 2\pi i\tau}$$

where $(\xi,\tau) \in \mathbb{R}^d \times \mathbb{R}$, with ξ dual to x, and τ dual to t.

[Hint: Use the two identities

$$\int_0^\infty e^{-4\pi^2|\xi|^2 t} e^{-2\pi i\tau t}\,dt = \frac{1}{4\pi^2|\xi|^2 + 2\pi i\tau} \qquad \text{for } |\xi| > 0$$

and

$$\int_{\mathbb{R}^d} \mathcal{H}_t(x) e^{-2\pi i x\cdot\xi}\,dx = e^{-4\pi^2|\xi|^2 t} \qquad \text{for } t > 0.]$$

22. Suppose f is a locally integrable function defined on $\mathbb{R}$, and let u be the function defined by $u(x,t) = f(x-t)$, for $(x,t) \in \mathbb{R}^2$. Verify that u, taken as a distribution, satisfies the wave equation

$$\frac{\partial^2 u}{\partial x^2} = \frac{\partial^2 u}{\partial t^2}.$$

More generally, let F be any distribution on $\mathbb{R}$. Construct U (in analogy to $f(x - t)$) as follows. If φ is in $\mathcal{D}(\mathbb{R}^2)$, $\mathbb{R}^2 = \{(x,t)\}$, set $U(\varphi) = \int_{\mathbb{R}} (F * \varphi(x,\cdot))(x)\,dx$. Then U satisfies

$$\frac{\partial^2 U}{\partial x^2} = \frac{\partial^2 U}{\partial t^2}.$$

Note that U is invariant under the translations (h,h), for $h \in \mathbb{R}$.

23. Show that in $\mathbb{R}^3$ the function

$$F(x) = \frac{-1}{4\pi|x|} e^{-|x|}$$

is a fundamental solution of the operator $\triangle - I$. The function F is the "Yukawa potential" in the theory of elementary particles. In contrast to the "Newtonian potential" $-1/(4\pi|x|)$, the fundamental solution of $\triangle$, the function F has a very rapid decay at infinity and it thus accounts for the short-range forces in the theory.

[Hint: Let F be the inverse Fourier transform of $-(1 + 4\pi^2|\xi|^2)^{-1}$. Going to polar coordinates in $\mathbb{R}^3$, one then uses the identity

$$\int_{|\xi|=1} e^{2\pi i\xi\cdot x}\,d\sigma(\xi) = \frac{2\sin(2\pi|x|)}{|x|},$$

together with the Fourier transform of the conjugate Poisson kernel, given by (18) of the previous chapter.]

24. The following statements deal with the uniqueness of the fundamental solutions of the Laplacian.

(a) Up to an additive constant, the unique fundamental solutions of $\triangle$ in $\mathbb{R}^d$, $d \geq 2$, that are rotationally invariant, are the ones given in Theorems 2.8 and 2.9.

(b) The unique fundamental solution of $\triangle$ in $\mathbb{R}^d$, $d \geq 3$, that vanishes at infinity is the one given in Theorem 2.8.

25. A distribution F defined on $\Omega \subset \mathbb{R}$ is **positive** if $F(\varphi) \geq 0$ for all $\varphi \in \mathcal{D}$ supported in Ω, with $\varphi \geq 0$. Show that F is positive if and only if $F(\varphi) = \int \varphi \, d\mu$ for some Borel measure $d\mu$ on Ω that is finite on compact subsets.

26. Recall that a real-valued function on (a, b) is **convex** if $f(x_0(1-t) + x_1 t) \leq (1-t)f(x_0) + tf(x_1)$, for $x_0, x_1 \in (a, b)$, $0 \leq t \leq 1$. (See also Problem 4 in Chapter 3, Book III.) A function f on $\Omega \subset \mathbb{R}^d$ is convex if the restriction of f to any line segment in Ω is convex.

(a) Suppose f is continuous on (a, b). Then it is convex if and only if the distribution $\frac{d^2 f}{dx^2}$ is positive.

(b) If f is continuous on $\Omega \subset \mathbb{R}^d$, it is convex if and only if for each $\xi = (\xi_1, \ldots, \xi_d) \subset \mathbb{R}^d$ the distribution $\sum_{1 \leq i,j, \leq d} \xi_j \xi_j \frac{\partial^2 f}{\partial x_i \partial x_j}$ is positive.

[Hint: For (a), let $\varphi \in \mathcal{D}$, $\varphi \geq 0$, $\int \varphi \, dx = 1$ and set $\varphi_\epsilon(x) = \epsilon^{-1}\varphi(x/\epsilon)$. Consider $f_\epsilon = f * \varphi_\epsilon$.]

27. Every distribution F of compact support in $\mathbb{R}^d$ is of **finite order** in the following sense: for each such F, there exists an integer M and continuous functions F_α of compact support, so that

$$F = \sum_{|\alpha| \leq M} \partial_x^\alpha F_\alpha.$$

Moreover if F is supported in C, then for every $\epsilon > 0$ we may take F_α to be supported in an ϵ-neighborhood of C. Prove this by carrying out the following three steps.

(a) Pick N so that $|F(\varphi)| \leq c\|\varphi\|_N$, for all $\varphi \in \mathcal{S}$, and choose M_0 so that $2M_0 > d + N$. Let Q be the inverse Fourier transform of $1/(1 + 4\pi^2|\xi|^2)^{M_0}$, and observe that Q is a fundamental solution of $(1 - \triangle)^{M_0}$, and Q is of class C^N.

(b) For each ϵ, construct Q_ϵ corresponding to Q, so that $(1-\triangle)^{M_0}Q_\epsilon = \delta + r_\epsilon$, where Q_ϵ is supported in the ϵ-neighborhood of the origin (as in Corollary 2.13). Prove that $F * Q_\epsilon$ is a continuous function, using the fact that $|F(\varphi)| \leq c\|\varphi\|_N$.

(c) Hence $F = (1-\triangle)^{M_0}(Q_\epsilon * F) - F * r_\epsilon$, and the result is proved with $M = 2M_0$.

28. One can characterize tempered distributions F whose Fourier transforms have compact support.

We already know by Proposition 1.6 that such an F must in fact be a function f that is C^∞ and slowly increasing. A precise characterization when $d = 1$ is given in the statement below.

The Fourier transform of a tempered distribution F is supported in the interval $[-M, M]$ if and only F equals a function f that is C^∞, slowly increasing, and having an analytic extension to the complex plane as an entire function of exponential type $2\pi M$; that is, for every $\epsilon > 0$, $|f(z)| \leq A_\epsilon e^{2\pi(M+\epsilon)|z|}$, where $z = x + iy$.

(An analogous assertion holds in higher dimensions.)

[Hint: Assume $\hat{F}$ is supported in $[-M, M]$. Using Exercise 27 allows us to write $\hat{F} = \sum_{|\alpha|N} \partial_x^\alpha(g_\alpha)$, where g_α are continuous and supported in $[-M-\epsilon, M+\epsilon]$, and thus reduce to the case when $\hat{F}$ is a continuous function.

To prove the converse, consider $f_\delta = f\gamma_\delta$ where $\gamma_\delta(x) = \frac{1}{\delta}\int e^{-2\pi i x\xi}\eta(\xi/\delta)\, d\xi$ with $\eta \in C^\infty$, supported in $|\xi| \leq 1$ and such that $\int \eta = 1$. Then $\gamma_\delta(z)$ is of exponential type $2\pi\delta$ and is rapidly decreasing on the real axis. Thus apply the simpler version of the result given in Theorem 3.3, Chapter 4 in Book II to the function f_δ, and let $\delta \to 0$.]

29. In this exercise, we consider the L^2 Sobolev spaces.

The space L^2_m consists of the functions $f \in L^2(\mathbb{R}^d)$ whose derivatives $\partial_x^\alpha f$ taken in the sense of distributions, are in $L^2(\mathbb{R}^d)$ for all $|\alpha| \leq m$. This space is sometimes denoted by $H_m(\mathbb{R}^d)$. Note that this is the special case for $p = 2$ of the Sobolev space given as an example in Section 3 of Chapter 1. However, here we use a slightly different (but equivalent) norm, which makes L^2_m into a Hilbert space.

On L^2_m we define the inner product

$$(f, g)_m = \sum_{|\alpha|\leq m} (\partial_x^\alpha f, \partial_x^\alpha g)_0,$$

with $(f, g)_0 = \int_{\mathbb{R}^d} f(x)\overline{g(x)}\, dx$. Then, L^2_m with the norm $\|f\|_{L^2_m} = (f, f)_m^{1/2}$ is a Hilbert space.

(a) Verify that $f \in L^2_m$ if and only if $\hat{f}(\xi)(1+|\xi|)^m \in L^2$, and that the norms $\|f\|_{L^2_m}$ and $\|\hat{f}(\xi)(1+|\xi|)^m\|_{L^2}$ are equivalent.

(b) If $m > d/2$, then f can be corrected on a set of measure zero, so that f becomes continuous and is in fact in C^k, for $k < m - d/2$. This is a version of the Sobolev embedding theorem.

[Hint: $\xi^\alpha \hat{f} \in L^1(\mathbb{R}^d)$ if $|\alpha| < m - d/2$.]

30. The following observation is useful in connection with the L^2 theory of Calderón-Zygmund distributions on $\mathbb{R}^d$.

(a) The Fourier transform of the distribution $\left[\frac{1}{|x|^d}\right]$ equals $c_1 \log|\xi| + c_2$, with $c_1 \neq 0$.

(b) Prove the following consequence of (a). Suppose k is a homogeneous function of degree $-d$ that is C^∞ away from the origin and with

$$\int_{|x|=1} k(x)\, d\sigma(x) \neq 0.$$

If K is any distribution that agrees with k away from the origin, then the Fourier transform of K is not a bounded function. Another way of stating this is that the operator T, defined by $T(\varphi) = K * \varphi$ initially defined for $\varphi \in \mathcal{D}$, does not extend to a bounded operator on $L^2(\mathbb{R}^d)$.

31. Suppose k is a C^∞ function homogenenous of degree $-d$, not identically equal to zero, and

$$\int_{|x|=1} k(x)\, d\sigma(x) = 0.$$

If K is the principal value distribution defined by k, that is, $K = \mathrm{pv}(k)$, then K is a Calderón-Zygmund distribution but the operator T given by $Tf = f * K$ is not bounded on L^1 or L^∞.

The special case of the Hilbert transform is in Exercise 7, Chapter 2.

[Hint: If $\varphi \in \mathcal{D}$, then $T\varphi(x) = ck(x) + O(|x|^{-d-1})$ as $|x| \to \infty$, where $c = \int \varphi$.]

32. The cancelation condition (21) for the Calderón-Zygmund distributions for some $n > 1$ implies the condition for $n = 1$. Show this by first proving the following fact: Whenever K satisfies (20) and (21) for some $n \geq 1$, then for every $1 \leq j \leq d$, the distribution $x_j \cdot K$ equals the locally integrable function $x_j k$.

[Hint: The distribution $x_j K - x_j k$ is supported at the origin. Then use Theorem 1.7 to test $x_j K - x_j k$ against φ_r as $r \to 0$ for suitable φ, to conclude that this difference vanishes. Next, write any $C^{(1)}$-normalized bump function as $\varphi(x) = \eta(x) + \sum_{j=1}^d x_j \varphi_j(x)$ where η and the φ_j are multiples of $C^{(n)}$ and $C^{(0)}$-normalized bump functions respectively, and use the above fact.]

33. Suppose k is a C^∞ function in $\mathbb{R}^d - \{0\}$, that satisfies the differential inequalities (20). Then there is a Calderón-Zygmund distribution K which has k as its associated function if and only if $\sup_{0<a<b} \left| \int_{a<|x|<b} k(x)\, dx \right| < \infty$.

[Hint: In one direction, note that $|K(\eta_b - \eta_a)| \leq 2A$, where $\eta(x) = 1$ if $|x| \leq 1/2$, and $\eta(x) = 0$ if $|x| \geq 1$, with $\eta \in C^\infty$. In the other direction, define

$$K(\varphi) = \int_{|x|\leq 1} k(x)(\varphi(x) - \varphi(0))\, dx + \int_{|x|\geq 1} k(x)\varphi(x)\, dx$$

and verify that the conditions (20) and (21) hold for K.]

34. Suppose K is a Calderón-Zygmund distribution and η belongs to $\mathcal{S}$. Verify that ηK is a Calderón-Zygmund distribution.

5 Problems

1. We consider **periodic distributions** and their Fourier series.

(a) The notion of a periodic distribution on $\mathbb{R}^d$ can be defined in two equivalent ways:

First, one can consider distributions F on $\mathbb{R}^d$ which are periodic in the sense that $\tau_h(F) = F$ for all $h \in \mathbb{Z}^d$;

Alternatively, one can consider the continuous linear functionals on $\mathcal{D}(\mathbb{T}^d)$, the space of C^∞ periodic functions on $\mathbb{R}^d$. (Here $\mathbb{T}^d = \mathbb{R}^d/\mathbb{Z}^d$ denotes the d-dimensional torus.)

(b) Note that if $\varphi \in \mathcal{D}(\mathbb{T}^d)$, then φ has a Fourier series expansion

$$\varphi(x) = \sum_n a_n e^{2\pi i n\cdot x},$$

where the Fourier coefficients $a_n = \int_{\mathbb{T}^d} f(x)e^{-2\pi i n\cdot x}\,dx$ are rapidly decreasing, that is, for every $N > 0$, $|a_n| \leq O(|n|^{-N})$ as $|n| \to \infty$.

Similarly, if F is a periodic distribution, and $a_n = F(e^{-2\pi i n x})$ denote its Fourier coefficients, then a_n are slowly increasing in the sense that for some $N > 0$, $|a_n| \leq O(|n|^N)$ as $|n| \to \infty$.

Moreover, the Fourier series $\sum a_n e^{2\pi i n x}$ converges to F in the sense of distributions.

[Hint: To prove the equivalence in (a), consider the "periodization" operator $P : \mathcal{D}(\mathbb{R}^d) \to \mathcal{D}(\mathbb{T}^d)$,

$$P(\varphi)(x) = \sum_{h\in\mathbb{Z}^d} \tau_h(\varphi)(x) = \sum_{h\in\mathbb{Z}^d} \varphi(x-h).$$

Then find $\gamma \in \mathcal{D}(\mathbb{R}^d)$ so that $P(\gamma) = 1$. This allows to prove that P is surjective, and that, in the same way, its dual $P^* : \mathcal{D}_2^* \to \mathcal{D}_1^*$ is also surjective. (Here $\mathcal{D}_1^*$ and $\mathcal{D}_2^*$ denote, respectively, the two spaces of distributions described in (a).) To construct γ, pick $\psi \in \mathcal{D}(\mathbb{R}^d)$ so that $\psi \geq 0$ and $\psi = 1$ on $\{0 \leq x_j < 1,\ \ 1 \leq j \leq d\}$, and let $\gamma = \psi/P(\psi)$.]

2. Suppose $Tf = f * K$ is a singular integral operator as in Theorem 3.2 of Section 3. Then the mapping $f \mapsto T(f)$ is bounded on the Hardy space $\mathbf{H}_r^1$, and in particular maps $\mathbf{H}_r^1$ to L^1.

[Hint: Consider first a 2-atom $\mathfrak{a}$ associated to the unit ball B. Then for an appropriate constant c (bounded independently of $\mathfrak{a}$) we have $T(\mathfrak{a}) = c(a_* + \Phi)$. Here a_* is a 2-atom for the ball $B^a st$, the double of B, and Φ satisfies $|\Phi(x)| \leq (1+|x|)^{-d-1}$, $\int_{\mathbb{R}^d} \Phi(x)\, dx = 0$. With this apply Exercise 21 in Chapter 2. Then obtain the analog for 2-atoms $\mathfrak{a}$, after rescaling and translation.]

3. Prove the following interior estimates for an elliptic operator L of order m with constant coefficients.

Suppose $\mathcal{O}$ and $\mathcal{O}_1$ are bounded subsets of $\mathbb{R}^d$ with $\overline{\mathcal{O}} \subset \mathcal{O}_1$. Assume u and f are L^p functions in $\mathcal{O}_1$ with $Lu = f$ in $\mathcal{O}_1$ in the sense of distributions. Then if $1 < p < \infty$ and k is a non-negative integer, we have

$$\sum_{|\alpha| \leq m+k} \|\partial_x^\alpha u\|_{L^p(\mathcal{O})} \leq c \left(\sum_{|\beta| \leq k} \|\partial_x^\beta f\|_{L^p(\mathcal{O}_1)} + \|u\|_{L^p(\mathcal{O}_1)} \right)$$

where the derivatives are taken in the sense of distributions.

[Hint: Consider the parametrix $Q_\epsilon = \eta_\epsilon Q$ given in Corollary 2.13 which is supported in $|x| \leq \epsilon$. Here ϵ is chosen so that $\overline{\mathcal{O}}_\epsilon \subset \mathcal{O}_1$, where $\mathcal{O}_\epsilon$ are the points of distance $\leq \epsilon$ from $\mathcal{O}$.

Set $U = \psi u$, with ψ a C^∞ function that is 1 near $\overline{\mathcal{O}}_\epsilon$ but vanishes outside $\mathcal{O}_1$. Then

$$L(U) = \psi L(u) + \sum_{|\gamma| < m} \psi_\gamma \partial_x^\gamma u,$$

and what is important is that the ψ_γ vanishes in $\overline{\mathcal{O}}_\epsilon$. Now $U + r_\epsilon * U = Q_\epsilon * L(U)$, where $r_\epsilon \in \mathcal{S}$. This gives

$$\psi u = Q_\epsilon * (\psi f) - r_\epsilon(\psi u) + \sum_r Q_\epsilon * (\psi_\gamma \partial_x^\gamma u).$$

As has been pointed out, $\partial_x^\alpha Q$ are Calderón-Zygmund distributions whenever $|\gamma| \leq m$, so the same is true for Q_ϵ. Then using Theorem 3.2, the result follows.]

4.* Let $P(x)$ be any real polynomial in $\mathbb{R}^d$, and k a homogeneous function of degree $-d$ with $\int_{|x|=1} k(x)\, d\sigma(x) = 0$.

(a) One can define the tempered distribution $\text{pv}\left(e^{iP(x)}k(x)\right) = K$ by

$$K(\varphi) = \lim_{\epsilon \to 0} \int_{|x| \geq \epsilon} e^{iP(x)} k(x) \varphi(x)\, dx.$$

(b) Then the Fourier transform of K is a bounded function (with bound independent of the coefficients of P).

5.* Let Q be a fixed real-valued polynomial on $\mathbb{R}^d$. Consider the distributions initially defined for $\text{Re}(s) > 0$ by

$$I(s)(\varphi) = \int_{Q(x)>0} |Q(x)|^s \varphi(x)\, dx, \qquad \text{where } \varphi \in \mathcal{S}.$$

Then $I(s)(\varphi)$ has a meromorphic continuation to the whole complex s-plane, with poles at most at $s = -k/m$, where m is a positive integer determined by Q, and k is any positive integer. The order of the poles do not exceed d.

6.* As a consequence of the results in Problem 5*, one may prove the following.

(a) Suppose $L = \sum_{|\alpha|\leq m} a_\alpha \partial_x^\alpha$ is a non-zero partial differential operator on $\mathbb{R}^d$ with a_α complex constants. Then L has a *tempered* fundamental solution. As an immediate corollary we also have:

(b) Suppose P is a complex-valued polynomial on $\mathbb{R}^d$. Then there exists a tempered distribution F that agrees with $1/P$ where $P(x) \neq 0$.

In fact, let P be the characteristic polynomial of L and apply the result of the previous problem to $Q = |P|^2$. Suppose $I(s)$ has a pole of order r at $s = 1$, then define the tempered distribution F by

$$F = \overline{P}\,\frac{1}{r!}\frac{d^r}{ds^r}(s+1)^r I(s)\Big|_{s=-1}.$$

Consequently, $PF = 1$, and the inverse Fourier transform of F is the desired fundamental solution of L.

7.* Suppose $L = \sum_{|\alpha|\leq m} a_\alpha \partial_x^\alpha$ is a partial differential operator on $\mathbb{R}^d$, with a_α complex constants. Then L is hypo-elliptic if and only if for each $\alpha \neq 0$

$$\frac{\partial_\xi^\alpha P(\xi)}{P(\xi)} \to 0 \qquad \text{as } |\xi| \to \infty,$$

where P denotes the characteristic polynomial of L.

8.* We describe several fundamental solutions of the wave operator

$$\Box = \frac{\partial^2}{\partial t^2} - \triangle_x,$$

where $(x,t) \in \mathbb{R}^d \times \mathbb{R}$ and $\triangle_x = \sum_{j=1}^d \frac{\partial^2}{\partial x_j^2}$.

We let Γ_+ be the forward open cone $= \{(x,t):\ t > |x|\}$, and $\Gamma_- = -\Gamma_+$, the backward cone. For each s with $\mathrm{Re}(s) > -1$ we define the function F_s by

$$F_s(x,t) = \begin{cases} a_s(t^2 - |x|^2)^{s/2}, & \text{if } (x,t) \in \Gamma_+ \\ 0 & \text{otherwise.} \end{cases} \tag{29}$$

Here $a_s^{-1} = 2^{s+d}\pi^{\frac{d-1}{2}}\Gamma\left(\frac{s+d+1}{2}\right)\Gamma(s/2+1)$. Then $s \mapsto F_s$ has an analytic continuation in the complex s plane as an entire (tempered) distribution-valued function. Moreover, one can prove that $F_+ = F_s|_{s=-d+1}$ is a fundamental solution of $\Box$.

Note that F_-, obtained from F_+ by mapping $t \mapsto -t$, is also a fundamental solution, and F_+ and F_- are supported in $\overline{\Gamma}_+$ and $\overline{\Gamma}_-$ respectively. In addition, if d is odd and $d \geq 3$, then a_s vanishes for $s = -d + 1$, so both F_+ and F_- are supported on the boundary of their cones, which is a reflection of the Huygens' principle.

Finally, a third fundamental solution F_0 of interest is given by

$$F_0^\wedge = \lim_{\epsilon \to 0,\, \epsilon > 0} \frac{1}{4\pi^2} \left(\frac{1}{|\xi|^2 - \tau^2 + i\epsilon} \right),$$

with the limit taken in the sense of distribution, and (ξ, τ) representing the dual variables to (x, t). The fundamental solutions F_+, F_-, and F_0 are each homogeneous of degree -2, and invariant under the Lorentz group of linear transformations of determinant 1 that preserves Γ_+. Also each fundamental solution of $\Box$ with these invariance properties can be written as $c_1 F_+ + c_2 F_- + c_3 F_0$, with $c_1 + c_2 + c_3 = 1$.

4 Applications of the Baire Category Theorem

> We see the profound difference that lies between sets of the two categories; this difference lies not within denumerability, nor within density, since a set of the first category can have the power of the continuum and can also be dense in any interval one considers; but it is in some sense a combination of these two preceding notions.
>
> *R. Baire,* 1899

In the late nineteenth century, Baire introduced in his doctoral dissertation a notion of size for subsets of the real line which has since provided many fascinating results. In fact, his careful study of functions led him to the definition of the first and second category of sets. Roughly speaking, sets of the first category are "small," while sets of the second category are "large." In this sense the complement of a set of the first category is "generic."

Over time the Baire category theorem has been applied to metric spaces in different and more abstract settings. Its noteworthy use has been to show that a number of phenomena in analysis, found first in specific counter-examples, are in fact generic occurrences.

This chapter is organized as follows. We begin by stating and proving the Baire category theorem, and then proceed with the presentation of a variety of interesting applications. We start with the result about continuous functions which Baire proved in his thesis: a pointwise limit of continuous functions has itself "many" points of continuity. Also, we shall prove the existence of a continuous but nowhere differentiable function, as well as the existence of a continuous function with Fourier series diverging at a point, by showing that the category theorem allows us to see that such functions are indeed generic. We also deduce from Baire's theorem two further general results, the open mapping and closed graph theorems, and provide in each case an example of their use. Finally, we apply the category theorem to show that a Besicovitch-Kakeya set is generic in a natural class of subsets of $\mathbb{R}^2$.

1 The Baire category theorem

Although Baire proved his theorem on the real line, his result actually holds in the more general setting of complete metric spaces. For the purpose of the applications we have in mind it is better to have access to this more general formulation right away. Fortunately, the proof of the theorem remains very simple and elegant.

To state the main result, we begin with a list of definitions. Let X be a metric space with metric d, carrying the natural topology induced by d. In other words, a set $\mathcal{O}$ in X is open if for every $x \in \mathcal{O}$ there exists $r > 0$ so that $B_r(x) \subset \mathcal{O}$, where $B_r(x)$ denotes the open ball centered at x and of radius r,

$$B_r(x) = \{y \in X : \ d(x, y) < r\}.$$

By definition, a set is closed if its complement is open.

We define the **interior** E° of a set $E \subset X$ to be the union of all open sets contained in E. Also, the **closure** $\overline{E}$ of E is the intersection of all closed sets containing E. Since one checks easily that the union of any collection of open sets is open, and the intersection of any collection of closed sets is closed, we see that E° is the "largest" open set contained in E, and $\overline{E}$ is the "smallest" closed set containing E.

Suppose E is a subset of X. We say that the set E is **dense** in X if $\overline{E} = X$. Also, the set E is **nowhere dense** if the interior of its closure is empty, $(\overline{E})^\circ = \emptyset$. For instance, any point in $\mathbb{R}^d$ is nowhere dense in $\mathbb{R}^d$. Also, the Cantor set is nowhere dense in $\mathbb{R}$, but the rationals $\mathbb{Q}$ are not since $\overline{\mathbb{Q}} = \mathbb{R}$. We note here that in general E is closed and nowhere dense if and only $\mathcal{O} = E^c$ is open and dense.

We now describe the central notion of category due to Baire, and the dichotomy it introduces.

- A set $E \subset X$ is of the **first category** in X if E is a countable union of nowhere dense sets in X. A set of the first category is sometimes said to be "meager." A set E that is not of the first category in X is referred to as being of the **second category** in X.

- A set $E \subset X$ is defined to be **generic** if its complement is of the first category.

Thus the idea of category is to describe "smallness" in purely topological terms (involving closures, interiors, etc.) It reflects the idea that elements of a set of the first category are to be thought of as "exceptional," while those of a generic set are to be considered "typical." Connected with this is the fact that a countable union of sets of the first category is of the first category, while the countable intersection of generic sets is a generic set. Also we record here the useful fact that any open dense set is generic (this follows from our remark earlier).

In general relying on one's intuition about the category of sets requires a little caution. For instance, there is no link between this notion and that of Lebesgue measure. Indeed, there are sets in $[0,1]$ of the first category that are of full measure, and hence uncountable and dense. By the same token, there are generic sets of measure zero. (Some examples are discussed in Exercise 1.)

The main result of Baire is that "the continuum is of the second category." The key ingredient used in his argument is the fact that the real line is complete. This is the main reason why his theorem immediately carries over to the case of a complete metric space.

Theorem 1.1 *Every complete metric space X is of the second category in itself, that is, X cannot be written as the countable union of nowhere dense sets.*

Corollary 1.2 *In a complete metric space, a generic set is dense.*

Proof of the theorem. We argue by contradiction, and assume that X is a countable union of nowhere dense sets F_n,

$$X = \bigcup_{n=1}^{\infty} F_n. \tag{1}$$

By replacing each F_n by its closure, we may assume that each F_n is closed. It now suffices to find a point $x \in X$ with $x \notin \bigcup F_n$.

Since F_1 is closed and nowhere dense, hence not all of X, there exists an open ball B_1 of some radius $r_1 > 0$ whose closure $\overline{B_1}$ is entirely contained in F_1^c.

Since F_2 is closed and nowhere dense, the ball B_1 cannot be entirely contained in F_2, otherwise F_2 would have a non-empty interior. Since F_2 is also closed, there exists a ball B_2 of some radius $r_2 > 0$ whose closure $\overline{B_2}$ is contained in B_1 and also in F_2^c. Clearly, we may choose r_2 so that $r_2 < r_1/2$.

Continuing in this fashion, we obtain a sequence of balls $\{B_n\}$ with the following properties:

(i) The radius of B_n tends to 0 as $n \to \infty$.

(ii) $B_{n+1} \subset B_n$.

(iii) $F_n \cap \overline{B_n}$ is empty.

Choose any point x_n in B_n. Then, $\{x_n\}_{n=1}^{\infty}$ is a Cauchy sequence because of properties (i) and (ii) above. Since X is complete, this sequence converges to a limit which we denote by x. By (ii) we see that $x \in \overline{B_n}$ for each n, and hence $x \notin F_n$ for all n by (iii). This contradicts (1), and the proof of the Baire category theorem is complete.

To prove the corollary, we argue by contradiction and assume that $E \subset X$ is generic but not dense. Then there exists a closed ball $\overline{B}$ entirely contained in E^c. Since E is generic we can write $E^c = \bigcup_{n=1}^{\infty} F_n$ where each F_n is nowhere dense, hence

$$\overline{B} = \bigcup_{n=1}^{\infty} (F_n \cap \overline{B}).$$

It is clear that $F_n \cap \overline{B}$ is nowhere dense, hence the above contradicts Theorem 1.1 applied to the complete metric space $\overline{B}$, and the corollary is proved.

The theorem actually extends to certain cases of metric spaces that are not complete, in particular to open subsets of a complete metric space. To be precise, suppose we are given a subset X_0 of a complete metric space X. Then X_0 is itself a metric space, inheriting its metric from X by restricting the metric on X to X_0. The fact is that if X_0 is an open subset of X, then the conclusion of the theorem holds for it; that is, X_0 cannot be written as a countable union of sets that are nowhere dense (in X_0). See Exercise 3. A simple example is given by the open interval $(0, 1)$ with the usual metric.

1.1 Continuity of the limit of a sequence of continuous functions

Suppose X is a complete metric space, $\{f_n\}$ is a sequence of continuous complex-valued functions on X, and that the limit

$$\lim_{n\to\infty} f_n(x) = f(x)$$

exists for each $x \in X$. It is well known that if the limit is uniform in x, then the limiting function f is also continuous. In general, when the limit is just pointwise, we may ask: must f have at least one point of continuity? We answer this question affirmatively with a simple application of the category theorem.

Theorem 1.3 *Suppose that $\{f_n\}$ is a sequence of continuous complex-valued functions on a complete metric space X, and*

$$\lim_{n\to\infty} f_n(x) = f(x)$$

exists for every $x \in X$. Then, the set of points where f is continuous is a generic set in X. In other words, the set of points where f is discontinuous is of the first category.

Therefore f is in fact continuous at "most" points of X.

To show that the set $\mathcal{D}$ of discontinuities of f is of the first category, we use a characterization of points of continuity of f in terms of its oscillations. More precisely, we define the **oscillation** of the function f at a point x by

$$\operatorname{osc}(f)(x) = \lim_{r\to 0} \omega(f)(r,x), \text{ where } \omega(f)(r,x) = \sup_{y,z\in B_r(x)} |f(y) - f(z)|.$$

The limit exists since the quantity $\omega(f)(r,x)$ decreases with r. In particular, we see that $\operatorname{osc}(f)(x) < \epsilon$ if there exists a ball B centered at x so that $|f(y) - f(z)| < \epsilon$ whenever $y, z \in B$. Two more observations are in order:

(i) $\operatorname{osc}(f)(x) = 0$ if and only if f is continuous at x.

(ii) The set $E_\epsilon = \{x \in X : \operatorname{osc}(f)(x) < \epsilon\}$ is open.

Property (i) follows immediately from the definition of continuity. For (ii), we note that if $x \in E_\epsilon$, there is an $r > 0$ so that $\sup_{y,z\in B_r(x)} |f(y) - f(z)| < \epsilon$. Consequently, if $x^* \in B_{r/2}(x)$, then $x^* \in E_\epsilon$ because

$$\sup_{y,z\in B_{r/2}(x^*)} |f(y) - f(z)| \le \sup_{y,z\in B_r(x)} |f(y) - f(z)| < \epsilon.$$

Lemma 1.4 *Suppose $\{f_n\}$ is a sequence of continuous functions on a complete metric space X, and $f_n(x) \to f(x)$ for each x as $n \to \infty$. Then, given an open ball $B \subset X$ and $\epsilon > 0$, there exists an open ball $B_0 \subset B$ and an integer $m \ge 1$ so that $|f_m(x) - f(x)| \le \epsilon$ for all $x \in B_0$.*

Proof. Let Y denote a closed ball contained in B. Note that Y is itself a complete metric space. Define

$$E_\ell = \{x \in Y : \sup_{j,k\ge \ell} |f_j(x) - f_k(x)| \le \epsilon\}.$$

Then, since $f_n(x)$ converges for every $x \in X$, we must have

$$Y = \bigcup_{\ell=1}^{\infty} E_\ell. \tag{2}$$

Moreover, each E_ℓ is closed since it is the intersection of sets of the type $\{x \in Y : |f_j(x) - f_k(x)| \leq \epsilon\}$ which are closed by the continuity of f_j and f_k. Therefore, by Theorem 1.1 applied to the complete metric space Y, some set in the union (2), say E_m, must contain an open ball B_0. By construction,

$$\sup_{j,k \geq m} |f_j(x) - f_k(x)| \leq \epsilon \qquad \text{whenever } x \in B_0,$$

and letting k tend to infinity we find that $|f_m(x) - f(x)| \leq \epsilon$ for all $x \in B_0$. This proves the lemma.

To finish the proof of Theorem 1.3, we define

$$F_n = \{x \in X : \ \operatorname{osc}(f)(x) \geq 1/n\},$$

in other words, $F_n = E_\epsilon^c$ with $\epsilon = 1/n$ in the notation of (ii) above.

Then, by our observation (i), we have

$$\mathcal{D} = \bigcup_{n=1}^{\infty} F_n,$$

where we recall that $\mathcal{D}$ is the set of discontinuities of f. The theorem will be proved if we can show that each F_n is nowhere dense.

Fix $n \geq 1$. Since F_n is closed, we must show that it has empty interior. Assume on the contrary, that B is an open ball with $B \subset F_n$. Then, if we set $\epsilon = 1/4n$ in the lemma, we find that there is an open ball $B_0 \subset B$, and an integer $m \geq 1$ so that

$$|f_m(x) - f(x)| \leq 1/4n, \qquad \text{for all } x \in B_0. \tag{3}$$

By the continuity of f_m, we may find a ball $B' \subset B_0$ so that

$$|f_m(y) - f_m(z)| \leq 1/4n, \qquad \text{for all } y, z \in B'. \tag{4}$$

Then, the triangle inequality implies

$$|f(y) - f(z)| \leq |f(y) - f_m(y)| + |f_m(y) - f_m(z)| + |f_m(z) - f(z)|.$$

If $y, z \in B'$, the first and third terms are bounded by $1/4n$ because of condition (3). The middle term is also bounded by $1/4n$ due to (4). Therefore

$$|f(y) - f(z)| \leq \frac{3}{4n} < \frac{1}{n} \quad \text{whenever } y, z \in B'.$$

Consequently, if x' denotes the center of B', we have $\text{osc}(f)(x') < 1/n$ which contradicts the fact that $x' \in F_n$. This concludes the proof of the theorem.

1.2 Continuous functions that are nowhere differentiable

Our next application of the category theorem is to the problem of the existence of a continuous function that is nowhere differentiable.

Our first answer to this question appeared in Chapter 4 of Book I where we showed that the complex-valued function f given by the following lacunary Fourier series

$$f(x) = \sum_{n=0}^{\infty} 2^{-n\alpha} e^{i2^n x} \quad \text{with } 0 < \alpha \leq 1$$

is continuous but nowhere differentiable. Moreover, a slight change in the proof shows that both the real and imaginary parts of f are also nowhere differentiable. Other examples arose in Chapter 7 of Book III, in the context of the von Koch and space-filling curves.

Here, we prove the existence of such functions by showing that they are generic in an appropriate complete metric space. The space we have in mind consists of all real-valued continuous functions on $[0, 1]$, which we denote by

$$X = C([0, 1]).$$

This vector space is equipped with the sup-norm

$$\|f\| = \sup_{x \in [0,1]} |f(x)|.$$

Together with this norm, $C([0, 1])$ is a complete normed vector space (a Banach space). The completeness follows because the uniform limit of a sequence of continuous functions is necessarily continuous. Finally, the metric d on X is chosen to be $d(f, g) = \|f - g\|$, and hence (X, d) is a complete metric space.

Theorem 1.5 *The set of functions in $C([0, 1])$ that are nowhere differentiable is generic.*

We must show that the set $\mathcal{D}$, of continuous functions in $[0,1]$ that are differentiable at least at one point, is of the first category. To this end, we let E_N denote the set of all continuous functions so that there exists $0 \le x^* \le 1$ with

$$|f(x) - f(x^*)| \le N|x - x^*|, \qquad \text{for all } x \in [0,1]. \tag{5}$$

These sets are related to $\mathcal{D}$ by the inclusion

$$\mathcal{D} \subset \bigcup_{N=1}^{\infty} E_N.$$

To prove the theorem it suffices to show that for each N, the set E_N is nowhere dense. This will be achieved by showing successively:

(i) E_N is a closed set.

(ii) the interior of E_N is empty.

Thus $\bigcup E_N$ is of the first category, hence so is the set $\mathcal{D}$.

Proof of property (i)

Suppose that $\{f_n\}$ is a sequence of functions in E_N so that $\|f_n - f\| \to 0$. We must show that $f \in E_N$. Let x_n^* be a point in $[0,1]$ for which (5) holds with f replaced by f_n. We may choose a subsequence $\{x_{n_k}^*\}$ that converges to a limit in $[0,1]$, which we denote by x^*. Then,

$$|f(x) - f(x^*)| \le |f(x) - f_{n_k}(x)| + |f_{n_k}(x) - f_{n_k}(x^*)| + |f_{n_k}(x^*) - f(x^*)|.$$

On the one hand, since $\|f_n - f\| \to 0$, we see that given $\epsilon > 0$, there exists $K > 0$ so that whenever $k > K$ the first and third terms together are $< \epsilon$. On the other hand, we may estimate the middle term by

$$|f_{n_k}(x) - f_{n_k}(x^*)| \le |f_{n_k}(x) - f_{n_k}(x_{n_k}^*)| + |f_{n_k}(x_{n_k}^*) - f_{n_k}(x^*)|.$$

Therefore, applying the fact that $f_{n_k} \in E_N$ twice yields

$$|f_{n_k}(x) - f_{n_k}(x^*)| \le N|x - x_{n_k}^*| + N|x_{n_k}^* - x^*|.$$

Putting all these estimates together, we obtain

$$|f(x) - f(x^*)| \le \epsilon + N|x - x_{n_k}^*| + N|x_{n_k}^* - x^*|$$

for all $k > K$. Letting k tend to infinity, and recalling that $x_{n_k}^* \to x^*$ we get

$$|f(x) - f(x^*)| \le \epsilon + N|x - x^*|.$$

Since ϵ is arbitrary, we conclude that $f \in E_N$, and (i) is proved.

Proof of property (ii)

To show that E_N has no interior, let $\mathcal{P}$ denote the subspace of $C([0,1])$ that consists of all continuous piecewise-linear functions. Also, for each $M > 0$, let $\mathcal{P}_M \subset \mathcal{P}$ denote the set of all continuous piecewise-linear functions, each of whose line segments have slopes either $\geq M$ or $\leq -M$. Functions in $\mathcal{P}_M$ are naturally called "zig-zag" functions. Note the key fact that $\mathcal{P}_M$ is disjoint from E_N if $M > N$.

Lemma 1.6 *For every $M > 0$, the set $\mathcal{P}_M$ of zig-zag functions is dense in $C([0,1])$.*

Proof. It is plain that given $\epsilon > 0$ and a continuous function f, there exists a function $g \in \mathcal{P}$ so that $\|f - g\| \leq \epsilon$. Indeed, since f is continuous on the compact set $[0,1]$ it must be uniformly continuous, and there exists $\delta > 0$ so that $|f(x) - f(y)| \leq \epsilon$ whenever $|x - y| < \delta$. If we choose n so large that $1/n < \delta$, and define g as a linear function on each interval $[k/n, (k+1)/n]$ for $k = 0, \ldots, n-1$ with $g(k/n) = f(k/n)$, $g((k+1)/n) = f((k+1)/n)$, we see at once that $\|f - g\| \leq \epsilon$.

It now suffices to see how to approximate g on $[0,1]$ by zig-zag functions in $\mathcal{P}_M$. Indeed, if g is given by $g(x) = ax + b$ for $0 \leq x \leq 1/n$, consider the two segments

$$\varphi_\epsilon(x) = g(x) + \epsilon \quad \text{and} \quad \psi_\epsilon(x) = g(x) - \epsilon.$$

Then, beginning at $g(0)$, we travel on a line segment of slope $+M$ until we intersect φ_ϵ. Then, we reverse direction and travel on a line segment of slope $-M$ until we intersect ψ_ϵ (see Figure 1).

We obtain $h \in \mathcal{P}_M$ so that

$$\psi_\epsilon(x) \leq h(x) \leq \varphi_\epsilon(x), \quad \text{for all } 0 \leq x \leq 1/n,$$

and therefore $|h(x) - g(x)| \leq \epsilon$ in $[0, 1/n]$.

Then, we begin at $h(1/n)$ and repeat this argument on the interval $[1/n, 2/n]$. Continuing in this fashion, we obtain a function $h \in \mathcal{P}_M$ with $\|h - g\| \leq \epsilon$. Hence $\|f - h\| \leq 2\epsilon$, and the lemma is proved.

We deduce at once from this lemma that E_N has no interior points. Indeed, given any $f \in E_N$ and $\epsilon > 0$, we first choose a fixed $M > N$. Then, there exists $h \in \mathcal{P}_M$ so that $\|f - h\| < \epsilon$, and moreover $h \notin E_N$ since $M > N$. Therefore, no open ball around f is entirely contained in E_N, which is the desired conclusion. Theorem 1.5 is proved.

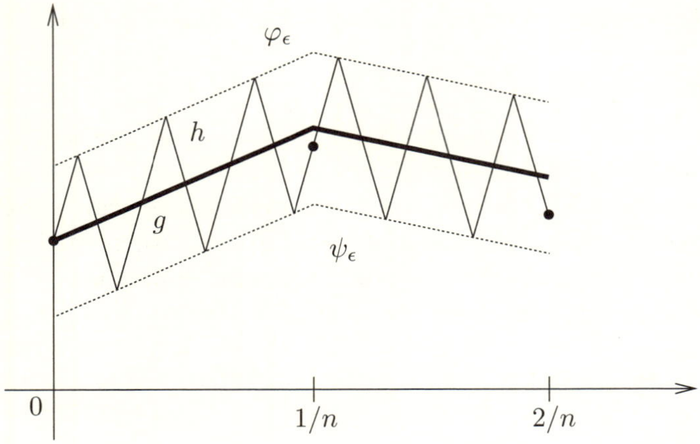

Figure 1. Approximation by $\mathcal{P}_M$

2 The uniform boundedness principle

Next, we turn to another corollary of Baire's theorem, one that itself has many applications. The main conclusion we find is that if a sequence of continuous linear functionals is pointwise bounded on a "large" set, then this sequence must in fact be bounded.

Theorem 2.1 *Suppose that $\mathcal{B}$ is a Banach space, and $\mathcal{L}$ is a collection of continuous linear functionals on $\mathcal{B}$.*

(i) *If* $\sup_{\ell \in \mathcal{L}} |\ell(f)| < \infty$ *for each* $f \in \mathcal{B}$, *then*

$$\sup_{\ell \in \mathcal{L}} \|\ell\| < \infty.$$

(ii) *This conclusion also holds if we only assume that* $\sup_{\ell \in \mathcal{L}} |\ell(f)| < \infty$ *for all* f *in some set of the second category.*

We note that the collection $\mathcal{L}$ need not be countable.

Proof. It suffices to show (ii) since by Baire's theorem, $\mathcal{B}$ is of the second category. So suppose that $\sup_{\ell \in \mathcal{L}} |\ell(f)| < \infty$ for all $f \in E$, where E is of the second category.

For each positive integer M, define

$$E_M = \{f \in \mathcal{B} : \sup_{\ell \in \mathcal{L}} |\ell(f)| \leq M\}.$$

Then, the hypothesis in the theorem guarantees that

$$E = \bigcup_{M=1}^{\infty} E_M.$$

Moreover, each E_M is closed, since it can be written as an intersection $E_M = \bigcap_{\ell \in \mathcal{L}} E_{M,\ell}$, where $E_{M,\ell} = \{f : |\ell(f)| \leq M\}$ is closed by the continuity of ℓ. Since E is of the second category, some E_M must have non-empty interior, say when $M = M_0$. In other words, there exists $f_0 \in \mathcal{B}$, and $r > 0$ so that $B_r(f_0) \subset E_{M_0}$. Hence for all $\ell \in \mathcal{L}$ we have

$$|\ell(f)| \leq M_0 \qquad \text{whenever } \|f - f_0\| < r.$$

As a result, for all $\|g\| < r$, and all $\ell \in \mathcal{L}$ we have

$$\|\ell(g)\| \leq \|\ell(g + f_0)\| + \|\ell(-f_0)\| \leq 2M_0,$$

and this implies the conclusion (ii) in the theorem.

2.1 Divergence of Fourier series

We now consider the problem of the existence of a continuous function whose Fourier series diverges at a point.

In Book I we gave an explicit construction of a function with this property. The main idea there was to break the symmetry inherent in the Fourier series $\sum_{|n| \neq 0} e^{inx}/n$ of the sawtooth function.

The solution we present here, which relies on a simple application of the uniform boundedness principle, provides only the *existence* of a continuous function with diverging Fourier series. However, we also learn that, in fact, a generic set of continuous functions have this property.

Let $\mathcal{B} = C([-\pi, \pi])$ be the Banach space of continuous complex-valued functions on $[-\pi, \pi]$ with the usual sup-norm $\|f\| = \sup_{x \in [-\pi,\pi]} |f(x)|$. The Fourier coefficients of $f \in \mathcal{B}$ are defined by

$$a_n = \hat{f}(n) = \frac{1}{2\pi} \int_{-\pi}^{\pi} f(x) e^{-inx} dx, \qquad \text{for all } n \in \mathbb{Z},$$

and the Fourier series of f is

$$f(x) \sim \sum_{n=-\infty}^{\infty} a_n e^{inx}.$$

Also, the Nth partial sum of this Fourier series is defined by

$$S_N(f)(x) = \sum_{n=-N}^{N} a_n e^{inx}.$$

We saw in Book I an elegant expression for these partial sums in terms of convolutions, namely

$$S_N(f)(x) = (f * D_N)(x)$$

where

$$D_N(x) = \sum_{n=-N}^{N} e^{inx} = \frac{\sin[(N+1/2)x]}{\sin(x/2)}$$

is the Dirichlet kernel, and

$$(f * g)(x) = \frac{1}{2\pi}\int_{-\pi}^{\pi} f(y)g(x-y)dy = \frac{1}{2\pi}\int_{-\pi}^{\pi} f(x-y)g(y)dy$$

is the convolution on the circle.

Theorem 2.2 *Let $\mathcal{B}$ denote the Banach space of continuous functions on $[-\pi, \pi]$ with the sup-norm.*

(i) *Given any point $x_0 \in [-\pi, \pi]$, there is a continuous function whose Fourier series diverges at x_0.*

(ii) *In fact, the set of continuous functions whose Fourier series diverge on a dense set in $[-\pi, \pi]$ is generic in $\mathcal{B}$.*

For a stronger version of these results, see Problem 3.

We begin with (i), and assume without loss of generality that $x_0 = 0$. Let ℓ_N denote the linear functional on $\mathcal{B}$ defined by

$$\ell_N(f) = S_N(f)(0) = \frac{1}{2\pi}\int_{-\pi}^{\pi} f(-y)D_N(y)\,dy.$$

If (i) were not true, then $\sup_N |\ell_N(f)| < \infty$ for every $f \in \mathcal{B}$. Moreover, if we knew that each ℓ_N is continuous, the uniform boundedness principle would then imply that $\sup_N \|\ell_N\| < \infty$. The proof of (i) will thus be complete if we can show that each ℓ_N is continuous yet $\|\ell_N\| \to \infty$ as N tends to infinity.

Now, ℓ_N is continuous for each N, since

$$\begin{aligned} |\ell_N(f)| &\leq \frac{1}{2\pi} \int_{-\pi}^{\pi} |f(-y)|\, |D_N(y)|\, dy \\ &\leq L_N \|f\|, \end{aligned}$$

where we have defined

$$L_N = \frac{1}{2\pi} \int_{-\pi}^{\pi} |D_N(y)|\, dy.$$

In fact, the norm of the linear functional ℓ_N is precisely equal to the integral L_N.

Lemma 2.3 *$\|\ell_N\| = L_N$ for all $N \geq 0$.*

Proof. We already know from the above that $\|\ell_N\| \leq L_N$. To prove the reverse inequality, it suffices to find a sequence of continuous functions $\{f_k\}$ so that $\|f_k\| \leq 1$, and $\ell_N(f_k) \to L_N$ as $k \to \infty$. To do so, first let g denote the function equal to 1 when D_N is positive and -1 when D_N is negative. Then g is measurable, $\|g\| \leq 1$, and

$$L_N = \frac{1}{2\pi} \int_{-\pi}^{\pi} g(-y) D_N(y)\, dy$$

where we used the fact that D_N is even, hence $g(y) = g(-y)$. Clearly, there exists a sequence of continuous functions $\{f_k\}$ with $-1 \leq f_k(x) \leq 1$ for all $-\pi \leq x \leq \pi$, and so that

$$\int_{-\pi}^{\pi} |f_k(y) - g(y)|\, dy \to 0 \qquad \text{as } k \to \infty.$$

As a result, we find that $\ell_N(f_k) \to L_N$ as $k \to \infty$, while $\|f_k\| \leq 1$, hence $\|\ell_N\| \geq L_N$, as desired.

The proof of part (i) in the theorem will be complete if we can show that $\|\ell_N\| = L_N$ tends to infinity as $N \to \infty$. This is precisely the content of our final lemma.

Lemma 2.4 *There is a constant $c > 0$ so that $L_N \geq c \log N$.*

Proof. Since $|\sin y|/|y| \leq 1$ for all y, and $\sin y$ is an odd function, we see that[1]

$$\begin{aligned} L_N &\geq c\int_0^{\pi} \frac{|\sin(N+1/2)y|}{|y|}\,dy \\ &\geq c\int_0^{(N+1/2)\pi} \frac{|\sin x|}{x}\,dx \\ &\geq c\sum_{k=0}^{N-1}\int_{k\pi}^{(k+1)\pi} \frac{|\sin x|}{x}\,dx \\ &\geq c\sum_{k=0}^{N-1}\frac{1}{(k+1)\pi}\int_{k\pi}^{(k+1)\pi} |\sin x|\,dx. \end{aligned}$$

However, for all k we have $\int_{k\pi}^{(k+1)\pi} |\sin x|\,dx = \int_0^{\pi} |\sin x|\,dx$, so that

$$L_N \geq c\sum_{k=0}^{N-1}\frac{1}{k+1} \geq c\log N,$$

as was to be shown.

The proof of (ii) in Theorem 2.2 is immediate. Indeed, part (ii) of the uniform boundedness principle, together with what we have just shown, guarantees that the set of continuous functions f for which $\sup_N |S_N(f)(0)| < \infty$ is of the first category, and consequently, the set of functions whose Fourier series converges at the origin is also of the first category. Therefore the set of functions whose Fourier series diverges at the origin is generic. Similarly, if $\{x_1, x_2, \dots\}$ is any countable collection of points in $[-\pi, \pi]$, then for each j, the set F_j of continuous functions whose Fourier series diverge at x_j is also generic. Hence the set $\bigcap_{j=1}^{\infty} F_{x_j}$ which consists of continuous functions whose Fourier series diverge at every point $x_1, x_2, \dots,$ is also generic, and the proof of the theorem is complete.

3 The open mapping theorem

Let X and Y be Banach spaces with norms $\|\cdot\|_X$ and $\|\cdot\|_Y$ respectively, and $T : X \to Y$ a mapping. Observe that T is continuous if and only if $\{x \in X : T(x) \in \mathcal{O}\}$ is open in X whenever $\mathcal{O}$ is open in Y. This holds regardless of whether T is linear or not. In particular, if T has an inverse $S : Y \to X$ that is also continuous, the above observation applied to S

[1] In this calculation, the value of the constant c may change from line to line.

shows that the image by T of any open set in X is open in Y. A mapping T that maps open sets to open sets is called an **open mapping**.

We recall that a mapping $T : X \to Y$ is **surjective** if $T(X) = Y$, and **injective** if $T(x) = T(y)$ implies $x = y$. Also, T is **bijective** if it is both surjective and injective.

A bijective mapping has an inverse $T^{-1} : Y \to X$ defined as follows: if $y \in Y$, then $T^{-1}(y)$ is the unique element $x \in X$ so that $T(x) = y$. This definition is unambiguous precisely because T is surjective and injective. In general, if T is linear, then the inverse T^{-1} is also linear, but T^{-1} need not be continuous. However, by the previous observation, we see that T^{-1} will be continuous if T is an open mapping. The next result says that surjectivity guarantees openness.

Theorem 3.1 *Suppose X and Y are Banach spaces, and $T : X \to Y$ is a continuous linear transformation. If T is surjective, then T is an open mapping.*

Proof. We denote by $B_X(x, r)$ and $B_Y(y, r)$ the open balls of radius r centered at $x \in X$ and $y \in Y$ respectively, and we write simply $B_X(r)$ and $B_Y(r)$ for the open balls centered at the origin. Since T is linear, it suffices to show that $T(B_X(1))$ contains an open ball centered at the origin.

First, we prove the weaker statement that $\overline{T(B_X(1))}$ contains an open ball centered at the origin. To see this, note that since T is surjective, we must have

$$Y = \bigcup_{n=1}^{\infty} T(B_X(n)).$$

By the Baire category theorem, not all the sets $T(B_X(n))$ can be nowhere dense, so for some n, the set $\overline{T(B_X(n))}$ must contain an interior point. As a result of the fact that T is linear, this implies that

$$\overline{T(B_X(1))} \supset B_Y(y_0, \epsilon)$$

for some $y_0 \in Y$, and $\epsilon > 0$. By definition of the closure, we may pick a point $y_1 = T(x_1)$ where $x_1 \in B_X(1)$ and $\|y_1 - y_0\|_Y < \epsilon/2$. Then, if $y \in B_Y(\epsilon/2)$, we find that $y - y_1$ belongs to $\overline{T(B_X(1))}$, and writing $y = T(x_1) + y - y_1$ we find that $y \in \overline{T(B_X(2))}$. Therefore, the ball $B_Y(\epsilon/2)$ is contained in $\overline{T(B_X(2))}$. Using once again the fact that T is linear, we see that $B_Y(\epsilon/4)$ is contained in $\overline{T(B_X(1))}$, and this proves the weaker claim. In fact, replacing T by $(4/\epsilon)T$, we may assume that

$$\overline{T(B_X(1))} \supset B_Y(1), \tag{6}$$

and consequently

$$\overline{T(B_X(2^{-k}))} \supset B_Y(2^{-k}), \qquad \text{for all } k. \tag{7}$$

Next, we strengthen the result and show that in fact

$$T(B_X(1)) \supset B_Y(1/2). \tag{8}$$

Indeed, let $y \in B_Y(1/2)$, and by (7) with $k = 1$, select a point $x_1 \in B_X(1/2)$ so that $y - T(x_1) \in B_Y(1/2^2)$. Then, by (7) again, applied with $k = 2$, we may find $x_2 \in B_X(1/2^2)$ so that $y - T(x_1) - T(x_2) \in B(1/2^3)$. Continuing this process, we obtain a sequence of points $\{x_1, x_2, \ldots\}$ so that $\|x_k\|_X < 1/2^k$. Since X is complete, the sum $x_1 + x_2 + \cdots$ converges to a limit $x \in X$ with $\|x\| < \sum_{k=1}^{\infty} 1/2^k = 1$. Moreover, since we have

$$y - T(x_1) - \cdots - T(x_k) \in B_Y(1/2^{k+1}),$$

and T is continuous, we find in the limit that $T(x) = y$. This implies (8), which then clearly implies that $T(B_X(1))$ contains an open ball centered at the origin.

We gather two interesting corollaries to this theorem.

Corollary 3.2 *If X and Y are Banach spaces, and $T : X \to Y$ is a continuous bijective linear transformation, then the inverse $T^{-1} : Y \to X$ of T is also continuous. Hence there are constants $c, C > 0$ with*

$$c\|f\|_X \leq \|T(f)\|_Y \leq C\|f\|_X \quad \textit{for all } f \in X.$$

This follows immediately from the discussion preceding Theorem 3.1.

Recall that two norms $\|\cdot\|_1$ and $\|\cdot\|_2$ on a vector space V are said to be equivalent, if there are constants $c, C > 0$ so that

$$c\|v\|_2 \leq \|v\|_1 \leq C\|v\|_2 \quad \text{for all } v \in V.$$

Corollary 3.3 *Suppose the vector space V is equipped with two norms $\|\cdot\|_1$ and $\|\cdot\|_2$. If*

$$\|v\|_1 \leq C\|v\|_2 \quad \textit{for all } v \in V,$$

and V is complete with respect to both norms, then $\|\cdot\|_1$ and $\|\cdot\|_2$ are equivalent.

Indeed, the hypothesis implies that the identity mapping $I : (V, \|\cdot\|_2) \to (V, \|\cdot\|_1)$ is continuous, and since it is clearly bijective, its inverse $I : (V, \|\cdot\|_1) \to (V, \|\cdot\|_2)$ is also continuous. Hence $c\|v\|_2 \leq \|v\|_1$ for some $c > 0$ and all $v \in V$.

3.1 Decay of Fourier coefficients of L^1-functions

We return to the Fourier series discussed in Section 2.1 for an interesting application of the open mapping theorem. Recall the Riemann-Lebesgue lemma, which states

$$\lim_{|n|\to\infty} |\hat{f}(n)| = 0,$$

if $f \in L^1([-\pi, \pi])$, where $\hat{f}(n)$ denotes the nth Fourier coefficient of f.[2] A natural question that arises is the following: given any sequence of complex numbers $\{a_n\}_{n\in\mathbb{Z}}$ that vanishes at infinity, that is, $|a_n| \to 0$ as $|n| \to \infty$, does there exist $f \in L^1([-\pi, \pi])$ with $\hat{f}(n) = a_n$ for all n?

To reformulate this question in terms of Banach spaces, we let $\mathcal{B}_1 = L^1([-\pi, \pi])$ equipped with the L^1-norm, and $\mathcal{B}_2$ denote the vector space of all sequences $\{a_n\}$ of complex numbers with $|a_n| \to 0$ as $|n| \to \infty$. The space $\mathcal{B}_2$ is equipped with the usual sup-norm $\|\{a_n\}\|_\infty = \sup_{n\in\mathbb{Z}} |a_n|$ which clearly makes $\mathcal{B}_2$ into a Banach space.

Then, we ask whether the mapping $T : \mathcal{B}_1 \to \mathcal{B}_2$ defined by

$$T(f) = \{\hat{f}(n)\}_{n\in\mathbb{Z}}$$

is surjective.

The answer to this is negative.

Theorem 3.4 *The mapping $T : \mathcal{B}_1 \to \mathcal{B}_2$ given by $T(f) = \{\hat{f}(n)\}$ is linear, continuous and injective, but not surjective.*

Therefore, there are sequences of complex numbers that vanish at infinity and that are not the Fourier coefficients of L^1-functions.

Proof. We first note that T is clearly linear, and also continuous with $\|T(f)\|_\infty \leq \|f\|_{L^1}$. Moreover, T is injective since $T(f) = 0$ implies that $\hat{f}(n) = 0$ for all n, which then implies[3] that $f = 0$ in L^1. If T were surjective, then Corollary 3.2 would imply that there is a constant $c > 0$ that satisfies

$$(9) \qquad c\|f\|_{L^1} \leq \|T(f)\|_\infty, \qquad \text{for all } f \in \mathcal{B}_1.$$

However, if we set $f = D_N$ the Nth Dirichlet kernel given by $D_N = \sum_{|n|\leq N} e^{inx}$, and recall from Lemma 2.4 that $\|D_N\|_{L^1} = L_N \to \infty$ as $N \to \infty$, we find that (9) is violated as N tends to infinity, which is our desired contradiction.

---

[2]See for instance Problem 1 in Chapter 2 of Book III.

[3]This result can be found in Theorem 3.1 in Chapter 4 of Book III.

4 The closed graph theorem

Suppose X and Y are two Banach spaces, with norms $\|\cdot\|_X$ and $\|\cdot\|_Y$ respectively, and $T : X \to Y$ is a linear map. The **graph** of T is defined as a subset of $X \times Y$ by

$$G_T = \{(x, y) \in X \times Y : \; y = T(x)\}.$$

The linear map T is **closed** if its graph is a closed subset in $X \times Y$. In other words, T is closed if whenever $\{x_n\} \subset X$ and $\{y_n\} \subset Y$ are two converging sequences in X and Y respectively, say $x_n \to x$ and $y_n \to y$, and if $T(x_n) = y_n$, then $T(x) = y$.

Theorem 4.1 *Suppose X and Y are two Banach spaces. If $T : X \to Y$ is a closed linear map, then T is continuous.*

Proof. Since the graph of T is a closed subspace of the Banach space $X \times Y$ with the norm $\|(x, y)\|_{X\times Y} = \|x\|_X + \|y\|_Y$, the graph G_T is itself a Banach space. Consider the two projections $P_X : G(T) \to X$ and $P_Y : G(T) \to Y$ defined by

$$P_X(x, T(x)) = x \quad \text{and} \quad P_Y(x, T(x)) = T(x).$$

The mappings P_X and P_Y are continuous and linear. Moreover, P_X is bijective, hence its inverse P_X^{-1} is continuous by Corollary 3.2. Since $T = P_Y \circ P_X^{-1}$, we conclude that T is continuous, as was to be shown.

4.1 Grothendieck's theorem on closed subspaces of L^p

As an application of the closed graph theorem, we prove the following result:

Theorem 4.2 *Let $(X, \mathcal{F}, \mu)$ be a finite measure space, that is, $\mu(X) < \infty$. Suppose that:*

(i) *E is a closed subspace of $L^p(X, \mu)$, for some $1 \leq p < \infty$, and*

(ii) *E is contained in $L^\infty(X, \mu)$.*

Then E is finite dimensional.

Since $E \subset L^\infty$, and X has finite measure, we find that $E \subset L^2$ with

$$\|f\|_{L^2} \leq C\|f\|_{L^\infty} \quad \text{whenever } f \in E.$$

The essential idea in the proof of the theorem is to reverse this inequality, and then use the Hilbert space structure of L^2.

Equipped with the L^p-norm, E is a Banach space since it is a closed subspace of $L^p(X, \mu)$. Let

$$I : E \to L^\infty(X, \mu)$$

denote the identity mapping $I(f) = f$. Then, I is linear and closed. Indeed, suppose that $f_n \to f$ in E and $f_n \to g$ in L^∞. Then, there exists a subsequence of $\{f_n\}$ that converges almost everywhere to f (see Exercise 5 in Chapter 1), and therefore $f = g$ almost everywhere, as desired. By the closed graph theorem there is an $M > 0$ so that

$$(10) \qquad \|f\|_{L^\infty} \leq M\|f\|_{L^p} \quad \text{for all } f \in E.$$

Lemma 4.3 *Under the assumptions of the theorem, there exists $A > 0$ so that*

$$\|f\|_{L^\infty} \leq A\|f\|_{L^2} \quad \textit{for all } f \in E.$$

Proof. If $1 \leq p \leq 2$, then Hölder's inequality with the conjugate exponents $r = 2/p$ and $r^* = 2/(2-p)$ yields

$$\int |f|^p \leq \left(\int |f|^2\right)^{p/2} \left(\int 1\right)^{\frac{2-p}{2}}.$$

Since X has finite measure, we see after taking p^{th} roots in the above, that there is some $B > 0$ so that $\|f\|_{L^p} \leq B\|f\|_{L^2}$ for all $f \in E$. Together with (10), this proves the lemma when $1 \leq p \leq 2$.

When $2 < p < \infty$, we note first that $|f(x)|^p \leq \|f\|_{L^\infty}^{p-2}|f(x)|^2$, and integrating this inequality gives

$$\|f\|_{L^p}^p \leq \|f\|_{L^\infty}^{p-2}\|f\|_{L^2}^2.$$

If we now use (10), and assume that $\|f\|_{L^\infty} \neq 0$, we find that for some $A > 0$, we have $\|f\|_{L^\infty} \leq A\|f\|_{L^2}$ whenever $f \in E$, and the proof of the lemma is complete.

We now return to the proof of Theorem 4.2. Suppose $f_1, \ldots, f_n$ is an orthonormal set in L^2 of functions in E, and let $\mathbb{B}$ denote the unit ball in $\mathbb{C}^n$,

$$\mathbb{B} = \{\zeta = (\zeta_1, \ldots, \zeta_n) \in \mathbb{C}^n : \sum_{j=1}^{n} |\zeta_j|^2 \leq 1\}.$$

For each $\zeta \in \mathbb{B}$, let $f_\zeta(x) = \sum_{j=1}^n \zeta_j f_j(x)$. By construction we have $\|f_\zeta\|_{L^2} \leq 1$, and the lemma gives $\|f_\zeta\|_{L^\infty} \leq A$. Hence for each ζ, there exists a measurable set X_ζ of full measure in X (that is, $\mu(X_\zeta) = \mu(X)$), so that

$$(11) \qquad |f_\zeta(x)| \leq A \quad \text{for all } x \in X_\zeta.$$

By first taking a countable dense subset of points in $\mathbb{B}$, and then using the continuity of the mapping $\zeta \mapsto f_\zeta(x)$, we see that (11) implies

$$(12) \qquad |f_\zeta(x)| \leq A \quad \text{for all } x \in X', \text{ and all } \zeta \in \mathbb{B}$$

where X' is a set of full measure in X. From this, we claim that

$$(13) \qquad \sum_{j=1}^n |f_j(x)|^2 \leq A^2 \quad \text{for all } x \in X'.$$

Indeed, it suffices to establish this inequality when the left-hand side is non-zero. Then, if we let $\sigma = (\sum_{j=1}^n |f_j(x)|^2)^{1/2}$, and set $\zeta_j = \overline{f_j(x)}/\sigma$, then by (12) we find that for all $x \in X'$

$$\frac{1}{\sigma}\sum_{j=1}^n |f_j(x)|^2 \leq A,$$

that is, $\sigma \leq A$, as we claimed.

Finally, integrating (13), and recalling that $\{f_1, \ldots, f_n\}$ is orthonormal, we find $n \leq A^2\mu(X)$, hence the dimension of E must be finite.

Remark. Problem 6 shows that the space L^∞ in the theorem cannot be replaced by any L^q for $1 \leq q < \infty$.

5 Besicovitch sets

In Section 4.4, Chapter 7 of Book III, we constructed an example of a **Besicovitch set** (or "Kakeya set") in $\mathbb{R}^2$, that is, a compact set with two-dimensional Lebesgue measure zero that contains a unit line segment in every direction. We recall that this set was obtained as a union of finitely many rotations of a specific set: one that is given as a union of line segments joining points from a Cantor-like set on the line $\{y = 0\}$ to another Cantor-like set on the line $\{y = 1\}$. Our goal here is to present an ingenious idea of Körner that proves the existence of Besicovitch sets using the Baire category theorem; in fact, it is shown that in the right metric space, such sets are generic.

The starting point of the analysis is an appropriate complete metric space of sets in $\mathbb{R}^2$. Suppose A is a subset of $\mathbb{R}^2$ and $\delta > 0$. We define the **δ-neighborhood** of A by

$$A^\delta = \{x : d(x, A) < \delta\}, \qquad \text{where } d(x, A) = \inf_{y\in A} |x - y|.$$

Then, if A and B are subsets of $\mathbb{R}^2$ we define the **Hausdorff distance**[4] between A and B by

$$\text{dist}(A, B) = \inf\{\delta : \ B \subset A^\delta \text{ and } A \subset B^\delta\}.$$

We shall restrict our attention to compact subsets of $\mathbb{R}^2$. The distance d then satisfies the following properties.

Suppose A, B and C are non-empty compact subsets of $\mathbb{R}^2$:

(i) $\text{dist}(A, B) = 0$ if and only if $A = B$.

(ii) $\text{dist}(A, B) = \text{dist}(B, A)$.

(iii) $\text{dist}(A, C) \leq \text{dist}(A, B) + \text{dist}(B, C)$.

(iv) The set of compact subsets of $\mathbb{R}^2$ equipped with the Hausdorff distance is a complete metric space.

Verification of (i), (ii), and (iii) can be left to the reader, while the proof of (iv), which is a little more intricate, is deferred to the end of this section.

We now restrict our attention to the compact subsets of the square $[-1/2, 1/2] \times [0, 1]$ which consist of a union of line segments joining points from $L_0 = \{-1/2 \leq x \leq 1/2, \quad y = 0\}$ to points on $L_1 = \{-1/2 \leq x \leq 1/2, \quad y = 1\}$ and spanning all possible directions. More precisely, let $\mathcal{K}$ denote the set of closed subsets K of the square $Q = [-1/2, 1/2] \times [0, 1]$ with the following properties:

(i) K is a union of line segments ℓ joining a point of L_0 to a point of L_1.

(ii) For every angle $\theta \in [-\pi/4, \pi/4]$ there exists a line segment ℓ in K making an oriented angle of θ with the y-axis.

Simple limiting arguments then show that $\mathcal{K}$ is a closed subset of the metric space of all compact subsets in $\mathbb{R}^2$ with the metric d, and consequently $\mathcal{K}$ with the Hausdorff distance is a complete metric space.

Our aim is to prove the following:

---

[4]Incidentally, this distance already arose in Chapter 7 of Book III.

Theorem 5.1 *The collection of sets in $\mathcal{K}$ of two-dimensional Lebesgue measure zero is generic.*

In particular, this collection is non-empty, and in fact dense.

Loosely stated, the key to the argument is to show that sets K in $\mathcal{K}$ whose horizontal slices $\{x : (x, y) \in K\}$ have "small" Lebesgue measure are generic. The argument is best carried out by using a "thickened" version K^η of K.

To this end, given $0 \leq y_0 \leq 1$ and $\epsilon > 0$ we define $\mathcal{K}(y_0, \epsilon)$ as the collection of all compact subsets K in $\mathcal{K}$ with the property that there exists $\eta > 0$ so that the η-neighborhood K^η satisfies: for every $y \in [y_0 - \epsilon, y_0 + \epsilon]$ the horizontal slice $\{x : (x, y) \in K^\eta\}$ has one-dimensional Lebesgue measure less than 10ϵ, that is,

$$(14) \qquad m_1(\{x : (x, y) \in K^\eta\}) < 10\epsilon, \qquad \text{for all } y \in [y_0 - \epsilon, y_0 + \epsilon].^5$$

Lemma 5.2 *For each fixed y_0 and ϵ, the collection of sets $\mathcal{K}(y_0, \epsilon)$ is open and dense in $\mathcal{K}$.*

To prove that $\mathcal{K}(y_0, \epsilon)$ is open, suppose $K \in \mathcal{K}(y_0, \epsilon)$ and pick η so that K^η satisfies the condition above. Suppose $K' \in \mathcal{K}$ with $\text{dist}(K, K') < \eta/2$. This means in particular that $K' \subset K^{\eta/2}$, and the triangle inequality then shows that $(K')^{\eta/2} \subset K^\eta$. Therefore

$$m_1(\{x : (x, y) \in (K')^{\eta/2}\}) \leq m_1(\{x : (x, y) \in K^\eta\}) < 10\epsilon,$$

and as a result $K' \in \mathcal{K}(y_0, \epsilon)$, as was to be shown.

To establish the rest of the lemma, we need to show that if $K \in \mathcal{K}$ and $\delta > 0$, there exists $K' \in \mathcal{K}(y_0, \epsilon)$ so that $\text{dist}(K, K') \leq \delta$. The set K' will be given as the union of two sets A and A'. The set A will be constructed by picking line segments ℓ in K, and looking at the corresponding angular sector obtained by rotating the line segment ℓ by a small angle around its intersection with $y = y_0$. This will result in two solid triangles with a vertex on $y = y_0$, and we shall try to control the length of the intersection of these triangles with any line segment parallel to the x-axis (Figure 2).

More precisely, if N is a positive integer, we can consider the partition of the interval $[-\pi/4, \pi/4]$ defined by

$$\theta_n = \frac{-\pi}{4} + \frac{n}{N}\frac{\pi}{2}, \qquad \text{for } n = 0, \ldots, N - 1.$$

[5] The choice of 10 for the constant appearing in (14) is of no particular significance; indeed, smaller constants would have done as well.

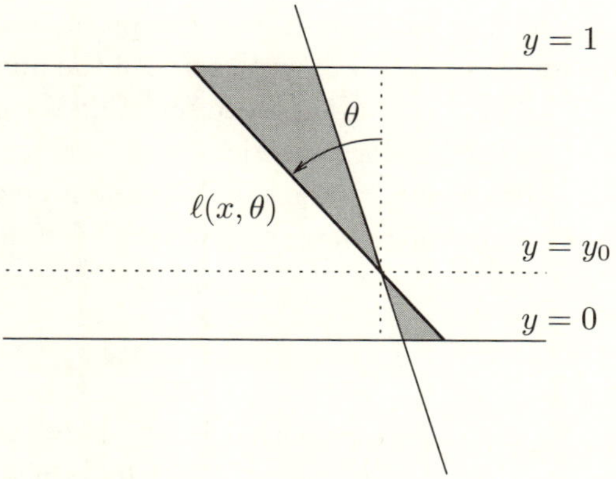

Figure 2. Rotation of $\ell(x, \theta)$ around its intersection with $y = y_0$

Then the angles θ_n are uniformly spaced in $[-\pi/4, \pi/4]$ and the N intervals defined by

$$I_n = [\theta_n, \theta_n + \pi/(2N)],$$

cover $[-\pi/4, \pi/4]$. Moreover each of these sub-intervals has length equal to $\pi/(2N)$.

If we use $\ell(x, \theta)$ to denote the line segment joining $\{y = 0\}$ to $\{y = 1\}$ that passes through the point (x, y_0) and which makes an oriented angle θ with the y-axis, then for each θ_n as defined above, by property (ii) of the set K there exists a number $-1/2 \leq x_n \leq 1/2$ so that $\ell(x_n, \theta_n) \in K$. For each $n = 0, \ldots, N$ consider the compact set

$$S_n = \bigcup_{\varphi \in I_n} \ell(x_n, \varphi).$$

Each S_n therefore consists of (at most) two closed triangles with vertex at the point (x_n, y_0). Now let

$$A = \bigcup_{n=0}^{N} S_n.$$

If $N \geq c/\delta$ (for a large enough constant c), then the sets S_n that are not entirely contained in the square Q can be translated slightly to the left or right so that the resulting set A belongs to Q, and moreover so that every point in A is at a distance less than δ from a point in K; that is $A \subset K^{\delta}$.

However it is not necessarily true that every point of K is close to A, since in defining A we have dealt only with some of the lines $\ell(x_n, \theta_n)$ that make up K. To remedy this we add a finite set of lines to obtain a set A' that is close to K in the Hausdorff metric. In more detail, recall that K is itself a union of lines, $K = \bigcup \ell$, and let ℓ^δ be the δ-neighborhood of ℓ. Then $\bigcup \ell^\delta$ is an open cover of K and thus we can select a finite subcover $\bigcup_{m=1}^M \ell_m^\delta$ of K. We define $A' = \bigcup_{m=1}^M \ell_m$ and set

$$K' = A \cup A'.$$

Observe first that $K' \in \mathcal{K}$. Note next that by its definition, $A' \subset K$, but $(A')^\delta \supset K$. Therefore $(K')^\delta \supset K$. Also $K^\delta \supset K'$, since $K^\delta \supset A$ as we have seen, and $K^\delta \supset K \supset A'$. This shows that $\mathrm{dist}(K', K) \leq \delta$.

We next estimate $m_1(\{x : (x, y) \in (K')^\eta\})$ for $y_0 - \epsilon \leq y \leq y_0 + \epsilon$, by adding the corresponding estimates with K' replaced by A and A'. Note that for fixed y the set $\{x : (x, y) \in A\}$ consists of N intervals arising from the intersection of the horizontal line at height y, with the N triangles that have their vertices at height y_0. By a simple trigonometric argument, since $|y - y_0| \leq \epsilon$ and the magnitudes of the angles at the vertices are $\pi/(2N)$, each corresponding interval of A^η has length $\leq 8\epsilon/N + 2\eta$. Thus

$$m_1(\{x : (x, y) \in (K')^\eta\}) \leq 8\epsilon + 2\eta N.$$

Next A' consists of M line segments, so the set $\{x : (x, y) \in A'\}$ consists of M points, and therefore the set $\{x : (x, y) \in (A')^\eta\}$ is the union of M intervals of length 2η; this has measure $\leq 2\eta M$. Altogether then $m_1(\{x : (x, y) \in (K')^\eta\}) \leq 8\epsilon + 2\eta(M + N)$ and we get estimate (14) for K' if we take $\eta < \epsilon/(M + N)$. This completes the proof of the lemma.

We can now proceed with the final argument in the proof of the theorem. For each m, consider the set

$$\mathcal{K}_M = \bigcap_{m=1}^{M} \mathcal{K}(m/M, 1/M).$$

Each $\mathcal{K}_M$ is open and dense, and moreover if $K \in \mathcal{K}_M$, each slice of K along any $0 \leq y \leq 1$ has one-dimensional Lebesgue measure that is $O(1/M)$. Since open dense sets are generic, and the countable intersection of generic sets is generic, the set

$$\mathcal{K}_* = \bigcap_{M=1}^{\infty} \mathcal{K}_M$$

is generic in $\mathcal{K}$, and by the above observation if $K \in \mathcal{K}_*$ then each slice $K^y = \{x : (x,y) \in K\}$ $(0 \leq y \leq 1)$ has Lebesgue measure 0, hence Fubini's theorem implies that K has two-dimensional Lebesgue measure equal to 0. This completes the proof of Theorem 5.1.

We conclude this section with the proof of property (iv) of the Hausdorff distance, the completeness of the metric.

Suppose $\{A_n\}$ is a sequence of (non-empty) compact subsets that is Cauchy with respect to the Haussdorff distance; let $\mathcal{A}_n = \overline{\bigcup_{k=n}^{\infty} A_k}$ and $\mathcal{A} = \bigcap_{n=1}^{\infty} \mathcal{A}_n$. We claim that $\mathcal{A}$ is non-empty, compact, and $A_n \to \mathcal{A}$.

Given $\epsilon > 0$ there exists N_1 so that $\operatorname{dist}(A_n, A_m) < \epsilon$ for all $n, m \geq N_1$. As a result, it is clear that whenever $n \geq N_1$, then $\bigcup_{k=n}^{\infty} A_k \subset (A_n)^{\epsilon}$, hence $\mathcal{A}_n \subset (A_n)^{2\epsilon}$. This implies

$$\mathcal{A} \subset (A_n)^{2\epsilon} \quad \text{whenever } n \geq N_1. \tag{15}$$

Since each $\mathcal{A}_n$ is non-empty and compact, and since $\mathcal{A}_{n+1} \subset \mathcal{A}_n$, it follows that $\mathcal{A}$ is non-empty and compact, and moreover $\operatorname{dist}(\mathcal{A}_n, \mathcal{A}) \to 0$. Indeed, if $\operatorname{dist}(\mathcal{A}_n, \mathcal{A})$ did not converge to zero, then there would exist $\epsilon_0 > 0$, an increasing sequence n_k of positive integers, and points $x_{n_k} \in \mathcal{A}_{n_k}$ so that $d(x_{n_k}, \mathcal{A}) \geq \epsilon_0$. Since $\{x_{n_k}\} \subset \mathcal{A}_1$, which is compact, we may assume (after picking a subsequence and relabeling if necessary) that $\{x_{n_k}\}$ converges to a limit, say x, which would clearly satisfy $d(x, \mathcal{A}) \geq \epsilon_0$. But for every M, we have $x_{n_k} \in \mathcal{A}_M$ for all sufficiently large n_k, and since $\mathcal{A}_M$ is compact, we must have $x \in \mathcal{A}_M$, thus $x \in \mathcal{A}$. This contradicts the fact that $d(x, \mathcal{A}) \geq \epsilon_0$, hence $\operatorname{dist}(\mathcal{A}_n, \mathcal{A}) \to 0$.

Returning to our proof of (iv), pick N_2 so that $\operatorname{dist}(\mathcal{A}_n, \mathcal{A}) < \epsilon$ for all $n \geq N_2$. This implies that $A_n \subset \mathcal{A}^{2\epsilon}$ for $n \geq N_2$, therefore

$$A_n \subset \mathcal{A}^{2\epsilon} \quad \text{whenever } n \geq N_2. \tag{16}$$

Combining (15) and (16) yields the inequality $\operatorname{dist}(A_n, \mathcal{A}) \leq 2\epsilon$ whenever $n \geq \max(N_1, N_2)$, which implies $A_n \to \mathcal{A}$, and that concludes the proof.

6 Exercises

1. Below are some examples of generic sets and sets of the first category.

(a) Let $\{x_j\}_{j=1}^{\infty}$ denote an enumeration of the rational numbers in $\mathbb{R}$, and consider the sets

$$U_n = \bigcup_{j=1}^{\infty} \left(x_j - \frac{1}{n2^j},\ x_j + \frac{1}{n2^j}\right), \qquad \text{and} \qquad U = \bigcap_{n=1}^{\infty} U_n.$$

Show that U is generic but has Lebesgue measure zero.

(b) Use a Cantor-like set (as described, for example, in Exercise 4, Chapter 1 of Book III) to give an example of a subset of the first category that has full Lebesgue measure in $[0,1]$. Note that automatically this subset will be uncountable and dense. Also, its complement is generic and has measure zero, giving an alternative to the set U in (a).

2. Suppose F is a closed subset and $\mathcal{O}$ an open subset of a complete metric space.

(a) Show that F is of the first category if and only if F has empty interior.

(b) Show that $\mathcal{O}$ is of the first category if and only if $\mathcal{O}$ is empty.

(c) Consequently, prove that F is generic if and only if $F = X$; and $\mathcal{O}$ is generic if and only if $\mathcal{O}^c$ contains no interior.

[Hint: For (a), argue by contradiction, assuming that a closed ball $\overline{B}$ is contained in F. Apply the category theorem to the complete metric space $\overline{B}$.]

3. Show that the conclusion of the Baire category theorem continues to hold if X_0 is a metric space that arises as an open subset of a complete metric space X. [Hint: Apply the Baire category theorem to the closure of X_0 in X.]

4. Prove that every continuous function on $[0,1]$ can be approximated uniformly by continuous nowhere differentiable functions. Do so by either:

(a) using Theorem 1.5.

(b) using only the fact that a continuous nowhere differentiable function exists.

5. Let X be a complete metric space. We recall that a set is a G_δ in X if it is a countable intersection of open sets. Also, a set is an F_σ in X if it is a countable union of closed sets.

(a) Show that a dense G_δ is generic.

(b) Hence a countable dense set that is not open is an F_σ, but not a G_δ.

(c) Prove the following partial converse to (a). If E is a generic set, then there exists $E_0 \subset E$ with E_0 a dense G_δ.

6. The function

$$f(x) = \begin{cases} 0 & \text{if } x \text{ is irrational} \\ 1/q & \text{if } x = p/q \text{ is rational and expressed in lowest form} \end{cases}$$

is continuous precisely at the irrationals. In contrast to this, prove that there is no function on $\mathbb{R}$ that is continuous precisely at the rationals.

[Hint: Show that the set of points where a function is continuous is a G_δ (see the proof of Theorem 1.3), and apply Exercise 5.]

7. Let E be a subset of $[0,1]$, and let I be any closed non-trivial interval in $[0,1]$.

(a) Suppose E is of the first category in $[0,1]$. Show that for every I, the set $E \cap I$ is of the first category in I.

(b) Suppose E is generic in $[0,1]$. Show that for every I, the set $E \cap I$ is generic in I.

(c) Construct a set E in $[0,1]$ so that for all I, the set $E \cap I$ is neither of the first category nor generic in I.

[Hint: Consider the Cantor set in $[0,1]$; then in each open interval of its complement place a scaled copy of the Cantor set; continue this process indefinitely. For a related measure theoretic result, see Exercise 36 in Chapter 1 of Book III.]

8. A **Hamel basis** for a vector space X is a collection $\mathcal{H}$ of vectors in X, such that any $x \in X$ can be written as a unique *finite* linear combination of elements in $\mathcal{H}$.

Prove that an infinite-dimensional Banach space cannot have a countable Hamel basis.

[Hint: Show that otherwise the Banach space would be of the first category in itself.]

9. Consider $L^p([0,1])$ with Lebesgue measure. Note that if $f \in L^p$ with $p > 1$, then $f \in L^1$. Show that the set of $f \in L^1$ so that $f \notin L^p$, is generic.

A more general result can be found in Problem 1.

[Hint: Consider the set $E_N = \{f \in L^1 : \int_I |f| \leq Nm(I)^{1-1/p} \text{ for all intervals } I\}$. Note that each E_N is closed and that $L^p \subset \bigcup_N E_N$. Finally, show that E_N is nowhere dense by considering $f_0 + \epsilon g$ where $g(x) = x^{-(1-\delta)}$ with $0 < \delta < 1 - 1/p$.]

10. Consider $\Lambda^\alpha(\mathbb{R})$, with $0 < \alpha < 1$. Show that the set of nowhere differentiable functions is a generic set in $\Lambda^\alpha(\mathbb{R})$.

Note however that functions corresponding to the case $\alpha = 1$, that is, Lipschitz functions, are almost everywhere differentiable. (See Exercise 32 in Chapter 3 of Book III.)

11. Consider the Banach space $X = C([0,1])$ over the reals, with the sup-norm on X. Let $\mathcal{M}$ be the collection of functions that are not monotonic (increasing or decreasing) in any interval $[a,b]$, where $0 \leq a < b \leq 1$. Prove that $\mathcal{M}$ is generic in X.

[Hint: Let $\mathcal{M}_{[a,b]}$ denote the subset of X consisting of functions that are not monotonic in $[a,b]$. Then $\mathcal{M}_{[a,b]}$ is dense in X, while $\mathcal{M}^c_{[a,b]}$ is closed.]

12. Suppose X, Y and Z are Banach spaces, and $T : X \times Y \to Z$ is a mapping such that:

(i) For each $x \in X$, the mapping $y \mapsto T(x, y)$ is linear and continuous on Y.

(ii) For each $y \in Y$, the mapping $x \mapsto T(x, y)$ is linear and continuous on X.

Prove that T is (jointly) continuous on $X \times Y$, and in fact,

$$\|T(x, y)\|_Z \leq C \|x\|_X \|y\|_Y$$

for some $C > 0$ and all $x \in X$ and $y \in Y$.

13. Let $(X, \mathcal{F}, \mu)$ be a measure space, and let $\{f_n\}$ a sequence of functions in $L^p(X, \mu)$. We know from Exercise 12 in Chapter 1, that if $1 < p < \infty$, and $\sup_n \|f_n\|_{L^p} < \infty$, then some subsequence of $\{f_n\}$ converges weakly in L^p. In other words, there exist a subsequence $\{f_{n_k}\}$ of $\{f_n\}$, and an $f \in L^p$, so that if q denotes the conjugate exponent of p, that is $1/p + 1/q = 1$, then

$$\int_X f_{n_k}(x) g(x)\, d\mu(x) \to \int_X f(x) g(x)\, d\mu(x) \qquad \text{for every } g \in L^q.$$

More generally, we say that a sequence $\{f_n\}$ in L^p is **weakly bounded** if

$$\sup_n \left| \int_X f_n(x) g(x)\, d\mu(x) \right| < \infty \qquad \text{for all } g \in L^q.$$

Prove that if $1 < p < \infty$, and $\{f_n\}$ is a sequence of functions in L^p that is weakly bounded, then

$$\sup_n \|f_n\|_{L^p} < \infty.$$

In particular this holds if $\{f_n\}$ converges weakly in L^p.
[Hint: Apply the uniform boundedness principle to $\ell_n(g) = \int_X f_n(x) g(x)\, d\mu(x)$.]

14. Suppose X is a complete metric space with respect to a metric d, and $T : X \to X$ a continuous function. An element x^* in X is **universal** for T if the orbit set $\{T^n(x^*)\}_{n=1}^{\infty}$ is dense in X. Here $T^n = T \circ T \circ \cdots \circ T$ denotes n compositions of T.

Show that the set of universal elements for T in X is either empty or generic.

[Hint: Suppose x^* is universal for T, let $\{x_j\}$ be a dense set of elements in X, and let $F_{j,k,N} = \{x \in X : \ d(T^n x, y_j) < 1/k \ \text{ for some } n \geq N\}$. Show that $F_{j,k,N}$ is open and dense.]

15. Let $\overline{B}$ denote the closure of the unit ball in $\mathbb{R}^d$, and consider the metric space $\mathcal{C}$ of compact subsets of $\overline{B}$ with the Hausdorff distance. (See Section 5.) Show that the following two collections are generic.

(a) The subsets of Lebesgue measure zero.

(b) The subsets that are nowhere dense.

[Hint: For (a) show that the collection of sets C so that $m(C) < 1/n$ is open and dense. In fact for such a set, $C^c \supset \bigcup_{j=1}^{M} Q_j$, where Q_j are disjoint open cubes so that $\sum |Q_j| > 1 - 1/n$. Now shrink the Q_j. For (b) fix an open set $\mathcal{O}$ and show that the collection $\mathcal{C}_{\mathcal{O}}$ of sets in $\mathcal{C}$ that contain $\mathcal{O}$ is closed and nowhere dense.]

7 Problems

1. Let $T : \mathcal{B}_1 \to \mathcal{B}_2$ be a bounded linear transformation of a Banach space $\mathcal{B}_1$ to a Banach space $\mathcal{B}_2$.

(a) Prove that either T is surjective, or the image $T(\mathcal{B}_1)$ is of the first category in $\mathcal{B}_2$.

(b) As a consequence, prove the following: Suppose (X, μ) is a finite measure space, and $1 \leq p_1 < p_2 \leq \infty$. One has of course $L^{p_2}(X) \subset L^{p_1}(X)$. Show that $L^{p_2}(X)$ is a set of the first category in $L^{p_1}(X)$ (except in the trivial case for which each element of L^{p_1} belongs to L^{p_2}).

[Hint: For (a), assume that $T(\mathcal{B}_1)$ is of the second category and use an argument similar to the proof of Theorem 3.1 to show that the image under T of a ball centered at the origin of $\mathcal{B}_1$ contains a ball centered at the origin in $\mathcal{B}_2$.]

2. For each integer $n \geq 2$, let Λ_n denote the set of real numbers x so that there exists infinitely many distinct fractions p/q so that

$$|x - p/q| \leq 1/q^n.$$

Show that:

(a) Λ_n is a generic set in $\mathbb{R}$.

(b) However, the Hausdorff dimension of Λ_n equals $2/n$.

(c) Hence $m(\Lambda_n) = 0$, if $n > 2$, where m denotes the Lebesgue measure.

The elements of $\Lambda = \bigcap_{n \geq 2} \Lambda_n$ are called the **Liouville numbers**. While it is not difficult to see that every element of Λ is transcendental, it is a deeper fact that the same holds for each element of Λ_n when $n > 2$. (Note that in the case $n = 2$, the set Λ consists of the irrationals.)

3. Consider the Banach space $\mathcal{B}$ of continuous functions on the circle (with the sup-norm). Prove that the set of f in $\mathcal{B}$ whose Fourier series diverges in a generic set on the circle, is itself a generic set in $\mathcal{B}$.

[Hint: Choose $\{x_i\}$ dense in $[0, 1]$, let $E_i = \{f \in \mathcal{B} : \sup_N |S_N(f)(x_i)| = \infty\}$, and $E = \cap E_i$. Then E is generic. For each $f \in E$, define $\mathcal{O}_n = \{x : |S_N(f)(x)| > n \text{ some } N\}$. Show that $\cap \mathcal{O}_n$ is generic.]

4. Let $\mathbb{D}$ denote the open unit disc in the complex plane, and let $\mathcal{A}$ be the Banach space of all continuous complex-valued functions on $\overline{\mathbb{D}}$ that are holomorphic on $\mathbb{D}$, equipped with the sup-norm. Then, the space of functions in $\mathcal{A}$ which cannot be extended analytically past any point of the boundary of $\mathbb{D}$ is generic. To prove this statement establish the following:

(a) The set $\mathcal{A}_N = \{f \in \mathcal{A} : \ |f(e^{i\theta}) - f(1)| \leq N|\theta|\}$ is closed.

(b) $\mathcal{A}_N$ is nowhere dense.

[Hint: For (b) use the function $f_0(z) = (1-z)^{1/2}$ and consider $f + \epsilon f_0$.]

5. Let $I = [0, 1]$ denote the unit interval, and $C^\infty(I)$ the vector space of all smooth functions on I equipped with the metric d given by

$$d(f, g) = \sum_{n=0}^{\infty} \frac{1}{2^n} \frac{\rho_n(f-g)}{1 + \rho_n(f-g)},$$

where $\rho_n(h) = \sup_{x \in I} |h^{(n)}(x)|$. A function $f \in C^\infty(I)$ is analytic at a point $x_0 \in I$, if its Taylor series

$$\sum_{n=0}^{\infty} \frac{f^{(n)}(x_0)}{n!}(x - x_0)^n$$

converges in a neighborhood of x_0 to the function f. The function f is said to be singular at x_0 if its Taylor series diverges at x_0.

(a) Show that $(C^\infty(I), d)$ is a complete metric space.

(b) Prove that the set of functions in $C^\infty(I)$ that are singular at every point is generic.

[Hint: For (b), consider the set F_K of smooth functions f that satisfy $|f^{(n)}(x^*)|/n! \leq K^n$ for some x^* and all n, and show that F_K is closed and nowhere dense.]

6. The space L^∞ in Theorem 4.2 cannot be replaced by any L^q, with $1 \leq q < \infty$. In fact there exists a closed infinite dimensional subspace of $L^1([0, 1])$ consisting of functions that belong to L^q for all $1 \leq q < \infty$.

[Hint: One may use Exercise 19 in the next chapter.]

7.* As an application of Exercise 14, let $\mathcal{H}$ denote the vector space of entire functions, that is, the set of functions that are holomorphic in all of $\mathbb{C}$. Given a compact subset K of the complex plane and $f \in \mathcal{H}$, let $\|f\|_K = \sup_{z \in K} |f(z)|$. If K_n denotes the closed disc centered at the origin and of radius n, define

$$d(f, g) = \sum_{n=1}^{\infty} \frac{1}{2^n} \frac{\|f-g\|_{K_n}}{1 + \|f-g\|_{K_n}} \qquad \text{whenever } f, g \in \mathcal{H}.$$

Then d is a metric, and $\mathcal{H}$ is a complete metric space with respect to d. Also, $d(f_n, f) \to 0$ if and only if f_n converges to f uniformly on every compact subset of $\mathbb{C}$.

Birkhoff's theorem (Problem 5, Chapter 2, Book II) states that there exists an entire function F so that the set $\{F(z+n)\}_{n=1}^{\infty}$ is dense in $\mathcal{H}$. Also, MacLane's theorem (see the end of the same problem in Book II) says that there is an entire function G so that the set of its derivatives $\{G^{(n)}(z)\}_{n=1}^{\infty}$ is dense in $\mathcal{H}$.

By Exercise 14, the set of functions in $\mathcal{H}$ with either of these properties is generic in $\mathcal{H}$, hence the set of entire functions with *both* properties is also generic.

5 Rudiments of Probability Theory

> The whole of my work in probability theory together with Khinchin, in general the whole first period of my work in this theory was marked by the fact that we employed methods worked out in the metric theory of functions. Such topics as conditions for the applicability of the law of large numbers or a condition for convergence of a series of independent random variables essentially involved methods forged in the general theory of trigonometric series...
>
> *A. N. Kolmogorov, ca. 1987*

> One owes to Steinhaus the definition of independent functions, whether there are finitely or infinitely many. It follows from this definition, first published here, that certain systems of orthogonal functions... (including) those of Rademacher, consist of independent functions.
>
> *M. Kac, 1936*

The simplest way to introduce the basic concepts of probability theory is to begin by considering Bernoulli trials (for example, coin flips) and inquire as to what happens in the limit as the number of trials tends to infinity. Essential here is the idea of independent events that is subsumed in the more elaborate notion of mutually independent random variables.[1]

The case of Bernoulli trials where each flip has probability $1/2$ can be translated as the study of the Rademacher functions. As we will see, the properties of these mutually independent functions lead to some remarkable consequences for random series. In particular, when a formal Fourier series is randomized by the Rademacher functions there is then the following striking instance of the "zero-one law": either almost every

[1] We prefer to use the terminology "function" instead of "random variable" in much of what follows.

resulting series corresponds to an L^p function for every $p < \infty$, or almost none is the Fourier series of an L^1 function.

From this special set of independent functions we turn to the aspects of the general theory, and our focus is on the behavior of sums of more general independent functions. In the first instance, when these functions are identically distributed (and square integrable) we obtain the "central limit theorem" in this more extended setting. We also see that there is a close link with the ergodic theorem, and this allows us to prove one form of the "law of large numbers."

Next we consider independent functions that are not necessarily identically distributed. Here the main property that is exploited is that the corresponding sums form a "martingale sequence." In fact, an interesting case of this was seen in the analysis of sums involving Rademacher functions. Of importance at this point is the maximal theorem for martingale sequences, akin to the maximal theorem in Chapter 2.

We conclude this chapter by returning to Bernoulli trials, now interpreted as a random walk on the line. It is natural to consider the analogous random walks in d dimensions. For these we find some striking differences between the cases $d \leq 2$ and $d \geq 3$, in terms of their recurrence properties.

1 Bernoulli trials

An examination of some questions related to coin flips give the easiest examples of some of the concepts of probability theory.

1.1 Coin flips

We begin by considering the simplest gambling game. Two players, A and B, decide to flip a fair coin N times. Each time the coin comes up "heads" player A wins one dollar; each time the coin comes up "tails" player A loses a dollar. Since each flip has two possible outcomes, there are 2^N possible sequences of outcomes for their game. If we take into account the resulting possibilities, a question that arises is: what are (say) player A's chances of winning, and in particular, his chances of winning k dollars, for some k?

To answer this question we first formalize the above situation and introduce some terminology whose more general usage will occur later. The 2^N possible scenarios (or "outcomes") under consideration can be thought of as points in $\mathbb{Z}_2^N$, the N-fold product of the two-point space

$\mathbb{Z}_2 = \{0, 1\}$, with 0 standing for heads and 1 for tails. That is,

$$\mathbb{Z}_2^N = \{x = (x_1, \ldots, x_N), \quad \text{with } x_j = 0 \text{ or } 1 \text{ for each } j,\ 1 \leq j \leq N\}.$$

If we assume that flipping heads or tails, at the nth flip, is equally probable (and hence each has probability $1/2$) for every n, we are then quickly led to the following definitions: The space $\mathbb{Z}_2^N$ is our underlying "probability space"; on it there is a measure m, the "probability measure" which assigns measure 2^{-N} to each point of $\mathbb{Z}_2^N$, and $m(\mathbb{Z}_2^N) = 1$. We note that if E_n denotes the collection of events for which the nth flip is heads, $E_n = \{x \in \mathbb{Z}_2^N : \ x_n = 0\}$, then $m(E_n) = 1/2$ for all $1 \leq n \leq N$; also $m(E_n \cap E_m) = m(E_n)m(E_m)$, for all n, m with $n \neq m$. The latter identity reflects the fact that the outcomes of the nth and m^{th} flips are "independent."

We also need to consider certain functions on our probability space. (In the parlance of probability theory, functions on probability spaces are often referred to as **random variables**; we prefer to retain the designation "functions.") We define the function r_n to be the amount player A wins (or losses) at the nth flip, that is, $r_n(x) = 1$ if $x_n = 0$, and $r_n(x) = -1$ if $x_n = 1$, where $x = (x_1, \ldots, x_n)$. The sum

$$S_N(x) = S(x) = \sum_{n=1}^{N} r_n(x)$$

gives the total winnings (or losses) of player A after N flips.

Next, let us get an idea of what is the probability that $S(x) = k$, for a given integer k. If a given point $x \in \mathbb{Z}_2^N$ has N_1 zeroes and N_2 ones among its coordinates, (that is, player A has N_1 wins and N_2 losses), then of course $S(x) = k$ means $k = N_1 - N_2$, while $N_1 + N_2 = N$. Thus

$$N_1 = (N + k)/2 \quad \text{and} \quad N_2 = (N - k)/2,$$

and k has the same parity as N. To proceed further, we assume that N is even; the case of N odd is similar. (See Exercise 1.)

Thus in our probability space one has as many points x for which $S(x) = k$ as ways one can choose N_1 zeroes when making N choices among either 0 or 1. This number is the binomial coefficient

$$\binom{N}{N_1} = \frac{N!}{N_1!(N - N_1)!} = \frac{N!}{\left(\frac{N+k}{2}\right)!\left(\frac{N-k}{2}\right)!}.$$

As a result, since each point carries measure 2^{-N}, we have that

$$m(\{x:\ S(x)=k\}) = 2^{-N}\frac{N!}{\left(\frac{N+k}{2}\right)!\left(\frac{N-k}{2}\right)!}. \tag{1}$$

What can we say about the relative size of these numbers as k varies from $-N$ to N, (with k even)? The smallest values of (1) are attained at the end-points, $k=-N$ or $k=N$, with $m(\{x:\ S(x)=N\}) = m(\{x:\ S(x)=-N\}) = 2^{-N}$. As k varies from $-N$ to 0 (with k even), $m(\{x:\ S(x)=k\})$ increases, and then decreases as k increases from 0 to N. This is because

$$\frac{m(\{x:\ S(x)=k+2\})}{m(\{x:\ S(x)=k\})} = \frac{N-k}{N+k+2},$$

and the right-hand side is greater than 1 or less than 1 according to whether $k \le -2$ or $k \ge 0$, respectively. Thus clearly (1) attains its maximum value at $k=0$, and this is

$$2^{-N}\frac{N!}{((N/2)!)^2}.$$

By Stirling's formula (more about this below), this quantity is approximately $\frac{2}{\sqrt{2\pi}}N^{-1/2}$, which is much larger than the minimum value 2^{-N}.

With this, we leave these elementary considerations and begin to deal with the questions of probability theory that arise when we pass to the limiting situation $N \to \infty$.

1.2 The case $N = \infty$

Here we take our probability space to be the infinite product of copies of $\mathbb{Z}_2$, which is written as $\mathbb{Z}_2^\infty$, and which we denote more simply as X. That is,

$$X = \{x = (x_1, \ldots, x_n, \ldots),\quad \text{each } x_n = 0 \text{ or } 1 \text{ for all } n \ge 1\}.$$

The space X inherits the natural product measure from each of the measures of the partial products $\mathbb{Z}_2^N$ (in turn from each of the factors $\mathbb{Z}_2$) above as follows. A set E is a **cylinder set** in X whenever there is a (finite) N and a set $E' \in \mathbb{Z}_2^N$, so that $x \in E$ if and only if $(x_1, \ldots, x_N) \in E'$. With this definition the collection of cylinder sets together with their finite unions and intersections, and complements, forms an algebra on X. The main point now is that the function m defined first on these sets by $m(E) = m_N(E')$, (where $m_N = m$ is the measure on $\mathbb{Z}_2^N$ described in the previous section) extends to a measure on the σ-algebra of sets generated by the cylinder sets. Clearly $m(X) = 1$. (In this connection, the reader may consult Exercises 14 and 15 in Chapter 6 of Book III.)

More generally consider a pair (X, m), where we are given a σ-algebra of subsets of X (the "measurable" sets, or "events") and a measure m on this σ-algebra, with $m(X) = 1$. Adopting the terminology used previously, we refer to X as a **probability space** and m as a **probability measure**. In this context, one uses the terminology "almost surely" to mean "almost everywhere."

Returning to the case $X = \mathbb{Z}_2^\infty$ with the product measure defined above, we can extend to it the functions r_n, for all $1 \leq n < \infty$. This means that we take $r_n(x) = 1 - 2x_n$, where $x = (x_1, \ldots, x_n, \ldots)$ and $x_n = 0$ or 1, for each n. These functions may also be viewed as setting up a correspondence between X and the interval $[0, 1]$, with the measure m then identified with Lebesgue measure on this interval. In fact, consider the mapping $D : X \to [0, 1]$ given by

$$D : (x_1, \ldots, x_n, \ldots) \mapsto \sum_{j=1}^{\infty} \frac{x_j}{2^j} = t \in [0, 1]. \tag{2}$$

The correspondence D becomes a bijection from X to $[0, 1]$ if we remove the denumerable sets Z_1 and Z_2 respectively from X and $[0, 1]$, with Z_1 consisting of all points in X whose coordinates are all 0 or all 1 after a finite number of places; and Z_2 consists of all dyadic rationals (points in $[0, 1]$ of the form $\ell/2^m$, with ℓ and m integers). Moreover, note that if $E \subset X$ is the cylinder set $E = \{x : \; x_j = a_j, \; 1 \leq j \leq N\}$ where the a_j are a given finite set of 0's and 1's, then $m(E) = 2^{-N}$. Moreover, D maps E to the dyadic interval $\left[\frac{\ell}{2^N}, \frac{\ell+1}{2^N}\right]$, with $\ell = \sum_{j=1}^{N} 2^{N-j} a_j$. Of course this interval has Lebesgue measure 2^{-N}. From this observation, the assertions about the correspondence of X with $[0, 1]$ follow easily.

The identification of X with $[0, 1]$ allows us to write the functions r_n also as functions of $t \in [0, 1]$ (each undefined on a finite set); thus we shall write $r_n(x)$ or $r_n(t)$ interchangeably (with $x \in X$, or $t \in [0, 1]$). Note that $r_1(t) = 1$ for $0 < t < 1/2$, and $r_1(t) = -1$, for $1/2 < t < 1$. Also if we extend r_1 to $\mathbb{R}$ by making it periodic of period 1, then $r_n(t) = r_1(2^{n-1}t)$. The functions $\{r_n\}$ on $[0, 1)$ are the **Rademacher functions**.

The critical property enjoyed by these functions is their mutual independence, defined as follows. Given a probability space (X, m), we say

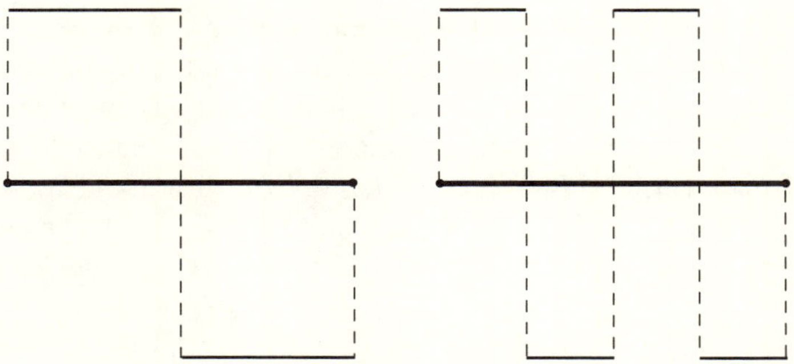

Figure 1. The Rademacher functions r_1 and r_2

that a sequence $\{f_n\}_{n=1}^{\infty}$ of real-valued measurable[2] functions on X are **mutually independent** if for any sequence of Borel sets B_n in $\mathbb{R}$

$$(3)\qquad m\left(\bigcap_{n=1}^{\infty}\{x:\ f_n(x)\in B_n\}\right)=\prod_{n=1}^{\infty}m(\{x:\ f_n(x)\in B_n\}).$$

Similarly, we say that a collection of sets $\{E_n\}$ are mutually independent if their characteristic functions are mutually independent. There is of course a similar definition of mutual independence if we are given only a finite collection $f_1,\ldots,f_N$ of functions or a finite collection $E_1,\ldots,E_N$ of sets. Note that for a pair of sets E_n and E_m, this notion coincides with what has been previously encountered. However a collection of functions (or sets) need not be mutually independent, even if they are pair-wise so. (See Exercise 2.) Also, note that if $f_1,\ldots,f_n$ are (say) bounded and mutually independent functions, then the integral of their product equals the product of their integrals,

$$(4)\qquad \int_X f_1(x)\cdots f_n(x)\,dm=\left(\int_X f_1(x)\,dm\right)\cdots\left(\int_X f_n(x)\,dm\right).$$

This follows by first verifying the identity directly when the f's are finite linear combinations of characteristic functions and then passing to the limit.

A general way that independent functions arise is as follows. Suppose our probability space (X,m) is a product of probability spaces (X_n,m_n),

[2]All functions (and sets) that arise are henceforth assumed to be measurable. Also we keep to the assumption that our functions (random variables) are real-valued, except in Section 1.7 and Section 2.6 onwards.

$n = 1, 2, \ldots$ with m equal to the product measure of the m_n. Assume that the function $f_n(x)$, defined for $x \in X$, depends only on the nth coordinate of x, that is $f_n(x) = F_n(x_n)$, where each F_n is given on X_n, and $x = (x_1, x_2, \ldots, x_n, \ldots)$. Then the functions $\{f_n\}$ are mutually independent. To see this set $E_n = \{x : \; f_n(x) \in B_n\}$ with $E_n \subset X$, similarly $E'_n = \{x_n : F_n(x_n) \in B_n\}$ with $E'_n \subset X_n$. Then $E_n = \{x : \; x_n \in E'_n\}$ is a cylinder set with $m(E_n) = m_n(E'_n)$. Hence it is clear that for each N

$$m\left(\bigcap_{n=1}^{N} E_n\right) = \prod_{n=1}^{N} m_n(E'_n) = \prod_{n=1}^{N} m(E_n).$$

Letting $N \to \infty$ gives (3), proving our assertion. This obviously applies to the Rademacher functions, showing their mutual independence.

Incidentally, this example of mutually independent random variables in a way represents the general situation. (See Exercise 6.)

1.3 Behavior of S_N as $N \to \infty$, first results

After these preliminaries we are ready to consider the behavior of

$$S_N(x) = \sum_{n=1}^{N} r_n(x),$$

which represents player A's winnings after N flips. It turns out that the order of magnitude of S_N, as $N \to \infty$, is essentially much smaller than N. A hint of what is to be expected comes from the following observation.

Proposition 1.1 *For each integer $N \geq 1$,*

$$\|S_N\|_{L^2} = N^{1/2}. \tag{5}$$

This proposition follows from the fact that $\{r_n(t)\}$ is an orthonormal system on $L^2([0,1])$. Indeed, we have that $\int_0^1 r_n(t)\,dt = 0$ because each r_n is equal to 1 on a set of measure $1/2$, and equal to -1 on a set of measure also $1/2$. Moreover, by their mutual independence and (4), we have

$$\int_0^1 r_n(t) r_m(t)\,dt = 0 \quad \text{if } n \neq m.$$

In addition, we obviously have $\int_0^1 r_n^2(t)\,dt = 1$. Therefore

$$\left\| \sum_{n=1}^{N} a_n r_n \right\|_{L^2}^2 = \sum_{n=1}^{N} |a_n|^2,$$

and the assertion follows by taking $a_n = 1$ for $1 \leq n \leq N$.

Note: The sequence $\{r_n\}$ is far from complete in $L^2([0,1])$. See Exercises 13 and 16.

As an immediate consequence we have the convergence of the averages S_N/N to 0 "in probability." The relevant definition is as follows. One says that a sequence of functions $\{f_n\}$ **converges to f in probability**, if for every $\epsilon > 0$,

$$m(\{x: \ |f_N(x) - f(x)| > \epsilon\}) \to 0 \qquad \text{as } N \to \infty.^3$$

Corollary 1.2 *S_N/N converges to 0 in probability.*

In fact,

$$m(\{|S_N(x)/N| > \epsilon\}) = m(\{|S_N(x)| > \epsilon N\}) \leq \frac{1}{\epsilon^2 N^2} \int |S_N(x)|^2 \, dm,$$

by Tchebychev's inequality. Hence $m(\{x : |S_N(x)/N| > \epsilon\}) \leq 1/(\epsilon^2 N)$, and the corollary is proved. It is to be noted that by the same argument one gets the better result that $S_N/N^\alpha \to 0$ in probability as $N \to \infty$, as long as $\alpha > 1/2$. A stronger version of this conclusion is given in Corollary 1.5 below.

1.4 Central limit theorem

The identity (5) suggests that the way to look more carefully at S_N for large N is to normalize it and consider instead $S_N/N^{1/2}$. Studying the limit of this quantity in the appropriate sense leads us to the **central limit theorem**. This is expressed in terms of the notion of **distribution measure** of a function, defined as follows. Whenever f is a (real-valued) function on a probability space (X, m), its distribution measure is defined to be the unique (Borel) measure $\mu = \mu_f$ on $\mathbb{R}$ that satisfies

$$\mu(B) = m(\{x: \ f(x) \in B\}) \qquad \text{for all Borel sets } B \subset \mathbb{R}.$$

Note that a distribution measure is automatically a probability measure on $\mathbb{R}$, since $\mu(\mathbb{R}) = 1$. Incidentally the distribution measure is closely related to the distribution function λ that appeared in Section 4.1 of Chapter 2, because

$$\lambda_f(\alpha) = m(\{x: \ |f(x)| > \alpha\}) = \mu_{|f|}((\alpha, \infty)).$$

---

[3] In measure theory, this notion is usually referred to as "convergence in measure."

The argument used there to prove (29) can also be applied to establish the following assertions. First, f is integrable on X precisely when $\int_{-\infty}^{\infty} |t|\, d\mu(t) < \infty$, and then $\int_X f(x)\, dm = \int_{-\infty}^{\infty} t\, d\mu(t)$. Similarly, f is in $L^p(X, m)$ exactly when $\int_{-\infty}^{\infty} |t|^p\, d\mu(t)$ is finite and this quantity equals $\|f\|_{L^p}^p$.

More generally, if G is a non-negative continuous function on $\mathbb{R}$ (or continuous and bounded), then

$$\int_X G(f)(x)\, dm = \int_{\mathbb{R}} G(t)\, d\mu(t). \tag{6}$$

See Exercise 12.

We say (using the parlance of probability theory) that f **has a mean** if f is integrable, and its **mean** m_0 (also called its **expectation**) is defined as

$$m_0 = \int_X f(x)\, dm = \int_{-\infty}^{\infty} t\, d\mu(t).$$

If f is also square integrable on X, the we define its **variance** σ^2 by

$$\sigma^2 = \int_X (f(x) - m_0)^2\, dm.$$

In particular, if $m_0 = 0$, then

$$\sigma^2 = \|f\|_{L^2}^2 = \int_{-\infty}^{\infty} t^2\, d\mu(t).$$

A measure μ that arises naturally in this context is the **Gaussian** (or **normal distribution**), the measure on $\mathbb{R}$ whose density function is $\frac{1}{\sqrt{2\pi}} e^{-t^2/2}$, that is,

$$\nu((a,b)) = \int_a^b \frac{1}{\sqrt{2\pi}} e^{-t^2/2}\, dt.$$

More generally, the normal measure with variance σ^2 is the one given by

$$\nu_{\sigma^2}((a,b)) = \int_a^b \frac{1}{\sigma\sqrt{2\pi}} e^{-t^2/(2\sigma^2)}\, dt.$$

1.5 Statement and proof of the theorem

We can now come to De Moivre's theorem, the central limit theorem in the special context of coin flips. It states that the distribution measure of $S_N/N^{1/2}$ converges to the normal distribution in the following sense.

Theorem 1.3 *For each $a < b$, we have*

$$m(\{x : \ a < S_N(x)/N^{1/2} < b\}) \to \int_a^b \frac{e^{-t^2/2}}{\sqrt{2\pi}}\,dt, \qquad \text{as } N \to \infty.$$

In proving this result we consider first the case when we restrict ourselves to N even; the limit when N is restricted to be odd is, except for small changes, treated the same way. Joining the two cases will give the desired result.

Proof. According to (1), with $k = 2r$, r an integer, and $\alpha < \beta$,

$$m(\{x : \ \alpha < S_N(x) < \beta\}) = \sum_{\alpha<2r<\beta} P_r, \qquad \text{where } P_r = \tfrac{2^{-N}N!}{(N/2+r)!(N/2-r)!}.$$

Hence

$$m(\{x : \ a < S_N(x)/N^{1/2} < b\}) = \sum_{aN^{1/2}<2r<bN^{1/2}} P_r.$$

With a and b fixed, this means that the r's are restricted by $r = O(N^{1/2})$. We claim that under this restriction

$$(7) \qquad P_r = \frac{2}{\sqrt{2\pi}N^{1/2}} e^{-2r^2/N}\left(1 + O(1/N^{1/2})\right) \qquad \text{as } N \to \infty.$$

To verify this we use a version of Stirling's formula,[4] which we state as

$$N! = \sqrt{2\pi} N^{N+1/2} e^{-N}\left(1 + O(1/N^{1/2})\right) \qquad \text{as } N \to \infty.$$

It follows from this that

$$P_r = \frac{2}{\sqrt{2\pi}}\frac{1}{N^{1/2}}\frac{1}{(1+\frac{2r}{N})^{N/2+r+1/2}}\frac{1}{(1-\frac{2r}{N})^{N/2-r+1/2}}\left(1 + O(1/N^{1/2})\right).$$

Now $\log(1 + x) = x - x^2/2 + O(|x|^3)$, as $x \to 0$, so if

$$A_r = \left(\frac{N}{2} + r + \frac{1}{2}\right)\log(1 + 2r/N),$$

[4]See for instance Theorem 2.3 in Appendix A, Book II. The error terms $O(1/N^{1/2})$ can be improved; but even a weaker bound would suffice for our purpose.

then

$$A_r = \left(\frac{N}{2} + r + \frac{1}{2}\right)\left(\frac{2r}{N} - \frac{2r^2}{N^2}\right) + O(N^{-1/2}) \quad \text{since } r = O(N^{1/2}).$$

Hence $A_r + A_{-r} = \frac{2r^2}{N} + O(N^{-1/2})$, and because

$$\left[\left(1 + \frac{2r}{N}\right)^{N/2+r+1/2}\left(1 - \frac{2r}{N}\right)^{N/2-r+1/2}\right] = e^{-A_r - A_{-r}},$$

we have the asserted result (7).

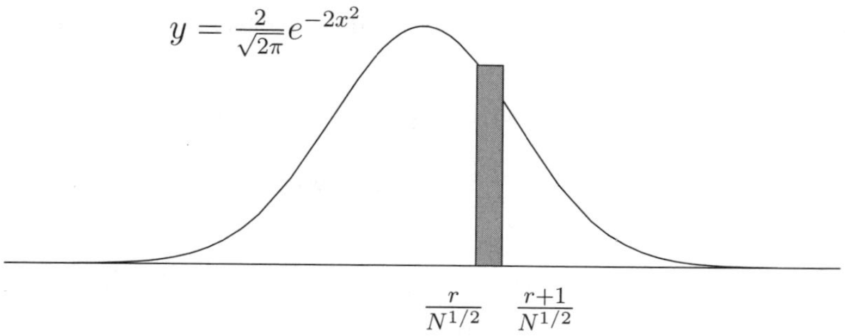

Figure 2. Approximating the integral of a Gaussian

Now $e^{-2r^2/N} - e^{-2t^2} = O(e^{-2t^2}/N^{1/2})$ if $t \in [r/N^{1/2}, (r+1)/N^{1/2}]$, again because $r = O(N^{1/2})$. Therefore

$$\frac{1}{N^{1/2}}e^{-2r^2/N} = \int_{r/N^{1/2}}^{(r+1)/N^{1/2}} e^{-2t^2}\, dt(1 + O(N^{-1/2})).$$

Taking (7) into account we see that as a result

$$\begin{aligned} m(\{x : \ a < S_N(x)/N^{1/2} < b\}) &= \sum_{aN^{1/2} < 2r < bN^{1/2}} P_r \\ &= \int_{a/2}^{b/2} \frac{2}{\sqrt{2\pi}} e^{-2t^2}\, dt + O(N^{-1/2}) \\ &= \int_a^b \frac{1}{\sqrt{2\pi}} e^{-t^2/2}\, dt + O(N^{-1/2}), \end{aligned}$$

upon making the change of variables $t \to t/2$. Letting $N \to \infty$ gives our desired conclusion.

1.6 Random series

A striking illustration of randomness inherent in the Rademacher functions is the observation that, although the series $\sum_{n=1}^{\infty} 1/n$ diverges, the series $\sum_{n=1}^{\infty}(\pm)1/n$ converges for "almost all" choices of $\pm$ signs, where the $\pm$ signs for the different n's are chosen independently and with equal probability.

A precise and more general formulation is as follows.

Theorem 1.4

(a) *Suppose* $\sum_{n=1}^{\infty} |a_n|^2 < \infty$. *Then for almost every* $t \in [0,1]$, *the series* $\sum_{n=1}^{\infty} a_n r_n(t)$ *converges.*

(b) *However if* $\sum_{n=1}^{\infty} |a_n|^2$ *diverges, then* $\sum_{n=1}^{\infty} a_n r_n(t)$ *diverges for almost all* $t \in [0,1]$.

Note. The fact that these conclusions must hold almost everywhere (if they hold on sets of positive measure) is a particular case of the "zero-one law." More about this in Section 2.3.

To prove the theorem, recall that $\{r_n\}$ is an orthonormal sequence in $L^2([0,1])$. Thus if $\sum_{n=1}^{\infty} |a_n|^2 < \infty$, the sequence $\{\sum_{n=1}^{N} a_n r_n(t)\}$ converges in the L^2 norm, as $N \to \infty$, to a function $f \in L^2([0,1])$. For this f it is convenient to write

$$f \sim \sum_{n=1}^{\infty} a_n r_n, \qquad \text{and set} \qquad S_N(f) = \sum_{n=1}^{N} a_n r_n.$$

To prove the almost everywhere convergence of the S_N, we bring in averaging operators that average over dyadic intervals, which are defined as follows. For each positive integer n the **dyadic intervals** of length 2^{-n} are the 2^n sub-intervals of $[0,1]$ of the form $\left(\frac{\ell}{2^n}, \frac{\ell+1}{2^n}\right]$ with $0 \leq \ell < 2^n$. These obviously form a disjoint covering of $[0,1]$ (except for the origin). Now for each f that is integrable over $[0,1]$, and every n, set

$$\mathbb{E}_n(f)(t) = \frac{1}{m(I)} \int_I f(s)\, ds$$

when $t \in I$, and I is a dyadic interval of length n. (Note that $\mathbb{E}_n(f)(t)$ is not defined for $t = 0$, but this is immaterial.)

For the functions f that arise as above (as L^2 limits of finite linear combinations of the r_n), there is the basic identify

$$\mathbb{E}_N(f) = S_N(f) \quad \text{for all } N. \tag{8}$$

To prove this note first that $\mathbb{E}_N(r_n) = r_n$ if $N \geq n$. In fact, when $N \geq n$, each r_n is constant on every dyadic interval of length 2^{-N}. Also $\mathbb{E}_N(r_n) = 0$ if $n > N$, since the integral of r_n on each dyadic interval of length 2^{-N} vanishes. These facts are easily reduced to the case $n = 1$ by using the identity $r_n(t) = r_1(2^{n-1}t)$. Thus we have shown that (8) holds for any finite linear combination of the Rademacher functions. Hence $S_N(f) = \mathbb{E}_N(S_n(f))$, if $n \geq N$, and a passage to the limit, as $n \to \infty$, establishes (8).

Now by the Lebesgue differentiation theorem[5] $\lim_{N\to\infty} \mathbb{E}_N(f)(t)$ exists and equals $f(t)$ at all points of the Lebesgue set of f, and hence almost everywhere. Thus by (8) the series converges almost everywhere, and part (a) is proved.

Before we turn to the converse, part (b), we digress to strengthen the conclusion obtained in Section 1.3. There we considered the sums $S_N(t) = \sum_{n=1}^N r_n(t)$ and showed that $S_N/N \to 0$ in probability. This initial conclusion is itself implied by the "strong law of large numbers," which in this case takes the following form.

Corollary 1.5 *Let $S_N(t) = \sum_{n=1}^N r_n(t)$. Then $S_N(t)/N \to 0$, as $N \to \infty$ for almost every t. In fact, if $\alpha > 1/2$, then $S_N(t)/N^\alpha \to 0$ for almost every t.*

Proof. Fix $1/2 < \beta < \alpha$, and let $a_n = n^{-\beta}$ and $b_n = n^\beta$. Clearly $\sum a_n^2 < \infty$. Set $\tilde{S}_N(t) = \sum_{n=1}^N a_n r_n(t)$. Then, by summation by parts, setting $\tilde{S}_0 = 0$, we get

$$\begin{aligned} S_N(t) &= \sum_{n=1}^N r_n = \sum_{n=1}^N a_n r_n b_n \\ &= \sum_{n=1}^N (\tilde{S}_n - \tilde{S}_{n-1}) b_n \\ &= \tilde{S}_N b_N + \sum_{n=1}^{N-1} \tilde{S}_n (b_n - b_{n+1}). \end{aligned}$$

However $|b_n - b_{n+1}| = b_{n+1} - b_n$, and $\sum_{n=1}^{N-1}(b_{n+1} - b_n) = b_N - 1 = O(N^\beta)$ while the convergence of the series $\sum_{n=1}^\infty a_n r_n(t)$ for almost all t guarantees that $|\tilde{S}_n(t)| = O(1)$ for almost every t. As a result, for those t, $S_N(t) = O(N^\beta)$ and this implies $S_N(t)/N^\alpha \to 0$ for almost all t, proving the corollary.

[5] See for example Theorem 1.3 and its corollaries in Chapter 3, Book III.

We now turn to the proof of part (b) of the theorem. It is based on the following lemma.

Lemma 1.6 *Suppose E is a subset of $[0,1]$ with $m(E) > 0$. Then there is a $c > 0$ and a positive integer N_0 so that if F is any finite sum of the form*

$$F(t) = \sum_{n \geq N_0} a_n r_n(t)$$

then

$$\int_E |F(t)|^2 \, dt \geq c \sum_{n \geq N_0} a_n^2.$$

Besides the orthogonality of the $\{r_n\}$ already used, the proof requires a stronger orthogonality, which again exploits the mutual independence of the Rademacher functions.

For each ordered pair (n, m), with $n < m$, define $\varphi_{n,m}(t) = r_n(t) r_m(t)$. Then the collection $\{\varphi_{n,m}\}$ is an orthonormal sequence in $L^2([0,1])$. To see this, consider $\int_0^1 \varphi_{n,m}(t)\varphi_{n',m'}(t)\, dt$. When $(n,m) = (n', m')$ the integral clearly equals 1. Now if $(n,m) \neq (n', m')$, but n or m equals n' or m' (in any order), then we see that the integral vanishes by the orthogonality of the $\{r_n\}$. Finally, if neither n or m equals n' or m', then we apply (4) to the four mutually independent functions r_n, r_m, $r_{n'}$ and $r_{m'}$, establishing the assertion.

Assuming that F is any finite sum of the form $\sum_n a_n r_n(t)$, we have

$$(F(t))^2 = \sum_n a_n^2 r_n^2(t) + 2 \sum_{n<m} a_n a_m r_n(t) r_m(t),$$

hence

$$\int_E (F(t))^2 \, dt = m(E) \sum_n a_n^2 + 2 \sum_{n<m} a_n a_m \gamma_{n,m}, \tag{9}$$

with $\gamma_{n,m} = \int_E r_n(t) r_m(t)\, dt = \int_0^1 \chi_E(t) \varphi_{n,m}(t)\, dt$. Thus by the orthogonality of the $\{\varphi_{n,m}\}$ and Bessel's inequality,[6] $\sum_{n,m} \gamma_{n,m}^2 \leq m(E) \leq 1$. Hence for any fixed $\delta > 0$ (δ will be chosen momentarily), there is an N_0 so that $\sum_{N_0 \leq n < m} \gamma_{n,m}^2 \leq \delta$. We apply this with Schwarz's inequality to the last term on the right-hand side of (9), restricting ourselves to F's of

[6]For Bessel's inequality, see Section 2.1 in Chapter 4 of Book III.

the form $F(t) = \sum_{n \geq N_0} a_n r_n(t)$. The result is that this term is bounded by

$$2\left(\sum_{N_0 \leq n < m} (a_n a_m)^2\right)^{1/2} \delta^{1/2} \leq 2\delta^{1/2} \sum_{n \geq N_0} a_n^2.$$

If we choose δ so that $2\delta^{1/2} \leq m(E)/2$, then from (9) we get

$$\int_E |F(t)|^2 \, dt \geq \frac{1}{2} m(E) \sum_{n \geq N_0} a_n^2,$$

and the lemma is proved with $c = m(E)/2$.

To conclude the proof of part (b) of Theorem 1.4 we suppose the opposite, that $\{S_N(t)\}$ converges in a set of positive measure. Then this sequence is uniformly bounded on a set of positive measure, and that means that there is an M and a set E, with $m(E) > 0$, so that $|S_N(t)| \leq M$ for all N if $t \in E$. As a result there is an M' so that, for all $N \geq N_0$, one has $\left|\sum_{N_0 \leq n \leq N} a_n r_n(t)\right| \leq M'$ whenever $t \in E$.

The lemma guarantees that $\sum_{N_0 \leq n \leq N} a_n^2 \leq c^{-1}(M')^2$ for all N, and letting $N \to \infty$ gives us that $\sum a_n^2$ converges. This establishes the contradiction and finishes the proof of the theorem.

1.7 Random Fourier series

The ideas above can also be used to obtain remarkable results about random Fourier series, that is, Fourier series on $[0, 2\pi]$ of the form

$$\sum_{n=-\infty}^{\infty} \pm c_n e^{in\theta}.$$

To parametrize the choices of $\pm$ signs in terms of the Rademacher functions, we need to re-index these functions so that their indices range over $\mathbb{Z}$. For this reason, it is convenient to change notation and write ρ_n for the functions defined by $\rho_n(t) = r_{2n+1}(t)$, if $n \geq 0$, and $\rho_n(t) = r_{-2n}(t)$, if $n < 0$, with $n \in \mathbb{Z}$. We allow the coefficients c_n to be complex, so that here we deal with complex-valued functions.

Theorem 1.7

(a) *If* $\sum_{n=-\infty}^{\infty} |c_n|^2 < \infty$, *then for almost every* $t \in [0, 1]$ *the function*

$$f_t(\theta) \sim \sum_{n=-\infty}^{\infty} \rho_n(t) c_n e^{in\theta} \tag{10}$$

belongs to $L^p([0, 2\pi])$ *for every* $p < \infty$.

(b) *If* $\sum_{n=-\infty}^{\infty} |c_n|^2 = \infty$, *then for almost every* $t \in [0, 1]$ *the series* (10) *is not the Fourier series of an integrable function.*

The proof is based on Khinchin's inequality, which like Lemma 1.6 is a further exploitation of the independence of the Rademacher functions.

Suppose $\{a_n\}$ are complex numbers with $\sum_{n=-\infty}^{\infty} |a_n|^2 < \infty$. Let $F(t) = \sum_{n=-\infty}^{\infty} a_n \rho_n(t)$, with F taken as the L^2 limit on $L^2([0, 1])$ of the partial sums.

Lemma 1.8 *For each* $p < \infty$ *there is a bound* A_p *so that*

$$\|F\|_{L^p} \leq A_p \|F\|_{L^2},$$

for all $F \in L^p([0, 1])$ *of the form* $F(t) = \sum_{n=-\infty}^{\infty} a_n \rho_n(t)$.

It clearly suffices to prove the corresponding statement when the a_n are assumed real and have been normalized so that $\|F\|_{L^2}^2 = \sum_{-\infty}^{\infty} a_n^2 = 1$.

Now observe that the defining property (3) shows that whenever $\{f_n\}$ is a sequence of mutually independent (real-valued) functions, so is the sequence $\{\Phi_n(f_n)\}$, with $\{\Phi_n\}$ any sequence of continuous functions from $\mathbb{R}$ to $\mathbb{R}$. As a result the functions $\{e^{a_n \rho_n(t)}\}$ are mutually independent. Thus if $F_N(t) = \sum_{|n| \leq N} a_n \rho_n(t)$, then

(11)

$$\int_0^1 e^{F_N(t)}\, dt = \int_0^1 \left(\prod_{n=-N}^{N} e^{a_n \rho_n(t)} \right) dt = \prod_{n=-N}^{N} \left(\int_0^1 e^{a_n \rho_n(t)}\, dt \right).$$

However, $\int_0^1 e^{a_n \rho_n(t)}\, dt = \cosh(a_n)$, since each ρ_n takes values $+1$ or -1 on sets of measure $1/2$ respectively. Also, $\cosh(x) \leq e^{x^2}$ for real x, as a comparison of their power series clearly shows. Hence

$$\int_0^1 e^{F_N(t)}\, dt \leq \prod_{n=-N}^{N} e^{a_n^2} \leq e^{\sum a_n^2} \leq e.$$

A similar inequality holds with the a_n replaced by $-a_n$. Altogether then

$$\int_0^1 e^{|F_N(t)|}\, dt \leq 2e.$$

A simple passage to the limit, as $N \to \infty$, then gives that $e^{|F(t)|}$ is integrable over $[0, 1]$, and $\int_0^1 e^{|F(t)|}\, dt \leq 2e$. However for each p there is a constant c_p so that $u^p \le c_p e^u$ for all $u \ge 0$. Thus $\|F\|_{L^p}^p \le 2ec_p$, and the lemma is proved with $A_p = (2ec_p)^{1/p}$.

We turn now to the proof of part (a) of the theorem. We may assume that $\sum_{n=-\infty}^{\infty} |c_n|^2 = 1$, and set $F(t) = f_t(\theta)$, with $a_n = c_n e^{in\theta}$, and θ fixed. Now

$$\int_0^1 |F(t)|^p \, dt = \int_0^1 |f_t(\theta)|^p \, dt \le A_p^p$$

by the lemma. Thus integrating over $\theta \in [0, 2\pi]$ gives

$$\int_0^{2\pi} \int_0^1 |f_t(\theta)|^p \, dt \, d\theta \le 2\pi A_p^p,$$

and by Fubini's theorem,

$$\int_0^{2\pi} |f_t(\theta)|^p \, d\theta < \infty,$$

for almost every $t \in [0, 1]$, and this is what was to be proved.

To prove the converse, part (b) in the theorem, suppose that for a set $E_1 \subset [0, 1]$ of positive measure we have $f_t(\theta) \in L^1([0, 2\pi])$, whenever $t \in E_1$. Since every function in $L^1([0, 2\pi])$ has a Fourier series that is Cesàro summable almost everywhere, it then follows that there is a set $\tilde{E} \subset [0, 1] \times [0, 2\pi]$ of positive two-dimensional measure, and an M so that

$$\sup_N |\sigma_N(f_t)(\theta)| \le M \qquad \text{for each } (t, \theta) \in \tilde{E}. \tag{12}$$

Here σ_N is the Cesàro sum given by $\sigma_N(f_t)(\theta) = \sum_{|n| \le N} \rho_n(t) c_n e^{in\theta}(1 - |n|/N)$. However, by Fubini's theorem, (12) holds for at least one θ_0, and all $t \in E$, where $m(E) > 0$. Now write $c_n e^{in\theta_0} = \alpha_n + i\beta_n$, with α_n and β_n real, then apply Lemma 1.6. Thus there is an M' and an N_0 so that

$$\sup_{N_0 \le |n| \le N} \sum \alpha_n^2 \le M',$$

and letting $N \to \infty$ shows that $\sum_{-\infty}^{\infty} \alpha_n^2$ converges. Similarly $\sum_{-\infty}^{\infty} \beta_n^2$ converges and the theorem is proved.

1.8 Bernoulli trials

Many of the results in Sections 1.1 to 1.5 that were proved above continue to hold in modified form when the equal probabilities of heads and tails are replaced by probabilities p and q, with $p+q=1$ and $0<p<1$. This more general situation is often referred to as that of **Bernoulli trials**.

To consider it we begin by replacing the probability measure on $\mathbb{Z}_2^\infty$ by the measure m_p that arises as the product measure on $\mathbb{Z}_2^\infty$ where for each factor $\mathbb{Z}_2=\{0,1\}$ the point 0 is assigned measure p and the point 1 is assigned measure q. (Incidentally, when $p\neq 1/2$, then under the mapping $D:\mathbb{Z}_2^\infty\to[0,1]$, the measure m_p now corresponds to a singular measure $d\mu_p$ on $[0,1]$. For this, see Problem 1.)

In this setting the law of large numbers takes the form that $S_N/N\to p-q$ as in Corollaries 1.2 and 1.5. The proof of the analog of the first corollary can be carried out in much the same way as before. The variant of the second corollary requires some further ideas and is dealt with in a general context in the next section. In addition, a modification of the proof of Theorem 1.3 gives its analog

$$m_p(\{x:\ a<\frac{S_N(x)-N(p-q)}{N^{1/2}}<b\})\to\frac{1}{\sigma\sqrt{2\pi}}\int_a^b e^{-t^2/(2\sigma^2)}\,dt$$

as $N\to\infty$, where $\sigma^2=1-(p-q)^2$.

This result is subsumed in the general form of the central limit theorem proved in the last part of the next section.

2 Sums of independent random variables

Our aim in this section is to put in a more general and abstract form some of the results for coin flips and Bernoulli trials dealt with in the first section. To begin with, we shall present a version of the law of large numbers.

2.1 Law of large numbers and ergodic theorem

Here we deduce a general form of this law from the ergodic theorem.[7] Another version, derived from the theory of martingales, will be presented in Section 2.2 below.

A sequence $(f_0,f_1,\ldots,f_n,\ldots)$ of functions is said to be **identically distributed** if the distribution measures μ_n of f_n (as defined in Section 1.4) are independent of n, that is, the measures $m(\{x:\ f_n(x)\in B\})$ are the same for all n for every Borel set B. If the sequence $\{f_n\}$ is

[7] A treatment of the ergodic theorem needed here can be found in Section 5* of Chapter 6 in Book III.

identically distributed and if f_0 has a mean (equal to m_0), then of course all f_n have a mean that equals m_0. The first main theorem is as follows.

Theorem 2.1 *Suppose $\{f_n\}$ is a sequence of functions that are mutually independent, are identically distributed, and have mean m_0. Then*

$$\frac{1}{N}\sum_{n=0}^{N-1} f_n(x) \to m_0 \quad \textit{for almost every } x \in X, \textit{ as } N \to \infty.$$

The possibility of reducing this theorem to the ergodic theorem depends on the device of replacing the sequence $\{f_n\}$ by another sequence that is "equimeasurable" with the first, in the following sense.

Given functions $f_1, \ldots, f_N$, their **joint distribution measure** is defined as the measure on $\mathbb{R}^N$ that satisfies for all Borel sets $B \subset \mathbb{R}^N$

$$\mu_{f_1,\ldots,f_N}(B) = m(\{x : (f_1(x), \ldots, f_N(x)) \in B\}).$$

Now suppose $\{g_n\}$ is a sequence on a (possibly different) probability space (Y, m^*). Then we say that $\{f_n\}$ and $\{g_n\}$ have the **same joint distribution** if for every N, we have

$$\mu_{f_1,\ldots,f_N}(B) = \mu_{g_1,\ldots,g_N}(B) \quad \text{for all Borel sets } B \subset \mathbb{R}^N.$$

With this definition in hand we come to the space Y that is relevant here. It is the infinite product $Y = R^\infty = \prod_{j=0}^\infty R_j$, where each R_j is $\mathbb{R}$. On each R_j we consider the measure μ, the common distribution measure of the f_n. Define m^* to be the corresponding product measure on Y.

We also consider the shift $\tau : Y \to Y$, given by $\tau(y) = (y_{n+1})_{n=0}^\infty$, if $y = (y_n)_{n=0}^\infty$. Finally we take for the $\{g_n\}$ the coordinate functions on Y given by $g_n(y) = y_n$, if $y = (y_n)_{n=0}^\infty$.

Everything will now be a consequence of the following four steps.

Observation 1. $g_n(\tau(y)) = g_{n+1}(y)$ for all $n \geq 0$; hence $g_n(y) = g_0(\tau^n y)$.

Observation 2. τ is measure-preserving and ergodic.

Conclusion 1. $\lim_{N\to\infty} \frac{1}{N}\sum_{n=0}^{N-1} g_n(y) = m_0$, for almost every $y \in Y$.

Conclusion 2. $\lim_{N\to\infty} \frac{1}{N}\sum_{n=0}^{N-1} f_n(x) = m_0$, for almost every $x \in X$.

The first observation is immediate.

That τ is measure preserving means that $m^*(\tau^{-1}(E)) = m^*(E)$ for every (measurable) set $E \subset Y$. Since Y is a product space, it suffices to verify this for all cylinder sets E, and then a simple limiting argument proves this for the general set E. If E is a cylinder set, E depends only on the first N coordinates, for some N. This means that $E = E' \times \prod_{j=N}^{\infty} R_j$, with E' a subset of $\prod_{j=0}^{N-1} R_j$, and $m^*(E) = \mu^{(N)}(E')$, where $\mu^{(N)}$ is the N-fold product of μ on the first N factors. However

$$\tau^{-1}(E) = R_0 \times E'' \times \prod_{j=N+1}^{\infty} R_j,$$

with $(y_1'', \ldots, y_N'') \in E''$ if and only if $(y_0', \ldots, y_{N-1}') \in E'$, where $y_{n+1}'' = y_n'$, for $0 \leq n \leq N-1$. Thus $m^*(\tau^{-1}(E)) = \mu^{(N+1)}(R_0 \times E'') = \mu^{(N)}(E')$ and the assertion $m^*(\tau^{-1}(E)) = m^*(E)$ is proved.

The ergodicity of τ follows from the fact that τ is **mixing**,[8] which means

$$\lim_{n\to\infty} m^*(\tau^{-n}(E) \cap F) = m^*(E)m^*(F) \tag{13}$$

for all pairs of $E, F \subset Y$.

To prove the mixing property it suffices, as before, to assume that both E and F are cylinder sets. So, for a sufficiently large N we have that $E = E' \times \prod_{j=N}^{\infty} R_j$ and $F = F' \times \prod_{j=N}^{\infty} R_j$, where both E' and F' are subsets of $\prod_{j=0}^{N-1} R_j$. Now, as above if $n \geq 1$,

$$\tau^{-n}(E) = \prod_{j=0}^{n-1} R_j \times E'' \times \prod_{j=N+n}^{\infty} R_j,$$

where E'' is the subset of $\prod_{j=n}^{N+n-1} R_j$ that corresponds to E'. Thus if $n > N$

$$\tau^{-n}(E) \cap F = F' \times \prod_{j=N}^{n-1} R_j \times E'' \times \prod_{j=N+n}^{\infty} R_j.$$

As a result $m^*(\tau^{-n}(E) \cap F) = m^*(E)m^*(F)$ whenever $n > N$ and (13) is established.

It follows immediately from (13), when taking $F = E$, that if E is an **invariant set**, that is $\tau^{-1}(E) = E$ almost everywhere, then $m^*(E) =$

---

[8]Also referred to as "strongly-mixing"; see Chapter 6 in Book III.

$(m^*(E))^2$, so $m^*(E) = 0$ or $m^*(E) = 1$. Thus there is no proper subset of X invariant under τ, and this means that τ is **ergodic**; so our second observation is established.

Now the function g_0 is integrable on Y since

$$\int_Y |g_0(y)|\, dm^*(y) = \int_{\mathbb{R}} |y_0|\, d\mu(y_0) = \int_X |f_0(x)|\, dm(x) < \infty,$$

because μ is the distribution measure of f_0 that is integrable. We can now apply the ergodic theorem in Corollary 5.6 of Chapter 6, Book III, which gives us the first conclusion with $m_0 = \int_Y g_0\, dm^* = \int_X f_0\, dm$.

To deduce the second conclusion we need the following lemmas.

Lemma 2.2 *If $\{f_N\}$ and $\{g_N\}$ have the same joint distribution, then so do the sequences $\{\Phi_N(f)\}$ and $\{\Phi_N(g)\}$. Here $\Phi_N(f) = \Phi_N(f_1, \ldots, f_N)$, $\Phi_N(g) = \Phi_N(g_1, \ldots, g_N)$, and each Φ_N is a continuous function from $\mathbb{R}^N$ to $\mathbb{R}$.*

To see this, note that if $B \subset \mathbb{R}^N$ is a Borel set, and $\Phi = (\Phi_1, \ldots, \Phi_N)$, then $B' = \Phi^{-1}(B)$ is also a Borel set in $\mathbb{R}^N$, so if $f = (f_1, \ldots, f_N)$ and $g = (g_1, \ldots, g_N)$, then $\mu_{\Phi(f)}(B) = \mu_f(B')$ and $\mu_{\Phi(g)}(B) = \mu_g(B')$. Since f and g have the same joint distribution we must have $\mu_f(B') = \mu_g(B')$, and the lemma is proved.

Lemma 2.3 *If $\{F_N\}$ and $\{G_N\}$ have the same joint distribution, then $F_N(x) \to m_0$ almost everywhere as $N \to \infty$ if and only if $G_N(y) \to m_0$ almost everywhere as $N \to \infty$.*

To prove this lemma, note that if we define $E_{N,k} = \{x : \sup_{r \geq N} |F_r(x) - m_0| \leq 1/k\}$, then $F_N \to m_0$ almost everywhere if and only if $m(E_{N,k}) \to 1$, as $N \to \infty$, for each k. If $E'_{N,k} = \{y : \sup_{r \geq N} |G_r(x) - m_0| \leq 1/k\}$, then $m(E_{N,k}) = m^*(E'_{N,k})$, and this leads to our desired result.

Once we take $\Phi_N(t_1, \ldots, t_N) = \frac{1}{N} \sum_{k=1}^N t_k$, $F_N(x) = \frac{1}{N} \sum_{k=0}^{N-1} f_k(x)$, and $G_N(y) = \frac{1}{N} \sum_{k=0}^{N-1} g_k(y)$, we see that the lemmas complete the proof of the theorem.

2.2 The role of martingales

We shall now look at sums of independent functions (random variables) from a different angle and relate these sums to the notion of martingales. The basic definition required is that of the conditional expectation of a function f with respect to a σ sub-algebra $\mathcal{A}$ of the σ-algebra $\mathcal{M}$ of measurable sets of X. In fact, for the sake of brevity of terminology, in what follows we drop the adjective "σ" and use "algebra" and sub-algebra to mean σ-algebra and σ sub-algebra, respectively.

Suppose $\mathcal{A}$ is a given such sub-algebra. We say that a function F on X is **measurable** with respect to $\mathcal{A}$ (or $\mathcal{A}$-measurable) if $F^{-1}(B) \in \mathcal{A}$ for all Borel subsets B of $\mathbb{R}$. The algebra $\mathcal{A}$ is said to be **determined** by F, sometimes written $\mathcal{A} = \mathcal{A}_F$, if $\mathcal{A}$ is the smallest algebra with respect to which F is measurable; that is, $\mathcal{A}_F = \{F^{-1}(B)\}$, as B ranges over the Borel sets of $\mathbb{R}$.

Given an integrable function f on X and a sub-algebra $\mathcal{A}$, then $\mathbb{E}_{\mathcal{A}}(f)$, also sometimes written as $\mathbb{E}(f|\mathcal{A})$, is the unique function F described by the proposition below. It is called the **conditional expectation** of f with respect to $\mathcal{A}$.

Proposition 2.4 *Given an integrable function f and a sub-algebra $\mathcal{A}$ of $\mathcal{M}$, there is a unique*[9] *function F so that:*

(i) *F is $\mathcal{A}$-measurable.*

(ii) *$\int_A F\,dm = \int_A f\,dm$ for any set $A \in \mathcal{A}$.*

In general, one may think of the conditional expectation as the "best guess" of the function f given the knowledge of $\mathcal{A}$. A simple example to keep in mind is $\mathbb{E}_{\mathcal{A}}(f) = \mathbb{E}_n(f)$ given in Section 1.6 above. In that case, $\mathcal{A}$ is the (finite) algebra generated by the dyadic intervals of length 2^{-n} on $[0,1]$.

Proof. We denote by m' the restriction of the measure m to $\mathcal{A}$. Define a (σ-finite) signed measure ν on $\mathcal{A}$ by $\nu(A) = \int_A f\,dm$, for $A \in \mathcal{A}$. Then since ν is clearly absolutely continuous with respect to m', the Lebesgue-Radon-Nikodym theorem[10] guarantees that there is a function F that is $\mathcal{A}$-measurable so that $\nu(A) = \int_A F\,dm' = \int_A F\,dm$. Given the definition of ν, the existence of the required F is therefore established. Its uniqueness is clear because if G is $\mathcal{A}$-measurable and $\int_A G\,dm = 0$ for every $A \in \mathcal{A}$, then necessarily $G = 0$.

Once the algebra $\mathcal{A}$ is fixed, we shall not always indicate the dependence of the conditional expectation on the algebra, but write it simply as $\mathbb{E}$ instead of $\mathbb{E}_{\mathcal{A}}$.

There are a number of elementary observations about conditional expectations $\mathbb{E}$ that are direct consequences of the defining proposition for $F = \mathbb{E}(f)$. We leave these for the reader to verify.

---

[9] Uniqueness, of course, means determined up to a set of measure zero.

[10] See for example Theorem 4.3 in Chapter 6 of Book III.

- The mapping $f \mapsto \mathbb{E}(f)$ is linear.
- $\int_X \mathbb{E}(f)\, dm = \int_X f\, dm$, and $\mathbb{E}(1) = 1$.
- $\mathbb{E}(f) \geq 0$ if $f \geq 0$, and $|\mathbb{E}(f_1)| \leq \mathbb{E}(f)$ if $|f_1| \leq f$.
- $\mathbb{E}^2 = \mathbb{E}$, and in particular $\mathbb{E}(f) = f$ if f is $\mathcal{A}$-measurable.
- $\mathbb{E}(gf) = g\mathbb{E}(f)$, if g is bounded and $\mathcal{A}$-measurable.

Two other noteworthy properties of $\mathbb{E}$ are contained in the following.

Lemma 2.5

(a) *If $f \in L^2$, then $\mathbb{E}(f) \in L^2$ and $\|\mathbb{E}(f)\|_{L^2} \leq \|f\|_{L^2}$.*

(b) *If $f, g \in L^2$, then $\int_X \mathbb{E}(f) g\, dm = \int_X f\mathbb{E}(g)\, dm$.*

Note. The conclusion (b) of the lemma, together with the property $\mathbb{E}^2 = \mathbb{E}$ shows that $\mathbb{E}$ is an orthogonal projection on the Hilbert space $L^2(X, m)$.

Proof. To establish (a) observe that if g is bounded and $\mathcal{A}$-measurable, then by the proposition above, $\int_X gf\, dm = \int_X \mathbb{E}(gf)\, dm = \int_X g\mathbb{E}(f)\, dm$. But

$$\|\mathbb{E}(f)\|_{L^2} = \sup_g \left| \int_X g\mathbb{E}(f)\, dm \right|,$$

where g ranges over bounded $\mathcal{A}$-measurable functions with $\|g\|_{L^2} \leq 1$ (see Lemma 4.2 in Chapter 1), because of the fact that $\mathbb{E}(f)$ is $\mathcal{A}$-measurable. Moreover $\left|\int_X gf\, dm\right| \leq \|f\|_{L^2}$ for such g, gives conclusion (a).

Next observe that $\int_X \mathbb{E}(g) f\, dm = \int_X \mathbb{E}(\mathbb{E}(g)f)\, dm = \int_X \mathbb{E}(g)\mathbb{E}(f)\, dm$, whenever g is bounded. By symmetry in f and g this gives conclusion (b) when both f and g are bounded, and by the continuity in (a) the result extends to f and g in L^2.

After these preliminaries, we are ready for the task at hand. We now assume we are given an increasing sequence of sub-algebras of $\mathcal{M}$, that is, we have

$$\mathcal{A}_0 \subset \mathcal{A}_1 \subset \cdots \subset \mathcal{A}_n \subset \cdots \subset \mathcal{M},$$

and to each sub-algebra we attach its conditional expectation,

$$\mathbb{E}_n = \mathbb{E}_{\mathcal{A}_n} \qquad \text{for } n = 0, 1, 2, \ldots.$$

The increasing character of the sequence $\mathcal{A}_n$ implies that expectation operators form an increasing sequence in the sense that

$$\mathbb{E}_n\mathbb{E}_m = \mathbb{E}_{\min(n,m)}, \qquad \text{for all } n, m.$$

Indeed, if $m \leq n$, then $\mathcal{A}_m \subset \mathcal{A}_n$, so $g = \mathbb{E}_m(f)$ is $\mathcal{A}_n$-measurable, and consequently $\mathbb{E}_n(g) = g$. In the other case, if $n \leq m$, and $A \in \mathcal{A}_n$, then

$$\int_A \mathbb{E}_n(f) = \int_A f = \int_A \mathbb{E}_m(f),$$

where the second equality follows from the fact that A is also $\mathcal{A}_m$-measurable. Therefore the definition of conditional expectation implies that $\mathbb{E}_n(\mathbb{E}_m(f)) = \mathbb{E}_n(f)$.

With this we arrive at the crucial definition. Having fixed our increasing family of algebras $\{\mathcal{A}_n\}$ and the resulting conditional expectations, we say that a sequence $\{s_n\}$ of integrable functions on X forms a **martingale sequence** if for all k and n,

$$s_k = \mathbb{E}_k(s_n), \qquad \text{whenever } k \leq n. \tag{14}$$

Note that by this definition, each s_k is automatically $\mathcal{A}_k$-measurable.

If the sequence is finite (and consists of $s_0, s_1, \ldots, s_m$) then this is equivalent with $s_k = \mathbb{E}_k(s_m)$, for all $k \leq m$. An important class of martingale sequences are those that are **complete**. This means that there is an integrable function s_∞ so that $s_k = \mathbb{E}_k(s_\infty)$ for all k.

The fundamental connection between sums of independent random variables and martingales is contained in the following assertion.

Proposition 2.6 *Suppose $\{f_k\}$ is a sequence of integrable functions that are mutually independent and each have mean zero. Then there is an increasing family $\mathcal{A}_n$ of sub-algebras so that with respect to these $s_n = \sum_{k=0}^{n} f_k$ is a martingale sequence.*

To see this, we require further terminology. Let $\{\mathcal{B}_n\}$ be a sequence of sub-algebras of $\mathcal{M}$ that are not assumed to be increasing. Then these are said to be **mutually independent** if for every N,

$$m\left(\bigcap_{j=0}^{N} B_j\right) = \prod_{j=0}^{N} m(B_j) \qquad \text{for all choices } B_j \in \mathcal{B}_j.$$

Notice that if $\mathcal{A}_{f_n}$ are the sub-algebras determined by the f_n, then the fact that $\{\mathcal{A}_{f_n}\}$ are mutually independent is equivalent to the functions $\{f_n\}$ being mutually independent, according to the definition given in (3).

Now starting with our independent functions $f_0, f_1, \ldots, f_n, \ldots$ we define $\mathcal{A}_n$ to be the algebra generated by $\mathcal{A}_{f_0} \cup \mathcal{A}_{f_1} \cup \cdots \cup \mathcal{A}_{f_n}$. It is useful to have the short-hand $\bigvee_{j=0}^{n} \mathcal{B}_j$ to denote the algebra generated by $\mathcal{B}_0 \cup \mathcal{B}_1 \cup \cdots \cup \mathcal{B}_n$. Thus we have set $\mathcal{A}_n = \bigvee_{j=0}^{n} \mathcal{A}_{f_j}$. Our claim is that $\bigvee_{j=0}^{n-1} \mathcal{A}_{f_j}$ and $\mathcal{A}_{f_n}$ are mutually independent. This is an immediate consequence of the following lemma.

Lemma 2.7 *Suppose $\mathcal{B}_0, \ldots, \mathcal{B}_n$ are mutually independent algebras. Then for each $k < n$, the algebras $\bigvee_{j=0}^{k} \mathcal{B}_j$ and $\mathcal{B}_n$ are mutually independent.*

See Exercise 7.

Now clearly $\{\mathcal{A}_n\}$ is an increasing sequence of algebras and $\mathbb{E}_k(f_\ell) = f_\ell$, if $k \geq \ell$, since each f_ℓ is also $\mathcal{A}_k$-measurable. We next observe that $\mathbb{E}_k(f_\ell) = 0$ if $k < \ell$. Indeed, recall first that $F = \mathbb{E}_k(f_\ell)$ is $\mathcal{A}_k$-measurable and

$$\int_{A_k} F \, dm = \int_{A_k} f_\ell \, dm, \qquad \text{for every set } A_k \in \mathcal{A}_k.$$

But

$$\int_{A_k} f_\ell \, dm = \int_X \chi_{A_k} f_\ell \, dm = m(\chi_{A_k}) \int_X f_\ell \, dm = 0,$$

by the independence of $\mathcal{A}_k$ and $\mathcal{A}_{f_\ell}$, and the fact that the mean of f_ℓ is zero. Hence $F = 0$. Finally for $k \leq n$

$$\begin{aligned}\mathbb{E}_k(s_n) &= \mathbb{E}_k(f_0 + f_1 + \cdots + f_k) + \mathbb{E}_k(f_{k+1} + \cdots + f_n) \\ &= f_0 + \cdots + f_k = s_k.\end{aligned}$$

Thus (14) holds and the proposition is proved.

Having reached this point, we are ready to use the ideas of martingales to extend the results of Section 1.6.

Theorem 2.8 *Suppose $f_0, \ldots, f_n, \ldots$ are independent functions that are square integrable, and that each has mean zero, and variance $\sigma_n^2 = \|f_n\|_{L^2}^2$. Assume that*

$$\sum_{n=0}^{\infty} \sigma_n^2 < \infty.$$

Then $s_n = \sum_{k=0}^{n} f_k$ converges (as $n \to \infty$) almost everywhere.

A corollary of this where we only assume that the $\{\sigma_n\}$ are bounded, gives the strong law of large numbers in this setting.

Corollary 2.9 *If* $\sup_n \sigma_n < \infty$, *then for each* $\alpha > 1/2$

$$\frac{s_n}{n^\alpha} \to 0 \qquad \textit{almost everywhere as } n \to \infty.$$

Note that here, unlike in Theorem 2.1, we have not assumed that the f_n are identically distributed. On the other hand, we have made a more restrictive assumption in requiring square integrability.

We begin the proof of the theorem by noting that under its assumptions the sequence $s_n = \sum_{k=0}^n f_k$ converges in the L^2 norm, as $n \to \infty$. Indeed, since the f_n are mutually independent and $\int_X f_n \, dm = 0$, then by (4) they are mutually orthogonal. Hence by Pythagoras' theorem, if $m < n$, $\|s_n - s_m\|_{L^2}^2 = \sum_{k=m+1}^n \|f_k\|^2 = \sum_{k=m+1}^n \sigma_k^2 \to 0$, as $n, m \to \infty$. Thus s_n converges to a limit (call it s_∞) in the L^2 norm. Using (14) and the fact that each $\mathbb{E}_n$ is continuous in the L^2 norm by Lemma 2.5, we arrive at

$$s_n = \mathbb{E}_n(s_\infty), \qquad \text{for all } n.$$

Our desired result now follows from a basic maximal theorem for martingales and its corollary, which gives convergence almost everywhere.

Theorem 2.10 *Suppose* s_∞ *is an integrable function, and* $s_n = \mathbb{E}_n(s_\infty)$, *where the* $\mathbb{E}_n$ *are conditional expectations for an increasing family* $\{\mathcal{A}_n\}$ *of sub-algebras of* $\mathcal{M}$. *Then:*

(a) $m(\{x : \sup_n |s_n(x)| > \alpha\}) \le \frac{1}{\alpha}\|s_\infty\|_{L^1}$ *for every* $\alpha > 0$.

(b) *If* s_n *converges in the* L^1 *norm as* $n \to \infty$, *then it also converges almost everywhere to the same limit.*

Note. The assumption in part (b) is in reality redundant because if $s_n = \mathbb{E}_n(s_\infty)$ with $s_\infty \in L^1$, then automatically s_n converges in the L^1 norm; but in general this limit need not be s_∞. (See Exercise 27.) However in the situation in which we apply the theorem, we know already that $s_n \to s_\infty$ in the L^2 norm, hence also in the L^1 norm.

For the proof of part (a) we may assume that s_∞ is non-negative, for otherwise we may proceed with $|s_\infty|$ instead of s_∞ and then obtain the result once we observe that $|\mathbb{E}_n(s_\infty)| \le \mathbb{E}_n(|s_\infty|)$. For fixed α, let $A = \{x : \sup_n s_n(x) > \alpha\}$. Then we can partition $A = \bigcup_{n=0}^\infty A_n$, where

A_n is the set where n is the first time that $s_n(x) > \alpha$. That is $A_n = \{x : s_n(x) > \alpha, \text{ but } s_k(x) \le \alpha, \text{ for } k < n\}$. Note that $A_n \in \mathcal{A}_n$. Also,

$$\int_A s_\infty \, dm = \sum_{n=0}^{\infty} \int_{A_n} s_\infty \, dm = \sum_{n=0}^{\infty} \int_{A_n} \mathbb{E}_n(s_\infty) \, dm \quad = \sum_{n=0}^{\infty} \int_{A_n} s_n \, dm$$
$$> \alpha \sum_n \int_{A_n} dm$$
$$= \alpha m(A).$$

The identity $\int_{A_n} \mathbb{E}_n(s_\infty) \, dm = \int_{A_n} s_\infty \, dm$ follows from the definition of the conditional expectation $\mathbb{E}_n(s_\infty)$. Thus

$$(15) \qquad m(A) \le \frac{1}{\alpha} \int_A s_\infty \, dm, \qquad \text{with } A = \{x : \sup_n s_n(x) > \alpha\},$$

and part (a) is proved. (The reader might find it instructive to compare (15) with a corresponding estimate for the Hardy-Littlewood maximal function in equation (28) of Chapter 2.)

To prove (b), assume first that $s_n \to s_\infty$ in the L^1 norm. Remark that we always have $s_n - s_\infty = \mathbb{E}_n(s_\infty - s_k) + s_k - s_\infty$ if $n \ge k$, because then $\mathbb{E}_n(s_k) = s_k$. We will show that if $A_\alpha = \{x : \limsup_{n\to\infty} |s_n(x) - s_\infty(x)| > 2\alpha\}$, then $m(A_\alpha) = 0$ for every $\alpha > 0$, and this assures our conclusion about the existence of the limit. Now with α given, let $\epsilon > 0$ be arbitrary. Then choose k so large that $\|s_k - s_\infty\|_{L^1} < \epsilon$. Then

$$\limsup_{n\to\infty} |s_n - s_\infty| \le \sup_{n \ge k} |\mathbb{E}_n(s_\infty - s_k)| + |s_k - s_\infty|.$$

If $A_\alpha^1 = \{x : \sup_n |\mathbb{E}_n(s_\infty - s_k)(x)| > \alpha\}$ and $A_\alpha^2 = \{x : |s_k(x) - s_\infty(x)| > \alpha\}$, then

$$m(A_\alpha) \le m(A_\alpha^1) + m(A_\alpha^2).$$

By part (a) applied to $s_\infty - s_k$ instead of s_∞, we get $m(A_\alpha^1) \le \epsilon/\alpha$. Also Tchebychev's inequality gives $m(A_\alpha^2) \le \epsilon/\alpha$. Altogether then $m(A_\alpha) \le 2\epsilon/\alpha$, and since ϵ was arbitrary we have $m(A_\alpha)$=0, which holds for every α, proving the result under the additional hypothesis that s_n converges to s_∞ in the L^1 norm. Dropping that assumption we can define s'_∞ to be the limit of the sequence $\{s_n\}$ in the L^1 norm which was assumed to exist. Then by (14) and the continuity of $\mathbb{E}_k$ on the L^1 norm, we get $s_k = \mathbb{E}_k(s'_\infty)$, and we are back to the previous situation with s'_∞ in place to s_∞. The theorem is therefore completely proved.

The corollary then follows by the same argument used in the proof of Corollary 1.5.

2.3 The zero-one law

The kernel of the idea is the observation that if $\mathcal{A}_1$ and $\mathcal{A}_2$ are two independent algebras, and the set A belongs simultaneously to both $\mathcal{A}_1$ and $\mathcal{A}_2$, then necessarily $m(A) = 0$ or $m(A) = 1$.

Indeed, in this situation, $m(A) = m(A \cap A) = m(A)m(A)$ by independence, which proves the assertion. This idea is elaborated in Kolmogorov's **zero-one law** that we now formulate.

Suppose $\mathcal{A}_0, \mathcal{A}_1, \ldots, \mathcal{A}_n, \ldots$ is a sequence of sub-algebras of $\mathcal{M}$, that are not necessarily increasing. With $\bigvee_{k=n}^{\infty} \mathcal{A}_k$ denoting the algebra[11] generated by $\mathcal{A}_n, \mathcal{A}_{n+1}, \ldots$, we define the **tail algebra** to be

$$\bigcap_{n=0}^{\infty} \bigvee_{k=n}^{\infty} \mathcal{A}_k.$$

Theorem 2.11 *If the algebras $\mathcal{A}_0, \mathcal{A}_1, \ldots, \mathcal{A}_n, \ldots$ are mutually independent then every element of the tail algebra has either measure zero or one.*

Proof. Let $\mathcal{B}$ denote the tail algebra. Note that $\mathcal{A}_r$ is automatically independent from $\bigvee_{k=r+1}^{\infty} \mathcal{A}_k$, by Lemma 2.7. Hence each $\mathcal{A}_r$ is independent of $\mathcal{B}$, and thus the algebras $\mathcal{B}$ and $\mathcal{B}$ are mutually independent! Therefore as observed above, every element of $\mathcal{B}$ has measure zero or one.

A simple consequence is the following.

Corollary 2.12 *Suppose $f_0, f_1, \ldots, f_n, \ldots$ are mutually independent functions. The set where $\sum_{k=0}^{\infty} f_k$ converges has measure zero or one.*

Proof. Set $\mathcal{A}_n = \mathcal{A}_{f_n}$. Then these algebras are independent. Now with $s_n = \sum_{k=0}^{n} f_k$, and a fixed positive integer n_0, we have by the Cauchy criterion that

$$\{x : \lim s_n(x) \text{ exists}\} = \bigcap_{\ell=1}^{\infty} \bigcup_{r=n_0}^{\infty} \{x : |s_n(x) - s_m(x)| < \frac{1}{\ell}, \text{ all } n, m \geq r\}.$$

Since $\{x : \ |s_n(x) - s_m(x)| < 1/\ell, \text{ all } n, m \geq r\} \in \bigvee_{k=n_0}^{\infty} \mathcal{A}_k$ whenever $r \geq n_0$ we conclude that the set of convergence is a tail set, as desired.

2.4 The central limit theorem

We generalize the special case of this theorem given in Section 1.4, connecting its proof in an elegant way with the Fourier transform.

---

[11] Recall that we are using "algebra" as a short-hand for "σ-algebra."

The setting is as follows. On our probability space (X, m) we are given a sequence $f_1, f_2, \ldots,$ of identically distributed, square integrable, and mutually independent functions (random variables) that each have mean m_0 and variance σ^2.

Theorem 2.13 *Let $S_N = \sum_{n=1}^N f_n$. Under the above conditions*

$$m(\{x : \; a < \frac{S_N - Nm_0}{N^{1/2}} < b\}) \to \frac{1}{\sigma\sqrt{2\pi}} \int_a^b e^{-t^2/(2\sigma^2)}\, dt$$

as $N \to \infty$, for each $a < b$.

In proving the theorem we can immediately reduce to the case where the mean m_0 is zero, by replacing f_n by $f_n - m_0$ for each n. Suppose now that μ is the common distribution measure of the f_n, that μ_N is the distribution measure of $S_N/N^{1/2}$ and ν_{σ^2} is the distribution measure of the Gaussian with mean zero and variance σ^2. We consider the Fourier transforms of these measures, called their **characteristic functions**. In the case of μ it is given by

$$\hat{\mu}(\xi) = \int_{-\infty}^{\infty} e^{-2\pi i \xi t}\, d\mu(t),$$

with similar formulas[12] for $\hat{\mu}_N$ and $\hat{\nu}_{\sigma^2}$.

Note first that $\hat{\nu}_{\sigma^2}$ can be computed explicitly. It is given by the formula[13]

$$\hat{\nu}_{\sigma^2}(\xi) = e^{-2\sigma^2\pi^2\xi^2}.$$

The proof of the theorem can now be presented in three relatively easy steps:

(i) The identity $\hat{\mu}_N(\xi) = \hat{\mu}(\xi/N^{1/2})^N$, for each N.

(ii) The fact that $\hat{\mu}_N(\xi) \to \hat{\nu}_{\sigma^2}(\xi)$, for each ξ, as $N \to \infty$.

(iii) The resulting consequence that $\mu_N((a,b)) \to \nu_{\sigma^2}((a,b))$, as $N \to \infty$ for all intervals (a,b).

---

[12] To be consistent with our previous usage of the Fourier transform, we have kept the factor 2π in the exponential, which is not the usual practice in probability theory.

[13] See for instance Chapter 5 in Book I.

Now if μ is the common distribution of the f_n, then as we noted in (6), for any $G : \mathbb{R} \to \mathbb{R}$ that is (say) continuous and bounded we have

$$\int_X G(f_n)(x)\,dm = \int_{-\infty}^{\infty} G(t)\,d\mu(t).$$

In particular, taking $G(t) = e^{-2\pi it\xi}$, with ξ real, we have

$$\hat{\mu}(\xi) = \int_X e^{-2\pi i\xi f_n(x)}\,dm.$$

Similarly $\hat{\mu}_N(\xi) = \int_X e^{-2\pi i\xi S_N(x)/N^{1/2}}\,dm$. However $S_N(x) = \sum_{n=1}^N f_n(x)$, thus by the mutual independence of the f_n

$$\int_X e^{-2\pi i\xi S_N(x)/N^{1/2}}\,dm = \prod_{n=1}^N \left(\int_X e^{-2\pi i\xi f_n(x)/N^{1/2}}\,dm\right) = \hat{\mu}(\xi/N^{1/2})^N.$$

(Note here the similarity with equation (11).) The identity (i) is therefore established.

To carry out the second step we prove the following.

Lemma 2.14 *$\hat{\mu}(\xi/N^{1/2}) = 1 - 2\sigma^2\pi^2\xi^2/N + o(1/N)$, as $N \to \infty$.*

Proof. Indeed, when ξ is fixed

$$e^{-2\pi i\xi t/N^{1/2}} = 1 - 2\pi i\xi t/N^{1/2} - 2\pi^2\xi^2t^2/N + E_N(t)$$

with $E_N(t) = O(t^2/N)$, but also $E_N(t) = O(t^3/N^{3/2})$. Integrating this in t, we get

$$\hat{\mu}(\xi/N^{1/2}) = 1 - \frac{2\pi^2\xi^2}{N}\sigma^2 + \int_{-\infty}^{\infty} E_N(t)\,d\mu(t),$$

because $m_0 = \int_{-\infty}^{\infty} t\,d\mu(t) = 0$, and $\sigma^2 = \int_{-\infty}^{\infty} t^2\,d\mu(t)$. The lemma will be proved as soon as we see that $\int_{-\infty}^{\infty} E_N(t)\,d\mu(t) = o(1/N)$. However, the integral in question can be divided into a part where $t^2 < \epsilon_N N$, and a complementary part where $t^2 \geq \epsilon_N N$. Here we choose ϵ_N to tend to zero as $N \to \infty$, while $\epsilon_N N \to \infty$ as $N \to \infty$; (for example, the choice $\epsilon_N = N^{-1/2}$ will do.) Now for the first part

$$\begin{aligned}\int_{t^2<\epsilon_N N} E_N(t)\,dt &= O\left(\int_{t^2<\epsilon_N N} t^3/N^{3/2}\,d\mu(t)\right)\\ &= O\left(\frac{\epsilon_N^{1/2}}{N}\int_{-\infty}^{\infty} t^2\,d\mu(t)\right)\\ &= o(1/N).\end{aligned}$$

In addition, for the second part we can estimate

$$\int_{t^2 \geq \epsilon_N N} E_N(t)\, dt = O\left(\frac{1}{N}\int_{t^2 \geq \epsilon_N N} t^2\, d\mu(t)\right) = o(1/N).$$

Having thus proved the lemma we see that

$$\hat{\mu}_N(\xi) = \hat{\mu}(\xi/N^{1/2})^N = \left(1 - 2\sigma^2\pi^2\xi^2/N + o(1/N)\right)^N,$$

and this converges to $e^{-2\sigma^2\pi^2\xi^2}$, completing step (ii).

To finish the proof of the theorem we need the following lemma. We say that a measure is **continuous** if each point has measure zero.

Lemma 2.15 *Suppose* $\{\mu_N\}$, $N = 1, 2, \ldots$, *and* ν *are non-negative finite Borel measures on* $\mathbb{R}$, *and that* ν *is continuous. Assume that* $\hat{\mu}_N(\xi) \to \hat{\nu}(\xi)$, *as* $N \to \infty$, *for each* $\xi \in \mathbb{R}$. *Then* $\mu_N((a,b)) \to \nu(a,b)$ *for all* $a < b$.

Proof. We prove first that

$$\mu_N(\varphi) \to \nu(\varphi) \quad \text{as } N \to \infty \tag{16}$$

for any φ that is C^∞ and has compact support, where we have used the notation $\mu_N(\varphi) = \int_{-\infty}^{\infty} \varphi(t)\, d\mu_N(t)$ and $\nu(\varphi) = \int_{-\infty}^{\infty} \varphi(t)\, d\nu(t)$.

Notice that since $\hat{\mu}_N(0) = \int_{-\infty}^{\infty} d\mu_N(t)$, then the convergent sequence $\int_{-\infty}^{\infty} d\mu_N(t)$ must be bounded. As a result, for some M we have $|\hat{\mu}_N(\xi)| \leq M$ for all N and also $|\hat{\nu}(\xi)| \leq M$.

Next, the function φ can be represented by its inverse Fourier transform $\varphi(t) = \int_{-\infty}^{\infty} e^{-2\pi i t}\varphi^\vee(\xi)\, d\xi$, where $\varphi^\vee(\xi) = \hat{\varphi}(-\xi)$ is necessarily in the Schwartz space $\mathcal{S}$. This shows that

$$\int_{\mathbb{R}} \varphi(t)\, d\mu_N(t) = \int_{\mathbb{R}} \varphi^\vee(\xi)\hat{\mu}_N(\xi)\, d\xi,$$

by applying Fubini's theorem to $\int\int e^{-2\pi i t\xi}\varphi^\vee(\xi)\, d\mu_N(t)\, d\xi$, which is justified by the rapid decrease of $\varphi^\vee$. Similarly, $\int \varphi\, d\nu = \int \varphi^\vee(\xi)\hat{\nu}(\xi)\, d\xi$. Then since $\hat{\mu}_N(\xi) \to \hat{\nu}(\xi)$ pointwise and boundedly we obtain (16).

Now for (a, b) fixed, let φ_ϵ be a sequence of positive C^∞ functions with $\varphi_\epsilon \leq \chi_{(a,b)}$, and $\varphi_\epsilon(t) \to \chi_{(a,b)}(t)$ for every t as $\epsilon \to 0$. Then

$$\mu_N((a,b)) \geq \mu_N(\varphi_\epsilon) \to \nu(\varphi_\epsilon) \quad \text{as } N \to \infty.$$

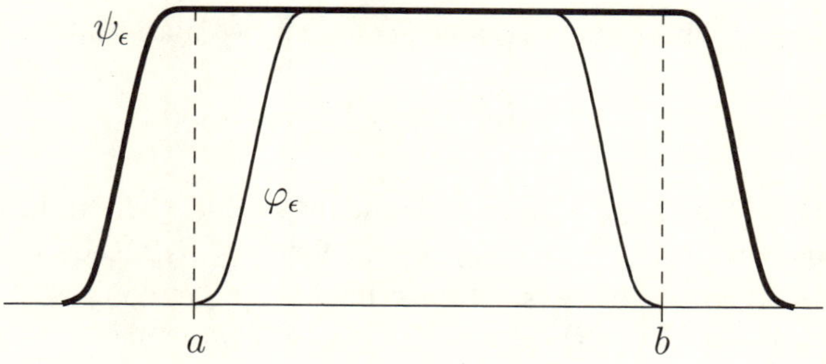

Figure 3. The functions φ_ϵ and ψ_ϵ in Lemma 2.15

As a result, $\liminf_{N\to\infty} \mu_N(a,b) \geq \nu(\varphi_\epsilon)$, and letting $\epsilon \to 0$ gives

$$\liminf_{N\to\infty} \mu_N((a,b)) \geq \nu((a,b)).$$

Similarly, let ψ_ϵ be a sequence of C^∞ functions so that $\psi_\epsilon \geq \chi_{[a,b]}$ and $\psi_\epsilon(t) \to \chi_{[a,b]}(t)$ for every t, as $\epsilon \to 0$. Then by the same reasoning, $\limsup_{N\to\infty} \mu_N((a,b)) \leq \nu([a,b]) = \nu((a,b))$, by the continuity of ν. Thus the lemma is proved, and with it the theorem is established, once we take $\nu = \nu_{\sigma^2}$.

Another way to put the conclusion of the theorem is in terms of weak convergence of measures. We say that a sequence of probability measures $\{\mu_N\}$ **converges weakly** to the probability measure ν, if (16) holds for all continuous functions φ that are bounded on $\mathbb{R}$.

Corollary 2.16 *If μ_N is the distribution measure of $(S_N - Nm_0)/N^{1/2}$, then μ_N converges weakly to $\nu = \nu_{\sigma^2}$.*

We note first that (16) holds for any function φ that is continuous and has compact support. Indeed such a φ can be uniformly approximated by a sequence $\{\varphi_\epsilon\}$ of C^∞ functions of compact support.[14] Now

$$\mu_N(\varphi) - \nu(\varphi) = \mu_N(\varphi - \varphi_\epsilon) - \nu(\varphi - \varphi_\epsilon) + (\mu_N - \nu)(\varphi_\epsilon).$$

Now the sum of the first two terms on the right-hand side is majorized by $2\sup_t |\varphi(t) - \varphi_\epsilon(t)|$, and this can be made small by choosing ϵ conveniently. Once ϵ is chosen we need only let $N \to \infty$ and apply (16) for φ_ϵ.

[14]See for example the proof of Lemma 4.10 in Chapter 5 of Book III.

To pass to φ whose support is not compact, we note that

$$\limsup_{N\to\infty} \mu_N(\chi_{(I_R)^c}) \le \epsilon(R), \tag{17}$$

where $\epsilon(R) \to 0$ as $R \to \infty$, and I_R is the interval $|t| \le R$. In fact if η_R is continuous with $0 \le \eta_R \le \chi_{I_R}$, $\eta_R(t) = 1$ for $|t| \le R/2$, then $\mu_N(\chi_{I_R}) \ge \mu_N(\eta_R) \to \nu(\eta_R)$, as $N \to \infty$. Hence $\liminf_{N\to\infty} \mu_N(\chi_{I_R}) \ge 1 - \nu(1 - \eta_R)$, but $\nu(1-\eta_R) = \epsilon(R) \to 0$, as $R \to \infty$ so (17) holds.

Now suppose φ is a given continuous and bounded function on $\mathbb{R}$. We can assume that $0 \le \varphi \le 1$. For each R, let φ_R be a continuous function with $\varphi_R(t) = \varphi(t)$ for $|t| \le R$, but $\varphi_R(t) = 0$ for $|t| \ge 2R$, while $0 \le \varphi_R(t) \le \varphi(t)$ everywhere.

Then $\varphi \le \varphi_R + \chi_{(I_R)^c}$, so

$$\mu_N(\varphi) \le \mu_N(\varphi_R) + \mu_N(\chi_{(I_R)^c}).$$

Therefore $\limsup_{N\to\infty} \mu_N(\varphi) \le \nu(\varphi) + \epsilon(R)$, and letting $R \to \infty$ gives $\limsup_{N\to\infty} \mu_N(\varphi) \le \nu(\varphi)$. However

$$\liminf_{N\to\infty} \mu_N(\varphi) \ge \lim_{N\to\infty} \mu_N(\varphi_R) = \nu(\varphi_R) \to \nu(\varphi) \qquad \text{as } R \to \infty.$$

Thus $\lim_{N\to\infty} \mu_N(\varphi) = \nu(\varphi)$, proving the corollary.

2.5 Random variables with values in $\mathbb{R}^d$

Up to this point, with the exception of Section 1.7, our functions have been assumed to be real-valued. However, for many purposes it is useful to extend the theory to the setting where the functions take their values in $\mathbb{R}^d$ (and in particular, to complex-valued functions, which corresponds to the case $d = 2$). Often this extension is rather routine. In what follows, we will limit ourselves to a formulation of the d-dimensional version of the central limit theorem. First, some notation.

Suppose f is an $\mathbb{R}^d$-valued function on (X, m). We write it in coordinates as $f = (f^{(1)}, f^{(2)}, \ldots, f^{(d)})$, where each $f^{(k)}$ is real-valued. The **distribution measure** of f is the non-negative Borel measure μ on $\mathbb{R}^d$ defined by

$$\mu(B) = m(f^{-1}(B)) = m(\{x : f(x) \in B\}), \qquad \text{for each Borel set } B \subset \mathbb{R}^d.$$

Of course $\mu(\mathbb{R}^d) = 1$, so μ is a probability measure.

The function f is said to the **integrable** if $|f| = \left(\sum_{k=1}^d |f^{(k)}|^2\right)^{1/2}$ is integrable. **Square integrability** of f is defined similarly. When

f is integrable then its mean (or expectation) is defined as the vector $m_0 = (m_0^{(k)})$, with $m_0^{(k)} = \int_X f^{(k)}(x)\, dm$.

If f is square integrable, the **covariance matrix** of f is the $d \times d$ matrix $\{a_{ij}\}$ with

$$a_{ij} = \int_X (f^{(i)}(x) - m_0^{(i)})(f^{(j)}(x) - m_0^{(j)})\, dm.$$

Note that $a_{ij} = \int_{\mathbb{R}^d} (t_i - m_0^{(i)})(t_j - m_0^{(j)})\, d\mu(t)$, and this matrix is symmetric and non-negative. It has a (unique) square root σ which is symmetric and non-negative, and thus we write σ^2 for the covariance matrix of f.

Next, we say that the sequence of $\mathbb{R}^d$-valued functions, $f_1, \ldots, f_n \cdots$ are mutually independent if the algebras

$$\mathcal{A}_n = \mathcal{A}_{f_n} = \{f_n^{-1}(B), \text{ all Borel sets } B \text{ in } \mathbb{R}^d\}$$

are mutually independent. Notice that this implies that for each vector $\xi = (\xi_1, \ldots, \xi_d) \in \mathbb{R}^d$ the scalar-valued functions $\xi \cdot f_1, \ldots, \xi \cdot f_n, \ldots$ where $\xi \cdot f_n = \xi_1 \cdot f_n^{(1)} + \xi_2 \cdot f_n^{(2)} + \cdots + \xi_d \cdot f_n^{(d)}$, are mutually independent.

Two other preliminary matters. Given an $\mathbb{R}^d$-valued random variable (function) f, its **characteristic function** is the d-dimensional Fourier transform $\hat{\mu}(\xi) = \int_{\mathbb{R}^d} e^{-2\pi i \xi \cdot t}\, d\mu(t)$, $\xi \in \mathbb{R}^d$, where μ is the distribution measure of f. Of course $\hat{\mu}(\xi) = \int_X e^{-2\pi i \xi \cdot f(x)}\, dm$.

Also adapting a previous terminology, if $\{\mu_N\}$ is a sequence of probability measures on $\mathbb{R}^d$, and ν is another probability measure on $\mathbb{R}^d$, then we say that $\mu_N \to \nu$ **weakly** if

$$\int_{\mathbb{R}^d} \varphi\, d\mu_N \to \int_{\mathbb{R}^d} \varphi\, d\nu \quad \text{as } N \to \infty,$$

for all continuous and bounded functions φ on $\mathbb{R}^d$.

We now come to the theorem. We suppose that our sequence $\{f_n\}$ of $\mathbb{R}^d$-valued functions are mutually independent, that they are identically distributed and are square integrable with mean zero. If σ^2 denotes the common covariance matrix, we assume that σ is invertible, and write σ^{-1} for its inverse.

Let μ_N be the distribution measure of $\frac{1}{N^{1/2}} \sum_{n=1}^{N} f_n$, and ν_{σ^2} be the measure on $\mathbb{R}^d$ given by

$$\nu_{\sigma^2}(B) = \frac{1}{(2\pi)^{d/2}(\det \sigma)} \int_B e^{-\frac{|\sigma^{-1}(x)|^2}{2}}\, dx$$

for all Borel sets $B \subset \mathbb{R}^d$.

Theorem 2.17 *Under the above conditions on* $\{f_n\}$, *the measures* μ_N *converge weakly to* ν_{σ^2} *as* $N \to \infty$.

The proof proceeds essentially as in the case of real-valued functions, showing first the analog of (16) for smooth functions with compact support, and then proceeding as in Corollary 2.16 for continuous functions. The calculation of the characteristic function of the Gaussian is given in Exercise 32.

Remark. The following generalization can be deduced by a slight modification of the proof of Theorem 2.17. Suppose $\{f_n\}$ satisfies the conditions of the theorem, $t > 0$, and define

$$S_{N,t} = \frac{1}{N^{1/2}} \sum_{n=1}^{[Nt]} f_n.$$

(Here $[x]$ denotes the integer part of x.) Then, the distribution measure of $S_{N,t}$ converges weakly to $\nu_{t\sigma^2}$ as $N \to \infty$. In fact, if $0 \leq s < t$, then the distribution measure of $S_{N,t} - S_{N,s}$ converges weakly to $\nu_{(t-s)\sigma^2}$ as $N \to \infty$.

2.6 Random walks

The coin tossing (or sums of Rademacher functions) considered in Section 1.1 can be thought of as representing a random walk on the real line. This walk can be described as follows.

One starts at the origin, then moves along a straight line with steps of unit length; each step taken has equal probability of going to the right or left, with different steps having independent probabilities. The position after the nth step is given by $s_n = \sum_{k=1}^{n} r_k$. Notice that the values of s_n are always integers.

In $\mathbb{R}^d$ we will consider a particular version of a **random walk**, giving the simplest generalization of the above. It starts at the origin, and the position of the nth step is obtained from the previous step by moving a unit length in a direction of one of the coordinate axes, and doing this with equal probability, (that is probability $1/(2d)$). The passage at each step will be assumed to be independent of the previous steps. We formalize this situation as follows.

Let $\mathbb{Z}_{2d}$ be the set of $2d$ points in $\mathbb{R}^d$ labeled by $\{\pm e_1, \pm e_2, \ldots, \pm e_d\}$, where $e_j = (0, \ldots, 0, 1, 0, \ldots, 0)$ with 1 in the jth coordinate and 0 elsewhere. Assign to $\mathbb{Z}_{2d}$ the measure that gives weight $1/(2d)$ to each of its points. Let $X = \mathbb{Z}_{2d}^{\infty}$ be the infinite product of copies of $\mathbb{Z}_{2d}$, endowed with the product measure, and call this measure m. Thus X consists of points $x = (x_n)_{n=1}^{\infty}$, where each $x_n \in \mathbb{Z}_{2d}$. Now define $\mathfrak{r}_n(x) = x_n$ for each n. So $\mathfrak{r}_n(x)$ is one of the $\pm e_j$, for each n, and therefore in fact takes its values in the lattice $\mathbb{Z}^d$ of $\mathbb{R}^d$. Also $\{\mathfrak{r}_n\}$ are mutually independent functions, since $\mathfrak{r}_n(x)$ depends only on the nth coordinate of x. Note finally that each $\mathfrak{r}_n$ has mean zero and the identity as its covariance matrix.

The sums

$$s_n(x) = \sum_{k=1}^{n} \mathfrak{r}_k(x)$$

represent our random walk, in that x labels a possible **path** and $s_n(x)$ gives the position of this path at the nth step. It is convenient to set $s_0(x) = 0$ for all x.

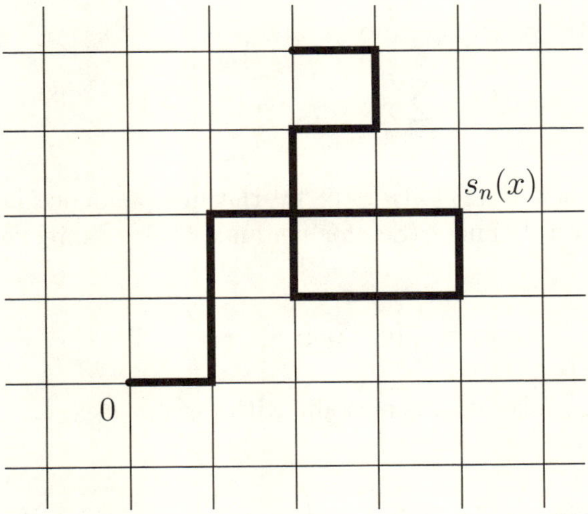

Figure 4. The random walk s_n in dimension two

Here we examine only one of the interesting properties exhibited by this random walk. It illustrates a significant dichotomy between the case of dimension $d \leq 2$ and $d \geq 3$.

Theorem 2.18 *For the above random walk:*

(a) *If $d = 1$ or 2, the random walk is* **recurrent** *in the sense that almost all paths return to the origin for infinitely many n.*

(b) *If $d \geq 3$, then almost every path returns to the origin at most a finite number of times. Moreover, there is a positive probability that the path never returns to the origin.*

In fact, when $d = 1$ or 2, the random walk visits almost surely every point of $\mathbb{Z}^d$ infinitely often. However, when $d \geq 3$, one has $\lim_{n\to\infty} |s_n| = \infty$ almost surely. The proofs of these further conclusions are outlined in Exercises 34 and 35.

Proof. Let μ be the common distribution measure of each of the $\mathfrak{r}_n$. Then μ is the measure on $\mathbb{R}^d$, concentrated at the points $\pm e_1, \pm e_2, \ldots, \pm e_d$, assigning measure $1/(2d)$ to each of these points. Let μ_n be the distribution measure of s_n. Like μ, the measure μ_n is clearly supported on $\mathbb{Z}^d$.

If

$$\hat{\mu}(\xi) = \sum_{k\in\mathbb{Z}^d} m(\{x : \mathfrak{r}_n(x) = k\})e^{-2\pi ik\xi}$$

is the characteristic function of μ, and

$$\hat{\mu}_n(\xi) = \sum_{k\in\mathbb{Z}^d} m(\{x : s_n(x) = k\})e^{-2\pi ik\xi}$$

that of μ_n, then $\hat{\mu}_n(\xi) = (\hat{\mu}(\xi))^n$ by the independence argument used previously several times. (See for instance (11).) Moreover, as is easily seen

$$\hat{\mu}(\xi) = \frac{1}{d}\left(\cos 2\pi\xi_1 + \cdots + \cos 2\pi\xi_d\right).$$

However $\hat{\mu}_n(\xi)$, like $\hat{\mu}(\xi)$, is periodic with periods $e_1, e_2, \ldots, e_d$, and thus for each n

$$(18) \qquad m(\{x : s_n(x) = 0\}) = \int_Q \hat{\mu}_n(\xi)\,d\xi = \int_Q (\hat{\mu}(\xi))^n\,d\xi,$$

where Q is the fundamental cube defined by $Q = \{\xi : -1/2 < \xi_j \leq 1/2, j = 1, \ldots, d\}$.

As a result of all of this we assert that,

$$(19) \qquad \sum_{n=0}^{\infty} m(\{x : s_n(x) = 0\}) = \int_Q \frac{d\xi}{1 - \hat{\mu}(\xi)}.$$

Note first that $\hat{\mu}(\xi) \leq 1$, so the integrand on the right-hand side is always non-negative (or $+\infty$). The claim is that both sides are simultaneously infinite, or finite and equal. In fact multiplying both sides of (18) by r^n, for $0 < r < 1$, and summing gives

$$\sum_{n=0}^{\infty} r^n m(\{x: \ s_n(x) = 0\}) = \int_Q \frac{d\xi}{1 - r\hat{\mu}(\xi)},$$

and letting $r \to 1$ then yields (19).

Now since

$$\begin{aligned} 1 - \hat{\mu}(\xi) &= 1 - \frac{1}{d}(\cos 2\pi\xi_1 + \cdots + \cos 2\pi\xi_d) \\ &= \frac{2\pi^2}{d}|\xi|^2 + O(|\xi|^4) \quad \text{as } \xi \to 0, \end{aligned}$$

and $1 - \hat{\mu}(\xi) \geq c_1$ if $c_2 \leq |\xi|$, $\xi \in Q$, for suitable positive constants c_1 and c_2, we can conclude that the integral

$$\int_Q \frac{d\xi}{1 - \hat{\mu}(\xi)}$$

diverges when $d = 1$ or $d = 2$, but converges when $d \geq 3$. This means that $\sum_{n=0}^{\infty} m(\{x: \ s_n(x) = 0\})$ diverges or converges depending on whether $d \leq 2$ or $d \geq 3$.

The above has the following interpretation. Let $A_n = \{x: \ s_n(x) = 0\}$, and χ_{A_n} its characteristic function. Then $\#(x) = \sum_{n=0}^{\infty} \chi_{A_n}(x)$ is the number of times the path x visits the origin. Thus $\int_X \#(x)\, dm$ is the expected number of times all paths visit the origin. However

$$\int_X \#(x)\, dm = \sum_{n=0}^{\infty} m(\{x: \ s_n(x) = 0\}) = \sum_{n=0}^{\infty} m(A_n),$$

so if $d \geq 3$ this expectation is finite, and therefore almost all paths return to the origin only a finite number of times, proving the first part of conclusion (b) of the theorem.

While the expectation is infinite when $d \leq 2$, this, by itself, does not show that almost all paths return infinitely often to the origin. That we will now see. To proceed we define F_k to be the set of paths where $s_k(x) = 0$ for the first time

$$F_k = \{x: \ s_k(x) = 0, \text{but } s_\ell(x) \neq 0 \text{ for } 1 \leq \ell < k\}.$$

(Here we set $F_1 = \emptyset$.) Since the F_k are disjoint, $\sum_{k=1}^{\infty} m(F_k) \leq 1$. We shall see that for $d = 1$ or 2 in fact $\sum_{k=1}^{\infty} m(F_k) = 1$, which means that almost all paths return to the origin at least once. This is in contrast with $d \geq 3$ where $\sum_{k=1}^{\infty} m(F_k) < 1$ which means that for a set of positive probability, the paths never return to the origin.

We prove these assertions by showing first

$$m(A_n) = \sum_{1 \leq k \leq n} m(F_k) m(A_{n-k}), \qquad \text{for all } n \geq 1. \tag{20}$$

In fact, $A_n = \bigcup_{1 \leq k \leq n} (F_k \cap A_n)$, where this union is disjoint. Therefore $m(A_n) = \sum_{1 \leq k \leq n} m(F_k \cap A_n)$. However

$$F_k \cap A_n = F_k \cap \{x : \; s_n(x) - s_k(x) = 0\}.$$

Hence $m(F_k \cap A_n) = m(F_k) m(\{x : \; s_n(x) - s_k(x) = 0\})$ since the sets F_k and $\{x : \; s_n(x) - s_k(x) = 0\} = \{x : \; \sum_{\ell=k+1}^{n} \mathfrak{r}_\ell(x) = 0\}$ are clearly independent. However

$$m(\{x : \; s_n(x) - s_k(x) = 0\}) = m(\{x : \; s_{n-k}(x) = 0\}) = m(A_{n-k}),$$

by the shift-invariance of the measure on the product space $\mathbb{Z}_{2d}^{\infty}$. (We have already observed this kind of invariance in Section 2.1 in a different setting.) Thus $m(F_k \cap A_n) = m(F_k) m(A_{n-k})$, giving us (20).

If we set $A(r) = \sum_{n=0}^{\infty} r^n m(A_n)$, $F(r) = \sum_{n=0}^{\infty} r^n m(F_n)$, $0 < r < 1$, then (20) can be interpreted to say $A(r) = A(r)F(r) + 1$, that is $F(r) = 1 - 1/A(r)$. First, when $d \leq 2$, since the series $\sum_{n=0}^{\infty} m(A_n)$ diverges, then $A(r) \to \infty$ as $r \to 1$, which gives $F(1) = \sum_{n=1}^{\infty} m(F_n) = 1$, and proves that almost all paths return to the origin at least once. Secondly, when $d \geq 3$, since the series $\sum_{n=0}^{\infty} m(A_n)$ converges, we deduce that $F(1) = \sum_{n=1}^{\infty} m(F_n) < 1$, hence there is a set of positive probability where paths never return to the origin.

For the case $d \leq 2$, to prove infinite recurrence we define for each $\ell \geq 1$

$$F_n^{(\ell)} = \{x : \; s_n(x) = 0, \text{ but } s_k(x) = 0, \; \ell - 1 \text{ times, for } 1 \leq k < n\}.$$

(Here we set $F_1^{(\ell)} = \emptyset$.) Note that $F_n^{(1)} = F_n$, and $\sum_{n=1}^{\infty} m(F_n^{(\ell)}) = 1$ means that almost every path returns to the origin at least ℓ times. Then by an argument very similar to that giving (20) we get when $\ell \geq 2$

$$m(F_n^{(\ell)}) = \sum_{1 \leq k \leq n} m(F_k^{(\ell-1)}) m(F_{n-k}^{(1)}).$$

So if $F^{(\ell)}(r)$ is defined by $\sum_{n=1}^{\infty} r^n m(F_n^{(\ell)})$, then

$$F^{(\ell)}(r) = F^{(\ell-1)}(r) F^{(1)}(r),$$

which by iteration yields $F^{(\ell)}(r) = (F^{(1)}(r))^\ell$. Letting $r \to 1$ then gives $\sum_{n=1}^\infty m(F_n^{(\ell)}) = 1$, so almost all paths return to the origin at least ℓ times. Since this holds for all $\ell \geq 1$, conclusion (a) of the theorem is also proved.

It is interesting to ask what happens to our random walk, when the time interval between successive steps is taken to be $1/n$, the paths are re-scaled by a factor $1/n^{1/2}$, and we then pass to the limit $n \to \infty$ in accordance with the central limit theorem. The answer is that in this way we are led to Brownian motion. This important topic will be our next subject.

3 Exercises

1. Consider $S_N(x) = \sum_{n=1}^N r_n(x)$, with N odd.

(a) Calculate $m(\{x:\ S_N(x) = k\})$, and show that as k varies over the integers, the maximum is attained at $k = -1$ and $k = 1$.

(b) Adapt the proof when N is even to show that for odd N,

$$m(\{x:\ a < \frac{S_N(x)}{N^{1/2}} < b\}) \to \frac{1}{\sqrt{2\pi}} \int_a^b e^{-t^2/2}\, dt, \qquad \text{as } N \to \infty.$$

2. Find three functions f_1, f_2, and f_3, so that any pair are mutually independent, but the three are not.

[Hint: Let $f_1 = r_1$, $f_2 = r_2$, and express f_3 in terms of r_1 and r_2.]

3. The collection $\{r_n\}$ of mutually independent functions on $[0, 1]$ cannot be much enlarged and still remain mutually independent. In fact, prove that if we adjoin a function f to the collection $\{r_n\}$, then the resulting collection is also mutually independent only when f is constant.

[Hint: See also Exercise 16.]

4. Suppose μ and ν are two finite measures on a space X that agree on a collection of sets $\mathcal{C}$. If $\mathcal{C}$ contains X and is closed under finite intersections, then show that $\mu = \nu$ on the σ-algebra generated by $\mathcal{C}$.

[Hint: The equality $\mu = \nu$ holds on finite unions of sets in $\mathcal{C}$ because

$$\mu(\bigcup_{j=1}^k C_j) = \sum_{j=1}^k \mu(C_j) - \sum_{i<j} \mu(C_i \cap C_j) + \\ + \sum_{i<j<\ell} \mu(C_i \cap C_j \cap C_\ell) + \cdots + (-1)^{k-1} \mu(\bigcap_{j=1}^k C_j).]$$

5. Prove that the $\mathbb{R}^d$-valued functions $f_1, \dots, f_n$ are mutually independent if and only if their joint distribution measure equals the product of the individual distribution measures:

$$\mu_{f_1,\dots,f_n} = \mu_{f_1} \times \cdots \times \mu_{f_n} \qquad \text{as measures on } \mathbb{R}^{nd} = \mathbb{R}^d \times \cdots \times \mathbb{R}^d.$$

[Hint: Check the equality on cylinder sets in $\mathbb{R}^{nd}$ and use the previous exercise.]

6. Suppose $\{f_n\}$ is a sequence of mutually independent functions on the probability space (X, m). Prove that there exists a probability space (X', m'), with X' an infinite product, $X' = \prod_{n=1}^{\infty} X_n$, (X_n, m_n) probability spaces, and m' the product measure of the m_n, so that the following holds: there are functions $\{g_n\}$ on X' so that $\{f_n\}$ and $\{g_n\}$ have the same joint distributions, but each function g_n depends only on the nth coordinate of X'.

[Hint: Take $(X_n, m_m) = (X, m)$ for each n and define g_n in terms of the f_n accordingly, and use the previous exercise.]

7. Show that if $\mathcal{B}_0, \dots, \mathcal{B}_n$ are mutually independent algebras, then for each $k < n$, the algebras $\bigvee_{j=0}^{k} \mathcal{B}_j$ and $\mathcal{B}_n$ are mutually independent.

Prove this by noting the following. First, use induction to show that if $B_j \in \mathcal{B}_j$, then $B_0 \cap \cdots \cap B_k$ is independent of $\mathcal{B}_n$. Now fix $B \in \mathcal{B}_n$, and consider the two finite measures $\mu(E) = m(E \cap B)$ and $\nu(E) = m(E)m(B)$, and the collection $\mathcal{C}$ of sets that are of the form $E = B_0 \cap \cdots \cap B_k$, where $B_j \in \mathcal{B}_j$. Then apply Exercise 4.

8. Verify the following further facts about probability distribution measures.

(a) Suppose $f = (f_1, \dots, f_k)$ with each f_j an $\mathbb{R}^d$-valued function. Let μ be the probability distribution measure of f, and let L be a linear transformation of $\mathbb{R}^{dk}$ to itself. Suppose that μ is the probability distribution measure of f. Then the distribution measure of $L(f)$ is μ_L, where $\mu_L(A) = \mu(L^{-1}A)$ for every Borel set $A \subset \mathbb{R}^{dk}$.

(b) Suppose the distribution measure of f_j is Gaussian with covariance matrix $\sigma_j^2 I$, $1 \le j \le k$. Assume also that the $\{f_j\}$ are mutually independent. Then the distribution measure of $c_1 f_1 + \cdots + c_k f_k$ is Gaussian with covariance matrix $(c_1^2\sigma_1^2 + \cdots + c_k^2\sigma_k^2)I$.

[Hint: For (b), compute the Fourier transforms (the characteristic functions) of the measures in question.]

9. Consider the space $L^2(\Omega, \mathbb{R}^d)$ of square-integrable $\mathbb{R}^d$-valued functions on the probability space (Ω, P). A closed subspace $\mathcal{G}$ of this space is called a **Gaussian subspace** if it is spanned by a sequence $\{f_n\}$ of mutually independent functions, each having a Gaussian distribution measure with mean zero, and covariance $\{\sigma_n^2 I\}$.

Prove that if $F_1, F_2, \dots, F_k$ are mutually orthogonal elements of $\mathcal{G}$, then they are mutually independent. Note that the converse of this is immediate.

[Hint: Consider the case when $\mathcal{G}$ is finite-dimensional, and $\mathcal{G}$ is spanned by $f_1, \ldots, f_N$. One may suppose, after multiplication by appropriate scalars, that the f_j and F_j each have L^2 norm equal to 1. So there exists an orthogonal linear transformation L so that $L(f_j) = F_j$. Then apply Exercises 5 and 8.]

10. Consider the following two types of convergence of a sequence $\{f_n\}$ to a limit f on a probability space:

(i) $f_n \to f$ almost everywhere,

(ii) $f_n \to f$ in terms of weak convergence of their distribution measures.

Prove that (i) implies (ii), but that this implication cannot be reversed.

[Hint: Recall that if φ is continuous and bounded, then $\int \varphi(f)\, d\mu = \int \varphi\, d\mu_f$, where μ_f is the distribution measure of f, and apply the dominated convergence theorem.]

11. On $[0, 1]$ with Lebesgue measure, construct a function f whose distribution measure is normal.

[Hint: Consider the "error function" $\text{Erf}(x) = \frac{1}{\sqrt{2\pi}} \int_{-\infty}^{x} e^{-t^2/2}\, dt$ and its inverse function.]

12. Prove the identity (6), which says that if G is a non-negative continuous function on $\mathbb{R}$ (or continuous and bounded), and f is a real-valued measurable function (on a probability space (X, m)) with distribution measure $\mu = \mu_f$, then

$$\int_X G(f)(x)\, dm = \int_{\mathbb{R}} G(t)\, d\mu(t).$$

[Hint: Note that if f is bounded, then $\sum_k G(k/n)\, m(\{k/n < f < (k+1)/n\})$ converges to both integrals as $n \to \infty$.]

13. The Rademacher sequence $\{r_n\}$ is far from complete on $L^2([0, 1])$. In fact it cannot be completed by adjoining any finite collection of functions. Prove this in two ways.

(a) By considering the functions $\{r_n r_m\}$ for $n < m$.

(b) By using the L^p inequality of Lemma 1.8.

See also Exercise 16.

14. Consider the power series

$$\sum_{n=1}^{\infty} \pm a_n z^n = \sum_{n=1}^{\infty} r_n(t) a_n z^n = F(z, t),$$

where $\sum |a_n|^2 = \infty$ and $\limsup |a_n|^{1/n} \le 1$.

Show that for almost every t, the function $F(z,t)$ cannot be analytically continued outside the unit disc.

[Hint: Argue as in Theorem 1.7 part (b) using Abel summation rather than Cesàro summation.]

15. Show that the $L^2([0,1])$ span of $\{r_n\}$ can be characterized as the subspace of L^2 consisting of those f for which

$$\mathbb{E}_N(f) = S_N(f), \qquad \text{for all } N,$$

where $\mathbb{E}_N$ are the conditional expectations corresponding to the dyadic intervals of length 2^{-N}.

16. A natural completion of the collection of Rademacher functions are the **Walsh-Paley functions**. One defines this collection on $[0,1]$, denoted by $\{w_n\}$, in the following way.

First one sets $w_0(t) = 1$, $w_1(t) = r_1(t)$, $w_2(t) = r_2(t)$ and $w_3(t) = r_1(t)r_2(t)$. More generally, if $n \geq 1$, $n = 2^{k_1} + 2^{k_2} + \cdots + 2^{k_\ell}$, where $0 \leq k_1 < \cdots < k_\ell$, then one defines

$$w_n(t) = \prod_{j=1}^{\ell} r_{k_j+1}(t).$$

In particular $w_{2^{k-1}} = r_k$.

(a) Prove that $\{w_n\}_{n=0}^{\infty}$ is a complete orthonormal system on $L^2([0,1])$.

(b) Verify the following additional interesting property of the Walsh-Paley functions: they are the continuous characters of the compact abelian group $\mathbb{Z}_2^{\infty}$ (thought of as the product of the two-point abelian groups $\mathbb{Z}_2$).

[Hint: Equip the group $\mathbb{Z}_2^{\infty}$ with the addition $x+y$ defined by $(x+y)_j = x_j + y_j$ mod 2 if $x = (x_j)$ and $y = (y_j)$. Then $r_k(x+y) = r_k(x)r_k(y)$.

Consider also the "Dirichlet kernel" $K_N(t) = \sum_{k=0}^{N-1} w_k(t)$, and show that if $N = 2^n$, then $K_N(t) = \prod_{j=1}^{n}(1 + r_j(t))$, hence $K_{2^n}(t) = 2^n$ if $0 \leq t \leq 2^{-n}$ and 0 otherwise. As a result, using the convolution $\int f(y)K_N(x+y)\,dy$, note that if $f \sim \sum a_k w_k$, then $\sum_{k<2^n} a_k w_k = \mathbb{E}_n(f)$, where $\mathbb{E}_n$ was defined in Section 1.6. See also Problem 2*.]

17. The inequality in Lemma 1.8 may be strengthened as follows. Let $F(t) = \sum_{n=1}^{\infty} a_n r_n(t)$, with a_n real and $\sum_{n=1}^{\infty} a_n^2 = 1$. Then

(a) $\int_0^1 e^{\mu|F(t)|}\,dt \leq 2e^{\mu^2}$, for all $0 \leq \mu$.

(b) As a result, for some $c > 0$, $\int_0^1 e^{c|F(t)|^2}\,dt < \infty$.

[Hint: Part (a) implies that $m(\{t : |F(t)| > \alpha\}) \leq 2e^{\mu^2 - \mu\alpha}$. Choose $\mu = \alpha/2$, and obtain (b) with $c < 1/4$.]

18. Prove that there exists an f in $L^p([0, 2\pi])$, for all $p < \infty$, $f \sim \sum_{-\infty}^{\infty} c_n e^{in\theta}$, so that $\sum |c_n|^q = \infty$, for all $q < 2$. Hence the Hausdorff-Young inequality in Section 2.1 of Chapter 2 fails for $p > 2$.

[Hint: Use Theorem 1.7.]

19. Suppose $F(t) = \sum_{n=1}^{\infty} a_n r_n(t)$ with $\sum a_n^2 < \infty$.

(a) Prove directly that there exists a constant A so that

$$\|F\|_{L^4} \leq A\|F\|_{L^2}.$$

(b) Show as a result that there is a constant A' so that $\|F\|_{L^2} \leq A'\|F\|_{L^1}$.

(c) Conclude that $\|F\|_{L^p} \leq A_p\|F\|_{L^1}$, for $1 \leq p < \infty$.

[Hint: For (a) write out $\int_0^1 F^4(t)\,dt$ as a sum and use the orthogonality of $r_n(t)r_m(t)$. For (b) use Hölder's inequality. For (c) use Lemma 1.8.]

20. Suppose $\{A_n\}$ is a sequence of subsets of the probability space X.

(a) If $\sum m(A_n) < \infty$, then $m(\limsup_{n\to\infty} A_n) = 0$, where $\limsup_{n\to\infty} A_n$ is defined as $\bigcap_{n=1}^{\infty} \bigcup_{k=n}^{\infty} A_k$.

(b) However if $\sum m(A_n) = \infty$, and the sets $\{A_n\}$ are mutually independent, then $m(\limsup_{n\to\infty} A_n) = 1$.

This dichotomy is often referred to as the Borel-Cantelli lemma. (See also Book III.)
[Hint: In the case (b), $m\left(\bigcap_{k=r}^{n} A_k^c\right) = \prod_{k=r}^{n}(1 - m(A_k))$.]

21. Except for a countable set (the dyadic rationals) it is possible to assign a unique dyadic expansion to each real number α in $[0, 1]$, that is,

$$\alpha = \sum_{j=1}^{\infty} \frac{x_j}{2^j} \qquad \text{with } x_j = 0 \text{ or } 1.$$

Given such a number α let $\#_N(\alpha)$ denote the number of 1's that appear among the first N terms in the dyadic expansion of α. We say that α is **normal**, if its dyadic expansion contains a density of 1's equal to the density of 0's, that is,

$$\lim_{N\to\infty} \frac{\#_N(\alpha)}{N} = 1/2.$$

(a) Prove that (with respect to the Lebesgue measure) almost every number in $[0, 1]$ is normal.

(b) More generally, given an integer $q \geq 2$ consider the q-expansion of a real number α in $[0, 1]$,

$$\alpha = \sum_{j=1}^{\infty} \frac{x_j}{q^j} \qquad \text{with } x_j = 0, 1, \ldots, q-1. \tag{21}$$

Again, ignoring a countable subset, this expansion is unique. For a given real number α and for each $0 \leq p \leq q-1$, define $\#_{p,N}(\alpha)$ to equal the number of j's with $0 \leq j \leq N$ for which $x_j = p$ in the q-expansion of α. A number that satisfies

$$\lim_{N\to\infty} \frac{\#_{p,N}(\alpha)}{N} = 1/q,$$

for every $0 \leq p \leq q-1$ is said to be **normal** to base q.

Show that almost all real numbers in $[0,1]$ satisfy this property.

[Hint: Consider the infinite product $\prod \mathbb{Z}_q$ with each factor given the uniform measure. Under (21) the product measure corresponds to the Lebesgue measure on $[0,1]$. Now apply the law of large numbers as in Theorem 2.1.]

22. A sequence $\{f_n\}_{n=0}^{\infty}$ of functions on X is called a (discrete) **stationary process** if for every N the joint probability distribution of $f_r, f_{r+1}, \ldots, f_{r+N}$ is independent of r.

Consider the probability space Y constructed as in the proof of Theorem 2.1. Show that whenever the sequence $\{f_n\}$ is a stationary process, then it has the same joint distribution as the sequence $\{g_0(\tau^n(y))\}$, where g_0 is a suitable function on Y and τ is the shift. Hence the ergodic theorem is equally applicable in this more general situation.

23. Prove that the conditions in Theorem 2.1 are sharp in the following sense. If $\{f_n\}_{n=0}^{\infty}$ are mutually independent and identically distributed, but $\int_X |f_0(x)|\,dm = \infty$, then for almost every x, the averages $\frac{1}{N}\sum_{n=0}^{N-1} f_n(x)$ fail to converge to a limit as $N \to \infty$.

[Hint: Let $A_n = \{x : \ |f_n(x)| > n\}$. The sets A_n are independent. However, $\sum_{n=0}^{\infty} m(A_n) = \sum_{n=0}^{\infty} m(\{x : \ |f_0(x)| > n\}) \approx \int_X |f_0(x)|\,dm = \infty$. Then use Exercise 20.]

24. The following are examples of conditional expectations.

(a) Suppose $X = \bigcup A_n$ is a finite (or countable) partition of X, with $m(A_n) > 0$ whenever A_n is non-empty. Let $\mathcal{A}$ be the algebra generated by the sets $\{A_n\}$. Then $\mathbb{E}_{\mathcal{A}}(f)(x) = \frac{1}{m(A_n)}\int_{A_n} f\,dm$ whenever $x \in A_n$.

(b) Let $X = X_1 \times X_2$, with the measure m on X being the product of the measures m_i on X_i. Let $\mathcal{A} = \{A \times X_2\}$, where A ranges over arbitrary measurable sets of X_1. Then $\mathbb{E}_{\mathcal{A}}(f)(x_1,x_2) = \int_{X_2} f(x_1,y)\,dm_2(y)$.

25. In the following four exercises $\{s_n\}$ will denote a martingale sequence corresponding to the increasing sequence of algebras $\mathcal{A}_n$ and their conditional expectations $\mathbb{E}_n$.

Suppose $s_n = \mathbb{E}_n(s_\infty)$ with $s_\infty \in L^2$. Then $\{s_n\}$ converges in L^2.

[Hint: Note that if $f_n = s_n - s_{n-1}$ then the f_n's are mutually orthogonal and $s_n - s_0 = \sum_{k=1}^{n} f_k$.]

26. Prove the following.

(a) If $s_\infty \in L^p$, then $s_n = \mathbb{E}_n(s_\infty) \in L^p$, and $\|s_n\|_{L^p} \leq \|s_\infty\|_{L^p}$ for all p with $1 \leq p \leq \infty$.

(b) Conversely, if $\{s_n\}$ is martingale and $\sup_n \|s_n\|_{L^p} < \infty$, then there exists $s_\infty \in L^p$, so that $s_n = \mathbb{E}_n(s_\infty)$, when $1 < p \leq \infty$.

(c) Show, however, that the conclusion in (b) may fail when $p = 1$.

[Hint: For (a) argue as in the proof of Lemma 2.5(a). For (b), use Lemma 2.5 and also the weak compactness of L^p, $p > 1$, as in Exercise 12 in Chapter 1. For (c), let $X = [0,1]$ with Lebesgue measure, and consider $s_n(x) = 2^n$ for $0 \leq x \leq 2^{-n}$, $s_n(x) = 0$ otherwise.]

27. Suppose that $s_n = \mathbb{E}_n(s_\infty)$, with s_∞ integrable on X.

(a) Show that s_n converges in the L^1 norm as $n \to \infty$.

(b) Moreover $s_n \to s_\infty$ in L^1 if and only if s_∞ is measurable with respect to the algebra $\mathcal{A}_\infty = \bigvee_{n=1}^{\infty} \mathcal{A}_n$.

[Hint: For (a) use Exercises 25 and 26 (a). Then $\lim s_n = \mathbb{E}_{\mathcal{A}_\infty}(s_\infty)$, and use the previous exercise.]

28. Suppose that $s_n = \mathbb{E}_n(s_\infty)$, and $s_\infty \in L^1$.

(a) Show that

$$m(\{x : \sup_n |s_n(x)| > \alpha\}) \leq \frac{1}{\alpha} \int_{|s_\infty(x)|>\alpha} |s_\infty(x)|\, dx.$$

(b) Prove as a result $\| \sup_n |s_n| \|_{L^p} \leq A_p \|s_\infty\|_{L^p}$ if $s_\infty \in L^p$ and $1 < p \leq \infty$.

[Hint: For (a), note that when $s_\infty \geq 0$ this is a consequence of (15). To deduce (b) adapt the argument in the proof of Theorem 4.1 in Chapter 2 for the maximal function f^*.]

29. The results for real-valued martingale sequences $\{s_n\}$ discussed in Section 2.2 go through if we assume that the s_n take their values in $\mathbb{R}^d$. Verify in particular that the following consequences of identity (14) hold:

(a) $|s_k| \leq \mathbb{E}_k(|s_n|)$, if $k \leq n$.

(b) $m(\{x : \sup_n |s_n(x)| > \alpha\}) \leq \frac{1}{\alpha} \int_{|s_\infty(x)|>\alpha} |s_\infty(x)|\, dx$.

Here $|\cdot|$ denotes the Euclidean norm in $\mathbb{R}^d$.

[Hint: To prove (a), note that $(s_k, v) = \mathbb{E}_k((s_n, v))$, where $(\cdot,\cdot)$ is the inner product on $\mathbb{R}^d$, and v is any fixed vector in $\mathbb{R}^d$. Then take the supremum over unit vectors v. The conclusion (b) is a consequence of (a) and part (a) of Exercise 28.]

30. The ideas regarding conditional expectations extend to spaces (X, m) whose total measures are not necessarily finite. Consider the following example: $X = \mathbb{R}^d$, with m the Lebesgue measure. For each $n \in \mathbb{Z}$, let $\mathcal{A}_n$ be the algebra generated by all dyadic cubes of side-length 2^{-n}. The dyadic cubes are the open cubes, whose vertices are points of $2^{-n}\mathbb{Z}^d$, and have side-length 2^{-n}. Clearly $\mathcal{A}_n \subset \mathcal{A}_{n+1}$ for all n. Let f be integrable on $\mathbb{R}^d$, and set $\mathbb{E}_n(f) = \mathbb{E}_{\mathcal{A}_n}(f)$, with

$$\mathbb{E}_n(f)(x) = \frac{1}{m(Q)} \int_Q f\, dm$$

whenever $x \in Q$, with Q a dyadic cube of length 2^{-n}.

(a) Show that the maximal inequality in Theorem 2.10 extends to this case.

(b) If $f \geq 0$, then $\sup_{n\in\mathbb{Z}} \mathbb{E}_n(f)(x) \leq cf^*(x)$ for an appropriate constant c, with f^* the Hardy-Littlewood maximal function discussed in Chapter 2.

(c) Show by example that the converse inequality $f^*(x) \leq c' \sup_{n\in\mathbb{Z}} \mathbb{E}_n(f)(x)$ fails. Prove however that a substitute result holds

$$m(\{x : f^*(x) > \alpha\}) \leq c_1 m(\{x : \sup_{n\in\mathbb{Z}} \mathbb{E}_n(f)(x) > c_2\alpha\})$$

for all $\alpha > 0$. Here c_1 and c_2 are appropriate constants.

31. Let $\{\mu_N\}_{N=1}^{\infty}$ and ν be probability measures on $\mathbb{R}^d$. Prove the following are equivalent as $N \to \infty$.

(a) $\hat{\mu}_N(\xi) \to \hat{\nu}(\xi)$, all $\xi \in \mathbb{R}^d$.

(b) $\mu_N \to \nu$ weakly.

(c) In $\mathbb{R}$, $\mu_N((a,b)) \to \nu((a,b))$ for all open intervals (a,b), if we assume the measure ν is continuous.

(d) In $\mathbb{R}^d$, $\mu_N(\mathcal{O}) \to \nu(\mathcal{O})$ for all open sets $\mathcal{O}$, if we assume the measure ν is absolutely continuous with respect to Lebesgue measure.

[Hint: In $\mathbb{R}$, the equivalence of (a), (b) and (c) is implicit in the argument given in the proofs of Lemma 2.15 and Corollary 2.16. To show that (a) implies (d) in the case when $\mathcal{O}$ is an open cube, generalize the argument given in the text to $\mathbb{R}^d$. Then, prove that the analog of (d) holds for closed cubes. Finally, use the fact that any open set is an almost disjoint union of closed cubes. To show that (d) implies (b), approximate a continuous function φ of compact support uniformly by step functions that are constant on cubes.]

32. The proof of Theorem 2.17 requires the following calculation. Suppose σ is a strictly positive definite symmetric matrix with σ^{-1} denoting its inverse. Let ν_{σ^2} be the measure on $\mathbb{R}^d$ with density equal to $\frac{1}{(2\pi)^{d/2}(\det\sigma)} e^{-\frac{|\sigma^{-1}(x)|^2}{2}}$, $x \in \mathbb{R}^d$. Then $\hat{\nu}_{\sigma^2}(\xi) = e^{-2\pi^2|\sigma(\xi)|^2}$.

[Hint: Verify this by making an orthogonal change of variables that puts σ in a diagonal form. This reduces the d-dimensional integral in question to a product of corresponding 1-dimensional integrals.]

33. For the d-dimensional random walk $\{s_n(x)\}$ considered in Section 2.6, find the limit of the distribution measures of $s_n(x)/n^{1/2}$ as $n \to \infty$.

34. If k is a lattice point in $\mathbb{Z}^d$ and $d = 1$ or 2, show that for almost every path, the random walk visits k infinitely often, that is,

$$m(\{x: \; s_n(x) = k \qquad \text{for infinitely many } n\} = 1.$$

[Hint: There exists ℓ_0 so that $m(\{s_{\ell_0} = -k\}) > 0$. If the conclusion fails, then there exists r_0 so that $m(\{s_n \neq k, \text{ for all } n \geq r_0\}) > 0$. Then note that

$$\{s_n \neq 0, \text{ all } n \geq \ell_0 + r_0\} \supset \{s_{\ell_0} = -k\} \cap \{s_n - s_{\ell_0} \neq k, \text{all } n \geq \ell_0 + r_0\},$$

and that the sets on the right-hand side are independent.]

35. Prove that if $d \geq 3$, then the random walk s_n satisfies $\lim_{n\to\infty} |s_n| = \infty$ almost everywhere.

[Hint: It is sufficient to prove that for any fixed $R > 0$ the set

$$B = \{x: \; \liminf_{n\to\infty} |s_n(x)| \leq R\}$$

has measure 0. To this end, for each lattice point k, define

$$B(k,\ell) = \{x: \; s_\ell(x) = k, \text{ and } s_n(x) = k \text{ for infinitely many } n\}.$$

Clearly, $B \subset \bigcup_{\ell,\ |k|\leq R} B(k,\ell)$. But $d \geq 3$, so $m(B(k,\ell)) = 0$).]

4 Problems

1. In the context of Bernoulli trials with probabilities $0 < p, q < 1$, where $p + q = 1$, let $D : \mathbb{Z}_2^\infty \to [0,1]$, be given by

$$D(x) = \sum_{n=1}^{\infty} x_n/2^n \qquad \text{if } x = (x_1, \ldots, x_n, \ldots).$$

Under this mapping the measure m_p goes to the measure μ_p that can be written symbolically as a "Riesz product," $\mu_p = \prod_{n=1}^{\infty}(1 + (p-q)r_n(t))\,dt$. The meaning of this is as follows. For each N, let

$$F_N(t) = \int_0^t \prod_{n=1}^{N} (1 + (p-q)r_n(s))\,ds.$$

Then one can show that:

(a) Each F_N is increasing on $[0, 1]$.

(b) $F_N(0) = 0$, $F_N(1) = 1$.

(c) F_N converges uniformly to a function F, as $N \to \infty$.

(d) $\mu_p = dF$, in the sense that $\mu_p((a, b)) = F(b) - F(a)$.

(e) If $p \neq 1/2$, then μ_p is completely singular (that is $dF/dt = 0$ almost everywhere.)

[Hint: Show that if $I = (a, b)$ is a dyadic interval of length 2^{-n}, $a = \ell/2^n$ and $b = (\ell + 1)/2^n$, and $N \geq n$, then

$$F_N(b) - F_N(a) = p^{n_0} q^{n_1},$$

where n_0 is the number of zeroes in the first n terms of the dyadic expansion of $\ell/2^n$, and n_1 is the number of 1's, with $n_0 + n_1 = n$.]

2.* There is an analogy between the Walsh-Paley expansion (see Exercise 16) and the Fourier expansion, that is, between $\{w_n\}_{n=0}^{\infty}$ and $\{e^{in\theta}\}_{n=-\infty}^{\infty}$. In this analogy the Rademacher functions $r_k = w_{2^{k-1}}$ correspond to the lacunary frequencies $\{e^{i2^k\theta}\}_{k=0}^{\infty}$. In fact, the following is known:

(a) If $\sum_{k=0}^{\infty} c_k e^{i2^k\theta}$ is an $L^2([0, 2\pi])$ function, then it belongs to L^p, for every $p < \infty$.

(b) If $\sum_{k=0}^{\infty} c_k e^{i2^k\theta}$ is the Fourier series of an integrable function, it belongs to L^2, and hence to L^p for every $p < \infty$.

(c) This function belongs to L^{∞} if and only if $\sum_{k=0}^{\infty} |c_k| < \infty$.

(d) From (c) it follows that the conclusion (a) of Theorem 1.7 does not necessarily extend to $p = \infty$.

3. The following is a general form of the central limit theorem. Suppose $f_1, \ldots, f_n, \ldots$ are square integrable mutually independent functions on X, and assume for simplicity that each has mean equal to 0. Let μ_n be the distribution measure of f_n, and σ_n^2 the variance. Set $S_n^2 = \sum_{k=1}^{n} \sigma_n^2$. The critical assumption is that for every $\epsilon > 0$

$$\lim_{n \to \infty} \frac{1}{S_n^2} \sum_{k=1}^{n} \int_{|t| \geq \epsilon S_n} t^2 \, d\mu_n(t) = 0.$$

Under these conditions the distribution measures of $\frac{1}{S_n} \sum_{k=1}^{n} f_k$ converge weakly to the normal distribution ν with variance 1.

4.* Suppose $\{f_n\}$ are identically distributed, square integrable, mutually independent, have each mean 0 and variance 1. Let $s_n = \sum_{k=1}^{n} f_k$. Then for a.e. x

$$\limsup_{n \to \infty} \frac{s_n(x)}{(2n \log \log n)^{1/2}} = 1.$$

This is the law of the "iterated logarithm."

5. An interesting variant of the random walk in $\mathbb{R}^d$ (often referred to as a "random flight") arises if the motion of unit distance at the nth step is allowed to be in *any* direction (of the unit sphere). More precisely

$$s_n = f_1 + \cdots + f_n,$$

where the f_n are mutually independent, and each f_n is uniformly distributed on the unit sphere $S^{d-1} \subset \mathbb{R}^d$. The underlying probability space is defined as the infinite product $X = \prod_{j=1}^{\infty} S_j$, where each $S_j = S^{d-1}$ with the usual surface measure normalized to have integral 1.

(a) If μ is the distribution measure of each f_n, connect $\hat{\mu}(\xi)$ with Bessel functions.

(b) What is the covariance matrix?

(c) What is the limiting distribution of $s_n(x)/n^{1/2}$?

[Hint: Show that $\hat{\mu}(\xi) = \Gamma(d/2)(\pi|\xi|)^{(2-d)/2} J_{(d-2)/2}(2\pi|\xi|)$ by using the formulas in Problem 2, Chapter 6 of Book I.]

6 An Introduction to Brownian Motion

> Norbert Wiener: a precocious genius... whose feeling for physics and appreciation of Lebesgue integration was so deep that he was the first to understand the necessity of and the proper context for a rigorous definition of Brownian motion, which he then devised, going on to initiate the fundamentally important theory of stochastic integrals; who, however, was so unfamiliar with the standard probability techniques even at elementary levels that his methods were so clumsily indirect that some of his own doctoral students did not realize that his Brownian motion process had independent increments; who was the first to offer a general definition of potential theoretic capacity; who, however, published his probabilistic and potential theoretic triumphs in a little-known journal, with the result that this work remained unknown until too late to have its deserved influence...
>
> *J. L. Doob,* 1992

Between the 19th and 20th centuries there was a change in the scientific view of the natural world. The belief in the ultimate regularity and predictability of nature gave way to the recognition of a degree of inherent irregularity, uncertainty, and randomness. No mathematical construct better encompasses this idea of randomness, nor has wider general interest, than the process of Brownian motion.

While there are different ways of constructing the Brownian motion process, the approach we have chosen attempts to see the Brownian paths in $\mathbb{R}^d$ as limits of random walk paths, appropriately rescaled. The analytic problem that then must be dealt with is the question of convergence of the measures induced by these random walks to the "Wiener measure" on the space $\mathcal{P}$ of paths.

A remarkable application of Brownian motion is to the solution of Dirichlet's problem in a general setting.[1] It is based on the following

[1]See also the previous discussion for the disc in terms of Fourier series in Book I, in relation to conformal mappings in Book II, and the use of Dirichlet's principle in Book III.

insight that goes back to Kakutani. Namely, whenever $\mathcal{R}$ is a bounded region in $\mathbb{R}^d$, x a fixed point in it, and E a subset of $\partial\mathcal{R}$, then the probability that a Brownian path starting at x exits $\mathcal{R}$ at E, is the "harmonic measure" of E with respect to x.

A key to understanding this approach is the notion of a "stopping time." The basic example here is the first time that the path starting at x hits the boundary. Incidentally, stopping times were already used implicitly in the proof of the martingale maximal theorem in the previous chapter.

One also needs to come to grips with the "strong Markov" property of Brownian motion, which essentially states that if the Brownian motion process is restarted after a stopping time, the result is an equivalent Brownian motion. The application of this Markov property is a little intricate, and it is best understood in terms of an identity that involves two stopping times.

1 The Framework

Here we begin by sketching the framework of our construction of Brownian motion. At first we describe the situation somewhat imprecisely, and postpone to Sections 2 and 3 below the exact definitions and statements.

We recall the random walk in $\mathbb{R}^d$ studied in the previous chapter (in Section 2.6). It is given by a sequence $\{s_n\}_{n=1}^{\infty}$ where

$$s_n = s_n(x) = \sum_{k=1}^{n} \mathfrak{r}_k(x),$$

with $s_n(x) \in \mathbb{Z}^d$ for each x in the probability space $\mathcal{Z}_{2d}^{\infty}$. This probability space carries the probability measure m, which is the product measure on $\mathcal{Z}_{2d}^{\infty}$. In this random walk we visit points in $\mathbb{Z}^d$ moving from a point to one of its neighbors in steps of unit "time" and "distance."

Next we consider the rectilinear paths obtained by joining these successive points, and then rescale both time and distance, so that between two consecutive steps the elapsed time is $1/N$ and the traversed distance is $1/N^{1/2}$, all in accordance with our experience with the central limit theorem. That is, for each N we consider

$$(1) \qquad S_t^{(N)}(x) = \frac{1}{N^{1/2}} \sum_{1 \le k \le [Nt]} \mathfrak{r}_k(x) + \frac{(Nt - [Nt])}{N^{1/2}} \mathfrak{r}_{[Nt]+1}(x).$$

Now for each N, $S_t^{(N)}$ is a **stochastic process**, that is, for $0 \le t < \infty$,

$S_t^{(N)}$ is a function (random variable) on a fixed probability space (here, $(\mathbb{Z}_{2d}^\infty, m)$).

Our goal is the proper formulation and proof of the assertion that we have the convergence

$$S_t^{(N)} \to B_t \quad \text{as } N \to \infty, \tag{2}$$

where B_t is the Brownian motion process in $\mathbb{R}^d$.

So to proceed we need first to set down the properties that characterize this process. Brownian motion B_t is defined in terms of a probability space (Ω, P), with P its probability measure and ω denoting a typical point in Ω. We suppose that for each t, $0 \leq t < \infty$, the function B_t is defined on Ω and takes values in $\mathbb{R}^d$. The **Brownian motion process** $B_t = B_t(\omega)$ is then assumed to satisfy $B_0(\omega) = 0$ almost everywhere and:

B-1 The increments are independent, that is, if $0 \leq t_1 < t_2 < \cdots < t_k$, then B_{t_1}, $B_{t_2} - B_{t_1}$, $\ldots$, $B_{t_k} - B_{t_{k-1}}$ are mutually independent.

B-2 The increments $B_{t+h} - B_t$ are Gaussian with covariance hI and mean zero,[2] for each $0 \leq t < \infty$. Here I is the $d \times d$ identity matrix.

B-3 For almost every $\omega \in \Omega$, the path $t \mapsto B_t(\omega)$ is continuous for $0 \leq t < \infty$.

Note that in particular, B_t is normally distributed with mean zero and covariance tI.

Now it will turn out that this process can be realized in a canonical way in terms of a natural choice of the probability space Ω. This probability space, denoted by $\mathcal{P}$, is the space of continuous paths in $\mathbb{R}^d$ starting at the origin: it consists of the continuous functions $t \mapsto \mathsf{p}(t)$ from $[0, \infty)$ to $\mathbb{R}^d$ with $\mathsf{p}(0) = 0$.

Since, by assumption B-3, for almost every $\omega \in \Omega$ the function $t \mapsto B_t(\omega)$ is such a continuous path, we get an inclusion $i : \Omega \to \mathcal{P}$ and then the probability measure P gives us, as we will see, a corresponding measure W (the "Wiener measure") on $\mathcal{P}$.[3]

One can in fact reverse the logic of these implications, starting with the space $\mathcal{P}$ and a probability measure W given on it. From this, one can define a process $\tilde{B}_t$ on $\mathcal{P}$ with

$$\tilde{B}_t(\mathsf{p}) = \mathsf{p}(t). \tag{3}$$

[2]In the notation of the previous chapter the increments have distribution ν_{hI}.

[3]More precisely, the inclusion i is defined on a subset of Ω of full measure.

We then say that the measure W on $\mathcal{P}$ is a **Wiener measure** if the process $\tilde{B}_t$ defined by (3) satisfies the properties of Brownian motion set down in B-1, B-2 and B-3 above. Thus the existence of a Wiener measure is tantamount to the existence of Brownian motion. In fact, we will focus on constructing a Wiener measure and then relabel $\tilde{B}_t$ and designate it by B_t. Moreover we will see that such a Wiener measure on $\mathcal{P}$ is unique, and so we speak of "the" Wiener measure.

Now returning to the random walks and their scalings given in (1), we have defined for each $x \in \mathbb{Z}_{2d}^{\infty}$ a continuous path $t \mapsto S_t^{(N)}(x)$ defined for $0 \leq t < \infty$. Thus the probability measure m on $\mathbb{Z}_{2d}^{\infty}$ induces a probability measure μ_N on $\mathcal{P}$ via

$$\mu_N(A) = m(\{x \in \mathbb{Z}_{2d}^{\infty} : S_t^{(N)}(x) \in A\}),$$

where A is any Borel subset of paths in $\mathcal{P}$. With this, our goal is the following assertion:

The measures μ_N converge weakly to the Wiener measure W as $N \to \infty$.

Notice that it is not claimed that the convergence in (2) is anything like pointwise almost everywhere, but only a statement essentially weaker in appearance in terms of convergence of induced measures.[4]

2 Technical Preliminaries

With $\mathcal{P}$ denoting the collection of continuous paths $t \mapsto \mathsf{p}(t)$ from $[0, \infty)$ to $\mathbb{R}^d$ such that $\mathsf{p}(0) = 0$, we endow $\mathcal{P}$ with a metric with respect to which convergence is equivalent to uniform convergence on compact subsets of $[0, \infty)$.

For two such paths, p and p' in $\mathcal{P}$, we set

$$d_n(\mathsf{p}, \mathsf{p}') = \sup_{0 \leq t \leq n} |\mathsf{p}(t) - \mathsf{p}'(t)|,$$

and

$$d(\mathsf{p}, \mathsf{p}') = \sum_{n=1}^{\infty} \frac{1}{2^n} \frac{d_n(\mathsf{p}, \mathsf{p}')}{1 + d_n(\mathsf{p}, \mathsf{p}')}.$$

Then it is easily verified that d is a metric on $\mathcal{P}$. We record here some simple properties of d, whose proofs may be left to the reader:

[4]Since $S_t^{(N)}$ and B_t are defined on different probability spaces, pointwise almost everywhere convergence would not be meaningful. It is also to be noted that the rectilinear paths corresponding to $S_t^{(N)}$ are a subset of zero W-measure of $\mathcal{P}$.

- We have $d(\mathsf{p}_k, \mathsf{p}) \to 0$, as $k \to \infty$, if and only if $\mathsf{p}_k \to \mathsf{p}$ uniformly on compact subsets of $[0, \infty)$.
- The space $\mathcal{P}$ is complete with respect to the metric d.
- $\mathcal{P}$ is separable.

(See also Exercise 2.)

We next consider the **Borel sets** $\mathcal{B}$ of $\mathcal{P}$, defined as the σ-algebra of subsets of $\mathcal{P}$ generated by the open sets. Since $\mathcal{P}$ is separable, the σ-algebra $\mathcal{B}$ is the same as the σ-algebra generated by the open balls in $\mathcal{P}$.

A useful class of elementary sets in $\mathcal{B}$ is that of the cylindrical sets, defined as follows. For each sequence $0 \leq t_1 \leq t_2 \leq \cdots \leq t_k$, and a Borel set A in $\mathbb{R}^{dk} = \mathbb{R}^d \times \cdots \times \mathbb{R}^d$ (that is, k factors $\mathbb{R}^d$), then

$$\{\mathsf{p} \in \mathcal{P} : (\mathsf{p}(t_1), \mathsf{p}(t_2), \ldots, \mathsf{p}(t_k)) \in A\}$$

is called a **cylindrical set**.[5] We denote by $\mathcal{C}$ the σ-algebra of $\mathcal{P}$ generated by these sets (as k ranges over all positive integers and A over all Borel sets in $\mathbb{R}^{dk}$).

Lemma 2.1 *The σ-algebra $\mathcal{C}$ is the same as the σ-algebra $\mathcal{B}$ of Borel sets.*

Proof. If $\mathcal{O}$ is an open set in $\mathbb{R}^{dk}$, then clearly

$$\{\mathsf{p} \in \mathcal{P} : (\mathsf{p}(t_1), \mathsf{p}(t_2), \ldots, \mathsf{p}(t_k)) \in \mathcal{O}\}$$

is open in $\mathcal{P}$, and hence this set belongs to $\mathcal{B}$. As a result, cylindrical sets are in $\mathcal{B}$, thus $\mathcal{C} \subset \mathcal{B}$.

To see the reverse inclusion, note that for any fixed n and a and a given path p_0, the set $\{\mathsf{p} \in \mathcal{P} : \sup_{0 \leq t \leq n} |\mathsf{p}(t) - \mathsf{p}_0(t)| \leq a\}$ is the same as the corresponding set where the supremum is restricted to the t in $[0, n]$ that are rational, and hence this set is in $\mathcal{C}$. It is then not too difficult to see that for any $\delta > 0$, the ball $\{\mathsf{p} \in \mathcal{P} : d(\mathsf{p}, \mathsf{p}_0) < \delta\}$ is in $\mathcal{C}$. Since open balls generate $\mathcal{B}$ we have $\mathcal{B} \subset \mathcal{C}$, and the lemma is established.

We will now consider probability measures on $\mathcal{P}$, and in what follows these will always be assumed to be **Borel measures**, that is, defined on the Borel subsets $\mathcal{B}$ of $\mathcal{P}$. For any such measure μ, and any choice

[5] This terminology is used to distinguish it from "cylinder sets" that appear in product spaces.

$0 \le t_1 \le t_2 \le \cdots \le t_k$, we define the **section** $\mu^{(t_1,t_2,\ldots,t_k)}$ of μ to be the measure on $\mathbb{R}^{dk}$ given by

(4) $$\mu^{(t_1,t_2,\ldots,t_k)}(A) = \mu(\{\mathsf{p} \in \mathcal{P} : \ (\mathsf{p}(t_1), \mathsf{p}(t_2), \ldots, \mathsf{p}(t_k)) \in A\})$$

for any Borel set A in $\mathbb{R}^{dk}$.

It follows from Lemma 2.1 and Exercise 4 in the previous chapter that two measures μ and ν on $\mathcal{P}$ are identical if $\mu^{(t_1,t_2,\ldots,t_k)} = \nu^{(t_1,t_2,\ldots,t_k)}$ for all $0 \le t_1 \le t_2 \le \cdots \le t_k$, since they then agree on all cylindrical sets (and the intersection of two cylindrical set is also a cylindrical set). The converse, that if $\mu = \nu$ then all their sections agree, is obviously true.

We will be concerned with a sequence $\{\mu_N\}$ of measures on $\mathcal{P}$, and the question whether this sequence **converges weakly**, that is, whether there exists another probability measure μ so that

(5) $$\int_{\mathcal{P}} f \, d\mu_N \to \int_{\mathcal{P}} f \, d\mu \quad \text{as } N \to \infty, \text{ for every } f \in C_b(\mathcal{P}).$$

Here $C_b(\mathcal{P})$ denotes the continuous bounded functions on $\mathcal{P}$.

A particular feature of our metric space $\mathcal{P}$ that does not allow certain compactness arguments to apply in regard to (5) is that $\mathcal{P}$ is not σ-compact. (See Exercise 3.) This is the reason for the significance of the following lemma of Prokhorov.

Suppose X is a metric space. Assume that $\{\mu_N\}$ is a sequence of probability measures on X, and that this sequence is **tight** in the sense that for each $\epsilon > 0$, there is a compact set $K_\epsilon \subset X$ so that

(6) $$\mu_N(K_\epsilon^c) \le \epsilon, \qquad \text{for all } N.$$

In other words, the measures μ_N assign a probability of at least $1 - \epsilon$ to K_ϵ for *all* N.

Lemma 2.2 *If $\{\mu_N\}$ is tight, then there is a subsequence $\{\mu_{N_k}\}$ that converges weakly to a probability measure μ on X.*

Proof. For each compact set $K_{1/m}$ arising in (6) with $\epsilon = 1/m$, we construct a countable collection of functions $\mathcal{D}_m \subset C_b(X)$ so that:

(i) The functions $g|_{K_{1/m}}$, with $g \in \mathcal{D}_m$, are dense in $C(K_{1/m})$.

(ii) $\sup_{x \in X} |g(x)| = \sup_{x \in K_{1/m}} |g(x)|$, if $g \in \mathcal{D}_m$.

The $\mathcal{D}_m$ can be obtained as follows. Since $K_{1/m}$ is compact, $K_{1/m}$ and $C(K_{1/m})$ are both separable. (See Exercise 4.) Now if $\{g'_\ell\}$ is a countable dense subset of $C(K_{1/m})$, we can extend each g'_ℓ defined on $K_{1/m}$ to a function g_ℓ defined on X by the Tietze extension principle. (See Exercise 5.) The resulting collection of functions is taken to be $\mathcal{D}_m$.

Now since $\mathcal{D} = \bigcup_{m=1}^{\infty} \mathcal{D}_m$ is a countable collection of functions in $C_b(X)$, we can use the usual diagonalization procedure to find a subsequence of the measures $\{\mu_N\}$, which we relabel as $\{\mu_N\}$, so that

$$\mu_N(g) = \int g \, d\mu_N$$

converges to a limit as $N \to \infty$, for each $g \in \mathcal{D}$.

Next we fix $f \in C_b(X)$, and write

$$\mu_N(f) = \mu_N(f - g) + \mu_N(g).$$

Now given any m we can find $g \in \mathcal{D}_m$, so that $|(f - g)(x)| \leq 1/m$ if $x \in K_{1/m}$. Therefore, with $\|\cdot\|$ denoting the sup-norm on X, we have

$$\begin{aligned} |\mu_N(f - g)| &\leq \int_{K_{1/m}} |f - g| \, d\mu_N + \int_{K_{1/m}^c} |f - g| \, d\mu_N \\ &\leq \frac{1}{m} + \frac{1}{m}\|f - g\| \\ &\leq \frac{1}{m} + \frac{1}{m}\left(2\|f\| + \frac{1}{m}\right), \end{aligned}$$

where we have used (ii) above. From this it is clear that

$$\limsup_{N\to\infty} \mu_N(f) - \liminf_{N\to\infty} \mu_N(f) = O(1/m),$$

and since m was arbitrary, the conclusion is that $\lim_{N\to\infty} \mu_N(f)$ exists. This defines a linear functional ℓ on $C_b(X)$ by

$$\ell(f) = \lim_{N\to\infty} \mu_N(f).$$

Now we note that ℓ satisfies the requirements of Theorem 7.4 in Chapter 1. In fact, given $\epsilon > 0$, if we choose K_ϵ as in the definition of tightness, then

$$|\mu_N(f)| \leq \int_{K_\epsilon} |f| \, d\mu_N + \int_{K_\epsilon^c} |f| \, d\mu_N,$$

so the inequality (6) implies

$$|\mu_N(f)| \leq \sup_{x \in K_\epsilon} |f(x)| + \epsilon \|f\|,$$

and thus the same estimate holds for $\ell(f)$, satisfying the hypothesis (21) of the relevant theorem in Chapter 1. This yields that the linear functional ℓ is representable by a measure μ, and since we then have $\mu_N(f) \to \mu(f)$ for all $f \in C_b(X)$, we see that $\mu_N \to \mu$ weakly.

Corollary 2.3 *Suppose the sequence of probability measures $\{\mu_N\}$ is tight, and for each $0 \leq t_1 \leq t_2 \leq \cdots \leq t_k$ the measures $\mu_N^{(t_1,\ldots,t_k)}$ converge weakly to a measure $\mu_{t_1,\ldots,t_k}$, as $N \to \infty$. Then the sequence $\{\mu_N\}$ converges weakly to a measure μ, and moreover $\mu^{(t_1,\ldots,t_k)} = \mu_{t_1,\ldots,t_k}$.*

Proof. First, by Lemma 2.2, there is a subsequence $\{\mu_{N_m}\}$ that converges weakly to a measure μ. Next, $\mu_{N_m}^{(t_1,\ldots,t_k)} \to \mu^{(t_1,\ldots,t_k)}$ weakly. In fact, if $\pi^{t_1,t_2,\ldots,t_k}$ is the continuous mapping from $\mathcal{P}$ to $\mathbb{R}^{kd}$ that assigns to $\mathsf{p} \in \mathcal{P}$ the point $(\mathsf{p}(t_1), \mathsf{p}(t_2), \ldots, \mathsf{p}(t_k)) \in \mathbb{R}^{kd}$, then, by definition, $\mu^{(t_1,\ldots,t_k)}(A) = \mu((\pi^{t_1,\ldots,t_k})^{-1}A)$ for any Borel set $A \subset \mathbb{R}^{dk}$. As a result

$$\int_{\mathbb{R}^{dk}} f\, d\mu^{(t_1,\ldots,t_k)} = \int_{\mathcal{P}} (f \circ \pi^{t_1,\ldots,t_k})\, d\mu$$

for any $f \in C_b(\mathbb{R}^{dk})$, with a similar identity with μ replaced by μ_{N_m}. From this, and the weak convergence of μ_{N_m} to μ, it follows that $\mu^{(t_1,\ldots,t_k)} = \mu_{t_1,\ldots,t_k}$.

We now observe that the full sequence $\{\mu_N\}$ must converge weakly to μ. Suppose the contrary. Then there is another sequence $\mu_{N'_n}$ and a bounded continuous function f on $\mathcal{P}$, so that $\int f\, d\mu_{N'_n}$ converges to a limit that is not equal to $\int f\, d\mu$. Now using Lemma 2.2 again, there is a further subsequence $\{\mu_{N''_n}\}$ and a measure ν, so that $\mu_{N''_n}$ converges weakly to ν, while $\nu \neq \mu$. However by the previous argument we have $\nu^{(t_1,\ldots,t_k)} = \mu^{(t_1,\ldots,t_k)}$ for all choices of $0 \leq t_1 \leq t_2 \leq \cdots \leq t_k$. Therefore $\nu = \mu$, and $\int f\, d\mu = \int f\, d\nu$. This contradiction completes the proof of the corollary.

In applying the lemma and its corollary it will be necessary to prove that appropriate subsets K of the path space $\mathcal{P}$ are compact. The following gives a sufficient condition for this when K is closed. (It can be shown to be necessary. See Exercise 6.)

Lemma 2.4 *A closed set $K \subset \mathcal{P}$ is compact if for each positive T there is a positive bounded function $h \mapsto w_T(h)$, defined for $h \in (0,1]$ with $w_T(h) \to 0$ as $h \to 0$, and so that*

$$\sup_{\mathsf{p}\in K}\sup_{0\leq t\leq T} |\mathsf{p}(t+h) - \mathsf{p}(t)| \leq w_T(h), \qquad \text{for } h \in (0,1]. \tag{7}$$

The condition (7) implies that the functions on K are equicontinuous on each interval $[0,T]$. The lemma then essentially follows from the Arzela-Ascoli criterion. (Recall, this criterion was used in a special setting in Section 3, Chapter 8 of Book II.)

3 Construction of Brownian motion

We now prove the existence of the probability measure W on $\mathcal{P}$ that satisfies the following: If we define the process B_t on the probability space $(\mathcal{P}, W)$ by

$$B_t(\mathsf{p}) = \mathsf{p}(t), \qquad \text{for } \mathsf{p} \in \mathcal{P},$$

then B_t verifies the defining properties B-1, B-2 and B-3 of Brownian motion set down at the beginning of this chapter (with $(\mathcal{P}, W)$ playing the role of (Ω, P)). Note that if we are assured of the existence of such a W, the measure $W^{(t_1,t_2,\ldots,t_k)}$ is the distribution measure of $(B_{t_1},\ldots,B_{t_k})$. Therefore, by Exercise 8 in Chapter 5, this distribution measure is determined by properties B-1 and B-2, hence with this data the Wiener measure W is uniquely determined, as in the remarks following the proof of Lemma 2.1.

To construct W we return to the random walk $\{s_n\}$ discussed at the beginning of this chapter, with its attached probability space $(\mathbb{Z}_{2d}^{\infty}, m)$. Now for each $x \in \mathbb{Z}_{2d}^{\infty}$ there is a path $t \mapsto S_t^{(N)}(x)$ given by (1). This gives an injection $i_N : \mathbb{Z}_{2d}^{\infty} \to \mathcal{P}$. If $\mathcal{P}_N$ denotes the image of i_N (the collection of random walk paths scaled by the factor $N^{-1/2}$) then $\mathcal{P}_N$ is clearly a closed subset of $\mathcal{P}$. Now via i_N, the product measure m on $\mathbb{Z}_{2d}^{\infty}$ induces a Borel probability measure μ_N on $\mathcal{P}$, which is supported on $\mathcal{P}_N$, by the identity $\mu_N(A) = m(i_N^{-1}(A \cap \mathcal{P}_N))$. (Note that $i_N^{-1}(A \cap \mathcal{P}_N)$ is a cylinder set in the product space $\mathbb{Z}_{2d}^{\infty}$ whenever A is a cylindrical set in $\mathcal{P}$.)

Theorem 3.1 *The measures μ_N on $\mathcal{P}$ converge weakly to a measure as $N \to \infty$. This limit is the Wiener measure W.*

There are two steps in the proof. The first, that the sequence μ_N satisfies the tightness condition, is a little intricate. The second, that then μ_N converges to the Wiener measure, is more direct. The second step is based on the central limit theorem.

For the first step, the following lemma is key. It is a consequence of the martingale properties of sums of independent random variables dealt with in the previous chapter. Consider the unscaled random walk

$$s_n(x) = \sum_{1\leq k\leq n} \mathfrak{r}_k(x).$$

This is $S_t^{(N)}$ in (1) with $N = 1$ and $t = n$.

Lemma 3.2 *We have as* $\lambda \to \infty$*,*

$$\sup_{n\geq 1} m(\{x : \sup_{k\leq n} |s_k(x)| > \lambda n^{1/2}\}) = O(\lambda^{-p}) \tag{8}$$

for every $p \geq 2$*.*

Remark. In the first application below it suffices to have the conclusion for some p such that $p > 2$.

To prove the lemma we apply the martingale maximal theorem of the previous chapter (Theorem 2.10, in the form that it takes in Exercise 29, part (b)) to the stopped sequence $\{s'_k\}$ defined as $s'_k = s_k$ if $k \leq n$, $s'_k = s_n$ if $k \geq n$, and $s'_\infty = s_n$. With $s^*_n = \sup_{k\leq n} |s_k| = \sup_k |s'_k|$ we then have

$$m(\{x :\ s^*_n > \alpha\}) \leq \frac{1}{\alpha}\int_{|s_n|>\alpha} |s_n|\, dm. \tag{9}$$

Multiplying both sides by $p\alpha^{p-1}$ and integrating, using an argument similar to the one used in Section 4.1 of Chapter 2 yields

$$\int (s^*_n)^p\, dm \leq \frac{p}{p-1}\int |s_n|^p\, dm.$$

Now, the Khinchin inequality of Lemma 1.8 in the previous chapter, applied to the more general setting described in Exercise 10 gives

$$\int |s_n|^p\, dm \leq A\left(\int |s_n|^2\, dm\right)^{p/2}.$$

Thus

$$m(\{s_n^* > \alpha\}) \leq \frac{1}{\alpha^p}\|s_n^*\|_{L^p}^p \leq \frac{A'}{\alpha^p}\|s_n\|_{L^2}^p.$$

Setting $\alpha = \lambda n^{1/2}$ and recalling that $\|s_n\|_{L^2} = n^{1/2}$ completes the proof of the lemma.

Let us now prove that the sequence $\{\mu_N\}$ converges weakly to a measure μ. For this we use Corollary 2.3, and begin by showing that the sequence $\{\mu_N\}$ is tight, that is, for every $\epsilon > 0$ there is a compact subset K_ϵ of $\mathcal{P}$ so that $\mu_N(K_\epsilon^c) \leq \epsilon$ for all N.

To this end we will invoke Lemma 2.4 and first consider the situation for $T = 1$. We fix $0 < a < 1/2$, throughout the rest of the proof of the theorem. Then with our given ϵ we will see that we can select a sufficiently large constant c_1 so that

$$(10) \quad m(\{x : \sup_{0\leq t\leq 1,\ 0\leq h\leq\delta} |S_{t+h}^{(N)} - S_t^{(N)}| > c_1\delta^a \text{ for some } \delta \leq 1\}) \leq \epsilon.$$

Therefore if we define

$$\mathcal{K}^{(1)} = \{x : \sup_{0\leq t\leq 1,\ 0\leq h\leq\delta} |S_{t+h}^{(N)} - S_t^{(N)}| \leq c_1\delta^a, \text{ all } \delta \leq 1\},$$

and

$$K^{(1)} = \{\mathsf{p} : \sup_{0\leq t\leq 1,\ 0\leq h\leq\delta} |\mathsf{p}(t+h) - \mathsf{p}(t)| \leq c_1\delta^a, \text{ all } \delta \leq 1\},$$

then $m((\mathcal{K}^{(1)})^c) = \mu_N((K^{(1)})^c) \leq \epsilon$. Note also that then (7) is satisfied for $K = K^{(1)}$, $T = 1$, and $w_1(\delta) = c_1\delta^a$, and hence $K^{(1)}$ is compact.

In proving (10), let us first consider the analog of this set but with δ fixed, and δ of the form $\delta = 2^{-k}$, with k a non-negative integer. We then decompose the interval $[0,1]$ via the $2^k + 1$ partition points $\{t_j\}$, where $t_j = j\delta = j2^{-k}$, with $0 \leq j \leq 2^k$. Next, observe that for any function f defined on $[0, 1+\delta]$, we have

$$\sup_{0\leq t\leq 1,\ 0\leq h\leq\delta} |f(t+h) - f(t)| \leq 2\max_j\{\sup_{0\leq h\leq\delta} |f(t_j+h) - f(t_j)|\}.$$

Thus with $f(t) = S_t^{(N)}$ and any fixed $\sigma > 0$,

$$m(\{\sup_{0\leq t\leq 1,\ 0\leq h\leq\delta} |S_{t+h}^{(N)} - S_t^{(N)}| > \sigma\}) \leq \sum_{j=0}^{2^k} m(\{\sup_{0\leq h\leq\delta} |S_{t_j+h}^{(N)} - S_{t_j}^{(N)}| > \frac{\sigma}{2}\}).$$

However $m(\{x: \sup_{0\leq h\leq\delta}|S^{(N)}_{t_j+h}-S^{(N)}_{t_j}|>\sigma/2\})$ equals the same quantity with t_j replaced by 0, that is, it equals

$$m(\{x: \sup_{0\leq h\leq\delta}|S^{(N)}_h|>\sigma/2\}),$$

and this itself equals $m(\{x: \sup_{n\leq\delta N}|s_n(x)|>(\sigma/2)N^{1/2}\})$. These assertions follow from the definition (1) and the "stationarity" of the random walk: the fact that the joint probability distribution of $(\mathfrak{r}_m,\mathfrak{r}_{m+1},\ldots,\mathfrak{r}_{m+n})$ is independent of m, for all $m\geq 1$ and $n\geq 0$. (Recall that $\{\mathfrak{r}_n\}$ are defined in Section 2.6 of the previous chapter.)

Thus by Lemma 3.2, if we take $\lambda=\sigma/(2\delta^{1/2})$, then $N^{1/2}\frac{\sigma}{2}=\lambda(\delta N)^{1/2}$, and

$$m(\{x: \sup_{0\leq t\leq 1,\ 0\leq h\leq\delta}|S^{(N)}_{t+h}-S^{(N)}_t|>\sigma\})=O\left(\frac{1}{\delta}\left(\frac{\sigma}{2\delta^{1/2}}\right)^{-p}\right).$$

Here p is at our disposal. We now set $\sigma=c_1\delta^a$, with a fixed $0<a<1/2$. Then the O term becomes $O(c_1^{-p}\delta^b)$, with $b=-1+(\frac{1}{2}-a)p$. Therefore, since $a<1/2$ we can make b strictly positive by choosing p large enough, and then fix p. To summarize, with $\delta=2^{-k}$ we have proved

$$m(\{x: \sup_{0\leq t\leq 1,\ 0\leq h\leq\delta}|S^{(N)}_{t+h}-S^{(N)}_t|\geq c_1\delta^a\})=O\left(c_1^{-p}2^{-kb}\right).$$

Now each δ, with $0<\delta\leq 1$, lies between 2^{-k+1} and 2^{-k} for some integer $k\geq 0$. Thus when we take the union of the corresponding sets and add their measures (summing over k) we get a total measure that is $O(c_1^{-p})$, and this is less than ϵ if c_1 is large enough. So we have obtained our desired conclusion (10).

In the same way we can prove the following analog of this conclusion: for any $T>0$, and $\epsilon_T>0$, there is a constant c_T sufficiently large so that $m((\mathcal{K}^{(T)})^c)\leq\epsilon_T$, where

$$\mathcal{K}^{(T)}=\{x: \sup_{0\leq t\leq T,\ 0\leq h\leq\delta}|S^{(N)}_{t+h}-S^{(N)}_t|\leq c_T\delta^a,\ \text{all }\delta\leq 1\}.$$

This can be restated as follows. If

$$K^{(T)}=\{\mathsf{p}\in\mathcal{P}: \sup_{0\leq t\leq T,\ 0\leq h\leq\delta}|\mathsf{p}(t+h)-\mathsf{p}(t)|\leq c_T\delta^a,\ \text{all }\delta\leq 1\},$$

then $\mu_N((K^{(T)})^c)=m((\mathcal{K}^{(T)})^c)\leq\epsilon_T$.

Therefore, if we let T range over the positive integers, set $\epsilon_n = \epsilon/2^n$, and $K = \bigcap_{n=1}^{\infty} K^{(n)}$, then we have $\mu_N(K^c) \leq \epsilon$, and thus by the compactness of K guaranteed by Lemma 2.4, the tightness of the sequence $\{\mu_N\}$ will be established.

Now to show that the measure converges weakly it suffices, by Corollary 2.3, to show that for each $0 \leq t_1 \leq t_2 \leq \cdots \leq t_k$, the measures $\mu_N^{(t_1,\ldots,t_k)}$ converge weakly to the putative measure $W^{(t_1,\ldots,t_k)}$. However the central limit theorem (Theorem 2.17 of the previous chapter together with Exercise 1 below) shows that the distribution measures of $S_{t_j}^{(N)} - S_{t_{j-1}}^{(N)}$ converge weakly to the Gaussian measure $\nu_{t_j - t_{j-1}}$ (see Exercise 1). Moreover, since

$$S_{t_\ell}^{(N)} = S_{t_1}^{(N)} + (S_{t_2}^{(N)} - S_{t_1}^{(N)}) + \cdots + (S_{t_\ell}^{(N)} - S_{t_{\ell-1}}^{(N)}),$$

Exercise 8 (a) in the previous chapter shows that the distribution measures of the vectors of random variables $(S_{t_1}^{(N)}, S_{t_2}^{(N)}, \ldots, S_{t_k}^{(N)})$ converge weakly to the presumed measure $W^{(t_1,\ldots,t_k)}$ as $N \to \infty$. Thus the sequence $\{\mu_N\}$ converges weakly to a measure and this measure is then the desired Wiener measure W, and this completes the proof of the theorem.

Our construction of Brownian motion was done in terms of the limit of scalings of the simple random walk treated in Section 2.6 of the previous chapter. However the Brownian motion process can also be obtained as a corresponding scaling limit of more general random walks, as follows.

Let $f_1, \ldots, f_n, \ldots$ be a sequence of identically distributed mutually independent square integrable $\mathbb{R}^d$-valued functions on a probability space (X, m), each having mean zero and the identity as its covariance matrix. Define, as in (1),

$$S_t^{(N)} = \frac{1}{N^{1/2}} \sum_{1 \leq k \leq [Nt]} f_k + \frac{(Nt - [Nt])}{N^{1/2}} f_{[Nt]+1},$$

and let μ_N be the corresponding measures on $\mathcal{P}$ induced via the measure m on X. The result is then that $\{\mu_N\}$ converges weakly to the Wiener measure W as $N \to \infty$.

In this general setting the result is known as the **Donsker invariance principle**. The modifications needed for a proof of this generalization are outlined in Problem 2. A particularly striking example of the convergence to the Brownian motion process then arises if we choose the $\{f_n\}$ occurring in the process of "random flight" discussed in Problem 5 in the previous chapter.

4 Some further properties of Brownian motion

We describe now several interesting properties enjoyed by the Brownian motion process. In general it is useful to think of this process as either in terms of an abstract realization B_t on (Ω, P) satisfying conditions B-1, B-2 and B-3, or its concrete realization on $(\mathcal{P}, W)$ with W the Wiener measure, given in terms of $B_t(\omega) = \mathsf{p}(t)$, where ω is identified with p. More about this identification can be found in Exercises 8 and 9 below. It will also be convenient to augment the Borel σ-algebra of $\mathcal{P}$ by all subsets of Borel sets of W-measure zero.[6]

We begin by observing three simple but significant invariance statements. (Another symmetry of Brownian motion is described in Exercise 13.)

Theorem 4.1 *The following are also Brownian motion processes:*

(a) $\delta^{-1/2}B_{t\delta}$ *for every fixed* $\delta > 0$.

(b) $\mathsf{o}(B_t)$ *whenever* o *is an orthogonal linear transformation on* $\mathbb{R}^d$.

(c) $B_{t+\sigma_0} - B_{\sigma_0}$ *whenever* $\sigma_0 \geq 0$ *is a constant.*

We need only check that these new processes satisfy the conditions B-1, B-2, and B-3 defining Brownian motion. Thus the assertion (a) of the theorem is clear once we observe that for any function f, the covariance matrix of $\delta^{-1/2}f$ is δ^{-1} times the covariance matrix of f. The assertion (b) is also obvious once we note that the covariance matrix of $\mathsf{o}(f)$ is the same as that of f; and that if $f_1, \ldots, f_n, \ldots$ are mutually independent so are $\mathsf{o}(f_1), \mathsf{o}(f_2), \ldots, \mathsf{o}(f_n), \ldots$. Finally (c) is immediate from the definition of Brownian motion.

The next result concerns the regularity of the paths of Brownian motion. The conclusion is that almost all paths satisfy a Hölder condition of exponent a, with $a < 1/2$; this fails however when $a > 1/2$. (This failure extends to the critical case $a = 1/2$, but this is discussed separately in Exercise 14.) Moreover, almost every path is nowhere differentiable. The conclusions are subsumed in the theorem below.

Theorem 4.2 *With W the Wiener measure on $\mathcal{P}$ we have:*

(a) *If* $0 < a < 1/2$ *and* $T > 0$, *then, with respect to* W *almost every path* p *satisfies*

$$\sup_{0 \leq t \leq T,\ 0 < h \leq 1} \frac{|\mathsf{p}(t+h) - \mathsf{p}(t)|}{h^a} < \infty.$$

[6]This is the completion of the measure space as outlined in Exercise 2, Chapter 6 of Book III.

(b) *On the other hand, if $a > 1/2$, then for almost every path* p

$$\limsup_{h\to 0} \frac{|\mathsf{p}(t+h) - \mathsf{p}(t)|}{h^a} = \infty, \qquad \textit{for every } t \geq 0.$$

The first conclusion is implicit in our construction of Brownian motion. Indeed, suppose $K^{(T)}$ is the set arising in the proof of Theorem 3.1. Then we have seen that $\mu_N(K^{(T)}) \geq 1 - \epsilon$ for every N. Thus the same holds for the weak limit of the $\{\mu_N\}$. Hence $W(K^{(T)}) \geq 1 - \epsilon$. But by the definition of $K^{(T)}$ we have the inequality in (a) for every $\mathsf{p} \in K^{(T)}$. Since ϵ is arbitrary, the first conclusion holds.

To prove the second conclusion we fix an $a > 1/2$, and a positive integer k, so that $dk(a - 1/2) > 1$.

Now, for any positive integer n, note that if there is a $t_0 \in [0, 1]$ so that

$$\sup_{0<h\leq (k+1)/n} \frac{|\mathsf{p}(t_0+h) - \mathsf{p}(t_0)|}{h^a} \leq \lambda, \tag{11}$$

then there is an integer j_0, $0 \leq j_0 \leq n-1$ so that

$$\max_{1\leq \ell\leq k} \left| \mathsf{p}\left(\frac{j_0+\ell+1}{n}\right) - \mathsf{p}\left(\frac{j_0+\ell}{n}\right)\right| \leq C_k \lambda n^{-a},$$

where $C_k = 2(k+1)^a$. By renaming λ, we may proceed assuming $C_k = 1$. Thus if we let E_n^λ denote the set of path p where (11) holds, then $E_n^\lambda \subset \tilde{E}_n^\lambda$ with

$$\tilde{E}_n^\lambda = \bigcup_{j_0=0}^{n-1} \left\{ \mathsf{p} \in \mathcal{P} : \max_{1\leq \ell\leq k} \left| \mathsf{p}\left(\frac{j_0+\ell+1}{n}\right) - \mathsf{p}\left(\frac{j_0+\ell}{n}\right)\right| \leq \lambda n^{-a} \right\}.$$

But the k sets $\left\{\mathsf{p} \in \mathcal{P} : \left|\mathsf{p}\left(\frac{j_0+\ell+1}{n}\right) - \mathsf{p}\left(\frac{j_0+\ell}{n}\right)\right| \leq \lambda n^{-a}\right\}$, $1 \leq \ell \leq k$ are mutually independent; also the measures of these sets are the same as ℓ and j_0 vary. Hence

$$W(\{\mathsf{p} \in \mathcal{P} : \max_{1\leq \ell\leq k} \left| \mathsf{p}\left(\frac{j_0+\ell+1}{n}\right) - \mathsf{p}\left(\frac{j_0+\ell}{n}\right)\right| \leq \lambda n^{-a}\}) =$$
$$= (W\{\mathsf{p} \in \mathcal{P} : |\mathsf{p}(1/n)| \leq \lambda n^{-a}\})^k.$$

Thus $W(E_n^\lambda) \leq W(\tilde{E}_n^\lambda) = n(W\{\mathsf{p} \in \mathcal{P} : |\mathsf{p}(1/n)| \leq \lambda n^{-a}\})^k$. However, by the scaling property (a) of the previous theorem

$$W\{\mathsf{p} \in \mathcal{P} : |\mathsf{p}(1/n)| \leq \lambda n^{-a}\} = W\{\mathsf{p} \in \mathcal{P} : |\mathsf{p}(1)| \leq \lambda n^{1/2-a}\}.$$

However $\mathsf{p}(1)$ has a Gaussian as its distribution measure. Thus the last quantity is $O(\lambda^d n^{d(1/2-a)})$ as $n \to \infty$. As a result

$$W(E_n^\lambda) = O(\lambda^{dk} n n^{dk(1/2-a)})$$

and this converges to zero as $n \to \infty$. Thus for every positive λ, the set of p where (11) holds has measure converging to zero as $n \to \infty$ because $a > 1/2$. This establishes conclusion (b) of the theorem.

At this point, it may be worthwhile to recall the variety of ways a nowhere differentiable function has arisen in different settings in these Volumes. First, as a specific example of a lacunary Fourier series in Book I; next as a von Koch fractal, in Book III; further as the generic continuous function via the Baire category theorem in Chapter 4; and now lastly as almost every Brownian path.

One last remark. Given our construction it is intuitively tempting to think of almost every Brownian path as the "limit" of an appropriate collection of random walk paths (paths in $\mathcal{P}_N$ with $N \to \infty$). However it is not clear how to make such an idea precise. Despite this, the following less satisfactory substitute is a direct consequence of Theorem 3.1.

Let $\mathsf{q} \in \mathcal{P}$ be any fixed path. Suppose $\epsilon > 0$ and $0 \le t_1 \le t_2 \le \cdots \le t_n$ are given. We consider the open set

$$\mathcal{O}_\epsilon = \{\mathsf{p} \in \mathcal{P} : \ |\mathsf{p}(t_j) - \mathsf{q}(t_j)| < \epsilon, \quad 1 \le j \le n\}$$

of paths close to q, and set $\mathcal{O}_\epsilon^{(N)} = \mathcal{O}_\epsilon \cap \mathcal{P}_N$, the bundle of corresponding random walk paths. Then

$$\text{(12)} \qquad m(\{x \in \mathbb{Z}_{2d}^\infty : \ S_t^{(N)}(x) \in \mathcal{O}_\epsilon^{(N)}\}) \to W(\mathcal{O}_\epsilon), \qquad \text{as } N \to \infty.$$

In fact, (12) is merely a restatement of the assertion $\mu_N(\mathcal{O}_\epsilon) \to W(\mathcal{O}_\epsilon)$ as $N \to \infty$. This follows because $\mu_N \to W$ weakly, using Exercise 7, since it is easily checked that $W(\overline{\mathcal{O}_\epsilon} - \mathcal{O}_\epsilon) = 0$.

5 Stopping times and the strong Markov property

The goal of the rest of this chapter is to exhibit the remarkable role of Brownian motion in the solution of the Dirichlet problem. A general setting for this problem is as follows.

We are given a bounded open set $\mathcal{R}$ in $\mathbb{R}^d$ and a continuous function f on the boundary $\partial\mathcal{R} = \overline{\mathcal{R}} - \mathcal{R}$. Then the issue is that of finding a function u, continuous on $\overline{\mathcal{R}}$, harmonic in $\mathcal{R}$, that is $\triangle u = 0$, and with the boundary condition $u|_{\partial\mathcal{R}} = f$.

The connection of this question with Brownian motion arises when we fix a point $x \in \mathcal{R}$ and consider Brownian motion *starting* at x, that is, the process $B_t^x = x + B_t$. Now for each $\omega \in \Omega$ we consider the first time $t = \tau(\omega) = \tau^x(\omega)$, when the Brownian motion path $t \mapsto B_t^x(\omega)$ exits $\mathcal{R}$ (in particular, $B_{\tau(\omega)}^x(\omega) = B_{\tau^x(\omega)}^x(\omega) \in \partial\mathcal{R}$).

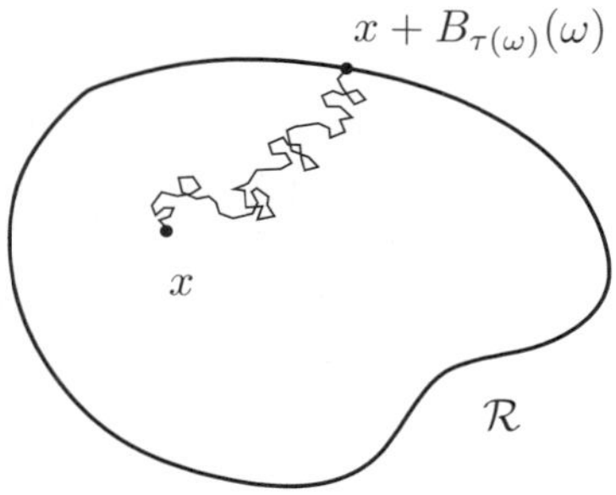

Figure 1. Path ω, exiting at time $\tau = \tau(\omega)$

Then the resulting induced measure $\mu^x = \mu$ on $\partial\mathcal{R}$, given by

$$\mu^x(E) = P(\{\omega : B_{\tau(\omega)}^x(\omega) \in E\})$$

(also called "harmonic measure") leads to the solution of the problem: under appropriate restrictions on the set $\mathcal{R}$

$$u(x) = \int_{\partial\mathcal{R}} f(y)\, d\mu^x(y), \qquad x \in \mathcal{R}$$

is the desired harmonic function.

Now the function $\omega \mapsto \tau(\omega)$ will be seen to be a "stopping time," and we begin by discussing this notion, which arose implicitly when we proved the maximal theorem for martingale sequences in Theorem 2.10 of the previous chapter.

5.1 Stopping times and the Blumenthal zero-one law

Suppose $\{s_n\}_{n=0}^{\infty}$ is a martingale sequence associated to the increasing sequence $\{\mathcal{A}_n\}_{n=0}^{\infty}$ of σ-algebras on the probability space (X, m). Then an integer-valued function $\tau : x \mapsto \tau(x)$ is a **stopping time** if $\{x : \tau(x) = n\} \in \mathcal{A}_n$ for all $n \geq 0$, or equivalently if $\{x : \tau(x) \leq n\} \in \mathcal{A}_n$ for all n.

We note here the basic fact that if (say) $\tau(x) \leq N < \infty$ for all x, then

$$(13) \qquad \int s_{\tau(x)}(x)\,dm = \int s_N(x)\,dm.$$

Indeed, the left-hand side is $\sum_{n=0}^{N} \int_{A_n} s_n(x)\,dx$, with $A_n = \{x : \tau(x) = n\}$. However, by the martingale property (that is, (14) in the previous chapter) $\int_{A_n} s_n(x)\,dx = \int_{A_n} s_N(x)\,dx$, and summing over n gives (13) above.

Similarly, for a subset A, we say that the integer-valued function $x \mapsto \tau(x)$ defined on A is a **stopping time relative to** A if $\{x \in A : \tau(x) = n\} \subset A_n$ for all n. In this case $\int_A s_{\tau(x)}(x)\,dx = \int_A s_N(x)\,dx$. When this is applied to $A = \{x : \sup_{n \leq N} s_n(x) > \alpha\}$, then this yields essentially the maximal inequality (15) in the previous chapter.

Martingales are relevant to Brownian motion because that process is a continuous version of a martingale in the following sense. For each $t \geq 0$, let $\mathcal{A}_t$ be the σ-algebra generated by all functions B_s, $0 \leq s \leq t$, that is, the smallest σ-algebra containing the $\mathcal{A}_{B_s}$ for all $0 \leq s \leq t$.[7] Then we have:

(a) For any sequence $0 \leq t_0 < t_1 < \cdots < t_n < \cdots$ the sequence $\{B_{t_n}\}_{n=0}^{\infty}$ is a martingale relative to the σ-algebras $\{\mathcal{A}_{t_n}\}_{n=0}^{\infty}$.

(b) For almost every ω, the path $B_t(\omega)$ is continuous in t.

Now (a) follows immediately from the proof of Proposition 2.6 in the previous chapter and the fact that the process B_t has independent increments, with each B_t having mean zero. Also, (b) is the condition B-3 arising in the definition of Brownian motion.

At this point, because it will be useful below, we remark that the maximal inequality (9) immediately leads to the Brownian motion inequality:

$$(14) \qquad P(\{\omega : \sup_{0 \leq t \leq T} |B_t(\omega)| > \alpha\}) \leq \frac{1}{\alpha}\|B_T\|_{L^1}$$

for all $T > 0$ and $\alpha > 0$.

In analogy with the discrete case above we say that a non-negative function $\omega \mapsto \tau(\omega)$ is a **stopping time** if $\{\omega : \tau(\omega) \leq t\} \in \mathcal{A}_t$ for every $t \geq 0$.

---

[7]To be precise, $\mathcal{A}_t$ is the σ-algebra generated by all functions B_s, $0 \leq s \leq t$, together with all subsets of sets of measure zero. See also the previous footnote.

Now suppose $\mathcal{R}$ is a bounded open set of $\mathbb{R}^d$ and define the first "exit time" for the path $B_t^x(\omega) = x + B_t(\omega)$ to be

$$\tau(\omega) = \tau^x(\omega) = \inf\{t \geq 0,\ B_t^x(\omega) \notin \mathcal{R}\}.$$

Also define the "strict" exit time $\tau_* = \tau_*^x$ by

$$\tau_*^x(\omega) = \inf\{t > 0,\ B_t^x(\omega) \notin \mathcal{R}\}.$$

Proposition 5.1 *Both τ^x and τ_*^x are stopping times.*

We note that both τ and τ_* are well-defined, that is, finite almost everywhere, because almost every path ultimately exits the bounded open set $\mathcal{R}$. (See Exercise 14.)

Proof. For simplicity of notation we take $x = 0$; we can then recover the general case by reducing to the situation where $\mathcal{R}$ is replaced by $\mathcal{R} - x$. Now for any open set $\mathcal{O}$ in $\mathbb{R}^d$ define $\tau_{\mathcal{O}}(\omega) = \inf\{t \geq 0,\ B_t(\omega) \in \mathcal{O}\}$. Then, up to a set of measure zero,

$$\{\tau_{\mathcal{O}}(\omega) < t\} = \bigcup_{r<t} \{B_r(\omega) \in \mathcal{O}\},$$

where the union is taken over all the indicated rationals r. This is because a continuous path is in $\mathcal{O}$ before time t if and only if it is in $\mathcal{O}$ at some rational time r, with $r < t$. Thus $\{\tau_{\mathcal{O}}(\omega) < t\} \in \mathcal{A}_t$. Next let $\mathcal{O}_n = \{x :\ d(x, \mathcal{R}^c) < 1/n\}$. If $t > 0$, then

$$\{\tau(\omega) \leq t\} = \bigcap_n \{\tau_{\mathcal{O}_n}(\omega) < t\}, \tag{15}$$

because a path exits $\mathcal{R}$ by time t, if and only if it is in $\mathcal{O}_n$ before time t, for every n. Therefore, for $t > 0$ we have $\{\tau(\omega) \leq t\} \in \mathcal{A}_t$. However $\{\tau(\omega) = 0\}$ is the empty set or Ω, depending on whether $x \in \mathcal{R}$ or not. Thus τ is a stopping time.

Note that $\tau_*^x(\omega) = \tau^x(\omega) > 0$ for all ω if $x \in \mathcal{R}$ while $\tau_*^x(\omega) = \tau^x(\omega) = 0$, for $x \notin \overline{\mathcal{R}}$. Therefore the only difference between τ_*^x and τ^x can occur when x is on the boundary, $\partial\mathcal{R} = \overline{\mathcal{R}} - \mathcal{R}$. We notice that as above, when $t > 0$,

$$\{\tau_*^x(\omega) \leq t\} \in \mathcal{A}_t.$$

But then $\{\tau_*^x(\omega) = 0\} \in \bigcap_t \mathcal{A}_t$. Given the increasing character of the σ-algebra $\mathcal{A}_t$, it is natural to denote $\bigcap_t \mathcal{A}_t$ by $\mathcal{A}_{0+}$. So the proposition follows from the lemma below.

Lemma 5.2 $\mathcal{A}_{0+} = \mathcal{A}_0$.

The proof of this simple looking fact is however a little indirect. The conclusion, that any set $A \in \bigcap_{t>0} \mathcal{A}_t$ is trivial (is either of measure 0 or 1), is referred to as Blumenthal's zero-one law. (A generalization is given in Exercise 16.)

As a result, for each x in the boundary of $\mathcal{R}$ we have the dichotomy: $\{\tau_*^x(\omega) = 0\}$ has measure 1 or 0. In the former case, the point x is called a **regular** point at the boundary. In brief, a boundary point is regular, if almost all paths starting at that point are outside $\mathcal{R}$ for arbitrarily small positive times. This property plays a crucial role in the Dirichlet problem for $\mathcal{R}$.

Proof of the lemma. Fix a bounded continuous function f on $\mathbb{R}^{kd}$, and a sequence $0 \leq t_1 < t_2 < \cdots < t_k$. For any $\delta > 0$, set

$$f_\delta = f(B_{t_1+\delta} - B_\delta, B_{t_2+\delta} - B_{t_1+\delta}, \ldots, B_{t_k+\delta} - B_{t_{k-1}+\delta}).$$

If A is any set in $\mathcal{A}_{0+}$, then $A \in \mathcal{A}_\delta$, for $\delta > 0$. Then by the independence of the above increments from B_δ, we see that

$$\int_A f_\delta \, dP = P(A) \int_\Omega f_\delta \, dP.$$

Thus by continuity of the paths we can let $\delta \to 0$ and obtain

$$\int_A f_0 \, dP = P(A) \int_\Omega f_0 \, dP.$$

Now any bounded continuous function g on $\mathbb{R}^{kd}$ can be written in the form $g(x_1, \ldots, x_k) = f(x_1, x_2 - x_1, \ldots, x_k - x_{k-1})$ where f is another such function. As a result

$$\int_A g(B_{t_1}, \ldots, B_{t_k}) \, dP = P(A) \int_\Omega g(B_{t_1}, \ldots, B_{t_k}) \, dP.$$

Hence by a passage to the limit, this holds if g is the characteristic function of a Borel set of $\mathbb{R}^{kd}$. Thus $P(A \cap E) = P(A)P(E)$ whenever E is a cylindrical set. From this, we deduce the same equality for any Borel set E by using Exercise 4 in the previous chapter. Therefore $P(A) = P(A)^2$, which implies $P(A) = 0$ or $P(A) = 1$. Since A was an arbitrary subset of $\mathcal{A}_{0+}$, the lemma, and also the proposition, are proved.

Note. Lastly, it will be important below to remark that the stopping time $\tau^x(\omega)$ is jointly measurable in x and ω. This follows from

$$\{(x, \omega) : \ \tau^x(\omega) > \rho\} = \bigcup_{n=1}^{\infty} \bigcap_{r \leq \rho, \, r \in \mathbb{Q}} \{\omega : \ x + B_r(\omega) \in \mathcal{R}_n\},$$

where $\mathcal{R}_n = \{x : d(x, \mathcal{R}^c) > 1/n\}$.

5.2 The strong Markov property

Suppose σ is a stopping time (relative to the σ-algebras $\{\mathcal{A}_t\}_{t\geq 0}$). We can define the collection $\mathcal{A}_\sigma$ to be the collection of sets A, such that $A \cap \{\sigma(\omega) \leq t\} \in \mathcal{A}_t$, for all $t \geq 0$. One notes that in fact $\mathcal{A}_\sigma$ is a σ-algebra; that $\mathcal{A}_\sigma = \mathcal{A}_{\sigma_0}$ if $\sigma(\omega)$ is constant and equals σ_0; and that σ is measurable with respect to $\mathcal{A}_\sigma$. (See also Exercise 18.)

In studying the Dirichlet problem we shall need, in addition to the stopping time τ (the first exit time from $\mathcal{R}$), another stopping time σ. What happens when Brownian motion is restarted after time σ is the subject of the "strong Markov property," one version of which is contained in the following.

Theorem 5.3 *Suppose B_t is a Brownian motion and σ is a stopping time. Then the process B_t^*, defined by*

$$B_t^*(\omega) = B_{t+\sigma(\omega)}(\omega) - B_{\sigma(\omega)}(\omega)$$

is also a Brownian motion. Moreover B_t^* is independent of $\mathcal{A}_\sigma$.

In other words, if a Brownian motion is stopped at time $\sigma(\omega)$, then the process which is appropriately restarted is also a Brownian motion that is now independent of the past $\mathcal{A}_\sigma$.[8]

Proof. We have already noted that if $\sigma(\omega)$ is a constant, $\sigma(\omega) = \sigma_0$, then $B_{t+\sigma_0} - B_{\sigma_0}$ is a Brownian motion (see Theorem 4.1), so the assertion in the theorem holds in this case.

Next assume that σ is discrete, that is, it takes on only a countable set of values $\sigma_1 < \sigma_2 < \cdots < \sigma_\ell < \ldots$. Also suppose $0 \leq t_1 < t_2 < \cdots < t_k$ are fixed. Let us use the temporary notation

$$\begin{aligned}
\mathbf{B} &= (B_{t_1}, B_{t_2}, \ldots, B_{t_k}) \\
\mathbf{B}^* &= (B_{t_1}^*, B_{t_2}^*, \ldots, B_{t_k}^*) \\
\mathbf{B}_\ell^* &= (B_{t_1+\sigma_\ell} - B_{\sigma_\ell}, B_{t_2+\sigma_\ell} - B_{\sigma_\ell}, \ldots, B_{t_k+\sigma_\ell} - B_{\sigma_\ell}),
\end{aligned}$$

with all these bold-face vectors taking values in $\mathbb{R}^{kd}$. Now if E is a Borel set in $\mathbb{R}^{kd}$, then

$$\{\omega : \mathbf{B}^* \in E\} = \bigcup_\ell \{\omega : \mathbf{B}_\ell^* \in E, \text{ and } \sigma = \sigma_\ell\}.$$

[8]A corresponding independence when σ is an arbitrary positive constant is characteristic of a "Markov" process.

So

$$\{\omega : \mathbf{B}^* \in E\} \cap A = \bigcup_\ell (\{\omega : \mathbf{B}^*_\ell \in E\} \cap A \cap \{\sigma = \sigma_\ell\}),$$

with the union clearly disjoint.

However if $A \in \mathcal{A}_\sigma$, then $A \cap \{\sigma = \sigma_\ell\} \in \mathcal{A}_{\sigma_\ell}$. By the special case when $\sigma = \sigma_\ell$ is constant throughout, we see that the measure of $\{\omega : \mathbf{B}^* \in E\} \cap A$ equals

$$\sum_\ell P(\mathbf{B}^*_\ell \in E) m(A \cap \{\sigma = \sigma_\ell\}),$$

because $A \cap \{\sigma = \sigma_\ell\} \in \mathcal{A}_{\sigma_\ell}$ and this set is independent of $\{\mathbf{B}^*_\ell \in E\}$. However $P(\mathbf{B}^*_\ell \in E) = P(\mathbf{B} \in E)$, and we obtain that

$$P(\{\omega : \mathbf{B}^* \in E,\ \omega \in A\}) = P(\{\omega : \ \mathbf{B} \in E\})P(A). \tag{16}$$

Now using (16) when $A = \Omega$ shows that $\mathbf{B}^*$ satisfies the conditions B-1 and B-2 of Brownian motion. Also B-3 is obvious. Finally, using (16) for any $A \subset \mathcal{A}_\sigma$ gives the desired independence of $\mathbf{B}^*$ from $\mathcal{A}_\sigma$.

Turning to the case of general stopping time σ, we approximate it by a sequence $\{\sigma^{(n)}\}$ of stopping times, so that each $\sigma^{(n)}$ takes on only a countable set of values as above, and

(i) $\sigma^{(n)}(\omega) \searrow \sigma(\omega)$, as $n \to \infty$, for every ω; and

(ii) $\mathcal{A}_\sigma \subset \mathcal{A}_{\sigma^{(n)}}$.

For each n define $\sigma^{(n)}(\omega) = k2^{-n}$ if $(k-1)2^{-n} < \sigma(\omega) \leq k2^{-n}$ for $k = 1, 2, \ldots$, and $\sigma^{(n)}(\omega) = 0$ if $\sigma(\omega) = 0$. Property (i) is obvious. Next, for each t there is a k so that $k2^{-n} \leq t < (k+1)2^{-n}$. Then $\{\sigma^{(n)} \leq t\} = \{\sigma \leq k2^{-n}\} \in \mathcal{A}_{k2^{-n}} \subset \mathcal{A}_t$. Thus $\sigma^{(n)}$ is a stopping time.

Also suppose that $A \subset \mathcal{A}_\sigma$, then $A \cap \{\sigma^{(n)} \leq t\} = A \cap \{\sigma \leq k2^{-n}\} \in \mathcal{A}_{k2^{-n}} \subset \mathcal{A}_t$, and hence $A \in \mathcal{A}_{\sigma^{(n)}}$. Thus (ii) is established.

Now let $B^{*(n)}_t$ be the analog of B^*_t, with σ replaced by $\sigma^{(n)}$, and let $\mathbf{B}^{*(n)} = (B^{*(n)}_{t_1}, \ldots, B^{*(n)}_{t_k})$. Suppose $A \subset \mathcal{A}_\sigma$ (then $A \subset \mathcal{A}_{\sigma^{(n)}}$). Then by what we have proved in the discrete case

$$P(\{\mathbf{B}^{*(n)} \in E,\ \omega \in A\}) = P(\mathbf{B} \in E)P(A).$$

A passage to the limit then shows that (16) holds for the general σ. This limiting argument is carried out in two steps using exercises from the previous chapter. First, by Exercises 10 and 31 part (d), since $\mathbf{B}^{*(n)}$

converges pointwise to $\mathbf{B}^*$, we have that (16) holds whenever E is an open set. To conclude that such equality holds for all Borel sets E, we apply Exercise 4 in the previous chapter.

For any given stopping time σ, let us write B_σ for the function $\omega \mapsto B_{\sigma(\omega)}(\omega)$. We note that the argument above, where we approximate the stopping time, also shows that B_σ is $\mathcal{A}_\sigma$-measurable. (See Exercise 18.)

5.3 Other forms of the strong Markov Property

Another version of the strong Markov property involves integration of functions defined on *all* paths. To describe this we need a little additional notation. We define $\tilde{\mathcal{P}}$ to be the space of all paths, that is, all continuous functions from $[0, \infty)$ to $\mathbb{R}^d$. The space $\tilde{\mathcal{P}}$ differs from the space $\mathcal{P}$ considered earlier, in that in the latter all paths start at the origin. We can write each $\tilde{\mathsf{p}}$ in $\tilde{\mathcal{P}}$ as a pair (p, x) with $\mathsf{p} \in \mathcal{P}$, $x \in \mathbb{R}^d$ where $\mathsf{p} = \tilde{\mathsf{p}} - \tilde{\mathsf{p}}(0)$, and $x = \tilde{\mathsf{p}}(0)$. So we have $\tilde{\mathcal{P}} = \mathcal{P} \times \mathbb{R}^d$, and every function f on $\tilde{\mathcal{P}}$ can be written as $f(\tilde{\mathsf{p}}) = f_1(\mathsf{p}, x)$, with f_1 a function on the product $\mathcal{P} \times \mathbb{R}^d$. Moreover, $\tilde{\mathcal{P}}$ inherits a metric from the metrics on $\mathcal{P}$ and $\mathbb{R}^d$, and a corresponding class of Borel subsets.

We shall also use the short-hand that the path $t \mapsto B_t(\omega)$ will be designated by $B_\cdot(\omega)$; similarly the path $t \mapsto B_{\sigma(\omega)+t}(\omega)$ will be written as $B_{\sigma(\omega)+\cdot}(\omega)$; also the paths $t \mapsto B_{\sigma(\omega)+t}(\omega) - B_{\sigma(\omega)}(\omega)$ that appear in Theorem 5.3 will be represented as $B_\cdot^*(\omega)$. With these definitions our result is as follows.

Theorem 5.4 *Let f be a bounded Borel function on the space $\tilde{\mathcal{P}}$ of all paths. Then*

(17)
$$\int_\Omega f\left(B_{\sigma(\omega)+\cdot}(\omega)\right) dP(\omega) = \iint_{\Omega\times\Omega} f\left(B_\cdot(\omega) + B_{\sigma(\omega')}(\omega')\right) dP(\omega)\, dP(\omega').$$

Proof. We write $f(\tilde{\mathsf{p}}) = f_1(\mathsf{p}, x)$ as above; then (17) becomes

(18)
$$\int_\Omega f_1\left(B_\cdot^*(\omega), B_{\sigma(\omega)}(\omega)\right) dP(\omega) =$$
$$\iint_{\Omega\times\Omega} f_1\left(B_\cdot(\omega), B_{\sigma(\omega')}(\omega')\right) dP(\omega)\, dP(\omega'),$$

since $B_{\sigma(\omega)+t}(\omega) = B_t^*(\omega) + B_{\sigma(\omega)}(\omega)$.

We consider first functions f_1 of the product form $f_1 = f_2 \cdot f_3$, with

$f_1(\mathsf{p}, x) = f_2(\mathsf{p}) f_3(x)$. Then the right-hand side of (18) is

$$\int_\Omega f_2(B_{\cdot}(\omega))\, dP(\omega) \times \int_\Omega f_3(B_{\sigma(\omega')}(\omega'))\, dP(\omega').$$

However $\int_\Omega f_2(B_{\cdot}(\omega))\, dP(\omega) = \int_\Omega f_2(B^*_{\cdot}(\omega))\, dP(\omega)$, since by Theorem 5.3 B^*_t is also a Brownian motion and so has the same distribution measures as B_t. Also, by the independence guaranteed by that theorem (and the fact that $B_{\sigma(\omega')}(\omega')$ is $\mathcal{A}_\sigma$-measurable) we see that the product

$$\int_\Omega f_2(B^*_{\cdot}(\omega))\, dP(\omega) \times \int_\Omega f_3(B_{\sigma(\omega')}(\omega'))\, dP(\omega')$$

equals

$$\int_\Omega f_2(B^*_{\cdot}(\omega)) f_3(B_{\sigma(\omega)}(\omega))\, dP(\omega),$$

which is the left-hand side of (18).

To pass to the case of general f we may argue as follows. Let μ and ν denote the measures on $\tilde{\mathcal{P}}$ defined by $\mu(E)$, (respectively $\nu(E)$), as the left-hand side, (respectively the right-hand side) of (18) whenever f is the characteristic function of E, with E any Borel set in $\tilde{\mathcal{P}}$. Then what we have already proved implies that $\mu(E) = \nu(E)$ for all Borel sets of the form $E = E_2 \times E_3$, with $E_2 \subset \mathcal{P}$ and $E_3 \subset \mathbb{R}^d$. According to Exercise 4 in the previous chapter, this identity then extends to the σ-algebra generated by these sets, and hence to all Borel sets of $\tilde{\mathcal{P}}$, because this σ-algebra contains the open sets. Finally, because any bounded Borel function on $\tilde{\mathcal{P}}$ is the bounded pointwise limit of finite linear combinations of characteristic functions of Borel sets, we see that (18) holds for all those $f_1 = f$, and the theorem is proved.

The final version of the strong Markov property we present is the statement closest to the immediate application to the Dirichlet problem. It involves two stopping times σ and τ, with $\sigma \leq \tau$, where τ is the exit time for the bounded open set $\mathcal{R}$. Let us recall that $B^y_t(\omega) = y + B_t(\omega)$, and $\tau^y(\omega) = \inf\{t \geq 0,\ B^y_t(\omega) \notin \mathcal{R}\}$. We define the **stopped process**

$$\hat{B}^y_t(\omega) = y + B_{t\wedge\tau^y(\omega)}(\omega)$$

where $t \wedge \tau^y(\omega) = \min(t, \tau^y(\omega))$. If $y = 0$ we drop the subscript y in the above definitions.

Theorem 5.5 *Suppose σ and τ are stopping times with $\sigma(\omega) \leq \tau(\omega)$ for all ω. If F is a bounded Borel function on $\mathbb{R}^d$, then for every $t \geq 0$*

$$(19) \quad \int_\Omega F\left(\hat{B}_{\sigma(\omega)+t}(\omega)\right) dP(\omega) = \iint_{\Omega\times\Omega} F\left(\hat{B}_t^{y(\omega')}(\omega)\right) dP(\omega)\, dP(\omega')$$

where $y(\omega') = \hat{B}_{\sigma(\omega')}(\omega')$.

Proof. Start with the left-hand side of (19). It equals

$$\int_{\tau(\omega)\geq\sigma(\omega)+t} F\left(\hat{B}_{\sigma(\omega)+t}(\omega)\right) dP(\omega) + \int_{\tau(\omega)<\sigma(\omega)+t} F\left(\hat{B}_{\sigma(\omega)+t}(\omega)\right) dP(\omega)$$
$$= \int_{\tau(\omega)\geq\sigma(\omega)+t} F\left(B_{\sigma(\omega)+t}(\omega)\right) dP(\omega) + \int_{\tau(\omega)<\sigma(\omega)+t} F\left(B_{\tau(\omega)}(\omega)\right) dP(\omega)$$
$$= I_1 + I_2.$$

We will first look at

$$I_1 = \int_\Omega F\left(B_{\sigma(\omega)+t}(\omega)\right) \chi_{\tau(\omega)\geq\sigma(\omega)+t}\, dP(\omega).$$

Consider the following real-valued function on paths:

$$f(\tilde{\mathsf{p}}) = F\left(\tilde{\mathsf{p}}(t)\right) \chi_{\tau(\mathsf{p})\geq t}.$$

Here we define for any path $\tilde{\mathsf{p}}$ the quantity $\tau(\tilde{\mathsf{p}}) = \inf\{s \geq 0 : \tilde{\mathsf{p}}(s) \notin \mathcal{R}\}$. In particular, note that if $\tilde{\mathsf{p}}(\cdot) = B_\cdot(\omega)$, then $\tau(\tilde{\mathsf{p}}) = \tau(\omega)$. Now, given ω set $\tilde{\mathsf{p}}(\cdot) = B_{\sigma(\omega)+\cdot}(\omega)$. Then

$$f(\tilde{\mathsf{p}}) = f\left(B_{\sigma(\omega)+\cdot}(\omega)\right) = F\left(B_{\sigma(\omega)+t}(\omega)\right) \chi_{\tau(\omega)-\sigma(\omega)\geq t}.$$

Indeed, note that

$$\tau(B_{\sigma(\omega)+\cdot}(\omega)) = \inf\{s \geq 0 :\ B_{\sigma(\omega)+s}(\omega) \notin \mathcal{R}\} = \tau(\omega) - \sigma(\omega).$$

This is true because the path $B_\cdot(\omega)$ exits at time $\tau(\omega)$, and therefore the path $B_{\sigma(\omega)+\cdot}(\omega)$ exits at time $\tau(\omega) - \sigma(\omega)$. Therefore

$$f\left(B_{\sigma(\omega)+\cdot}(\omega)\right) = F\left(B_{\sigma(\omega)+t}\right) \chi_{\tau(\omega)\geq\sigma(\omega)+t},$$

which is the integrand in I_1, so we can apply (17) to get

$$I_1 = \int_\Omega\int_\Omega f\left(B_{\sigma(\omega')}(\omega') + B_\cdot(\omega)\right) dP(\omega)\, dP(\omega').$$

But now note that the integrand in the above equals

$$F\left(B_{\sigma(\omega')}(\omega') + B_t(\omega)\right) \chi_{\tau\left(B_{\sigma(\omega')}(\omega')+B.(\omega)\right)\geq t}.$$

To conclude the calculation of I_1 it suffices to note that the quantity $\tau\left(B_{\sigma(\omega')}(\omega') + B.(\omega)\right)$ equals $\tau^{y(\omega')}(\omega)$, and so

$$\begin{aligned} I_1 &= \int_\Omega\int_\Omega F\left(B_{\sigma(\omega')}(\omega') + B_t(\omega)\right) \chi_{\tau^{y(\omega')}(\omega)\geq t}\, dP(\omega)\, dP(\omega') \\ &= \int_\Omega\int_\Omega F\left(B_t^{y(\omega')}(\omega)\right) \chi_{\tau^{y(\omega')}(\omega)\geq t}\, dP(\omega)\, dP(\omega') \\ &= \int_\Omega\int_\Omega F\left(\hat{B}_t^{y(\omega')}(\omega)\right) \chi_{\tau^{y(\omega')}(\omega)\geq t}\, dP(\omega)\, dP(\omega'). \end{aligned}$$

We now look at the second integral I_2 defined by

$$I_2 = \int_\Omega F\left(B_{\tau(\omega)}(\omega)\right) \chi_{\tau(\omega)<\sigma(\omega)+t}\, dP(\omega).$$

Here we define a real-valued function on paths

$$g(\tilde{\mathsf{p}}) = F\left(\tilde{\mathsf{p}}(\tau(\tilde{\mathsf{p}}))\right) \chi_{\tau(\tilde{\mathsf{p}})<t}.$$

Setting $\tilde{\mathsf{p}}(\cdot) = B_{\sigma(\omega)+\cdot}(\omega)$ gives

$$g(B_{\sigma(\omega)+\cdot}(\omega)) = F\left(B_{\tau(\omega)}(\omega)\right) \chi_{\tau(\omega)<\sigma(\omega)+t}.$$

For the characteristic function χ, the argument is the same as above. For the first part (that is, the component $F(\cdots)$), note that $\tau(\tilde{\mathsf{p}})$ gives the time of exit of $\mathcal{R}$ of the path $\tilde{\mathsf{p}}$, and $\mathsf{p}(\tau(\mathsf{p}))$ the value (in $\mathbb{R}^d$) where the path exits. Since both $B_{\sigma(\omega)+\cdot}(\omega)$ and $B.(\omega)$ exit at the same point in space (although at different times, namely, $\tau(\omega) - \sigma(\omega)$ and $\tau(\omega)$ respectively) we get the above. Therefore by (17)

$$I_2 = \int_\Omega\int_\Omega g\left(B_{\sigma(\omega')}(\omega') + B.(\omega)\right)\, dP(\omega)\, dP(\omega').$$

Now note that

$$g\left(B_{\sigma(\omega')}(\omega') + B.(\omega)\right) = F\left(B_{\sigma(\omega')}(\omega') + B_{\tau^{y(\omega')}(\omega)}(\omega)\right) \chi_{\tau^{y(\omega')}(\omega)<t}.$$

Hence

$$
\begin{aligned}
I_2 &= \int_\Omega \int_\Omega g\left(B_{\sigma(\omega')}(\omega') + B_\cdot(\omega)\right) dP(\omega)\, dP(\omega') \\
&= \int_\Omega \int_\Omega F\left(B_{\sigma(\omega')}(\omega') + B_{\tau^{y(\omega')}(\omega)}(\omega)\right) \chi_{\tau^{y(\omega')}(\omega)<t}\, dP(\omega)\, dP(\omega') \\
&= \int_\Omega \int_\Omega F\left(\hat{B}_t^{y(\omega')}(\omega)\right) \chi_{\tau^{y(\omega')}(\omega)<t}\, dP(\omega)\, dP(\omega').
\end{aligned}
$$

Therefore, putting the two integrals for I_1 and I_2 together yields

$$I_1 + I_2 = \int_\Omega \int_\Omega F\left(\hat{B}_t^{y(\omega')}(\omega)\right) dP(\omega)\, dP(\omega'),$$

which completes the proof of (19).

Final remark. With almost no change in the argument one can prove generalizations of the two theorems above in which the left-hand side of (17) and (19) are integrated over any set A in $\mathcal{A}_\sigma$, instead of Ω. The result corresponding to (17) may then be rephrased in terms of conditional expectations $\mathbb{E}_{\mathcal{A}_\sigma}$ to read:

$$\mathbb{E}_{\mathcal{A}_\sigma}(f(B_{\sigma(\omega)+\cdot})) = \int_\Omega f(B_\cdot(\omega) + x)\, dP(\omega)\bigg|_{x=B_{\sigma(\omega')}(\omega')}.$$

The conclusion corresponding to (19) is

$$\int_A F\left(\hat{B}_{\sigma(\omega)+t}(\omega)\right) dP(\omega) = \int_A \int_\Omega F\left(\hat{B}_t^{y(\omega')}(\omega)\right) dP(\omega)\, dP(\omega'),$$

whenever $A \in \mathcal{A}_\sigma$.

6 Solution of the Dirichlet problem

Recall the definitions given at the beginning of Section 5. Here $\mathcal{R}$ is a bounded open set in $\mathbb{R}^d$, and for each $x \in \mathcal{R}$, we define μ^x as the measure on the boundary $\partial\mathcal{R}$ of $\mathcal{R}$, given by

$$\mu^x(E) = P(\{\omega :\ B^x_{\tau^x(\omega)}(\omega) \in E\}),$$

with $\tau^x(\omega)$ the first exit time of the path $B_t^x(\omega)$. Here E ranges over the Borels sets of $\partial\mathcal{R}$, which itself is a compact subset of $\mathbb{R}^d$.

For a continuous function f on $\partial\mathcal{R}$ we defined

$$u(x) = \int_{\partial\mathcal{R}} f(y)\, d\mu^x(y), \qquad \text{when } x \in \mathcal{R}. \tag{20}$$

Observe that u is measurable (and in fact Borel measurable) since

$$u(x) = \int_\Omega f(x + B_{\tau^x(\omega)}(\omega))\, dP(\omega),$$

and $\tau^x(\omega)$ is jointly measurable, as noted at the end of Section 5.1.

The main theorem is as follows.

Theorem 6.1 *If u is defined by (20), then:*

(a) *u is a harmonic function in $\mathcal{R}$.*

(b) *$u(x) \to f(y)$, as $x \to y$, for $x \in \mathcal{R}$, if y is a regular point of $\partial\mathcal{R}$.*

Proof. To establish (a) we fix $x \in \mathcal{R}$ and let S denote a sphere centered at x together with its interior ball is contained in $\mathcal{R}$. We will prove the mean-value property

$$u(x) = \int_S u(y)\, dm(y), \tag{21}$$

where m is the standard measure on the sphere, normalized to have total mass 1. To prove (21) let σ be the stopping time defined as the first time $B_t^x(\omega)$ hits S.

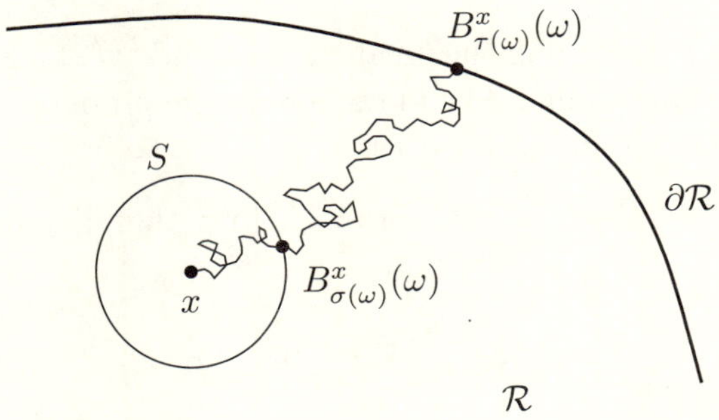

Figure 2. Brownian motion stopping on S and then $\partial\mathcal{R}$

We claim that for any continuous function G on S we have

$$\int_\Omega G(B^x_{\sigma(\omega_1)}(\omega_1))\, dP(\omega_1) = \int_S G(y)\, dm(y). \tag{22}$$

To see this, consider the case where $x = 0$, and note that the left-hand side defines a continuous linear functional on the continuous functions on S, and hence is of the form $\int_S G(y)\,d\mu(y)$, for some measure μ on S. By the rotation invariance of the Brownian motion it follows that μ is rotation-invariant and hence by Problem 4 in Chapter 6, Book III, we have $\mu = m$.

Suppose $\hat{B}_t^x = B_{t\wedge\tau^x}^x$ is the stopped process. Note that $\hat{B}^x_{\sigma(\omega_1)}(\omega_1) = B^x_{\sigma(\omega_1)}(\omega_1) = y(\omega_1) \in S$, because a path starting at x meets S before it meets $\partial\mathcal{R}$.

We now invoke (19). If we take F to be any continuous bounded extension of f to all of $\mathbb{R}^d$, and let $t \to \infty$ we obtain

(23)
$$\int\!\!\int_{\Omega\times\Omega} F(B^{y(\omega_1)}_{\tau^{y(\omega_1)}(\omega_2)}(\omega_2))\,dP(\omega_2)\,dP(\omega_1) = \int_\Omega F(B^x_{\tau^x(\omega)}(\omega))\,dP(\omega).$$

The right-hand side of (23) above equals $u(x)$, while the left-hand side equals

$$\int_\Omega u(y(\omega_1))\,dP(\omega_1).$$

Finally, since $B^x_{\sigma(\omega_1)}(\omega_1) = y(\omega_1)$ we can apply (22) with $u = G$ and deduce that

$$\int_\Omega u(y(\omega_1))\,dP(\omega_1) = \int_S u(y)\,dm(y).$$

This completes the proof of the mean-value identity (21), and from this it follows that u is harmonic. The ideas behind the proof of this well-known fact are summarized in Exercise 19.

To prove conclusion (b), we establish first that if $y \in \partial\mathcal{R}$, and y is regular, then

(24)
$$\lim_{x\to y,\ x\in\mathcal{R}} P(\{\tau^x > \delta\}) = 0, \qquad \text{for all } \delta > 0.$$

In fact, $P(\{B_t^x \in \mathcal{R},\ \text{all } \epsilon \le t \le \delta\})$, $\epsilon > 0$, is continuous in x, because at each ω for which B_t is continuous, the characteristic function of $\{B_t^x \in \mathcal{R},\ \text{all } \epsilon \le t \le \delta\}$ at ω converges to the characteristic function of $\{B_t^y \in \mathcal{R},\ \text{all } \epsilon \le t \le \delta\}$ at ω, as $x \to y$. However the functions $P(\{B_t^x \in \mathcal{R},\ \text{all } \epsilon \le t \le \delta\})$ are decreasing as $\epsilon \searrow 0$. The limit is

$$P(\{\omega :\ B_t^x(\omega) \in \mathcal{R},\ \text{all } 0 < t \le \delta\}) = P(\{\tau^x > \delta\})$$

and is thus upper semi-continuous in x. Hence $\limsup_{x\to y} P(\{\tau^x > \delta\}) \leq P(\{\tau^y > \delta\}) = 0$, since y is a regular point. Thus (24) is established. As a consequence we have for $s > 0$ and $\epsilon > 0$ given,

$$(25) \qquad P(\{\omega : \ |y - B^x_{\tau^x(\omega)}(\omega)| > s\}) < \epsilon$$

if x is sufficiently close to $y \in \partial\mathcal{R}$. In fact, by the maximal inequality (14) we can find a $\delta > 0$, so that $P(\{\omega : \ \sup_{t\leq\delta} |B_t(\omega)| > s/2\}) \leq \epsilon/2$, since $\|B_\delta\|_{L^1} = c\delta^{1/2}$. Also by (24), if x is sufficiently close to y, $P(\{\tau^x > \delta\}) \leq \epsilon/2$. As a result, if x is sufficiently close to y, (25) holds.

Now

$$u(x) - f(y) = \int_{\partial\mathcal{R}} (f(y') - f(y))\, d\mu^x(y') = \int_{\partial\mathcal{R}_1} + \int_{\partial\mathcal{R}_2} = I_1 + I_2.$$

Here $\partial\mathcal{R}_1$ is the set of y' in $\partial\mathcal{R}$ so that $|y' - y| \leq s$ and $\partial\mathcal{R}_2$ is the complementary set in $\partial\mathcal{R}$. Now the points $y' \in \partial\mathcal{R}$ are of the form $y' = B^x_{\tau(\omega)}(\omega)$, while $\mu^x(\partial\mathcal{R}_2) = P(\{\omega : \ |y - B^x_{\tau(\omega)}| > s\})$. Thus by (25) we see that $\mu^x(\partial\mathcal{R}_2) \leq \epsilon$ if x is sufficiently close to y. So the contribution of I_2 is majorized by $2\sup|f|\mu^x(\partial\mathcal{R}_2)\epsilon = O(\epsilon)$. Also $|f(y) - f(y')| < \epsilon$ if $|y - y'| \leq s$ and s is small enough, so the contribution of I_1 can be made less than ϵ. Altogether this shows that $u(x) - f(y)$ is majorized by a multiple of ϵ for x sufficiently close to y. Since ϵ was arbitrary, the second assertion of the theorem is proved.

Our final result is a very useful sufficient condition for the regularity of a boundary point. A **(truncated) cone** Γ is the open set

$$\Gamma = \{y \in \mathbb{R}^d : \ |y| < \alpha(y \cdot \gamma), \ |y| < \delta\}.$$

Here γ is a unit vector, $\alpha > 1$, $\delta > 0$ are fixed, and $y \cdot \gamma$ is the inner product between y and γ. The vector γ determines the direction of the cone, and the constant α gives the size of the aperture.

Proposition 6.2 *Suppose $x \in \partial\mathcal{R}$ and $x + \Gamma$ is disjoint from $\mathcal{R}$, for some truncated cone Γ. Then x is a regular point.*

Proof. We assume $x = 0$, and consider the set A of Brownian paths starting at the origin that enter Γ for an infinite sequence of times tending to zero. Let $A_n = \bigcup_{r_k<1/n}\{\omega : \ B_{r_k}(\omega) \in \Gamma\}$ where r_k is an enumeration of the positive rationals. Then $A = \bigcap_{n=1}^{\infty} A_n$. However $A_n \in \mathcal{A}_n$ for each n, and hence $A \in \mathcal{A}_{0+} = \mathcal{A}_0$, by the zero-one law. So $m(A) = 0$ or $m(A) = 1$, and we show that in fact $m(A) = 1$. Assume the contrary, that is $m(A) = 0$. By the rotation invariance of Brownian motion, the same

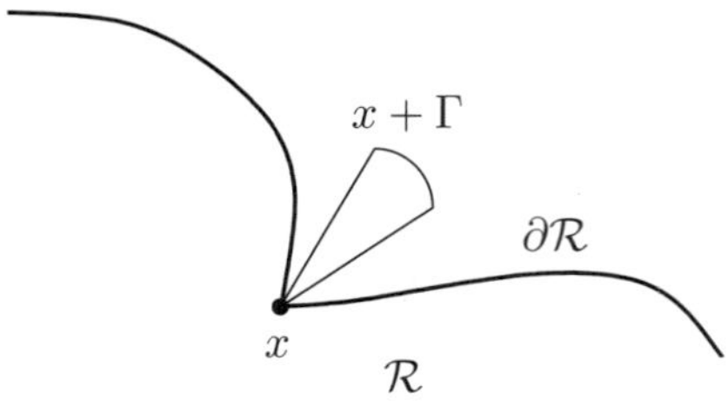

Figure 3. Truncated cone at x disjoint from $\mathcal{R}$

result would hold for any rotation of our truncated cone, and finitely many such rotations cover the ball of radius δ, with the origin excluded, while every path enters that ball at arbitrarily small times. This is a contradiction.

Now returning to our boundary point x, if $x + \Gamma$ is disjoint from $\mathcal{R}$, then there are, for each ω, arbitrarily small times for which $B_t(\omega) \in \Gamma$, and hence $B_t^x(\omega) \notin \mathcal{R}$. Thus x is regular.

In view of the above we say that a bounded open set $\mathcal{R}$ satisfies the **outside cone condition**, if whenever $x \in \partial\mathcal{R}$, there is a truncated cone Γ, so that $x + \Gamma$ is disjoint from $\mathcal{R}$. Our final result generalizes the theorem proved by very different methods in Chapter 5, Book III only for the special case of two dimensions.

Corollary 6.3 *Suppose the bounded open set $\mathcal{R}$ satisfies the outside cone condition. Assume f is a given continuous function on $\partial\mathcal{R}$. Then there is a unique function u that is continuous in $\overline{\mathcal{R}}$, harmonic in $\mathcal{R}$, and such that $u|_{\partial\mathcal{R}} = f$.*

Proof. Theorem 6.1 and Proposition 6.2 show that u is continuous in $\overline{\mathcal{R}}$ and $u|_{\partial\mathcal{R}} = f$. The uniqueness is a consequence of the well-known maximum principle.[9]

7 Exercises

1. Show that if $t > 0$, then the distribution measure of $S_t^{(N)}$ converges weakly to the Gaussian ν_t with mean zero and variance t as $N \to \infty$. More generally, if $t > s \geq 0$, then the distribution measure of $S_t^{(N)} - S_s^{(N)}$ converges weakly to the Gaussian ν_{t-s} with mean zero and covariance matrix $(t-s)I$.

[Hint: Using the notation in the remark following Theorem 2.17 in Chapter 5, and setting $f_k = \mathfrak{r}_k$, one has $S_t^{(N)} - S_{N,t} = \frac{(Nt-[Nt])}{N^{1/2}}\mathfrak{r}_{[Nt]+1}$.]

[9]See for example Corollary 4.4 in Chapter 5 of Book III.

2. Let $(\mathcal{P}, d)$ be the metric space defined in Section 2. Verify:

(a) The space is complete.

(b) The space is separable.

[Hint: For (b), let $e_1, \ldots, e_d$ be a basis for $\mathbb{R}^d$, and consider the polynomials $p(t) = e_1 p_1(t) + \cdots + e_d p_d(t)$, where the p_j have rational coefficients.]

3. Show that the metric space $(\mathcal{P}, d)$ is not σ-compact.

[Hint: Assume the contrary. Then the Baire category theorem implies that there exists a compact set that has a non-empty interior. As a result there exists an open ball whose closure is compact. However, consider for example the ball of radius 1 centered at 0, and a sequence of continuous piecewise linear functions $\{f_n\}$ with $f_n(0) = 1$, $f_n(x) = 0$, when $x \geq 1/n$.]

4. Suppose X is a compact metric space. Show that:

(a) X is separable.

(b) $C(X)$ is separable.

[Hint: For each m, find a finite collection $\mathcal{B}_m$ of open balls, each of radius $1/m$, so that the collection $\mathcal{B}_m$ covers X. For (a) take the centers of the balls in $\bigcup_{m=1}^{\infty} \mathcal{B}_m$. For (b), consider $\{\eta_k^{(m)}\}$ the partition of unity corresponding to the covering of X by $\mathcal{B}_m$ (as given, for example, in Chapter 1). Show that the finite linear combinations of the $\eta_k^{(m)}$ with rational coefficients are dense in $C(X)$.]

5. Let X be a metric space, $K \subset X$ a compact subset, and f a continuous function on K. There there is a continuous function F on X, so that

$$F|_K = f, \qquad \text{and} \qquad \sup_{x \in X} |F(x)| = \sup_{x \in K} |f(x)|.$$

[Hint: The argument given in Lemma 4.11, Chapter 5 of Book III for $X = \mathbb{R}^d$ can be copied over in this general setting.]

6. Suppose K is a compact subset of $\mathcal{P}$. Show that for each $T > 0$, there exists a function $w_T(h)$, defined for $h \in (0, 1]$ with $w_T(h) \to 0$ as $h \to 0$ and such that

$$\sup_{\mathsf{p} \in K} \sup_{0 \leq t \leq T} |\mathsf{p}(t+h) - \mathsf{p}(t)| \leq w_T(h), \qquad \text{for } h \in (0, 1].$$

[Hint: Fix $T > 0$ and $\epsilon > 0$. Each p is uniformly continuous on closed intervals, so there exists $\delta = \delta(\mathsf{p}) > 0$ so that $\sup_{0 \leq t \leq T} |\mathsf{p}(t+h) - \mathsf{p}(t)| \leq \epsilon$ whenever $0 < h \leq \delta$. Now use the fact that since K is compact, the covering $K \subset \bigcup_{\mathsf{p}} \{\mathsf{p}' \in \mathcal{P} : d(\mathsf{p}', \mathsf{p}) < \epsilon\}$ has a finite subcover.]

7. Suppose $\mu_N \to \mu$ weakly. Show as a result:

(a) $\liminf_{N\to\infty} \mu_N(\mathcal{O}) \geq \mu(\mathcal{O})$ for any open set $\mathcal{O}$.

(b) $\lim_{N\to\infty} \mu_N(\mathcal{O}) = \mu(\mathcal{O})$, $\mathcal{O}$ an open set, if $\mu(\overline{\mathcal{O}} - \mathcal{O}) = 0$.

[Hint: $\mu(\mathcal{O}) = \sup_f\{\int f\, d\mu$, where $0 \leq f \leq 1$ and $\mathrm{supp}(f) \subset \mathcal{O}\}$.]

8. Given the Wiener measure W in $\mathcal{P}$, we have a realization of Brownian motion (satisfying B-1, B-2 and B-3) with $B_t(\omega) = \mathsf{p}(t)$, $\Omega = \mathcal{P}$, and $P = W$. Conversely, suppose we start with $\{B_t\}$ satisfying B-1, B-2 and B-3. For any cylindrical set $C = \{\mathsf{p} \in \mathcal{P} : (\mathsf{p}(t_1), \ldots, \mathsf{p}(t_k)) \in A\}$ in $\mathcal{P}$, define $W^o(C) = P(\{\omega : (B_{t_1}(\omega), \ldots, B_{t_k}(\omega)) \in A\})$. Verify that W^o, initially defined on the cylindrical sets, extends to the Wiener measure on $\mathcal{P}$.

9. This exercise deals with the degree to which the Brownian motion process is uniquely determined by the properties B-1, B-2 and B-3.

Let us say that such a process is "strict" if in addition to the above it satisfies the following two conditions:

(i) $B_t(\omega_1) = B_t(\omega_2)$ for all t implies $\omega_1 = \omega_2$.

(ii) The collection of measurable sets of (Ω, P) is exactly $\mathcal{A}_\infty$, which is the σ-algebra generated by the $\mathcal{A}_t$, with $t < \infty$.

Now given any Brownian motion process B_t on (Ω, P) it induces a strict process $B_t^\#$ on $(\Omega^\#, P^\#)$ as follows: Let $\Omega^\#$ denote the collection of equivalence classes on Ω under the equivalence relation $\omega_1 \sim \omega_2$ if $B_t(\omega_1) = B_t(\omega_2)$ for all t. We also denote by $\{\omega\}$ the equivalence class to which ω belongs. On $\Omega^\#$ define $B_t^\#(\{\omega\}) = B_t(\omega)$, and $P^\#(\{A\}) = P(A)$ if $A \in \mathcal{A}_\infty$. Verify:

(a) $B_t^\#$ is a strict Brownian motion on $(\Omega^\#, P^\#)$.

(b) The process $(\mathcal{P}, W)$ constructed in Section 3 is a strict Brownian motion.

(c) If (B_t^1, Ω^1, P^1) and (B_t^2, Ω^2, P^2) are a pair of Brownian motion processes, then up to subsets of sets of measure zero, there is a bijection $\Phi : (\Omega^1)^\# \to (\Omega^2)^\#$, so that $(P^2)^\#(\Phi(A)) = (P^1)^\#(A)$ and $(B_t^2)^\#(\Phi(\omega)) = (B_t^1)^\#(\omega)$.

10. Prove the following version of Khinchin's inequality (Lemma 1.8 in the previous chapter). Suppose $\{f_n\}$ are identically distributed $\mathbb{R}^d$-valued functions that are bounded, have mean zero, and are mutually independent. Then for any $p < \infty$, we have

$$\|\sum a_n f_n\|_{L^p} \leq A_p \left(\sum |a_n|^2\right)^{1/2}.$$

[Hint: One can reduce to the case $d = 1$. Assuming $\sum |a_n|^2 \leq 1$, write $\int e^{\sum a_n f_n} = \prod_n \int e^{a_n f_n}$ and use $e^u = 1 + u + O(u^2)$ if $|u| \leq M$. The result is that the first integral above is majorized by $\prod(1 + M^2 a_n^2)$ if $|f_n| \leq M$ for all n.]

11. Prove the following variant of Lemma 3.2. Suppose $\{f_k\}_{k=1}^\infty$ is a sequence of identically distributed, mutually independent, $\mathbb{R}^d$-valued functions on a probability space (X, m) each having mean zero and the identity as its covariance matrix. If $s_n = \sum_{k=1}^n f_k$, then

$$\limsup_{n\to\infty} m(\{x : \sup_{1\le k\le n} |s_k(x)| > \lambda n^{1/2}\}) = O(\lambda^{-p}), \qquad \text{for } p > 0.$$

[Hint: If ν_n denotes the distribution measure of $s_n/n^{1/2}$ and $\alpha = \lambda n^{1/2}$, then the right-hand side of (9) equals $\frac{1}{\lambda}\int_{|t|>\lambda} |t|\,d\nu_n(t)$. With $\lambda \ge 1$ fix $M \ge 1$ and write this last integral as the sum of two terms $\lambda^{-1}\int_{|t|>\lambda^M} + \lambda^{-1}\int_{\lambda^M \ge |t| > \lambda}$. Using the fact that $\int \frac{|s_n|^2}{n}\,dm = 1$, the first term is $O(\lambda^{-1-M})$. By the central limit theorem $\lim_{n\to\infty} \lambda^{-1}\int_{\lambda^M\ge|t|>\lambda} |t|\,d\nu_n(t) = O\left(\lambda^{-1}\int_{|t|>\lambda} |t| e^{-|t|^2/2}\,dt\right)$ so the limit of the second term is also $O(\lambda^{-1-M})$.]

12. Prove that almost everywhere

$$|B_t(\omega)| = O(t^{1/2+\epsilon}), \qquad \text{as } t \to \infty,$$

for every $\epsilon > 0$. This is the analog of the strong law of large numbers given in Corollary 2.9 of the previous chapter.

[Hint: If $B_T^*(\omega)$ denotes $\sup_{0\le t\le T} |B_t(\omega)|$, then the maximal inequality (14) gives $W(\{B_T^* > \alpha\}) \le \frac{1}{\alpha}\|B_T\|_{L^1} = c'\frac{T^{1/2}}{\alpha}$. If $E_k = \{B_{2^k}^* > 2^{\frac{k}{2}(1+\epsilon)}\}$, then we have $\sum_{k\ge 0} W(E_k) = O\left(\sum_{k\ge0} 2^{-\frac{k}{2}\epsilon}\right) < \infty$.]

13. If B_t is a Brownian motion process then so is $B'_t = tB_{1/t}$.

[Hint: Note the continuity of almost all paths of B'_t at the origin follows from the previous exercise. To verify property B-2, use Exercise 29 in the previous chapter.]

14. Show that $\limsup_{t\to 0} \frac{|B_t(\omega)|}{t^{1/2}} = \infty$ almost everywhere; hence almost all Brownian paths are not Hölder 1/2.

Also show that $\limsup_{t\to\infty} \frac{|B_t(\omega)|}{t^{1/2}} = \infty$ almost everywhere; hence almost all Brownian paths exit every ball.

[Hint: By the previous exercise it suffices to check the result when $t \to 0$. Consider $d = 1$. Then

$$W(\{|B_\alpha - B_\beta| > \gamma\}) = \frac{1}{\sqrt{2\pi(\beta-\alpha)}}\int_{|u|>\gamma} e^{-\frac{u^2}{2(\beta-\alpha)}}\,du, \qquad \text{if } \beta > \alpha.$$

Thus

$$W(\{|B_{2^{-k}} - B_{2^{-k+1}}| > 2^{-k/2}\mu_k\}) \ge \frac{1}{\sqrt{2\pi}}\int_{|u|\ge\mu_k} e^{-u^2/2}\,du \ge c_1 e^{-c_2\mu_k^2}.$$

Now choose $\mu_k \to \infty$ so slowly that $\sum_{k\ge0} e^{-c_2\mu_k^2} = \infty$ and apply the Borel-Cantelli lemma (Exercise 20 in the previous chapter).]

15. Calculate the (joint) probability distribution measure of $(B_{t_1}, B_{t_2}, \ldots, B_{t_k})$.
[Hint: Use Exercise 8 (a) in the previous chapter.]

16. Show that the following generalization of the fact that $\mathcal{A}_{0+} = \mathcal{A}_0$ holds: if we define $\mathcal{A}_{t+}$ to be $\bigcap_{s>t} \mathcal{A}_s$, then $\mathcal{A}_{t+} = \mathcal{A}_t$.

17. The previous exercise gives the right-continuity of the collection $\{\mathcal{A}_s\}$. Prove the following left-continuity for every $t > 0$, $\mathcal{A}_t = \mathcal{A}_{t-}$, where $\mathcal{A}_{t-}$ is the σ-algebra generated by all $\mathcal{A}_s$ for $s < t$.
[Hint: Consider first cylindrical sets in $\mathcal{A}_t$.]

18. Let σ be a stopping time. Show that:

(a) σ is $\mathcal{A}_\sigma$-measurable.

(b) $B_{\sigma(\omega)}(\omega)$ is $\mathcal{A}_\sigma$-measurable.

(c) $\mathcal{A}_\sigma$ is the σ-algebra determined by the stopped process $\hat{B}_t$ with $\hat{B}_t(\omega) = B_{t\wedge\sigma(\omega)}(\omega)$.

[Hint: For (a), note that $\{\sigma(\omega) \leq \alpha\} \cap \{\sigma(\omega) \leq t\} = \{\sigma(\omega) \leq \min(\alpha, t)\}$. For (b), show first that for any Borel subset E of $\mathbb{R}^d$ and $t \geq 0$, one has $\{B_{\sigma(\omega)}(\omega) \in E\} \cap \{\sigma \leq t\} \in \mathcal{A}_t$ whenever σ takes on only discrete values. Then approximate σ by $\sigma^{(n)}$ as in the proof of Theorem 5.3.]

19. Let u be a bounded Borel measurable function on a bounded open set $\mathcal{R} \subset \mathbb{R}^d$. Suppose that u satisfies the mean-value property on spheres, that is, (21).

(a) Show that if B is a ball contained in $\mathcal{R}$ and centered at x, then

$$u(x) = \frac{1}{m(B)} \int_B u(y)\, dy,$$

where m is the Lebesgue measure on $\mathbb{R}^d$.

(b) As a result, the function u is continuous in $\mathcal{R}$ and the argument in Section 4.1, Chapter 5 of Book III shows that the function u is harmonic in $\mathcal{R}$.

[Hint: For (b), show that locally, $u(x) = (u * \varphi)(x)$, where φ is a smooth radial function supported on an appropriately small ball and with $\int \varphi = 1$.]

20. An bounded open set $\mathcal{R}$ has a **Lipschitz boundary** if $\partial\mathcal{R}$ can be covered by finitely many balls, so that for each such ball B, the set $\partial\mathcal{R} \cap B$ can (possibly after a rotation and translation) be written as $x_d = \varphi(x_1, \ldots, x_{d-1})$, where φ is a function that satisfies a Lipschitz condition.

Verify that if $\mathcal{R}$ has a Lipschitz boundary, then it satisfies the outside cone condition. Thus, in particular, if $\mathcal{R}$ is of class C^1 (in the sense of Section 4 in Chapter 7) then $\mathcal{R}$ satisfies the outside cone condition.

So in these cases the Dirichlet problem is uniquely solvable.

21. Suppose $\mathcal{R}_1$ and $\mathcal{R}_2$ are two open and bounded sets in $\mathbb{R}^d$, with $\overline{\mathcal{R}}_1 \subset \mathcal{R}_2$. Let μ_1^x and μ_2^x denote the harmonic measures of $\mathcal{R}_1$ and $\mathcal{R}_2$ respectively, as defined at the beginning of Section 5. Show that the following generalization of the mean-value property (21) holds: whenever $x \in \mathcal{R}_1$, then

$$\mu_2^x = \int_{\partial\mathcal{R}_1} \mu_2^y \, d\mu_1^x(y),$$

in the sense that $\mu_2^x(E) = \int_{\partial\mathcal{R}_1} \mu_2^y(E)\, d\mu_1^x(y)$ for any Borel set $E \subset \partial\mathcal{R}_2$.

8 Problems

1. The condition of continuity of Brownian paths B-3 is in effect a consequence of properties B-1 and B-2. This is implied by the following general theorem.

Suppose that for each $t \geq 0$, we are given an L^p function $F_t = F_t(x)$ on the space (X, m). Assume that $\|F_{t_1} - F_{t_2}\|_{L^p} \leq c|t_1 - t_2|^\alpha$, with $\alpha > 1/p$, and $1 \leq p \leq \infty$. Then there is a "corrected" $\tilde{F}_t$, so that for each t, $F_t = \tilde{F}_t$ (almost everywhere with respect to m), and so that $t \mapsto \tilde{F}_t(x)$ is continuous for all $t \geq 0$, for almost every $x \in X$. Moreover the functions $t \mapsto \tilde{F}_t(x)$ satisfy a Lipschitz condition of order γ if $\gamma < \alpha - 1/p$.

2. The proof of the Donsker invariance principle follows along the same lines as the proof of Theorem 3.1. Let $f_1, \ldots, f_n, \ldots$ be a sequence of identically distributed mutually independent square integrable $\mathbb{R}^d$-valued functions on a probability space (X, m), each having mean zero and the identity as its covariance matrix. Define

$$S_t^{(N)} = \frac{1}{N^{1/2}} \sum_{1 \leq k \leq [Nt]} f_k + \frac{(Nt - [Nt])}{N^{1/2}} f_{[Nt]+1},$$

and let $\{\mu_N\}$ be the corresponding measures on $\mathcal{P}$ induced via the measure m on X.

(a) Instead of Lemma 3.2 use Exercise 11 to show that for $T = 1$, $\eta > 0$ and $\sigma > 0$, there exists $0 < \delta < 1$ and an integer N_0 so that for all $0 \leq t \leq 1$ one has

$$m(\{x : \sup_{0<h<\delta} |S_{t+h}^{(N)} - S_t^{(N)}| > \sigma\}) \leq \delta\eta, \qquad \text{for all } N \geq N_0.$$

(b) Deduce from the above that for all $T > 0$, $\epsilon > 0$, and $\sigma > 0$ there is a $\delta > 0$ so that

$$m(\{x : \sup_{0 \leq t \leq T,\ 0<h<\delta} |S_{t+h}^{(N)} - S_t^{(N)}| > \sigma\}) \leq \epsilon, \qquad \text{for all } N \geq 1.$$

(c) Use the inequality in (b) to show that the sequence $\{\mu_N\}$ is tight.

(d) Conclude as before that $\{\mu_N\}$ converges weakly to W.

3. There are a number of other constructions of Brownian motion besides the one given in this chapter. A particularly elegant approach is based on simple Hilbert space ideas.

On (Ω, P) consider a sequence $\{f_n\}$ of independent, identically distributed $\mathbb{R}^d$-valued functions with Gaussian distribution of mean zero and covariance matrix equal to the identity. Observe that the sequence $\{f_n\}$ is an orthonormal sequence of $L^2(\Omega, \mathbb{R}^d)$. Let $\mathcal{H}$ denote the closed subspace of $L^2(\Omega, \mathbb{R}^d)$ spanned by $\{f_n\}$.

Observe that $\mathcal{H}$ is a separable infinite dimensional Hilbert space. Hence there is a unitary correspondence U between $L^2([0,\infty), dx)$ and $\mathcal{H}$. Let $B_t = U(\chi_t)$ where χ_t is the characteristic function of the interval $[0,t]$. Then, each B_t can be corrected as in Problem 1, so that the process $\{B_t\}$ becomes Brownian motion. In this connection see also Exercise 9 in Chapter 5.

Note that for instance, if $B_t = \sum c_n(t) f_n$, then $B_t - B_s = \sum [c_n(t) - c_n(s)] f_n$ with $\sum |c_n(t) - c_n(s)|^2 = t - s$.

4.* In the previous chapter, we noted that recurrence results for the (discrete) random walks depend on the dimension d, and in particular, whether $d \leq 2$ or $d \geq 3$ (see Theorem 2.18 in Chapter 5 and the remark that follows it).

One can establish the following results for the (continuous) Brownian motion B_t in $\mathbb{R}^d$.

(a) If $d = 1$, Brownian motion hits, almost surely, every point infinitely often, in the sense that for each $x \in \mathbb{R}$ and for any $t_0 > 0$,

$$P(\{\omega : B_t(\omega) = x \text{ for some } t \geq t_0\}) = 1.$$

Thus B_t is pointwise recurrent in $\mathbb{R}$.

(b) If $d \geq 2$, then for every point $x \in \mathbb{R}^d$, Brownian motion almost surely never hits that point, that is,

$$P(\{\omega : B_t(\omega) = x \text{ for some } t > 0\}) = 0.$$

So, in this case, Brownian motion is not pointwise recurrent.

(c) However if $d = 2$, then B_t is recurrent in every neighborhood of every point, that is, if D is any open disc with positive radius, and $t_0 > 0$, then

$$P(\{\omega : B_t(\omega) \in D \text{ for some } t \geq t_0\}) = 1.$$

(d) Finally, when $d \geq 3$, Brownian motion is transient, that is, it escapes to infinity in the sense that

$$P(\{\omega : \lim_{t\to\infty} |B_t(\omega)| = \infty\}) = 1.$$

5.* The law of the iterated logarithm describes the amplitude of the oscillations of Brownian motion as $t \to \infty$ and $t \to 0$: if B_t is an $\mathbb{R}$-valued Brownian motion process, then for almost all ω

$$\limsup_{t\to\infty} \frac{B_t(\omega)}{\sqrt{2t\log\log t}} \to 1, \qquad \liminf_{t\to\infty} \frac{B_t(\omega)}{\sqrt{2t\log\log t}} \to -1.$$

By Exercise 13, time inversion implies that for almost all ω

$$\limsup_{t\to 0} \frac{B_t(\omega)}{\sqrt{2t\log\log(1/t)}} \to 1, \qquad \liminf_{t\to 0} \frac{B_t(\omega)}{\sqrt{2t\log\log(1/t)}} \to -1.$$

6.* There is a converse to Theorem 6.1 when $d \geq 2$: if $u(x) \to f(y)$ as $x \to y$ with $x \in \mathcal{R}$, for each continuous function f, then y is a regular point.

[Hint: If y is not regular, then, using Problem 4* (b), show that $P(\{|B^y_{\tau^y} - y| > 0\}) = 1$, hence $P(\{|B^y_{\tau^y} - y| \geq \delta\}) > 1/2$ for some $\delta > 0$. If S_ϵ denotes the sphere centered at y of radius $\epsilon < \delta$, use the strong Markov property to prove that there exists $x_\epsilon \in S_\epsilon \cap \mathcal{R}$ so that $P(\{|B^{x_\epsilon}_{\tau^{x_\epsilon}} - y| \geq \delta\}) > 1/2$. Then, considering any continuous function $0 \leq f \leq 1$ on $\mathcal{R}$ with $f(y) = 1$, and $f(z) = 0$ whenever $|z - y| \geq \delta$, leads to a contradiction.]

7.* A simple example of a non-regular point arises when we remove from an open ball its center, with the center then becoming a non-regular point. A more interesting example of a non-regular point is given by Lebesgue's thorn with its cusp at the origin.

Suppose $d \geq 3$, and consider the ball $B = \{x \in \mathbb{R}^d : |x| < 1\}$ from which we remove the set

$$E = \{(x_1, \ldots, x_d) \in \mathbb{R}^d :\ 0 \leq x_1 \leq 1,\ x_2^2 + \cdots + x_d^2 \leq f(x_1)\}.$$

Here f is continuous and $f(x) > 0$ if $x > 0$. If $f(x)$ decreases sufficiently rapidly as $x \to 0$, then the origin is non-regular for the set $\mathcal{R} = B - E$. Clearly, $\mathcal{R}$ can be modified so that its boundary is smooth except at the origin.

7 A Glimpse into Several Complex Variables

> In dealing with the existence of solutions of partial differential equations it was customary during the nineteenth century and it still is today in many applications, to appeal to the theorem of Cauchy-Kowalewski, which guarantees the existence of analytic solutions for analytic partial differential equations. On the other hand a deeper understanding of the nature of solutions requires the admission of non-analytic functions in equations and solutions. For large classes of equations this extension of the range of equation and solution has been carried out since the beginning of this century. In particular much attention has been given to linear partial differential equations and systems of such. Uniformly the experience of the investigated types has shown that — speaking of existence in the local sense — there always were solutions, indeed, smooth solutions, provided the equations were smooth enough. It was therefore a matter of considerable surprise to this author, to discover that this inference is in general erroneous.
>
> *H. Lewy*, 1957

When we go beyond the introductory parts of the subject, what is striking is the extent to which the study of complex analysis in several variables differs from that of one variable. Among the new features that arise are: the automatic analytic continuation of functions from certain domains to larger domains; the crucial role of the tangential Cauchy-Riemann operators; and the significance of (complex) convexity properties of boundaries of domains.

Even though the subject has developed far exploiting these concepts, it is our purpose here to give the reader only a first look at these ideas.

1 Elementary properties

The definition and elementary properties of analytic (or "holomorphic") functions in $\mathbb{C}^n$ are straight-forward adaptations of the corresponding notions for the case $n = 1$. We start with a bit of notation. For any $z^0 = (z_1^0, \ldots, z_n^0) \in \mathbb{C}^n$ and $r = (r_1, \ldots, r_n)$ with $r_j > 0$, we denote by $\mathbb{P}_r(z^0)$ the **polydisc** given by the product

$$\mathbb{P}_r(z^0) = \{z = (z_1, \ldots, z_n) \in \mathbb{C}^n : \ |z_j - z_j^0| < r_j, \text{ for all } 0 \leq j \leq n\}.$$

We will also set $C_r(z^0)$ to be the corresponding product of boundary circles

$$C_r(z^0) = \{z = (z_1, \ldots, z_n) \in \mathbb{C}^n : \ |z_j - z_j^0| = r_j, \text{ all } 0 \leq j \leq n\}.$$

We also write z^α for the monomial $z_1^{\alpha_1} z_2^{\alpha_2} \cdots z_n^{\alpha_n}$, where $\alpha = (\alpha_1, \ldots, \alpha_n)$ with α_j non-negative integers.

We shall see below that for any continuous function f on an open set Ω, the following conditions, defining the analyticity of f, are equivalent:

(i) The function f satisfies the **Cauchy-Riemann equations**

$$(1) \qquad \frac{\partial f}{\partial \bar{z}_j} = 0, \qquad \text{for } j = 1, \ldots, n$$

(taken in the sense of distributions). Here

$$\frac{\partial f}{\partial \bar{z}_j} = \frac{1}{2}\left(\frac{\partial f}{\partial x_j} + i\frac{\partial f}{\partial y_j}\right), \qquad \text{and} \qquad z_j = x_j + iy_j, \text{ with } x_j, y_j \in \mathbb{R}.$$

(ii) For each $z^0 \in \Omega$ and $1 \leq k \leq n$, the function

$$g(z_k) = f(z_1^0, \ldots, z_{k-1}^0, z_k, z_{k+1}^0, \ldots, z_n^0)$$

is analytic in z_k (in the one-variable sense) for z_k in some neighborhood of z_k^0.

(iii) For any polydisc $\mathbb{P}_r(z^0)$ whose closure lies in Ω we have the **Cauchy integral representation**

$$(2) \qquad f(z) = \frac{1}{(2\pi i)^n} \int_{C_r(z^0)} f(\zeta) \prod_{k=1}^{n} \frac{d\zeta_k}{\zeta_k - z_k}, \qquad \text{for } z \in \mathbb{P}_r(z^0).$$

(iv) For each $z^0 \in \Omega$, the function f has a power series expansion $f(z) = \sum a_\alpha (z - z^0)^\alpha$ that converges absolutely and uniformly in a neighborhood of z^0.

Proposition 1.1 *For a continuous function f given in an open set Ω, the conditions* (i) *to* (iv) *above are equivalent.*

Proof. To see why (i) implies (ii), let $\triangle$ be the Laplacian on $\mathbb{C}^n$,

$$\triangle = \sum_{j=1}^{n} \left(\frac{\partial^2}{\partial x_j^2} + \frac{\partial^2}{\partial y_j^2} \right),$$

with $z_j = x_j + iy_j$, and where $\mathbb{C}^n$ is thus identified with $\mathbb{R}^{2n}$. Note that then

$$\triangle = 4 \sum_{j=1}^{n} \frac{\partial}{\partial z_j} \frac{\partial}{\partial \overline{z}_j},$$

where $\frac{\partial f}{\partial \overline{z}_j} = \frac{1}{2}\left(\frac{\partial f}{\partial x_j} + i\frac{\partial f}{\partial y_j} \right)$ and $\frac{\partial f}{\partial z_j} = \frac{1}{2}\left(\frac{\partial f}{\partial x_j} - i\frac{\partial f}{\partial y_j} \right)$, so if f satisfies (i) (in the sense of distributions), then in fact $\triangle f = 0$. From the ellipticity of the operator $\triangle$ and its resulting regularity (see Section 2.5 of Chapter 3) we see that f is in C^∞, and in particular in C^1. Thus the Cauchy-Riemann equations are satisfied in the usual sense and (ii) is established.

Now suppose $z \in \mathbb{P}_r(z^0)$, with $\overline{\mathbb{P}_r(z^0)} \subset \Omega$. Then if (ii) holds we can apply the one-variable Cauchy integral formula in the first variable, with $z_2, z_3, \ldots, z_n$ fixed, to obtain

$$f(z) = \frac{1}{2\pi i} \int_{|\zeta_1 - z_1^0| = r_1} f(\zeta_1, z_2, \ldots, z_n) \frac{d\zeta_1}{\zeta_1 - z_1}.$$

Next, using the Cauchy integral formula in the second variable to represent $f(\zeta_1, z_2, \ldots, z_n)$ with $\zeta_1, z_3, \ldots, z_n$ fixed, gives

$$f(z) = \frac{1}{(2\pi i)^2} \int_{|\zeta_1 - z_1^0| = r_1} \int_{|\zeta_2 - z_2^0| = r_2} \frac{f(\zeta_1, \zeta_2, \ldots, z_n)}{(\zeta_2 - z_2)(\zeta_1 - z_1)} \, d\zeta_2 \, d\zeta_1.$$

Continuing this way yields assertion (iii).

To obtain (iv) as a consequence of (iii), note that

$$\frac{1}{\zeta_k - z_k} = \frac{1}{\zeta_k - z_k^0 - (z_k - z_k^0)} = \sum_{m=0}^{\infty} \frac{(z_k - z_k^0)^m}{(\zeta_k - z_k^0)^{m+1}}.$$

This series converges for $z \in \mathbb{P}_r(z^0)$ and $\zeta \in C_r(z^0)$, since then $|z_k - z_k^0| < |\zeta_k - z_k^0| = r_k$ for all k. So if we take $\mathbb{P}_r(z^0)$ with $\overline{\mathbb{P}_r(z^0)} \subset \Omega$, and insert for each k the series in formula (2) we get $f(z) = \sum a_\alpha (z - z^0)^\alpha$ with

$$a_\alpha = \frac{1}{(2\pi i)^n} \int_{C_r(z^0)} f(\zeta) \prod_{k=1}^{n} \frac{d\zeta_k}{(\zeta_k - z_k^0)^{\alpha_k + 1}}.$$

As a result $|a_\alpha| \leq M r^{-\alpha}$, where $r^{-\alpha} = r_1^{-\alpha_1} r_2^{-\alpha_2} \cdots r_n^{-\alpha_n}$, and

$$M = \sup_{\zeta \in C_r(z^0)} |f(\zeta)|.$$

Thus the series converges uniformly and absolutely if $z \in \mathbb{P}_{r'}(z^0)$ and $r'_k < r_k$, for all $k = 1, \ldots, n$.

To complete the proof of the proposition, note that (iv) implies (i) as follows. If $\sum a_\alpha (z - z^0)^\alpha$ converges absolutely for all z near z^0, we can choose a z' near z^0, so that $z'_k - z_k^0 \neq 0$ for each k with $1 \leq k \leq n$, and thus $\sum |a_\alpha| \rho^\alpha$ converges with $\rho = (\rho_1, \ldots, \rho_n)$, $\rho_k = |z'_k - z_k^0| > 0$. Thus for any $z \in \mathbb{P}_\rho(z^0)$ we can differentiate the series term by term and see that in particular f is in C^1 in that polydisc and satisfies the usual Cauchy-Riemann equations there. Since this is valid for each $z^0 \in \Omega$, it follows that f is of class C^1 throughout Ω and satisfies (1) in the usual sense. A fortiori property (i) holds, and the proof of the proposition is concluded.

Two additional remarks are in order. First, the requirement in (i) that f be continuous can be weakened. In particular, if f is merely locally integrable and satisfies (i) in the sense of distributions then f can be corrected on a set of measure zero so as to become continuous (and thus by the above, analytic).

Second, a more difficult equivalence is that it suffices to have assertion (ii) without the a priori assumption that f be (jointly) continuous. See Problem 1*.

Another aspect of analysis in $\mathbb{C}^n$ that is essentially unchanged from the case of one variable is the following feature of analytic identity.

Proposition 1.2 *Suppose f and g are a pair of holomorphic functions in a region*[1] *Ω, and f and g agree in a neighborhood of a point $z^0 \in \Omega$. Then f and g agree throughout Ω.*

Proof. We may assume that $g = 0$. If we fix any point $z' \in \Omega$, it suffices to prove that $f(z') = 0$. Using the pathwise connectedness of Ω we can find a sequence of points $z^1, \ldots, z^N = z'$ in Ω and polydiscs $\mathbb{P}_{r_k}(z^k)$, for $0 \leq k \leq N$, so that

[1] Recall that a region is defined to be an open and connected set.

(a) $\mathbb{P}_{r_k}(z^k) \subset \Omega$,

(b) $z^{k+1} \in \mathbb{P}_{r_k}(z^k)$, for $0 \leq k \leq N-1$.

Now if f vanishes in a neighborhood of z^k, it must necessarily vanish in all of $\mathbb{P}_{r_k}(z^k)$. (This little fact is established in Exercise 1.) Thus f vanishes in $\mathbb{P}_{r_0}(z^0)$, and by (b), it vanishes in $\mathbb{P}_{r_{k-1}}(z^{k+1})$ if it vanishes in $\mathbb{P}_{r_k}(z^k)$. Hence, by an induction on k, we arrive at the conclusion that the function f vanishes on $\mathbb{P}_{r_N}(z^N)$, and therefore $f(z') = 0$, and the proposition is proved.

2 Hartogs' phenomenon: an example

As soon as we get past the elementary properties of holomorphic functions of several variables, we find new phenomena for which there are no analogs in the case of one variable. This is highlighted by the following striking example.

We let Ω be the region in $\mathbb{C}^n$, $n \geq 2$, lying between two concentric spheres; take in particular $\Omega = \{z \in \mathbb{C}^n,\ \rho < |z| < 1\}$, for some fixed $0 < \rho < 1$.

Theorem 2.1 *Suppose F is holomorphic in $\Omega = \{z \in \mathbb{C}^n,\ \rho < |z| < 1\}$, for some fixed ρ, $0 < \rho < 1$. Then F can be analytically continued into the ball $\{z \in \mathbb{C}^n :\ |z| < 1\}$.*

Here we give a simple and elementary proof of this. Using more sophisticated arguments we shall see below that this property of "automatic" continuation holds under very general circumstances.

The quick proof we have in mind is based on a primitive example of this continuation, which we give in the case of $\mathbb{C}^2$. Suppose

$$K_1 = \{(z_1, z_2) :\ |z_1| \leq a, \text{ and } |z_2| = b_1\}$$

and

$$K_2 = \{(z_1, z_2) :\ |z_1| = a, \text{ and } b_2 \leq |z_2| \leq b_1\}.$$

Lemma 2.2 *If the function F is holomorphic in a region $\mathcal{O}$ that contains the union $K_1 \cup K_2$ then F extends analytically to an open set $\tilde{\mathcal{O}}$ containing the product set*

$$\{(z_1, z_2) :\ |z_1| \leq a,\ b_2 \leq |z_2| \leq b_1\}. \tag{3}$$

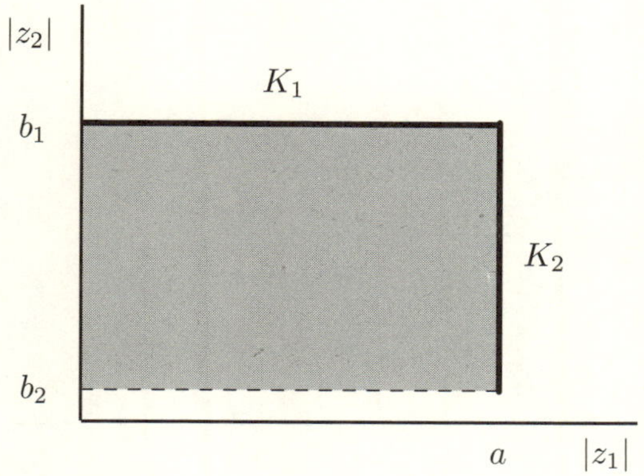

Figure 1. $\tilde{\mathcal{O}}$ contains the shaded region

See Figure 1 for an illustration of the sets K_1, K_2 and their product.

Proof. Consider the integral

$$I(z_1, z_2) = \frac{1}{2\pi i} \int_{|\zeta_1|=a+\epsilon} \frac{F(\zeta_1, z_2)}{\zeta_1 - z_1} \, d\zeta_1,$$

which is well-defined for small positive ϵ, when (z_1, z_2) is in a neighborhood $\tilde{\mathcal{O}}$ of the product set (3). In fact then the variable of integration ranges over a neighborhood of K_2, where F is analytic and hence continuous. Moreover $I(z_1, z_2)$ is analytic in $\tilde{\mathcal{O}}$, since it is visibly analytic in z_1 for fixed z_2 when $|z_1| < a + \epsilon$, and z_2 is near the set $b_2 \leq |z_2| \leq b_1$; also it is analytic in z_2 (for fixed z_1) in that set, by virtue of the analyticity of F. Finally when (z_1, z_2) is near the set K_1, then $I(z_1, z_2) = F(z_1, z_2)$ by the Cauchy integral formula, and thus I provides the desired continuation of F.

We give the proof of the theorem in the case $n = 2$, and start when $\rho < 1/\sqrt{2}$. Here we let $K_1 = \{|z_1| \leq a_1, \ |z_2| = b_1\}$ and $K_2 = \{|z_1| = a_1, \ b_2 \leq |z_2| \leq b_1\}$ with $a_1 = b_1$, $\rho < a_1, b_1 < 1/\sqrt{2}$, and $b_2 = 0$. (See Figure 2.)

Then K_1 and K_2 both belong to Ω, and according to the lemma, F continues to the product $\{|z_1| \leq 1/\sqrt{2}, \ |z_2| \leq 1/\sqrt{2}\}$, which together with Ω covers the entire unit ball.

When $1/\sqrt{2} \leq \rho < 1$, we use the same idea, but now carry out the argument by descending in a finite number of steps the staircase in the $(|z_1|, |z_2|)$ plane whose corners are denoted by (α_k, β_k). (See Figure 3.)

We take $\beta_1 = \rho$, $\alpha_1 = (1 - \beta_1^2)^{1/2} = (1 - \rho^2)^{1/2}$, and more generally

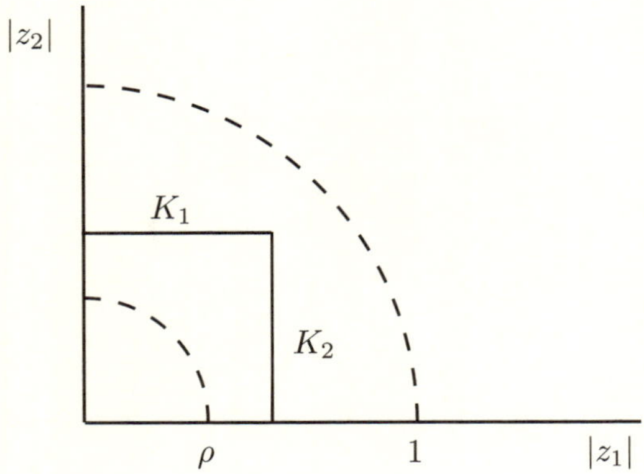

Figure 2. The case where $\rho < 1/\sqrt{2}$

$\beta_{k+1}^2 = \rho^2 - \alpha_k^2, \alpha_{k+1}^2 = 1 - \beta_{k+1}^2$. Hence $\beta_k^2 = 1 - k(1-\rho^2)$, $\alpha_k^2 = k(1 - \rho^2)$.

We start at $k = 1$ and stop as soon as $1 - k(1-\rho^2) < 0$ for $k = N$, with N the smallest integer $> 1/(1-\rho^2)$. With this we choose (a_k, b_k) so that $a_k < \alpha_k$, $b_k > \beta_k$ with (a_k, b_k) near (α_k, β_k), yet $a_N = 1$, $b_N = 0$.

Now let $\mathcal{R}_k = \{\rho < |z| < 1\} \cup \{|z| < 1;\ b_k \leq |z_2|\}$. As above, the lemma gives a continuation of F into a neighborhood of $\mathcal{R}_1$. Using the lemma again (this time with $a = a_k$, $b_1 = b_k$, $b_2 = b_{k+1}$) gives a continuation of F from a neighborhood of $\mathcal{R}_k$ to a neighborhood of $\mathcal{R}_{k+1}$. Now $\mathcal{R}_N = \{|z| < 1\}$, and so we are done.

The corresponding argument in dimension ≥ 3 is similar to that of $n = 2$, and is left to the interested reader to work out.

We mention one immediate application of the previous theorem: a holomorphic function in $\mathbb{C}^n$, $n > 1$, cannot have an isolated singularity; nor can it have an isolated zero. In fact we need only apply Theorem 2.1 to an appropriate pair of concentric balls, centered at the purported singularity. The fact that a zero of f cannot be isolated follows from the previous conclusion applied to the function $1/f$. A more extensive assertion holds, namely if f is holomorphic in Ω and vanishes somewhere, its zero set must reach the boundary of Ω. (See Exercise 4.) Also the nature of the zero set of f near a point where f vanishes can be described quite precisely by the Weierstrass preparation theorem, discussed in Problem 2*.

Finally, notice that holomorphic functions inside the unit ball $\{|z| < 1\}$ cannot necessarily be extended outside the ball, as the simple example

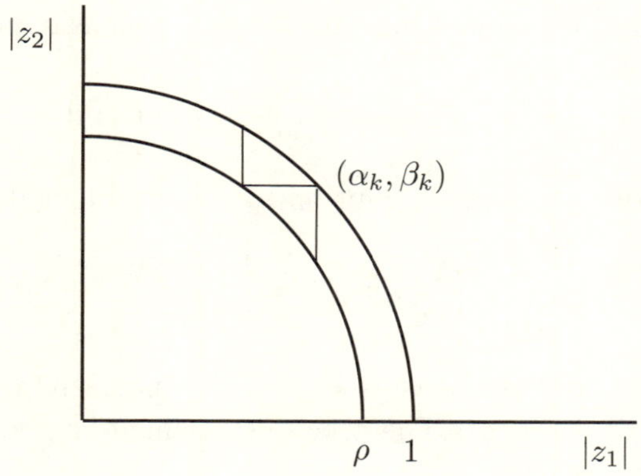

Figure 3. Staircase

$f(z) = 1/(z_1 - 1)$ shows. In fact, we shall see later that the "convexity" of the boundary of Ω plays a crucial role in determining whether a function can be extended past its boundary.

3 Hartogs' theorem: the inhomogeneous Cauchy-Riemann equations

Having seen some simple examples of automatic analytic continuation, we now come to the general situation. The method that will be used here, and that turns out to be useful in a number of questions in complex analysis, is the study of solutions of the system of inhomogeneous Cauchy-Riemann equations

$$\frac{\partial u}{\partial \overline{z}_j} = f_j \quad \text{for } j = 1, \ldots, n, \tag{4}$$

where the f_j are given functions.

The wide applicability of solutions of these equations results from the following necessity. Often one wishes to construct a holomorphic function F with certain desired properties. A first approximation F_1 can be found that enjoys these properties, but with that function not usually holomorphic. The extent to which it fails to satisfy that requirement is given by the non-vanishing of $\partial F_1/\partial \overline{z}_j = f_j$, for $1 \leq j \leq n$. Now if we could find an appropriately well-chosen u that solves $\partial u/\partial \overline{z}_j = f_j$, then we could correct our F_1 by subtracting u from it. In the case below, the "good" choice of u will be the one that has compact support (assuming the f_j have compact support).

In considering (4), we look first at the one-dimensional case, which is

$$(5)\quad \frac{\partial u}{\partial \bar{z}}(z) = f(z), \qquad \text{where } \frac{\partial}{\partial \bar{z}} = \frac{1}{2}\left(\frac{\partial}{\partial x} + i\frac{\partial}{\partial y}\right) \text{ and } z = x + iy \in \mathbb{C}^1.$$

One can state right away a solution to this problem. It is given by

$$(6)\qquad u(z) = \frac{1}{\pi}\int_{\mathbb{C}^1} \frac{f(\zeta)}{z-\zeta}\, dm(\zeta) = \frac{1}{\pi}\int_{\mathbb{C}^1} \frac{f(z-\zeta)}{\zeta}\, dm(\zeta)$$

with $dm(\zeta)$ the Lebesgue measure in $\mathbb{C}^1$. Alternatively, we can write $u = f * \Phi$, with $\Phi(z) = 1/(\pi z)$. The precise statement regarding (5) and (6) is the following assertion.

Proposition 3.1 *Suppose f is continuous and has compact support on $\mathbb{C}$. Then:*

(a) *u given by* (6) *is also continuous and satisfies* (5) *in the sense of distributions.*

(b) *If f is in the class C^k, $k \geq 1$, then so is u, and u satisfies* (5) *in the usual sense.*

(c) *If u is any C^1 function of compact support, then u is already of the form* (6)*; in fact*

$$u = \frac{\partial u}{\partial \bar{z}} * \Phi.$$

Proof. Note first that

$$u(z+h) - u(z) = \frac{1}{\pi}\int_{\mathbb{C}^1} f(z+h-\zeta) - f(z-\zeta)\, \frac{d\zeta}{\zeta},$$

and that this tends to zero as $h \to 0$, by the uniform continuity of f and the fact that the function $1/\zeta$ is integrable over compact sets in $\mathbb{C}^1$. If f is in the class C^k, $k \geq 1$, an easy elaboration of this shows that we can differentiate under the integral sign in (6) and find that any partial derivative of u of order $\leq k$ is represented in the same way in terms of partial derivatives of f.

Next we use the fact that $\Phi(z) = 1/(\pi z)$ is a fundamental solution of the operator $\partial/\partial\bar{z}$. This means that in the sense of distributions $\frac{\partial}{\partial \bar{z}}\Phi = \delta_0$, with δ_0 the Dirac delta function at the origin. (See Exercise 16 in Chapter 3.) So using the formalism of distributions, as in Chapter 3, we have

$$\frac{\partial}{\partial \bar{z}}(f * \Phi) = f * \left(\frac{\partial \Phi}{\partial \bar{z}}\right) = \left(\frac{\partial f}{\partial \bar{z}}\right) * \Phi.$$

The first set of equalities means that $\partial u/\partial\bar{z} = f$, since $f * \delta_0 = f$, and so assertions (a) and (b) are now proved. Using the equality of the second and third members above (with u in place of f) gives $u = u * \delta_0 = \frac{\partial u}{\partial \bar{z}} * \Phi$, and this is assertion (c).

When we turn to the inhomogeneous Cauchy-Riemann equations (5) for $n \geq 2$, there is an immediate difference that is obvious: the f_j's cannot be given "arbitrarily" but must satisfy a necessary consistency condition

$$\frac{\partial f_j}{\partial \bar{z}_k} = \frac{\partial f_k}{\partial \bar{z}_j}, \qquad \text{for all } 1 \leq j, k \leq n. \tag{7}$$

Moreover, it turns out that now the assumption that the f_j have compact support implies the existence of a solution of compact support. The result is contained in the following proposition.

Proposition 3.2 *Suppose $n \geq 2$. If f_j, $1 \leq j \leq n$, are functions of class C^k of compact support that satisfy* (7), *then there exists a function u of class C^k and of compact support that satisfies the inhomogeneous Cauchy-Riemann equations* (4).[2]

Proof. Write $z = (z', z_n)$, where $z' = (z_1, \ldots, z_{n-1}) \in \mathbb{C}^{n-1}$ and set

$$u(z) = \frac{1}{\pi}\int_{\mathbb{C}^1} f_n(z', z_n - \zeta)\,\frac{dm(\zeta)}{\zeta}. \tag{8}$$

Then by the previous proposition $\partial u/\partial \bar{z}_n = f_n$. However by differentiating under the integral sign (which is easily justified) we see that for $1 \leq j \leq n-1$

$$\begin{aligned}
\frac{\partial u}{\partial \bar{z}_j} &= \frac{1}{\pi}\int_{\mathbb{C}^1} \frac{\partial f_n}{\partial \bar{z}_j}(z', z_n - \zeta)\,\frac{dm(\zeta)}{\zeta} \\
&= \frac{1}{\pi}\int_{\mathbb{C}^1} \frac{\partial f_j}{\partial \bar{z}_n}(z', z_n - \zeta)\,\frac{dm(\zeta)}{\zeta} \\
&= f_j(z', z_n).
\end{aligned}$$

The next-to-last step results from the consistency condition (7), and the last step is a consequence of part (c) of Proposition 3.1. Therefore u solves (4).

Next, since the f_j have compact support, there is a fixed R, so that the f_j vanish when $|z| > R$ for all j. Thus by Proposition 1.1, u is holomorphic in $|z'| > R$, so by (8), u also vanishes there. Since the latter

---

[2]In the case $k = 0$, the identities (7) and (4) are taken in the sense of distributions.

is an open subset of the connected set $|z| > R$, Proposition 1.2 implies that u vanishes when $|z| > R$, and all our assertions are proved.

A few remarks may help clarify the nature of the solutions provided by the previous propositions.

- As opposed to the higher-dimensional case, when $n = 1$ it is not possible in general to solve (4) with a function u of compact support, given f of compact support. In fact it is easily seen that a necessary condition for the existence of such a solution is that $\int_{\mathbb{C}^1} f(z)\, dm(z) = 0$. The full necessary and sufficient conditions are described in Exercise 7.
- When $n \geq 2$, the solution given by (8) is the unique solution which has compact support. This is evident because the difference of two solutions is a holomorphic function on all of $\mathbb{C}^n$. Similarly, when $n = 1$, the solution u given by (6) is the unique one for which $u(z) \to 0$, as $|z| \to \infty$.

The simple facts that we have proved about solutions of the inhomogeneous Cauchy-Riemann equations in the whole space $\mathbb{C}^n$ allow us to obtain a general form of Hartog's principle illustrated by Theorem 2.1. This can be formulated as follows.

Theorem 3.3 *Suppose Ω is a bounded region in $\mathbb{C}^n$, $n \geq 2$, and K is a compact subset of Ω such that $\Omega - K$ is connected. Then any function F_0 analytic in $\Omega - K$ has an analytic continuation into Ω.*

This means that there is an analytic function F on Ω, so that $F = F_0$ on $\Omega - K$.

To prove the theorem observe first that there exists $\epsilon > 0$, so that the open set $\mathcal{O}_\epsilon = \{z : \; d(z, \Omega^c) < \epsilon\}$ is at a positive distance from K. Note that then $(\Omega \cap \mathcal{O}_\epsilon) \subset (\Omega - K)$. Next we can construct a C^∞ cut-off function[3] η so that $\eta(z) = 0$ for z in a neighborhood of K, while $\eta(z) = 1$ for $z \in \mathcal{O}_\epsilon$. With this function we define F_1 in Ω by

$$F_1(z) = \begin{cases} \eta(z)F_0(z) & \text{for } z \in \Omega - K \\ 0 & \text{for } z \in K. \end{cases}$$

The function F_1 is C^∞ in Ω. While F_1 gives an extension to Ω of F_0, this extension is of course not analytic. But by how much does it fail to have this property? To answer this, we define f_j by

$$f_j = \frac{\partial F_1}{\partial \overline{z}_j}, \quad \text{for } j = 1, \ldots, n. \tag{9}$$

---

[3]Note that C^2 instead of C^∞ would do for the rest of this proof.

Note that the f_j are C^∞ functions in Ω, and automatically satisfy the consistency conditions (7) there. Moreover the f_j vanish near the boundary of Ω (in particular for $z \in \mathcal{O}_\epsilon \cap \Omega$) because of the analyticity of F_0. Thus the f_j can be extended to be zero outside Ω so that now the extended f_j are C^∞ and satisfy (7) in the whole of $\mathbb{C}^n$. We call the extended f_j by the same name. We now correct the error given by (9) using Proposition 3.2 to find a function u of compact support so that $\partial u/\partial \overline{z}_j = f_j$ for all j, and take $F = F_1 - u$.

Note that F is holomorphic in Ω (since $\partial F/\partial \overline{z}_j = 0$, $1 \leq j \leq n$, there). We will next see that F agrees with F_0 in an appropriate open subset of $\Omega - K$, which is the same as saying that u vanishes in that open set.

To describe the open set in question we find the smallest R so that $\Omega \subset \{|z| \leq R\}$. Then clearly there is a $z^0 \in \partial\Omega$ with $|z^0| = R$. We set $B_\epsilon = B_\epsilon(z^0) = \{z : \ |z - z^0| < \epsilon\}$, and will see that $\Omega \cap B_\epsilon$ is an open set in $\Omega - K$ where u vanishes. (See Figure 4.)

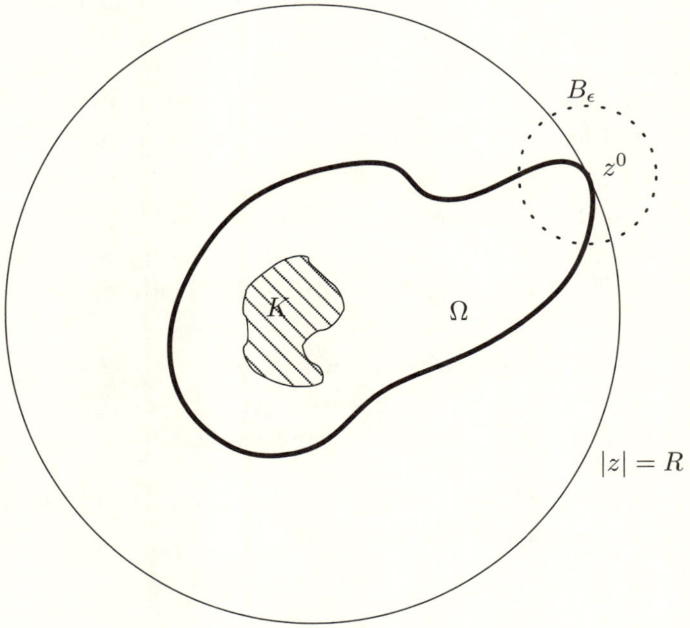

Figure 4. The function u vanishes in $\Omega \cap B_\epsilon$

The fact that $\Omega \cap B_\epsilon$ is an open non-empty set in $\Omega - K$ is immediate since $B_\epsilon \subset \mathcal{O}_\epsilon$ and hence B_ϵ is disjoint from K; also if $\Omega \cap B_\epsilon$ were empty, z^0 could not be a boundary point of Ω. In addition, u is holomorphic in B_ϵ (more generally in $\mathcal{O}_\epsilon$), since the f_j vanish there. Moreover, u is zero in $\{|z| > R\}$ since u is analytic there, this set is connected, and u vanishes outside a compact set. Finally, $B_\epsilon \cap \{|z| > R\}$ is clearly a non-empty open set of B_ϵ. Therefore u vanishes throughout B_ϵ and in particular in $\Omega \cap B_\epsilon$. This shows that F and F_0 agree on an open set of $\Omega - K$, and since the latter set is connected, they agree throughout $\Omega - K$. The theorem is therefore proved.

4 A boundary version: the tangential Cauchy-Riemann equations

We have just seen that if a holomorphic function F_0 is given in a (connected) neighborhood of the boundary of a region Ω in $\mathbb{C}^n$, $n \geq 2$, then it extends to the whole region. Since the neighborhood on which F_0 is given can in principle be arbitrarily narrow, it is natural to ask what happens in the limiting situation where F_0 is given only on the boundary $\partial\Omega$ of Ω. To answer this we must answer the question: what functions F_0 given only on $\partial\Omega$ extend to holomorphic functions in all of Ω?

We shall formulate this problem precisely and solve it in the context of regions with sufficiently smooth boundaries. We begin by reviewing the relevant definitions and elementary background facts that are needed for this.

We start in the setting of $\mathbb{R}^d$ and later pass to $\mathbb{C}^n$ by identifying the latter space with the former when $d = 2n$. Now suppose we are given a region Ω in $\mathbb{R}^d$. A **defining function** ρ of Ω is a real-valued function on $\mathbb{R}^d$ so that

$$\begin{cases} \rho(x) < 0, & \text{when } x \in \Omega, \\ \rho(x) = 0, & \text{when } x \in \partial\Omega, \\ \rho(x) > 0, & \text{when } x \in \overline{\Omega}^c. \end{cases}$$

For any integer $k \geq 1$ the boundary of Ω is said to be of **class** C^k if Ω has a defining function ρ which satisfies

- $\rho \in C^k(\mathbb{R}^d)$;
- $|\nabla\rho(x)| > 0$, whenever $x \in \partial\Omega$.

The boundary $\partial\Omega$ is an example of a hypersurface of class C^k. More generally we shall say that M is a (local) **hypersurface** of class C^k if there is a real-valued C^k function ρ, defined on a ball $B \subset \mathbb{R}^d$, so that $M = \{x \in B : \rho(x) = 0\}$, and $|\nabla\rho(x)| > 0$ whenever $x \in M$.

For a region Ω whose boundary is of class C^k one knows that near any boundary point $\partial\Omega$ can be realized as a "graph." More precisely, fixing any point of reference $x^0 \in \partial\Omega$ and making an appropriate affine-linear change of coordinates (in fact a translation and rotation of $\mathbb{R}^d$) then, by the implicit function theorem, we can achieve the following: With the new coordinate system written as $x = (x', x_d)$ where $x' \in \mathbb{R}^{d-1}$ and $x_d \in \mathbb{R}$, the initial reference point x^0 corresponds to $(0, 0)$ and near $x^0 = (0, 0)$ the region Ω and its boundary are given by

$$\begin{cases} \Omega: & x_d > \varphi(x'), \\ \partial\Omega: & x_d = \varphi(x'). \end{cases} \tag{10}$$

Here φ is a C^k function defined near the origin in $\mathbb{R}^{d-1}$. We can also arrange matters so that (in addition to $\varphi(0) = 0$), one has $\nabla_{x'}(\varphi)(x')|_{x'=0} = 0$, which means that the tangent plane to $\partial\Omega$ at the origin is the hyperplane $x_d = 0$. (See Figure 5.)

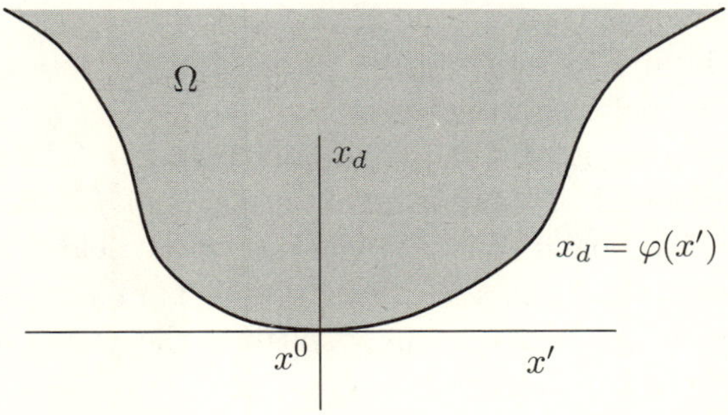

Figure 5. The set Ω and its boundary in the coordinate system (x', x_d)

In this coordinate system, because $\rho(x', \varphi(x')) = 0$, we have

$$\begin{aligned} \rho(x) &= \rho(x', x_d) - \rho(x', \varphi(x')) \\ &= \int_0^1 \frac{\partial}{\partial t}\rho(x', tx_d + (1-t)\varphi(x'))\, dt \\ &= (\varphi(x') - x_d)a(x), \end{aligned}$$

with $a(x) = -\int_0^1 \frac{\partial \rho}{\partial x_d}(x', tx_d + (1-t)\varphi(x'))\, dt$. In other words, $\rho(x) = a(x)(\varphi(x') - x_d)$, where a is a C^{k-1} function. Also $a(x) > 0$ if x is sufficiently close to the reference point x^0, since then $\frac{\partial \rho}{\partial x_d} < 0$, in view of the fact that $\partial/\partial x_d$ points "inwards" with respect to Ω.

Now suppose $\tilde{\rho}$ is another C^k defining function for Ω. Then near x^0 we again have $\tilde{\rho}(x) = \tilde{a}(x)(\varphi(x') - x_d)$ and thus

$$\tilde{\rho} = c\rho, \quad \text{where } c(x) > 0 \tag{11}$$

and c is of class C^{k-1}.

Next we recall that a vector field X on $\mathbb{R}^d$ can be viewed as a first-order linear differential operator of the form

$$X(f) = \sum_{j=1}^{d} a_j(x) \frac{\partial f}{\partial x_j}$$

with $(a_1(x), a_2(x), \ldots, a_d(x))$ the "vector" corresponding to the point $x \in \mathbb{R}^d$. This vector field is **tangential** at $\partial\Omega$, if

$$X(\rho) = \sum_{j=1}^{d} a_j(x) \frac{\partial \rho}{\partial x_j} = 0, \qquad \text{whenever } x \in \partial\Omega.$$

Because of (11) and Leibnitz's rule, this definition does not depend on the choice of the defining function of Ω.

Next we fix an ℓ with $\ell \leq k$. Then, any function f_0 defined on $\partial\Omega$ is said to be of **class** C^ℓ if there is an extension f of f_0 to $\mathbb{R}^d$ so that f is of class C^ℓ on $\mathbb{R}^d$. Now if X is a tangential vector field and f and f' are any two extensions of f_0, then as is easily seen $X(f)|_{\partial\Omega} = X(f')|_{\partial\Omega}$. (See Exercise 8.) So in this sense we may speak of the action of a tangential vector field on functions defined only on $\partial\Omega$.

We now pass to the complex space $\mathbb{C}^n$ that we identify with $\mathbb{R}^d$, $d = 2n$. We do this by writing $z \in \mathbb{C}^n$, $z = (z_1, \ldots, z_n)$, $z_j = x_j + iy_j$, $1 \leq j \leq n$, and then setting $x = (x_1, \ldots, x_{2n}) \in \mathbb{R}^{2n}$ with x_j, $1 \leq j \leq n$, as before, and $x_{j+n} = y_j$, for $1 \leq j \leq n$. Vector fields on $\mathbb{C}^n$ can now be written as

$$\sum_{j=1}^{n} \left(a_j(z) \frac{\partial}{\partial \overline{z}_j} + b_j(z) \frac{\partial}{\partial z_j} \right).$$

(Here it is necessary to allow the coefficients to be complex-valued.) Such a vector field is called a **Cauchy-Riemann vector field**, if $b_j = 0$ for all j, that is, X is of the form

$$X = \sum_{j=1}^{n} a_j(z) \frac{\partial}{\partial \overline{z}_j}.$$

Equivalently, X is a Cauchy-Riemann vector field if it annihilates all holomorphic functions.

Given a region Ω (with C^k boundary) then the above Cauchy-Riemann vector field X is **tangential** if

$$\sum_{j=1}^{n} a_j(z)\rho_j(z) = 0, \qquad \text{where } \rho_j(z) = \frac{\partial \rho}{\partial \overline{z}_j}.$$

Now near any fixed $z^0 \in \partial\Omega$, at least one of the $\rho_j(z^0)$, $1 \leq j \leq n$, must be non-zero, since $|\nabla\rho(z^0)| > 0$; for simplicity we may assume $j = n$. Then the $n-1$ vector fields

$$\rho_n \frac{\partial}{\partial \overline{z}_j} - \rho_j \frac{\partial}{\partial \overline{z}_n}, \qquad 1 \leq j \leq n-1 \tag{12}$$

are linearly independent and span the tangential Cauchy-Riemann vector fields near z^0 (up to multiplication by functions).

Without making the particular choice $j = n$ one notes that the $n(n-1)/2$ vector fields

$$\rho_k \frac{\partial}{\partial \overline{z}_j} - \rho_j \frac{\partial}{\partial \overline{z}_k}, \qquad 1 \leq j < k \leq n \tag{13}$$

span the tangential Cauchy-Riemann vector fields (globally), but of course are not linearly independent.

There is a way of expressing this neatly by using the language of differential forms. Suppose u is a complex-valued function. Then we can abbreviate the equations $\frac{\partial u}{\partial \overline{z}_j} = f_j$, for $1 \leq j \leq n$, by

$$\overline{\partial} u = f,$$

with $\overline{\partial} u$ and f the "one-forms"[4] defined by $\sum_{j=1}^{n} \frac{\partial u}{\partial \overline{z}_j} d\overline{z}_j$ and $\sum_{j=1}^{n} f_j d\overline{z}_j$, respectively. Now for any one-form $w = \sum_{j=1}^{n} w_j d\overline{z}_j$, we define the two-form $\overline{\partial} w$ by

$$\begin{aligned}
\overline{\partial} w &= \sum_{j=1}^{n} \overline{\partial} w_j \wedge d\overline{z}_j \\
&= \sum_{1 \leq k,j \leq n} \frac{\partial w_j}{\partial \overline{z}_k}\, d\overline{z}_k \wedge d\overline{z}_j \\
&= \sum_{1 \leq k < j \leq n} \left(\frac{\partial w_j}{\partial \overline{z}_k} - \frac{\partial w_k}{\partial \overline{z}_j} \right) d\overline{z}_k \wedge d\overline{z}_j,
\end{aligned}$$

[4]More precisely, $(0,1)$-forms.

since $d\overline{z}_k \wedge d\overline{z}_j = -d\overline{z}_j \wedge d\overline{z}_k$ in this formalism.

With this notation the inhomogeneous Cauchy-Riemann equations (4) can be written as $\overline{\partial} u = f$, and the consistency condition (7) is the same as $\overline{\partial} f = 0$. Moreover a function F_0 is annihilated by the tangential Cauchy-Riemann vector fields ((12) or (13)) exactly when

$$\overline{\partial} F_0 \wedge \overline{\partial} \rho|_{\partial\Omega} = 0. \tag{14}$$

So whenever F_0 is the restriction to $\partial\Omega$ of a function of class $C^1(\overline{\Omega})$ that is holomorphic in Ω, it must satisfy these tangential Cauchy-Riemann equations. The remarkable fact is that, broadly speaking, the converse of this holds. This is the thrust of Bochner's theorem.

Theorem 4.1 *Assume Ω is a bounded region in $\mathbb{C}^n$, whose boundary is of class C^3, and suppose the complement of $\overline{\Omega}$ is connected. If F_0 is a function of class C^3 on $\partial\Omega$ that satisfies the tangential Cauchy-Riemann equations, then there is a holomorphic function F in Ω that is continuous in $\overline{\Omega}$, so that $F|_{\partial\Omega} = F_0$.*

The fact that some connectedness property is required for both this and the previous theorem can be seen in Exercise 10.

The proof of this theorem is in the same spirit as the previous one, but the details are different. The function F_0 of class $C^3(\partial\Omega)$ can, by definition, be thought of as a function of class C^3 on the whole space. Now F_0 satisfies the tangential Cauchy-Riemann equations, and we can modify it (without changing its restriction to $\partial\Omega$), so that the modified function F_1 is of class C^2 and

$$\overline{\partial} F_1|_{\partial\Omega} = 0. \tag{15}$$

This modification is achieved by taking $F_1 = F_0 - a\rho$, where a is a suitable C^2 function. Indeed, F_1 already satisfies the tangential Cauchy-Riemann equations. An independent Cauchy-Riemann vector field (that is not tangential) is given by N, with

$$N(f) = \sum_{j=1}^{n} \overline{\rho}_j \frac{\partial f}{\partial \overline{z}_j}.$$

In fact, we note that

$$N(\rho) = \sum_{j=1}^{n} \left| \frac{\partial \rho}{\partial \overline{z}_j} \right|^2 = \frac{1}{4} |\nabla \rho|^2 > 0.$$

Thus if we set $a = N(F_0)/N(\rho)$ near the boundary of Ω and extend a strictly away from the boundary to be zero, then (15) is achieved because of (14).

We now define the one-form f in Ω by $f = \overline{\partial} F_1$. Then f is continuous on $\overline{\Omega}$, is of class $C^1(\overline{\Omega})$, vanishes on $\partial\Omega$, and satisfies $\overline{\partial} f = 0$ in the interior of Ω. We can now extend f to $\mathbb{C}^n$ (keeping the same name) so that $f = 0$ outside of Ω. Then f satisfies $\overline{\partial} f = 0$ in $\mathbb{C}^n$ (at least in the sense of distributions). This would be evident if we supposed that F_0 and $\partial\Omega$ were of class C^4 instead of class C^3. In the latter case an additional argument is needed (see Exercise 6 in Chapter 3). We can now invoke Proposition 3.2 to obtain a continuous function u so that $\overline{\partial} u = f$ and moreover u has compact support. Since u is holomorphic on $\overline{\Omega}^c$ and this set is connected, it follows that u vanishes throughout $\overline{\Omega}^c$ and by continuity it vanishes on $\partial\Omega$. Finally, take $F = F_1 - u$, then F is holomorphic in Ω, continuous in $\overline{\Omega}$ and $F|_{\partial\Omega} = F_1|_{\partial\Omega} = F_0|_{\partial\Omega}$, completing the proof of the theorem.

In the case $n = 1$, there are no tangential Cauchy-Riemann equations and the conditions on F_0 are global in nature. See Exercise 12.

By a different argument one can reduce the degree of regularity involved on F_0. See Problem 3*.

Given the nature of the conditions that are sufficient when $n > 1$, it is natural to ask if there is in fact a "local" version of the extension theorem just proved. For this to be possible, the formulation of such a result must distinguish on which "side" of the boundary this continuation holds. The example of the "inside" of the sphere, where continuation takes place as opposed to the "outside" where it fails, suggests that a convexity property might be involved. This is indeed the case because of the complex structure of $\mathbb{C}^n$, as we will see when we examine the local nature of the boundary of a region.

5 The Levi form

Let us briefly glance back to the situation in $\mathbb{R}^d$. We will see that near any boundary point x^0 the region Ω can be put in a very simple canonical form. We already noted earlier that near x^0, in the appropriate coordinates, we can represent Ω as $\{x_d > \varphi(x')\}$. Now if we introduce new coordinates $(\overline{x}_1, \overline{x}_2, \ldots, \overline{x}_d)$, by $\overline{x}_d = x_d - \varphi(x')$, $\overline{x}_j = x_j$, $1 \le j < d$ (with inverse $x_d = \overline{x}_d + \varphi(\overline{x}')$, $x_j = \overline{x}_j$, $1 \le j < d$) we obtain that locally Ω is now represented by the half-space $\overline{x}_d > 0$, and $\partial\Omega$ by the hyperplane $\overline{x}_d = 0$.

However to be applicable to the study of holomorphic functions in

$\mathbb{C}^n$, the new coordinates that we can allow (that is, the change of variables that is permissible) must be given by holomorphic functions, so our choices are more restricted. The coordinates that result from such changes of variables (starting with the standard coordinates about a fixed point z^0) will be called **holomorphic coordinates**. Here we assume that $\partial\Omega$ is of class C^2, and use the notation $z_j = x_j + iy_j$.

Proposition 5.1 *Near any point $z^0 \in \partial\Omega$ we can introduce holomorphic coordinates $(z_1, \ldots, z_n)$ centered at z^0 so that*

$$\Omega = \{\mathrm{Im}(z_n) > \sum_{j=1}^{n-1} \lambda_j |z_j|^2 + E(z)\}. \tag{16}$$

Here the λ_j are real numbers, and $E(z) = x_n \ell(z') + Dx_n^2 + o(|z|^2)$, as $z \to 0$; [5] *also $\ell(z')$ is a linear function of $x_1, \ldots, x_{n-1}, y_1, \ldots, y_{n-1}$, and D is a real number.*

A few remarks may help to clarify the nature of the cannonical representation (16).

- By making a further change of scale $z_j \to \delta_j z_j$, $\delta_j \neq 0$, we can set the λ_j to be either 1, -1 or 0.

- The number of λ_j that are positive, negative or zero (the **signature** of the quadratic form) is a holomorphic invariant as we will see below.

- It can be seen from (16) that it is natural to assign the variables $z_1, \ldots, z_{n-1}$ "weight 1" and the variable z_n "weight 2," which, disregarding the error term, makes the expression homogeneous of weight 2. This homogeneous version of (16) gives us the "half-space" $\mathcal{U}$ that we consider further in the Appendix to this chapter.

- If we had assumed that $\partial\Omega$ was of class C^3, then the error estimate $o(|z|^2)$ would be improved to $O(|z|^3)$, as $z \to 0$.

Proof of the proposition. As in (10), we see that we can introduce complex coordinates (with an affine complex linear change of variables) so that near z^0 the set Ω is given by

$$\mathrm{Im}(z_n) > \varphi(z', x_n)$$

[5] $f(z) = o(|z|^2)$ as $z \to 0$ means that $|f(z)|/|z|^2 \to 0$, as $|z| \to 0$.

with $z = (z', z_n)$, $z' = (z_1, \dots, z_{n-1})$, and $z_j = x_j + iy_j$. We can also arrange matters so that $\varphi(0,0) = 0$ and

$$\frac{\partial}{\partial x_j}\varphi|_{(0,0)} = \frac{\partial}{\partial y_j}\varphi|_{(0,0)} = \frac{\partial}{\partial x_n}\varphi|_{(0,0)}, \quad 1 \le j \le n-1.$$

Using Taylor's expansion of φ at the origin up to order 2 we see that

$$\begin{aligned}\varphi = \sum_{1\le j,k\le n-1} & (\alpha_{jk}z_j z_k + \overline{\alpha}_{jk}\overline{z}_j\overline{z}_k) + \\ & + \sum_{1\le j,k\le n-1} \beta_{jk} z_j \overline{z}_k + x_n \ell'(z') + \\ & + Dx_n^2 + o(|z|^2), \quad \text{as } z \to 0.\end{aligned}$$

Here $\beta_{jk} = \overline{\beta}_{kj}$ and ℓ' is a (real) linear function of the variables $x_1, \dots,$ x_{n-1} and $y_1, \dots, y_{n-1}$, with D a real number.

Next we introduce the (global) holomorphic change of coordinates $\zeta_n = z_n - 2i\sum_{1\le j,k\le n-1}\alpha_{jk}z_j z_k$, and $\zeta_k = z_k$, for $1 \le k \le n-1$. Then $\text{Im}(\zeta_n) = \text{Im}(z_n) - \sum_{1\le j,k\le n-1}(\alpha_{jk}z_j z_k + \overline{\alpha}_{jk}\overline{z}_j\overline{z}_k)$, and thus in these new coordinates (where we immediately relabel the ζ's as z's) the function φ becomes $\sum_{1\le j,k\le n-1}\beta_{jk}z_j\overline{z}_k + x_n\ell'(z') + Dx_n^2 + o(|z|^2)$.

Next, a unitary mapping (in the $z_1, \dots, z_{n-1}$ variables) allows us to diagonalize the Hermitian form and φ becomes

$$\sum_{j=1}^{n-1}\lambda_j|z_j|^2 + x_n\ell(z') + Dx_n^2 + o(|z|^2) \tag{17}$$

with $\lambda_1, \dots, \lambda_{n-1}$, the eigenvalues of the quadratic form. This proves the proposition.

The Hermitian matrix $\left\{\frac{\partial^2\varphi}{\partial z_j\partial\overline{z}_k}\right\}_{1\le j,k\le n-1}$ that appears implicitly above, or its diagonalized version the form in (16), $\sum_{j=1}^{n-1}\lambda_j|z_j|^2$, is referred to as the **Levi form** of Ω (at the boundary point z^0.) A more intrinsic definition comes about by noticing that the vectors $\partial/\partial\overline{z}_j$, $1 \le j \le n-1$, are tangent to $\partial\Omega$ at z_0. If $\rho(z) = \varphi(z', x_n) - y_n$, then the corresponding quadratic form is

$$\sum_{1\le j,k\le n}\frac{\partial^2\rho}{\partial z_j\partial\overline{z}_k}\overline{a}_j a_k, \tag{18}$$

restricted to the vectors $\sum_{k=1}^{n} a_k\partial/\partial\overline{z}_k$ that are tangential at z^0. Note also that these tangent vectors form a complex subspace (of complex dimension $n-1$) of the full tangent space (which has real dimension $2n-1$).

Now let ρ' be another defining function for Ω. Then $\rho' = c\rho$ with $c > 0$, and we assume c is of class C^2. Then by Leibniz's rule,

$$\sum \frac{\partial^2 \rho'}{\partial z_j \partial \overline{z}_k} \overline{a}_j a_k = c \sum \frac{\partial^2 \rho}{\partial z_j \partial \overline{z}_k} \overline{a}_j a_k \qquad \text{on } \partial\Omega,$$

since $\rho = 0$ there, and also $\sum_{k=1}^n a_k \frac{\partial \rho}{\partial \overline{z}_k} = 0$ because $\sum_{k=1}^n a_k \frac{\partial}{\partial \overline{z}_k}$ is tangential. Thus the signature of the form (18) is independent of the choice of defining function.

Finally let $z \mapsto \Phi(z) = w$ be a biholomorphic mapping defined near the origin (with $\Phi(0) = 0$), giving us a new holomorphic coordinate system $(w_1, \ldots, w_n)$ in the neighborhood of z^0. Then by holomorphicity the differential of Φ maps tangent vectors at z^0 of the form $\sum_{k=1}^n a_k \frac{\partial}{\partial \overline{z}_k}$ to tangent vectors of the form $\sum_{k=1}^n a'_k \frac{\partial}{\partial \overline{w}_k}$. Now if ρ' is a defining function of $\Phi(\Omega)$ then $\rho'(\Phi(z)) = \rho''(z)$ is another defining function of Ω near z^0 and we can conclude by the above that the signature of (18) is invariant under holomorphic bijections.

With regard to the above, one says that a boundary point $z^0 \in \partial\Omega$ is **pseudo-convex** if the Levi form is non-negative, and **strongly pseudo-convex** if that form is strictly positive definite. A region Ω is pseudo-convex if every boundary point of Ω has this property.

A good illustration is given by the unit ball $\{|z| < 1\}$. If we take $\rho(z) = |z|^2 - 1$ to be its defining function, we see that at every boundary point the Levi form corresponds to the identity matrix, and hence the unit ball is strongly pseudo-convex.

Pseudo-convexity may be thought of as the complex analytic analog for $n > 1$ of the standard (real) convexity in $\mathbb{R}^d$; for the latter see Exercise 26 in Chapter 3 and the problems in Chapter 3 of Book III. The nature of the Levi form at z^0 turns out to have important implications for the behavior of holomorphic functions defined in Ω near z^0. In particular, we shall next see some interesting consequences that follow if one of the eigenvalues of the Levi form is strictly positive.

6 A maximum principle

A noteworthy implication of the partial positivity of the Levi form is the following "local" maximum principle in $\mathbb{C}^n$, $n \geq 2$, which has no analog in the case $n = 1$.

Suppose we are given a region Ω with boundary of class C^2, and B is an open ball centered at some point $z^0 \in \partial\Omega$. Assume that at each $z \in \partial\Omega \cap B$ at least one eigenvalue of the Levi form is strictly positive.

Theorem 6.1 *In the above circumstances there exists a (smaller) ball $B' \subset B$, centered at z^0, so that whenever F is a holomorphic function on $\Omega \cap B$ that is continuous on $\overline{\Omega} \cap B$, then*

$$\sup_{z \in \Omega \cap B'} |F(z)| \leq \sup_{z \in \partial\Omega \cap B} |F(z)|. \tag{19}$$

A counter-example of assertion (19) in the case $n = 1$ is outlined in Exercise 16.

Proof. We consider first the special situation when $z^0 = 0$ and Ω is given in the canonical form (16). We may assume that $\lambda_1 > 0$.

We write $z = (z_1, z'', z_n)$, where $z'' = (z_2, \ldots, z_{n-1}) \in \mathbb{C}^{n-2}$, and we consider points of the form $(0, 0, iy_n)$. We denote by $B = B_r$ the ball of radius r centered at the origin and prove that whenever $0 < y_n \leq cr^2$, with r sufficiently small, then at these special points we have the preliminary conclusion

$$|F(0, 0, iy_n)| \leq \sup_{z \in \partial\Omega \cap B_r} |F(z)|. \tag{20}$$

Here c is a constant to be chosen below ($c = \min(1, \lambda_1/2)$ will do).

This will be proved by considering the complex one-dimensional slice passing through the point $(0, 0, iy_n)$. Indeed, let $\Omega_1 = \{z_1 : (z_1, 0, iy_n) \in \Omega \cap B_r\}$. It is obvious that Ω_1 is an open set containing the point $(0, 0, iy_n)$. We note the following key fact: if r is sufficiently small, then

$$\text{If } z_1 \in \partial\Omega_1 \text{ then } (z_1, 0, iy_n) \in \partial\Omega \cap B_r. \tag{21}$$

Indeed, if z_1 is on the boundary of the slice Ω_1, then either $(z_1, 0, iy_n)$ is on the boundary of Ω, or $(z_1, 0, iy_n)$ is on the boundary of B_r (or both alternatives hold). In fact the second alternative is not possible, because if it held, then it would imply that $|z_1|^2 + y_n^2 = r^2$. Since $y_n \leq cr^2$ this yields $|z_1|^2 \geq r^2 - c^2r^4 \geq 3r^2/4$, if we take $c \leq 1$ and $r \leq 1/2$. Moreover since any such point must be in $\overline{\Omega}$ we must have that $y_n \geq \lambda_1|z_1|^2 + o(|z_1|^2)$ and therefore $cr^2 \geq \lambda_1 3r^2/4 + o(r^2)$, which is not possible if we take $c \leq \lambda_1/2$ and r is sufficiently small. Since now the second alternative has been ruled out, we have established (21).

Now for y_n fixed, we define $f(z_1) = F(z_1, 0, iy_n)$. Then f is a holomorphic function in z_1 on the slice Ω_1 and is continuous on $\overline{\Omega}_1$. Since

$0 \in \Omega_1$, the usual maximum principle implies

$$|F(0,0,iy_n)| = |f(0)| \leq \sup_{z_1 \in \overline{\Omega}_1} |f(z_1)| = \sup_{z_1 \in \partial\Omega_1} |f(z_1)| \leq \sup_{z \in \partial\Omega \cap B_r} |F(z)|,$$

because of (21). Therefore the claim in (20) is established.

We will pass from this particular estimate to the general situation by showing that for every point $z \in \Omega$ sufficiently close to the boundary of Ω, we can find an appropriate coordinate system so that with respect to it the point z is given by $(0,0,iy_n)$, and thus the conclusion (20) holds for z. This is done as follows.

First, for every point $z \in \Omega$ sufficiently close to $\partial\Omega$ there is a (unique) point $\pi(z) \in \partial\Omega$ which is nearest to z and moreover, the vector from $\pi(z)$ to z is perpendicular to the tangent plane at $\pi(z)$. Now at each $\pi(z) \in \partial\Omega$ we can introduce a coordinate system leading to the description (17) of Ω near $\pi(z)$. We also observe that the mapping from the initial ambient coordinates of $\mathbb{C}^n$ to those appearing in (17) is affine linear and preserves Euclidean distances. Because of the orthgonality of the vector from $\pi(z)$ to z to the tangent plane, the point z has coordinates $(0,0,iy_n)$ in this coordinate system, and in fact $|z - \pi(z)| = y_n$.

With B the initial ball centered at z^0, we will define $B' = B_\delta(z^0)$ to be the ball of radius δ centered at z^0. That radius will be determined by another radius r, so that $\delta = c_* r^2$, with the constant c_* specified below. We will have $0 < c_* \leq 1$, and ultimately take r (and hence δ) sufficiently small.

We can assume that λ_1 is the largest eigenvalue appearing in (17) and since $\partial\Omega$ is of class C^2, the quantity λ_1 varies continuously with the base point $\pi(z)$. We denote by λ_* the infimum of these λ_1, and in parallel with the special case treated above we set $c_* = \min(1, \lambda_*/2)$.

We then note that if $z \in \Omega \cap B_\delta$ and we take r sufficiently small, then:

- $|z - \pi(z)| < \delta$, and;
- $B_r(\pi(z)) \subset B$.

In fact if $z \in B_\delta(z^0)$, then $z^0 \in \partial\Omega$ implies that $d(z, \partial\Omega) < \delta$, which gives $|z - \pi(z)| < \delta$.

Secondly

$$|\zeta - z^0| \leq |\zeta - \pi(z)| + |\pi(z) - z| + |z - z^0|,$$

so if $\zeta \in B_r(\pi(z))$, then $|\zeta - \pi(z)| < r$ while $|z - \pi(z)| < \delta$, and $|z - z^0| < \delta$ (since $z \in B_\delta$). This means that $|\zeta - z^0| \leq r + 2\delta$, and hence $\zeta \in B$, if r (and then $\delta = c_* r^2$) are sufficiently small.

We can now return to the argument leading to the proof of the special case (20). With the ball $B_r(\pi(z))$ playing the role of B_r above, we see as before that we obtain (20) by the maximum principle, because for $z \in \Omega \cap B$ we have $y_n > \lambda_*|z_1|^2 + o(|z|^2)$, $z_1 \to 0$ with an "o" term that is uniform as z (and hence $\pi(z)$) varies. (This uniformity is a consequence of the fact that the corresponding "o" term in the Taylor development of φ in (17) is uniform, by virtue of the fact that φ is of class C^2.)

All this shows that if we take r sufficiently small, and $\delta = c_* r^2$, then for $z \in B_\delta(z_0) = B'$ the conclusion of the theorem holds.

The implication of the theorem, and its proof, are valid in a more general setting where the boundary $\partial\Omega$ is replaced by a local hypersurface. This can be formulated as follows.

Suppose M is a local C^2 hypersurface given in a ball B with a defining function ρ, so that $M = \{z \in B : \rho(z) = 0\}$. Set $\Omega_- = \{z \in B : \rho(z) < 0\}$.

Corollary 6.2 *Suppose the Levi form, as given by* (18), *has at least one strictly positive eigenvalue for each* $z \in M$. *Under these circumstances, for every* $z^0 \in M$ *there is a ball* B' *centered at* z^0 *so that whenever* F *is holomorphic in* Ω_- *and continuous in* $\Omega_- \cup M$ *we have*

$$\sup_{z \in \Omega_- \cap B'} |F(z)| \leq \sup_{z \in M} |F(z)|. \tag{22}$$

The theorem we have just proved tells us that when an eigenvalue of the Levi form is positive, the control of the restriction of a holomorphic function to a small piece of the boundary gives us a corresponding control of the function in an interior region. This is a strong hint that for such boundaries a local version of Bochner's theorem (Theorem 4.1) should be valid. Our proof of this will be based on a remarkable extension of the Weierstrass approximation theorem, to which we now turn.

7 Approximation and extension theorems

The classical Weierstrass approximation theorem can be restated to assert: given a continuous function f on a compact segment of the real axis in $\mathbb{C}^1$, then f can be uniformly approximated by polynomials in $z = x + iy$. The general question we will deal with is as follows. Suppose M is a (local) hypersurface in $\mathbb{C}^n$. Given a continuous function F on M, can F be approximated on M by polynomials P_ℓ in $z_1, z_2, \ldots, z_n$?

Note that if $n > 1$, the restriction to M of each P_ℓ necessarily satisfies the tangential Cauchy-Riemann equations, and so F would necessarily have to satisfy these equations in at least some "weak" sense. We shall now see that this necessary condition is indeed sufficient. That is the thrust of the Baouendi-Treves approximation theorem stated below.

We suppose we are given a C^2 local hypersurface M in $\mathbb{C}^n$, defined near $z^0 \in M$, which after a complex affine-linear change of coordinates, the point z^0 has been brought to the origin and M is represented near z^0 as a graph

$$M = \{z = (z', z_n) : \operatorname{Im}(z_n) = \varphi(z', x_n)\}. \tag{23}$$

If we set $\rho(z) = \varphi(z', x_n) - y_n$, with $y_n = \operatorname{Im}(z_n)$, the tangential Cauchy-Riemann vector fields are spanned by

$$\rho_n \frac{\partial}{\partial \overline{z}_j} - \rho_j \frac{\partial}{\partial \overline{z}_n}, \qquad 1 \leq j \leq n-1,$$

with $\rho_j = \partial\rho/\partial\overline{z}_j$, and in particular $\rho_n = \frac{1}{2}(\varphi_{x_n} - i)$, where we define $\varphi_{x_n} = \partial\varphi/\partial x_n$. Thus we can write the corresponding tangential Cauchy-Riemann equations as

$$L_j(f) = 0, \qquad 1 \leq j \leq n-1,$$

with

$$L_j(f) = \frac{\partial f}{\partial \overline{z}_j} - a_j \frac{\partial f}{\partial \overline{z}_n}, \qquad \text{where } a_j = \rho_j/\rho_n. \tag{24}$$

In the coordinates (z', x_n) on M, these become $L_j(f) = \frac{\partial f}{\partial \overline{z}_j} - \frac{a_j}{2}\frac{\partial f}{\partial x_n}$.

Next, we define the transpose of L_j, namely, L_j^t, by

$$L_j^t(\psi) = -\left(\frac{\partial \psi}{\partial \overline{z}_j} - \frac{1}{2}\frac{\partial(a_j\psi)}{\partial x_n}\right),$$

so that

$$\int_{\mathbb{C}^{n-1}\times\mathbb{R}} L_j(f)\psi \, dz'\, dx_n = \int_{\mathbb{C}^{n-1}\times\mathbb{R}} f L_j^t(\psi)\, dz'\, dx_n$$

whenever both f and ψ are C^1 functions, with one of them having compact support. (We use the shorthand $dz'dx_n$ to designate Lebesgue measure on $\mathbb{C}^{n-1} \times \mathbb{R}$.) In view of the above we say that a continuous function f satisfies the tangential Cauchy-Riemann equations in the **weak sense** if

$$\int_{\mathbb{C}^{n-1}\times\mathbb{R}} f L_j^t(\psi)\, dz' dx = 0$$

for all ψ that are in C^1 and whose support is sufficiently small. Our theorem is then as follows:

Theorem 7.1 *Suppose $M \subset \mathbb{C}^n$ is a hypersurface of class C^2 as above. Given a point $z^0 \in M$, there are open balls B' and B, centered at z^0, with $\overline{B}' \subset B$, so that: if F is a continuous function in $M \cap B$ that satisfies the tangential Cauchy-Riemann equations in the weak sense, then F can be uniformly approximated on $M \cap \overline{B}'$ by polynomials in $z_1, z_2, \ldots, z_n$.*

Two remarks may help to clarify the nature of the conclusion asserted above.

- The theorem holds for all $n \geq 1$. In the case $n = 1$ there are of course no tangential Cauchy-Riemann equations so the conclusion is valid without further assumptions on F. Note however that in general the scope of this theorem must be local in nature. A simple illustration of this arises already when $n = 1$ and M is the boundary of the unit disc. See also Exercise 12.

- Note that for $n > 1$, there are no requirements on a Levi form related to M.

Proof. We shall first take B small enough so that in B, the hypersurface M has been represented by $M = \{y_n = \varphi(z', x_n)\}$ where z^0 corresponds to the origin. Besides $\varphi(0,0) = 0$, we can also suppose that the partial derivatives $\frac{\partial\varphi}{\partial x_j}$, $1 \leq j \leq n$, and $\frac{\partial\varphi}{\partial y_j}$, $1 \leq j \leq n-1$, vanish at the origin.

Now for each $u \in \mathbb{R}^{n-1}$, sufficiently close to the origin we define the slice M_u of M to be the n-dimensional sub-manifold given by

$$M_u = \{z : \ y_n = \varphi(z', x_n), \text{ with } z' = x' + iu\}.$$

We let $\Phi = \Phi^u$ be the mapping identifying the neighborhood of the origin $\mathbb{R}^n$ with M_u given by $\Phi(x) = (x' + iu, x_n + i\varphi(x' + iu, x_n))$ with $x = (x', x_n) \in \mathbb{R}^{n-1} \times \mathbb{R} = \mathbb{R}^n$. Observe that M is fibered by the collection $\{M_u\}_u$. Now for fixed u, the Jacobian of the mapping $x \mapsto \Phi(x)$, that is, $\frac{\partial\Phi}{\partial x}$, is the complex $n \times n$ matrix given by $I + A(x)$, where the entries of $A(x)$ are zero, except in the last row, and in that row we have the vector $\left(i\frac{\partial\varphi}{\partial x_1}, i\frac{\partial\varphi}{\partial x_2}, \ldots, i\frac{\partial\varphi}{\partial x_n}\right)$. So $A(0) = 0$, and $\det\left(\frac{\partial\Phi}{\partial x}\right) = 1 + i\frac{\partial\varphi}{\partial x_n}$. We shall need to shrink the ball B further so that $\|A(x)\| \leq 1/2$, on this ball, where $\|\cdot\|$ denotes the matrix-norm.

Now with u fixed, the map Φ carries the Lebesgue measure on $\mathbb{R}^n$ to a measure (with complex density) $dm_u(z) = \mathcal{J}(x)\,dx$ on M_u defined by

$$\int_{M_u} f(z)\,dm_u(z) = \int_{\mathbb{R}^n} f(\Phi(x))\mathcal{J}(x)\,dx, \qquad \text{where } \mathcal{J}(x) = \det\left(\tfrac{\partial\Phi(x)}{\partial x}\right)$$

for every continuous function f with sufficiently small support.

Next take B' any ball with the same center as B but strictly interior to it. Define χ to be a smooth (say C^1) cut-off function which is 1 on a neighborhood of B', and vanishes when $x \notin B$. With this, define for each $u \in \mathbb{R}^{n-1}$ (close to the origin), and $\epsilon > 0$, the function F_ϵ^u by

$$F_\epsilon^u(\zeta) = \frac{1}{\epsilon^{n/2}} \int_{M_u} e^{-\frac{\pi}{\epsilon}(z-\zeta)^2} F(z)\chi(z)\, dm_u(z). \tag{25}$$

Here we use the shorthand $w^2 = w_1^2 + \cdots + w_n^2$ if $w = (w_1, \ldots, w_n) \in \mathbb{C}^n$. We should remark at this point that, like the classical approximation theorem, the argument below comes down to the fact that the functions $\epsilon^{-n/2} e^{\frac{-\pi}{\epsilon}x^2}$ form an "approximation to the identity" in $\mathbb{R}^n$.[6]

The F_ϵ^u have the following three properties:

(i) Each $F_\epsilon^u(\zeta)$ is an entire function of $\zeta \in \mathbb{C}^n$.

(ii) Whenever $\zeta \in M_u$ and $\zeta \in \overline{B}'$, the $F_\epsilon^u(\zeta)$ converge uniformly to $F(\zeta)$, as $\epsilon \to 0$.

(iii) For each u, $\lim_{\epsilon\to 0} F_\epsilon^u(\zeta) - F_\epsilon^0(\zeta) = 0$, uniformly for $\zeta \in \overline{B}'$.

The first property is clear, since $e^{-\frac{\pi}{\epsilon}(z-\zeta)^2}$ is an entire function in ζ, and the integration in z is taken over a compact set.

For the second property note that $z \in M_u$, and $\zeta = \xi + i\eta \in M_u$, if $z = \Phi(x)$ and $\zeta = \Phi(\xi)$, with $\Phi = \Phi^u$. Therefore

$$\begin{aligned}(z-\zeta)^2 = (\Phi(x) - \Phi(\xi))^2 &= \left(\frac{\partial \Phi}{\partial \xi}(\xi)(x-\xi)\right)^2 + O(|x-\xi|^3)\\ &= ((I + A(\xi))(x-\xi))^2 + O(|x-\xi|^3).\end{aligned}$$

Now making our initial ball B smaller if necessary (which of course decreases the size of B'), we can guarantee that whenever z and ζ are in B

$$\mathrm{Re}(z-\zeta)^2 \geq c|x-\xi|^2, \qquad c > 0, \tag{26}$$

once we take into account that $\|A(\xi)\| \leq 1/2$. Thus the exponential appearing in (25) can be written as $e^{-\frac{\pi}{\epsilon}((1+A(\xi))(x-\xi))^2} + O\left(\frac{|x-\xi|^3}{\epsilon} e^{-\frac{c'|x-\xi|^2}{\epsilon}}\right)$. Thus $F_\epsilon^u(\zeta) = I + II$, with

$$I = \epsilon^{-n/2} \int_{\mathbb{R}^n} e^{-\frac{\pi}{\epsilon}((I+A(\xi))(x-\xi))^2} f(x)\, dx$$

[6]For the classical theorem, see for instance Theorem 1.13 in Chapter 5 of Book I.

and

$$II = O\left(\epsilon^{-n/2}\int_{\mathbb{R}^n}\frac{|v^3|}{\epsilon}e^{-c'|v|^2/\epsilon}\,dv\right),$$

with $f(x) = F(\Phi(x))\chi(\Phi(x))\det(I + A(x))$, where $\frac{\partial\Phi}{\partial x} = I + A(x)$. Now after a change of variables $v = x - \xi$, the first integral is handled by the following observation.

Lemma 7.2 *If A is an $n \times n$ complex matrix with constant coefficients and $\|A\| < 1$ then for every $\epsilon > 0$*

$$\text{(27)} \qquad \frac{1}{\epsilon^{n/2}}\det(I + A)\int_{\mathbb{R}^n} e^{-\frac{\pi}{\epsilon}((I+A)v)^2}\,dv = 1.$$

Corollary 7.3 *If f is a continuous function of compact support, then*

$$\frac{\det(I + A)}{\epsilon^{n/2}}\int_{\mathbb{R}^n} e^{-\frac{\pi}{\epsilon}((I+A)v)^2} f(\xi + v)\,dv \to f(\xi)$$

uniformly in ξ as $\epsilon \to 0$.

To prove the lemma note that $\mathrm{Re}(((I + A)v)^2) \geq |v|^2 - \|A\||v|^2 \geq c|v|^2$, with $c > 0$, so that the integral in (27) converges. A change of scale reduces the identity to the case $\epsilon = 1$. Now if A is real, a further change of variables $v' = (I + A)v$ (which is invertible since $\|A\| < 1$) reduces this case to the standard Gaussian integral. Finally, we pass to the general situation by analytic continuation, noting that the left-hand side of (27) is holomorphic in the entries of A, whenever $\|A\| < 1$. The corollary then follows from the usual arguments about approximations of the identity as in Section 4, Chapter 2 in Book I and Section 2 in Chapter 3 of Book III.

Now the term II is dominated by a multiple of $\int_{\mathbb{R}^n}\epsilon^{1/2}|v|^3 e^{-c'|v|^2}\,dv = c\epsilon^{1/2}$, as is seen by a change of scale. Thus property (ii) is proved.

Up to this point, we have not used the fact that F satisfies the tangential Cauchy-Riemann equations. It is in the proof of property (iii) that this is crucial. We begin by considering the case where F is assumed to be in class C^1. Later we will see how to lift this restriction. We recall that the tangential Cauchy-Riemann vector field L_j is given by (24).

Lemma 7.4 *Suppose f is a C^1 function on M. Then*

$$\text{(28)} \qquad \frac{\partial}{\partial u_j}\left(\int_{M_u} f(z)\,dm_u(z)\right) = \frac{2}{i}\int_{M_u} L_j(f)\,dm_u(z),$$

for all $1 \leq j \leq n - 1$.

Proof. Recall that $\Phi(x) = \Phi^u(x) = (x' + iu, x_n + i\varphi(x' + iu, x_n))$ and from before we have that $\det\left(\frac{\partial\Phi}{\partial x}\right) = 1 + i\varphi_{x_n}$. Also, recall that $\rho(z) = \varphi(z', x_n) - y_n$, hence, for $1 \le j \le n-1$, one has

$$L_j = \frac{\partial}{\partial \overline{z}_j} - \frac{\rho_j}{\rho_n}\frac{\partial}{\partial \overline{z}_n} = \frac{\partial}{\partial \overline{z}_j} + \frac{2}{i}\frac{\frac{\partial \varphi}{\partial \overline{z}_j}}{(1 + i\varphi_{x_n})}\frac{\partial}{\partial \overline{z}_n},$$

and therefore

$$\begin{aligned}\frac{2}{i}\int_{M_u} L_j(f)\,dm_u(z) &= \frac{2}{i}\int_{\mathbb{R}^n}(L_j f)(\Phi)(1 + i\varphi_{x_n})\\ &= \frac{2}{i}\int_{\mathbb{R}^n}\frac{\partial f}{\partial \overline{z}_j}(1 + i\varphi_{x_n}) - 4\int_{\mathbb{R}^n}\frac{\partial \varphi}{\partial \overline{z}_j}\frac{\partial f}{\partial \overline{z}_n},\end{aligned}$$

where we simplify the writing by sometimes omitting Φ from the formulas. Now, starting from the left-hand side of (28)

$$\begin{aligned}\frac{\partial}{\partial u_j}\left(\int_{M_u} f(z)\,dm_u(z)\right) &= \frac{\partial}{\partial u_j}\left(\int_{\mathbb{R}^n} f(\Phi)(1 + i\varphi_{x_n})\right)\\ &= \int\left(\frac{\partial f}{\partial u_j} + \varphi_{u_j}\frac{\partial f}{\partial y_n}\right)(1 + i\varphi_{x_n}) - i\int\varphi_{u_j}\left(\frac{\partial f}{\partial x_n} + \varphi_{x_n}\frac{\partial f}{\partial y_n}\right),\end{aligned}$$

where we have used an integration by parts and the fact that f has compact support to obtain the second integral on the right-hand side. Using the fact that f has compact support again, we also note that

$$\begin{aligned}0 &= \int_{\mathbb{R}^n}\frac{\partial}{\partial x_j}[f(\Phi)(1 + i\varphi_{x_n})]\\ &= \int_{\mathbb{R}^n}\left(\frac{\partial f}{\partial x_j} + \varphi_{x_j}\frac{\partial f}{\partial y_n}\right)(1 + i\varphi_{x_n}) - i\int_{\mathbb{R}^n}\varphi_{x_j}\left(\frac{\partial f}{\partial x_n} + \varphi_{x_n}\frac{\partial f}{\partial y_n}\right),\end{aligned}$$

where once again we have integrated by parts to obtain the last integral. Combining the two results above we find that

$$\begin{aligned}&\frac{\partial}{\partial u_j}\left(\int_{M_u} f(z)\,dm_u(z)\right) =\\ &\quad= -2i\int_{\mathbb{R}^n}\left(\frac{\partial f}{\partial \overline{z}_j} + \frac{\partial \varphi}{\partial \overline{z}_j}\frac{\partial f}{\partial y_n}\right)(1 + i\varphi_{x_n})\\ &\qquad\qquad - 2\int_{\mathbb{R}^n}\frac{\partial \varphi}{\partial \overline{z}_j}\left(\frac{\partial f}{\partial x_n} + \varphi_{x_n}\frac{\partial f}{\partial y_n}\right)\\ &\quad= \frac{2}{i}\int_{\mathbb{R}^n}\frac{\partial f}{\partial \overline{z}_j}(1 + i\varphi_{x_n}) - 4\int_{\mathbb{R}^n}\frac{\partial \varphi}{\partial \overline{z}_j}\frac{\partial f}{\partial \overline{z}_n}\\ &\quad= \frac{2}{i}\int_{M_u} L_j(f)\,dm_u(z),\end{aligned}$$

which is (28).

Now set $f(z) = \epsilon^{-n/2} e^{-\frac{\pi}{\epsilon}(z-\zeta)^2} F(z)\chi(z)$. Then

$$\begin{aligned} F_\epsilon^u - F_\epsilon^0 &= \int_0^1 \frac{\partial}{\partial s} F_\epsilon^{us}\, ds \\ &= \int_0^1 \sum_{j=1}^n u_j \frac{\partial}{\partial(u_j s)} \left(\int_{M_{us}} f(z)\, dm_{us}(z) \right) ds \\ &= \int_0^1 \sum_{j=1}^n u_j \frac{2}{i} \left(\int_{M_{us}} L_j(f)\, dm_{us}(z) \right) ds, \end{aligned}$$

because of Lemma 7.4. Now $L_j(f) = \epsilon^{-n/2} e^{-\pi(z-\zeta)^2/\epsilon} F L_j(\chi)$, since $e^{-(z-\zeta)^2/\epsilon}$ is holomorphic in z, and $L_j(F) = 0$ by assumption. However $L_j(\chi)$ is supported at a positive distance from B'. So if $\zeta \in B'$, the inequality (26) guarantees that

$$|F_\epsilon^u - F_\epsilon^0| = O(\epsilon^{-n/2} e^{-c'/\epsilon}) \qquad \text{as } \epsilon \to 0$$

for some $c' > 0$, and the property (iii) is established, under the assumption that $F \in C^1$.

To complete the proof of the theorem note that a combination of (ii) and (iii) shows that F_ϵ^0 converges uniformly to F when $\zeta \in M \cap \overline{B}'$. Now each F_ϵ^0, being an entire function of ζ, can be uniformly approximated by polynomials in ζ for ζ in the compact set $\overline{B}'$. Altogether then, F can be uniformly approximated by polynomials on $M \cap \overline{B}'$ and the theorem is proved in that case.

To pass to the general case note that what we have shown in (28) is that when f is of class C^1, $u = (0, \dots, 0, u_j, 0, \dots, 0)$, and $v = (0, \dots, 0, v_j, 0, \dots, 0)$ then

$$F_\epsilon^u - F_\epsilon^v = \frac{2}{i} \int_{v_j}^{u_j} \int_{\mathbb{R}^n} L_j(f) \mathcal{J}(x)\, dx\, dy_j. \tag{29}$$

To extend (29) to the case where f is merely continuous, and $L_j(f)$ (taken in the sense of distributions) is also continuous, a limiting argument with (29), as it stands, will not suffice. This is because the "weak" definition of $L_j(f)$ requires an integration over $\mathbb{R}^n \times \mathbb{R}^{n-1}$, while in (29) we only integrate over $\mathbb{R}^n \times \mathbb{R}$. To get around this we observe first (still assuming $f \in C^1$ and has compact support) that (28) implies

$$(30)\quad -\int_{\mathbb{R}^n\times\mathbb{R}^{n-1}} f(\Phi^{y'}(x))\frac{\partial\psi}{\partial y_j}(y')\mathcal{J}(x)\,dx\,dy' = \\ = \frac{2}{i}\int_{\mathbb{R}^n\times\mathbb{R}^{n-1}} f(\Phi^{y'}(x))L_j^t[\psi(y')\mathcal{J}(x)]\,dx\,dy'$$

for any C^1 function ψ on $\mathbb{R}^{n-1}$ having compact support. Now at this stage we can pass to an arbitrary continuous f of compact support (by approximating such f uniformly by C^1 functions) and see that (30) holds for f that are merely continuous and of compact support.

As a result we have that

$$(31)\quad -\int_{\mathbb{R}^n\times\mathbb{R}^{n-1}} f(\Phi^{y'}(x))\frac{\partial\psi}{\partial y_j}(y')\mathcal{J}(x)\,dx\,dy' = \\ = \frac{2}{i}\int_{\mathbb{R}^n\times\mathbb{R}^{n-1}} L_j(f)\psi(y')\mathcal{J}(x)\,dx\,dy',$$

where $L_j(f)$ is taken in the sense of distributions (assuming that $L_j(f)$ is continuous).

Now set $\psi(y') = \psi_\delta(y_j)\tilde{\psi}_\delta(\tilde{y})$, where $\tilde{y}$ is defined by $\tilde{y} = (y_1, \ldots, y_{j-1}, 0, y_{j+1}, \ldots, y_{n-1})$. Here $\psi_\delta(y_j) = 1$ if $v_j \le y_j \le u_j$, and vanishes if $y_j \le v_j - \delta$ or $y_j \ge u_j + \delta$; in addition $\left|\frac{\partial\psi_\delta(y_j)}{\partial y_j}\right| \le c\delta^{-1}$. As a result note that for any continuous function g

$$-\int g(y_j)\frac{\partial\psi_\delta}{\partial y_j}\,dy_j = g(u_j) - g(v_j) \quad \text{as } \delta \to 0,$$

since $\frac{\partial\psi_\delta}{\partial y_j}$ is the difference of two approximations to the identity centered at u_j and v_j, respectively.

Also $\tilde{\psi}_\delta(\tilde{y}) = \delta^{-n+2}\tilde{\psi}(\tilde{y}/\delta)$, where $\int_{\mathbb{R}^{n-2}} \tilde{\psi}(\tilde{y})\,d\tilde{y} = 1$, making $\{\tilde{\psi}_\delta\}$ an approximation to the identity in $\mathbb{R}^{n-2}$. Inserting these in (31) and letting $\delta \to 0$ shows that the left-hand side of (31) converges to $F_\epsilon^u - F_\epsilon^v$, while the right-hand side converges to $\frac{2}{i}\int_{v_j}^{u_j}\int_{\mathbb{R}^n} L_j(f)\,dx\,dy_j$, and (29) is proved. The rest of the argument then continues as before, and the proof of the theorem is now complete.

The approximation theorem just proved, together with the maximum principle in Section 6 lead directly to the famous Lewy extension theorem. Here again M is a C^2 hypersurface given in a ball B, with $M = \{z \in B,\ \rho(z) = 0\}$. As before we set $\Omega_- = \{x \in B,\ \rho(z) < 0\}$.

Theorem 7.5 *Suppose that the Levi form* (18) *has at least one strictly positive eigenvalue for each* $z \in M$. *Then for each* $z^0 \in M$, *there is a ball* B' *centered at* z^0 *so that whenever* F_0 *is a continuous function on* M *that satisfies the tangential Cauchy-Riemann equations in the weak sense, there exists an* F *which is holomorphic in* $\Omega_- \cap B'$, *continuous in* $\overline{\Omega}_- \cap B'$ *and so that* $F(z) = F_0(z)$ *for* $z \in M \cap B'$.

To prove the theorem we first use Theorem 7.1 to find a ball B_1 centered at z_0 so that F_0 can be uniformly approximated (on $M \cap B_1$) by polynomials $\{p_n(z)\}$. Then we invoke the corollary to Theorem 6.1 to find a ball B' so that (22) holds (with B_1 in place of B). Therefore the p_n also converge uniformly in $\Omega_- \cap B'$. The limit of this sequence, F, is then holomorphic there, continuous in $\overline{\Omega}_- \cap B'$, and gives the desired extension of F_0.

8 Appendix: The upper half-space

In this appendix we want to illustrate some of the concepts discussed in the present chapter, as viewed in terms of a special model region. We will only sketch the proofs of the results, leaving the details to the interested reader, and providing some further relevant ideas in Exercises 17 to 19.

The region we have in mind is the **upper half-space** $\mathcal{U}$ in $\mathbb{C}^n$ given by

$$\mathcal{U} = \{z \in \mathbb{C}^n : \ \text{Im}(z_n) > |z'|^2\},$$

and its boundary

$$\partial\mathcal{U} = \{z \in \mathbb{C}^n, \ \text{Im}(z_n) = |z'|^2\}, \tag{32}$$

with $z = (z', z_n)$, and $z' = (z_1, \ldots, z_{n-1})$. It is prompted by the canonical form (16). The region $\mathcal{U}$ in $\mathbb{C}^n$, $n > 1$, plays a role similar to the upper half-plane in $\mathbb{C}^1$. The definitions suggest that z_n can be thought of as the "classical" variable, while z' is the "new" variable that comes about when $n > 1$. As in the case $n = 1$, the region $\mathcal{U}$ is holomorphically equivalent with the unit ball $\{w \in \mathbb{C}^n : \ |w| < 1\}$ via a fractional linear transformation, namely

$$w_n = \frac{i - z_n}{i + z_n} \qquad w_k = \frac{2iz_k}{i + z_n}, \qquad k = 1, \ldots, n-1,$$

as the reader may easily verify.

This mapping also extends to a correspondence of the boundaries, except that the "south-pole" of the unit ball $(0, \ldots, 0, -1)$ corresponds to the point at infinity of $\partial\mathcal{U}$. The analysis of the region $\mathcal{U}$ is enriched by a number of symmetries it enjoys.

The boundary of $\mathcal{U}$, which by (32) is parametrized by $(z', x_n) \in \mathbb{C}^{n-1} \times \mathbb{R}$, carries a natural measure $d\beta = dm(z', x_n)$, with the latter being Lebesgue measure on $\mathbb{C}^{n-1} \times \mathbb{R}$. More precisely, if F_0 is a function on $\partial\mathcal{U}$, and $F_0^\sharp$ designates the corresponding function on $\mathbb{C}^{n-1} \times \mathbb{R}$,

$$F_0(z', x_n + i|z'|^2) = F_0^\sharp(z', x_n),$$

then by definition

$$\int_{\partial\mathcal{U}} F_0 \, d\beta = \int_{\mathbb{C}^{n-1}\times\mathbb{R}} F_0^\sharp \, dm.$$

8.1 Hardy space

In analogy with $\mathbb{C}^1$, we consider the **Hardy space** $H^2(\mathcal{U})$, which consists of all functions F, holomorphic in $\mathcal{U}$, that satisfy

$$\sup_{\epsilon>0} \int_{\partial\mathcal{U}} |F(z', z_n + i\epsilon)|^2 \, d\beta < \infty.$$

For those F the number $\|F\|_{H^2(\mathcal{U})}$ is defined as the square root of the above supremum. It will be convenient to abbreviate $F(z', z_n + i\epsilon)$ by $F_\epsilon(z)$, and sometimes also use the same symbol for the restriction of F_ϵ to $\partial\mathcal{U}$.

Theorem 8.1 *Suppose $F \in H^2(\mathcal{U})$. Then, when restricted to $z \in \partial\mathcal{U}$, the limit*

$$\lim_{\epsilon\to 0} F_\epsilon = F_0$$

exists in the $L^2(\partial\mathcal{U}, d\beta)$ norm. Also

$$\|F\|_{H^2(\mathcal{U})} = \|F_0\|_{L^2(\partial\mathcal{U})}.$$

For several arguments below we use the following observation.

Lemma 8.2 *Suppose B_1 and B_2 are two open balls in $\mathbb{C}^{n-1}$, with $\overline{B}_1 \subset B_2$. Then, whenever f is holomorphic in $\mathbb{C}^{n-1}$*

$$\sup_{z'\in B_1} |f(z')|^2 \le c \int_{B_2} |f(w')|^2 \, dm(w').$$

Indeed for sufficiently small δ, whenever $z' \in B_1$ then $B_\delta(z') \subset B_2$, so since f is harmonic in $\mathbb{R}^{2n-2}$, the mean-value property and the Cauchy-Schwarz inequality gives

$$|f(z')|^2 \le \frac{1}{m(B_\delta)} \int_{B_\delta(z')} |f(w')|^2 \, dm(w'),$$

proving the claim.

The proof of the theorem can be given by the Fourier transform representation of each $F \in H^2(\mathcal{U})$ in analogy with the case $n = 1$ treated in Chapter 5 of Book III.

We define the space $\mathcal{H}$ of functions $f(z', \lambda)$, with $(z', \lambda) \in \mathbb{C}^{n-1} \times \mathbb{R}^+$, that are jointly measurable, holomorphic in $z' \in \mathbb{C}^{n-1}$ for almost every λ, and for which

$$\|f\|_{\mathcal{H}}^2 = \int_0^\infty \int_{\mathbb{C}^{n-1}} |f(z', \lambda)|^2 e^{-4\pi\lambda|z'|^2} \, dm(z') \, d\lambda < \infty.$$

One can show that with this norm the space $\mathcal{H}$ is complete and hence a Hilbert space (see Exercises 18 and 19). With this, every $F \in H^2(\mathcal{U})$ can be represented as

$$F(z', z_n) = \int_0^\infty f(z', \lambda) e^{2\pi i \lambda z_n} \, d\lambda, \qquad \text{with } f \in \mathcal{H}. \tag{33}$$

Proposition 8.3 *If $f \in \mathcal{H}$, then the integral in* (33) *converges absolutely and uniformly for (z', z_n) lying in compact subsets of $\mathcal{U}$, and $F \in H^2(\mathcal{U})$. Conversely any $F \in H^2(\mathcal{U})$ can be written as* (33) *for some $f \in \mathcal{H}$.*

In fact if (z', z_n) belongs to a compact subset of $\mathcal{U}$, we may suppose that $\mathrm{Im}(z_n) > |z'|^2 + \epsilon$, for some $\epsilon > 0$. We will also restrict z' to range in a ball B_1, with $\overline{B}_1 \subset B_2$, and take the radius B_2 so small that $\mathrm{Im}(w_n) > |w'|^2 + \epsilon/2$, if $w' \in B_2$.

Now by the Cauchy-Schwarz inequality the absolute value of the integral in (33) is estimated by

$$\left(\int_0^\infty |f(z', \lambda)|^2 e^{-4\pi\lambda(y_n - \epsilon/2)} \, d\lambda \right)^{1/2} \left(\int_0^\infty e^{-4\pi\lambda\epsilon/2} \, d\lambda \right)^{1/2}.$$

Invoking the lemma we get as an estimate for this

$$c \left(\int_0^\infty \int_{\mathbb{C}^{n-1}} |f(w', \lambda)|^2 e^{-4\pi\lambda|w'|^2} \, dm(w') \, d\lambda \right)^{1/2} c' \epsilon^{-1/2} = c'' \epsilon^{-1/2} \|f\|_{\mathcal{H}}.$$

This shows that the integral converges absolutely and uniformly when $z' \in B_1$ and $\mathrm{Im}(z_n) > |z'|^2 + \epsilon$, and thus uniformly on any compact subset of $\mathcal{U}$. Thus F is holomorphic in $\mathcal{U}$. Observe next that for F given by (33), $F_\epsilon(z) = F(z', z_n + i\epsilon)$ is given in terms of f_ϵ, with $f_\epsilon(z', \lambda) = f(z', \lambda) e^{-2\pi\lambda\epsilon}$. Now for fixed z', Plancherel's theorem in the x_n variable shows that

$$\int_{\mathbb{R}} |F_\epsilon(z', x_n + i|z'|^2)|^2 \, dx_n = \int_0^\infty |f_\epsilon(z', \lambda) e^{-2\pi\lambda|z'|^2}|^2 \, d\lambda.$$

Integrating in z' gives

$$\int_{\partial\mathcal{U}} |F_\epsilon|^2 \, d\beta = \|f_\epsilon\|_{\mathcal{H}}^2 \leq \|f\|_{\mathcal{H}}^2.$$

By the same token, $\int_{\partial\mathcal{U}} |F_\epsilon - F_{\epsilon'}|^2 \, d\beta = \|f_\epsilon - f_{\epsilon'}\|_{\mathcal{H}}^2 \to 0$ as $\epsilon, \epsilon' \to 0$. Thus F_ϵ converges in $L^2(\partial\mathcal{U}, d\beta)$ to a limit F_0 given by (33) with $y_n = |z'|^2$. Moreover

$$\|F_0\|_{L^2(\partial\mathcal{U})} = \|F\|_{H^2(\mathcal{U})} = \|f\|_{\mathcal{H}}. \tag{34}$$

Conversely, suppose $F \in H^2(\mathcal{U})$. One observes that whenever z' is restricted to a compact subset of $\mathbb{C}^{n-1}$,

$$|F(z', z_n + i\epsilon)| \leq \frac{c}{\epsilon^{1/2}} \|F\|_{H^2}.$$

(Here we use Lemma 8.2 and also follow the reasoning used in the case $n = 1$ to study $H^2(\mathbb{R}^2_+)$ in Section 2 of Chapter 5 in Book III.) We set $F^\delta_\epsilon(z) = F(z', z_n + i\epsilon)(1 - i\delta z_n)^{-2}$. Then for each z', the function $F^\delta_\epsilon(z', z_n)$ is in H^2 of the half-space $\{\text{Im}(z_n) > |z'|^2\}$. So we may define $f^\delta_\epsilon(z', \lambda)$ by

$$f^\delta_\epsilon(z', \lambda) = \int_{\mathbb{R}} e^{-2\pi i\lambda(x_n + iy_n)} F^\delta_\epsilon(z', z_n)\, dx_n,$$

noting that the right-hand side is independent of y_n, if $y_n > |z'|^2$, by Cauchy's theorem. Also then F^δ_ϵ is represented by (33) with f^δ_ϵ in place of f and $f^\delta_\epsilon \in \mathcal{H}$.

Now letting $\delta \to 0$ and using (34) we see that $F_\epsilon(z)$ is given by (33), with $f_\epsilon = f^\delta_\epsilon|_{\delta=0}$ in place of f, and that $f_\epsilon \in \mathcal{H}$. Finally, since $F_\epsilon(z) = F(z', z_n + i\epsilon)$, we have that $f_\epsilon(z', \lambda) = f(z', \lambda)e^{-2\pi\lambda\epsilon}$, and using (34) again with $\epsilon \to 0$ gives us the representation (33) for our given $F \in H^2(\mathcal{U})$. The theorem is thus proved.

Remark. By the completeness of $\mathcal{H}$ given in Exercise 19 we see that $H^2(\mathcal{U})$ is also a Hilbert space.

We now ask:

Which $F_0 \in L^2(\partial\mathcal{U})$ arise as $\lim_{\epsilon\to 0} F_\epsilon$ for $F \in H^2(\mathcal{U})$?

When $n > 1$ the tangential Cauchy-Riemann operators provide the answer. If F_0 is given on $\partial\mathcal{U}$, recall that $F^\sharp_0(z', x_n) = F_0(z', x_n + i|z'|^2)$ is the corresponding function on $\mathbb{C}^{n-1} \times \mathbb{R}$. In this setting the vector fields L_j, given by

$$L_j = \frac{\partial}{\partial \overline{z}_j} - iz_j \frac{\partial}{\partial x_n}, \qquad j = 1, \ldots, n-1,$$

form a basis for the tangential Cauchy-Riemann vector fields, as is given by (24), with $\rho(z) = |z'|^2 - \text{Im}(z_n)$. Note that in this case $L^t_j = -L_j$. So here a function $G \in L^2(\mathbb{C}^{n-1} \times \mathbb{R})$ satisfies the tangential Cauchy-Riemann equations $L_j(G) = 0$, $j = 1, \ldots, n-1$, in the weak sense, if

$$(35) \qquad \int_{\mathbb{C}^{n-1}\times\mathbb{R}} G(z', x_n) L^t_j(\psi)(z', x_n)\, dm(z', x_n) = 0, \qquad 1 \leq j \leq n-1,$$

for all ψ that are (say) C^∞ and have compact support.

Proposition 8.4 *An F_0 in $L^2(\partial\mathcal{U})$ arises from an $F \in H^2(\mathcal{U})$ as in Theorem 8.1 if and only if $F^\sharp_0$ satisfies the tangential Cauchy-Riemann equations in the weak sense.*

Proof. First, assume that $F \in H^2(\mathcal{U})$. Then since F_ϵ is holomorphic in a neighborhood of $\overline{\mathcal{U}}$, the function $F^\sharp_\epsilon$ satisfies $L_j(F^\sharp_\epsilon) = 0$ in the usual sense. The fact that $F_\epsilon \to F_0$ in the $L^2(\partial\mathcal{U})$ norm (which is the same as $F_\epsilon^\sharp \to F_0^\sharp$ in $L^2(\mathbb{C}^{n-1} \times \mathbb{R})$) then implies that $F_0^\sharp$ satisfies (35) with $G = F_0^\sharp$.

Conversely, suppose G is in $L^2(\mathbb{C}^{n-1} \times \mathbb{R})$, and set

$$g(z', \lambda) = \int_{\mathbb{R}} e^{-2\pi i \lambda x_n} G(z', x_n)\, dx_n. \tag{36}$$

Also choose $\psi(z', x_n) = \psi_1(z')\psi_2(x_n)$. Then by Plancherel's theorem in the x_n variable,

$$\int_{\mathbb{R}} G(z', x_n) \frac{\partial \psi_2}{\partial x_n}(x_n)\, dx_n = -\int_{\mathbb{R}} g(z', \lambda) 2\pi i \lambda \hat{\psi}_2(-\lambda)\, d\lambda$$

for almost every z'. Integrating in z' then shows that

$$\int_{\mathbb{C}^{n-1}\times\mathbb{R}} G(z', x_n) L_j^t(\psi(z', x_n))\, dm(z', x_n) =$$

$$= -\int_{\mathbb{C}^{n-1}} \int_{\mathbb{R}} g(z', \lambda) \left(\frac{\partial \psi_1}{\partial \overline{z}_j}(z') - 2\pi\lambda z_j \psi_1(z') \right) \hat{\psi}_2(-\lambda) d\lambda\, dm(z').$$

So if G satisfies (35) it follows that

$$\int_{\mathbb{C}^{n-1}} g(z', \lambda) \left(\frac{\partial \psi_1}{\partial \overline{z}_j}(z') - 2\pi\lambda z_j \psi_1(z') \right) dm(z') = 0$$

for almost every λ, and this means that

$$\int_{\mathbb{C}^{n-1}} f(z', \lambda) \frac{\partial(\psi_1(z') e^{-2\pi |z'|^2 \lambda})}{\partial \overline{z}_j}(z')\, dm(z') = 0,$$

where $f(z', \lambda) = g(z', \lambda) e^{2\pi\lambda |z'|^2}$, which itself implies that $f(z', \lambda)$ satisfies the Cauchy-Riemann equations in $\mathbb{C}^{n-1}$ in the weak sense, for almost every λ. But we saw in Section 1 that this shows that the functions $f(z', \lambda)$ are holomorphic in z'. Now (36) and the Fourier inversion formula shows that

$$\int_{\mathbb{R}} \int_{\mathbb{C}^{n-1}} |g(z', \lambda)|^2\, dm(z')\, d\lambda = \int_{\mathbb{R}} \int_{\mathbb{C}^{n-1}} |f(z', \lambda)|^2 e^{-4\pi\lambda |z'|^2}\, dm(z')\, d\lambda$$

are both finite. Also, with F given by (33), we have $G(z', x_n) = F(z', x_n + i|z'|^2)$. Finally, because $\int_{\mathbb{C}^{n-1}} |f(z', \lambda)|^2 e^{-4\pi\lambda|z'|^2}\, dm(z') < \infty$ for almost every λ, then necessarily $f(z', \lambda) = 0$ for those λ that are negative. Thus we have given G as $F_0^\sharp$, with F as in (33), and $f \in \mathcal{H}$. The proposition is therefore proved.

8.2 Cauchy integral

The Cauchy integral[7] in $\mathcal{U}$ can be defined as follows. For each $z, w \in \mathbb{C}^n$ we set

$$r(z, w) = \frac{i}{2}(\overline{w}_n - z_n) - z' \cdot \overline{w}'$$

---

[7] Also referred to as the Cauchy-Szegö integral.

with $z = (z', z_n)$, $w = (w', w_n)$ and

$$z' \cdot \overline{w}' = z_1\overline{w}_1 + \cdots + z_{n-1}\overline{w}_{n-1}.$$

Note that $r(z, w)$ is holomorphic in z, conjugate holomorphic in w, and $r(z, z) = \text{Im}(z_n) - |z'|^2 = -\rho(z)$, with ρ the defining function for $\mathcal{U}$ used earlier.

Next, we define

$$S(z, w) = c_n r(z, w)^{-n}, \qquad \text{where } c_n = \tfrac{(n-1)!}{(4\pi)^n}.$$

Observe that $S(z, w) = \overline{S(w, z)}$, and that for each $w \in \mathcal{U}$, the function $z \mapsto S(z, w)$ is in $H^2(\mathcal{U})$. Also for each $z \in \mathcal{U}$, the function $w \mapsto S(z, w)$ is in $L^2(\partial\mathcal{U})$. We define the **Cauchy integral** $C(f)$ of a function f on $\mathcal{U}$ by

(37) $$C(f)(z) = \int_{\partial\mathcal{U}} S(z, w) f(w)\, d\beta(w), \qquad z \in \mathcal{U}.$$

The reproducing property of C is what interests us here.

Theorem 8.5 *Suppose $F \in H^2(\mathcal{U})$, and let $F_0 = \lim_{\epsilon \to 0} F_\epsilon$ as in Theorem 8.1. Then*

(38) $$C(F_0)(z) = F(z).$$

The key lemma used is an observation giving a reproducing identity for a related space of entire functions on $\mathbb{C}^{n-1}$. We consider the holomorphic functions f on $\mathbb{C}^{n-1}$ for which

$$\int_{\mathbb{C}^{n-1}} |f(z')|^2 e^{-4\pi\lambda|z'|^2}\, dm(z') < \infty,$$

where $\lambda > 0$ is fixed.

Lemma 8.6 *For f as above, we have*

(39) $$f(z') = \int_{\mathbb{C}^{n-1}} K_\lambda(z', w') f(w') e^{-4\pi\lambda|w'|^2}\, dm(w')$$

with $K_\lambda(z', w') = (4\lambda)^{n-1} e^{4\pi\lambda z' \cdot \overline{w}'}$.

Proof. In fact, consider first the case when $4\lambda = 1$, and $z' = 0$. Then (39), which states $f(0) = \int_{\mathbb{C}^{n-1}} f(w') e^{-\pi|w'|^2}\, dm(w')$, is a simple consequence of the mean-value property of f (taken on spheres in $\mathbb{C}^{n-1}$ centered at the origin) and the fact that $\int_{\mathbb{C}^{n-1}} e^{-\pi|z'|^2}\, dm(z') = 1$.

We now apply this identity to $w' \mapsto f(z' + w') e^{-\pi\overline{z}' \cdot w'}$ for fixed z'. The result is then (39) when $4\lambda = 1$. A simple rescaling argument then gives (39) in general.

Turning to the proof of the theorem, we observe that

$$S(z, w) = \int_0^\infty \lambda^{n-1} e^{-4\pi\lambda r(z,w)}\, d\lambda,$$

since $\int_0^\infty \lambda^{n-1}e^{-A\lambda}\,d\lambda = (n-1)!A^{-n}$, whenever $\mathrm{Re}(A) > 0$. So, at least formally,

$$\int_{\partial\mathcal{U}} S(z,w)F_0(w)\,d\beta = \\ = \int_0^\infty \int_{\partial\mathcal{U}} F_0(w', u_n + i|w'|^2)\lambda^{n-1}e^{-4\pi\lambda r(z,w)}\,dm(w', u_n)\,d\lambda.$$

But as we have seen

$$\int_{\mathbb{R}} F_0(w', u_n + iv_n)e^{-2\pi i\lambda(u_n+iv_n)}\,du_n = f(w',\lambda).$$

Now insert this in the above, recalling that $r(z,w) = -\frac{\overline{w}_n - z_n}{2i} - z'\cdot\overline{w}'$, and that $(4\lambda)^{n-1}\int_{\mathbb{C}^{n-1}} f(w',\lambda)e^{-4\pi\lambda|w'|^2}\,dm(w') = f(z',\lambda)$. The result is that

$$\int_{\partial\mathcal{U}} S(z,w)F_0(w)\,d\beta(w) = \int_0^\infty f(z',\lambda)e^{2\pi i\lambda z_n}\,d\lambda,$$

which by (33) is what we want to obtain.

To make this argument rigorous, we proceed as in the proof of Theorem 8.1, with the improved function F_ϵ^δ in place of F. Then all the integrals in question converge absolutely, and therefore the interchanges of integration are justified. This gives the reproducing property (38) for F_ϵ^δ instead of F. Then we let $\delta \to 0$, and next $\epsilon \to 0$, giving (38) for any $F \in H^2(\mathcal{U})$.

8.3 Non-solvability

We will use the Cauchy integral C to illuminate a basic example of Lewy of a non-solvable partial differential equation.

Here we look at $\mathcal{U}$ in $\mathbb{C}^2$, with its boundary parametrized by $\mathbb{C}\times\mathbb{R}$. We consider the tangential Cauchy-Riemann vector field $L = L_1 = \frac{\partial}{\partial \overline{z}_1} - iz_1\frac{\partial}{\partial x_2}$, and show that in order for $L(U) = f$ to be even locally solvable, the function f must satisfy a stringent necessary condition. For purposes of the statement of the result, it will be more convenient to deal with

$$\overline{L} = \frac{\partial}{\partial z_1} + i\overline{z}_1\frac{\partial}{\partial x_2}$$

instead of L. (To revert back to L then one needs only to replace f by its conjugate.)

We consider the Cauchy integral (37), written now as acting on functions on $\mathbb{C}\times\mathbb{R}$, identified with $\partial\mathcal{U}$ in $\mathbb{C}^2$. If f is such a function then (37) takes the form

$$\int_{\mathbb{C}\times\mathbb{R}} S(z, u_2 + i|w_1|^2)f(w_1,u_2)\,dm(w_1,u_2). \tag{40}$$

We can extend (40) to define the Cauchy integral when f is a distribution (say of compact support), by setting

$$C(f)(z) = \langle f, S(z, u_2 + i|w_1|^2)\rangle, \qquad z \in \mathcal{U}.$$

Here $\langle\cdot,\cdot\rangle$ is a pairing between the distribution f and the C^∞ function $(w_1, u_2) \mapsto S(z, u_2 + i|w_1|^2)$, with z fixed. The necessary condition is then:

(41) $\quad C(f)(z)$ *has an analytic continuation to a neighborhood of* 0.

Note that this property depends only on the behavior of f near the origin. Indeed, if f_1 agrees with f near the origin, then $C(f - f_1)$ is automatically holomorphic near the origin, because visibly $S(z, w)$ is holomorphic for z in a small neighborhood of the origin, with w staying outside a given neighborhood of the origin in $\mathbb{C}^n$.

Theorem 8.7 *Suppose U is a distribution defined on $\mathbb{C} \times \mathbb{R}$, so that $\overline{L}(U) = f$ in a neighborhood of the origin. Then* (41) *must hold.*

Proof. Assume first that U has compact support, and $\overline{L}(U) = f$ everywhere. Then

$$\begin{aligned} C(f)(z) = \langle f, S(z, u_2 + i|w_1|^2)\rangle &= \langle \overline{L}(U), S(z, u_2 + i|w_1|^2)\rangle \\ &= -\langle U, \overline{L}(S(z, u_2 + i|w_1|^2))\rangle \\ &= 0, \end{aligned}$$

since $\overline{L}(S(z, u_2 + i|w_1|^2)) = 0$, because $w \mapsto S(z, w)$ is conjugate holomorphic. Thus trivially $C(f)(z)$ is holomorphic everywhere.

If U does not have compact support and $\overline{L}(U) = f$ only in a neighborhood of the origin, then replace U by ηU, with η a C^∞ cut-off function that is 1 near the origin. With $U' = \eta U$, then $\overline{L}(U') = f'$ everywhere, so $C(f') = 0$ but $C(f - f')$ is analytic near the origin because $f - f'$ vanishes near the origin of $\mathbb{C} \times \mathbb{R}$. Therefore (41) holds.

We give a particular example. Take the function

$$F(z_1, z_2) = e^{-(z_2/2)^{1/2}} e^{-(i/z_2)^{1/2}} = F(z_2).$$

It is easy to verify that F is holomorphic in the half-plane $\mathrm{Im}(z_2) > 0$, continuous (in fact C^∞) in the closure, and rapidly decreasing as a function of $(z_1, z_2) \in \overline{\mathcal{U}}$. However it is clearly not holomorphic in a neighborhood of the origin.

Now set $f = F|_{\partial\mathcal{U}}$, that is, in the $\mathbb{C} \times \mathbb{R}$ coordinates, $f(z_1, x_2) = F(x_2 + i|z_1|^2)$. However $C(f) = F$ by Theorem 8.5.

Thus we have reached the conclusion that $\overline{L}(U) = f$ is not locally solvable near the origin, even though this particular f is a C^∞ function.

9 Exercises

1. Suppose f is holomorphic in a polydisc $\mathbb{P}_r(z^0)$, and assume that f vanishes in a neighborhood of z^0. Then $f = 0$ throughout $\mathbb{P}_r(z^0)$.

[Hint: Expand $f(z) = \sum a_\alpha (z - z^0)^\alpha$ in $\mathbb{P}_r(z^0)$, using Proposition 1.1, and note that all a_α are zero.]

2. Show that:

(a) If f is holomorphic in a pair $\mathbb{P}_\sigma(z^0)$ and $\mathbb{P}_\tau(z^0)$ of polydiscs centered at z^0 with $\sigma = (\sigma_1, \ldots, \sigma_n)$ and $\tau = (\tau_1, \ldots, \tau_n)$, then f extends to be holomorphic in $\mathbb{P}_r(z^0)$, wherever $r = (r_1, \ldots, r_n)$ and $r_j \leq \sigma_j^{1-\theta}\tau_j^\theta$, $1 \leq j \leq n$, for some $0 \leq \theta \leq 1$.

(b) If $S = \{s = (s_1, \ldots, s_n),\ \ s_j = \log r_j,\ \ \text{where } f \text{ is holomorphic in } \mathbb{P}_r(z^0)\}$, then S is a convex set.

[Hint: Consider $\sum a_\alpha(z - z^0)$ that represents f both in $\mathbb{P}_\sigma(z^0)$ and $\mathbb{P}_\tau(z^0)$.]

3. Given Ω any open subset of $\mathbb{C}^1$, construct a holomorphic function f in Ω that cannot be continued analytically outside Ω.

[Hint: Given any sequence of points $\{z_j\}$ in Ω, which does not have a limit point in Ω, there exists an analytic function in Ω vanishing exactly at those z_j.]

4. Suppose Ω is a bounded region in $\mathbb{C}^n$, $n > 1$, and f is holomorphic in Ω. Suppose Z, the zero set of f, is non-empty. Then $\overline{Z}$ intersects $\partial\Omega$, that is, $\overline{Z} \cap \partial\Omega$ is not empty.

[Hint: Let w be a point in $\overline{\Omega}^c$. Let $z^0 \in Z$ be a point furthest from w. Define γ to be the unit vector in the direction from z_0 to w, and let ν be another unit vector so that both ν and $i\nu$ are perpendicular to γ. Consider the one-variable function $h_\epsilon(\zeta)$ given by $h_\epsilon(\zeta) = f(z^0 - \epsilon\gamma + \zeta\nu)$. Then for $\epsilon > 0$, the function $h_\epsilon(\zeta)$ does not vanish in a fixed neighborhood of $\zeta = 0$.]

5. Suppose f is continuous and has compact support in $\mathbb{C}^1$.

(a) Show that $u = f * \Phi$ in Proposition 3.1 belongs to $\text{Lip}(\alpha)$, for every $\alpha < 1$.

(b) Show that u is not necessarily in C^1.

[Hint: For (b) consider $f(z) = z(\log(1/|z|))^\epsilon$ but modified away from the origin to have compact support.]

6. Verify the identity in $\mathbb{C}^1$

$$F(z) = \frac{1}{2\pi i}\int_{\partial\Omega}\frac{F(\zeta)}{\zeta - z}\,d\zeta - \frac{1}{\pi}\int_\Omega \frac{(\partial F/\partial\overline{\zeta})(\zeta)}{(\zeta - z)}\,dm(\zeta)$$

for appropriate regions Ω and C^1 functions F. Use this identity to give an alternative proof of Proposition 3.1.

7. Prove the following. The necessary and sufficient condition that the solution $u(z) = \frac{1}{\pi}\int \frac{f(\zeta)}{\zeta - z}\,dm(\zeta)$ of $\partial u/\partial\overline{z} = f$ in $\mathbb{C}^1$, have compact support when f has compact support, is that

$$\int_{\mathbb{C}} \zeta^n f(\zeta)\,dm(\zeta) = 0, \qquad \text{for all } n \geq 0.$$

[Hint: In one direction, note that $\frac{\partial}{\partial \overline{z}}(z^n u(z)) = z^n f(z)$. For the converse, observe that for large z, $u(z) = \sum_{n=0}^{\infty} a_n z^{-n-1}$, with $a_n = \frac{1}{\pi}\int \zeta^n f(\zeta)\, dm(\zeta)$.]

8. Suppose Ω is a region in $\mathbb{R}^d$ with a defining function ρ that is of class C^k.

(a) If F is a C^k function defined on $\mathbb{R}^d$ and $F = 0$ on $\partial\Omega$, show that $F = a\rho$, with $a \in C^{k-1}$.

(b) Suppose $F_1 = F_2$ on $\partial\Omega$. Show that if X is any tangential vector field then

$$X(F_1)|_{\partial\Omega} = X(F_2)|_{\partial\Omega}.$$

[Hint: Write $F_1 - F_2 = a\rho$.]

9. Verify that the extension F given by Theorem 4.1 is the unique solution to the Dirichlet problem for Ω with boundary data F_0.

10. Use the region $\{z \in \mathbb{C}^n : \rho < |z| < 1\}$ to show that the connectedness hypotheses in Theorem 3.3 and Theorem 4.1 are necessary.

11. That the connectedness properties in the hypotheses of Theorems 3.3 and 4.1 are related can be seen as follows. Suppose Ω is a bounded region with C^1 boundary. For $\epsilon > 0$, let Ω_ϵ be the "collar" defined by $\{z : d(z, \partial\Omega) < \epsilon\}$, and let $\Omega_\epsilon^- = \Omega_\epsilon \cap \Omega$. Then for sufficiently small ϵ the following are equivalent:

(i) $\overline{\Omega}^c$ is connected,

(ii) Ω_ϵ is connected,

(iii) Ω_ϵ^- is connected.

[Hint: For instance to see why (ii) or (iii) implies (i), suppose P_1 and P_2 are two points in $\overline{\Omega}^c$, and let Γ_1 and Γ_2 denote the connected components of $\overline{\Omega}^c$ which contain P_1 and P_2 respectively. Connect P_1 to a point Q_1 on $\partial\Omega \cap \overline{\Gamma}_1$, and P_2 to a point Q_2 on $\partial\Omega \cap \overline{\Gamma}_2$. Since Ω_ϵ^- is connected one can then connect Q_1 to Q_2 by a path in $\overline{\Omega}^c$.

Conversely, to show that (i) implies (iii) for example, let A be a point in Ω and B a point in $\overline{\Omega}^c$. If P_0 and P_1 belong to $\partial\Omega$, let γ_0 be any path starting at A traveling in Ω, passing through P_0, then traveling in Ω^c ending at B. Similarly, let γ_1 be path connecting A to B passing through P_1. These paths can be constructed because both Ω and $\overline{\Omega}^c$ are connected. Then, since $\mathbb{C}^n$ is simply-connected, deform the path γ_0 into γ_1, and denote such transformation by $s \mapsto \gamma_s$ with $0 \le s \le 1$. To conclude, consider the intersection of γ_s with $\partial\Omega$.]

12. Let Ω be a simply connected bounded region in $\mathbb{C}^1$ with a boundary of class C^1. Suppose F_0 is a given continuous function on $\partial\Omega$. Show that a necessary and sufficient condition that there is an F, holomorphic in Ω, continuous on $\overline{\Omega}$ so that $F = F_0$ on $\partial\Omega$ is that $\int_{\partial\Omega} z^n F_0(z)\, dz = 0$, for $n = 0, 1, 2, \ldots$.

[Hint: One direction is clear from Cauchy's theorem. For the converse define $F^{\pm}(z) = \frac{1}{2\pi i}\int_{\partial\Omega}\frac{F_0(\zeta)}{\zeta - z}\,d\zeta$, according to whether $z \in \Omega$ or $z \in \overline{\Omega}^c$. Now the hypothesis implies that $F^+(z) = 0$, $z \in \overline{\Omega}^c$. Also $F^-(z) - F^+(\tilde{z}) \to F_0(\zeta)$, $z \to \zeta$ if $\zeta \in \partial\Omega$, $z \in \Omega$, the segment $[z, \zeta]$ is normal to the tangent line of $\partial\Omega$ at ζ, and $\tilde{z}$ is the reflection of z across that line. That is, $\frac{\tilde{z}+z}{2} = \zeta$, $\tilde{z} \in \overline{\Omega}^c$. The convergence asserted is related to the expression of the delta function given by $i\pi\delta = \frac{1}{2}\left(\frac{1}{x-i0} - \frac{1}{x+i0}\right)$, in Section 2 of Chapter 3.]

13. Show that with an additional change of variables, that is, introducing complex coordinates, the canonical representations (16) and (17) of the boundary can be simplified to state

$$y_n = \sum_{j=1}^{n-1} \lambda_j |z_j|^2 + o(|z'|^2), \qquad \text{for } z' \to 0.$$

[Hint: Consider the change of variables $z_n \mapsto z_n - z_n(c_1 z_1 + \cdots + c_{n-1}z_{n-1} + Dz_n)$, $z_j \mapsto z_j$, $1 \le j \le n-1$, for suitable constants $c_1, \ldots, c_{n-1}$.]

14. The fact that when $n = 1$ there are no local holomorphic invariants at boundary points is indicated by the following fact. Suppose γ is a C^k curve in $\mathbb{C}^1$. Then for every $z^0 \in \gamma$, there is a holomorphic bijection Φ of a neighborhood of z^0 to a neighborhood of the origin, so that $\Phi(\gamma)$ is the curve $\{y = \varphi(x)\}$, with $\varphi(x) = o(x^k)$ as $x \to 0$.

[Hint: Suppose $y = a_2x^2 + \cdots + a_kx^k + o(x^k)$ as $x \to 0$, and consider Φ^{-1} defined by $\Phi^{-1}(z) = z + i\left(\sum_{j=2}^k a_j z^j\right)$.]

15. Consider the hypersurface M in $\mathbb{C}^3$ given by $M = \{\mathrm{Im}(z_3) = |z_1|^2 - |z_2|^2\}$. Show that M has the remarkable property that any holomorphic function F defined in a neighborhood of M continues analytically into all of $\mathbb{C}^3$.

[Hint: Use Theorem 7.5 to find a fixed ball B centered at the origin so that F continues into all of B. Then rescale.]

16. That the maximum principle of Theorem 6.1 does not hold in the case $n = 1$ can be seen as follows. Start with $f(e^{i\theta}) \in C^\infty$, so that $f \ge 0$, $f(e^{i\theta}) = 0$ for $|\theta| \le \pi/2$, $f(e^{i\theta}) = 1$ for $3\pi/4 \le |\theta| \le \pi$. Write $f(e^{i\theta}) = \sum_{n=0}^\infty a_n e^{in\theta} + \sum_{-\infty}^{n=-1} \overline{a}_n e^{in\theta}$, $G(z) = \sum_{n=0}^\infty a_n z^n$, and $F_N(z) = e^{NG(z)}$. Verify that F_N is continuous in the closed disc $|z| \le 1$, $|F_N(e^{i\theta})| = 1$, for $|\theta| \le \pi/2$ but $|F_N(z)| \ge c_1 e^{c_2 N(1-|z|)}$ in the closed disc, for two positive constants c_1 and c_2.

[Hint: $G(z) = u + iv$ where $u(r,\theta) = f * P_r$, with P_r the Poisson kernel.]

17. Verify the following:

(a) The inverse of the mapping of $\mathcal{U}$ to the unit ball given in the Appendix is $z_n = i\left(\frac{1-w_n}{1+w_n}\right)$, and $z_k = \frac{w_k}{1+w_n}$, $k = 1, \ldots, n-1$.

(b) For each $(\zeta, t) \in \mathbb{C}^{n-1} \times \mathbb{R}$ consider the following "translation" on $\mathbb{C}^n$, $r_{(\zeta,t)}$ given by

$$r_{(\zeta,t)}(z', z_n) = (z' + \zeta, z_n + t + 2i(z' \cdot \overline{\zeta}) + i|\zeta|^2).$$

Then $r_{(\zeta,t)}$ maps $\mathcal{U}$ and $\partial\mathcal{U}$ to themselves, respectively. Composing these mappings leads to the composition formula

$$(\zeta, t) \cdot (\zeta', t') = (\zeta + \zeta', t + t' + 2\mathrm{Im}(\zeta \cdot \overline{\zeta}')).$$

Under this law $\mathbb{C}^{n-1} \times \mathbb{R}$ becomes the "Heisenberg group."

(c) $\mathcal{U}$ (as well as $\partial\mathcal{U}$) is invariant under the "non-isotropic" dilations $(z', z_n) \to (\delta z', \delta^2 z_n)$, $\delta > 0$.

(d) Both $\mathcal{U}$ and $\partial\mathcal{U}$ are invariant under the mappings $(z', z_n) \mapsto (u(z'), z_n)$, where u is a unitary mapping of $\mathbb{C}^{n-1}$.

18. Define $\mathcal{H}_\lambda$ to be the space of functions f holomorphic in $\mathbb{C}^{n-1}$, for which

$$\int_{\mathbb{C}^{n-1}} |f(z)|^2 e^{-4\pi\lambda|z|^2}\, dm(z) = \|f\|^2_{\mathcal{H}_\lambda} < \infty.$$

Show that:

(a) $\mathcal{H}_\lambda$ is trivial if $\lambda \leq 0$.

(b) $\mathcal{H}_\lambda$ is complete in the indicated norm, so $\mathcal{H}_\lambda$ is a Hilbert space.

(c) Define $P_\lambda(f)(z) = \int_{\mathbb{C}^{n-1}} f(w) K_\lambda(z, w) e^{-4\pi\lambda|w|^2}\, dm(w)$, where $K_\lambda(z, w) = (4\lambda)^{n-1} e^{4\pi\lambda z \cdot \overline{w}}$.

Then P_λ is the orthogonal projection of $L^2(e^{-4\pi\lambda|w|^2} dm(w))$ to $\mathcal{H}_\lambda$.

[Hint: Show that convergence in the norm $\mathcal{H}_\lambda$ implies uniform convergence on compact subsets of $\mathbb{C}^{n-1}$, using Lemma 8.2.]

19. Prove:

(a) The space $\mathcal{H}$ in Section 8.1, is complete, and hence is a Hilbert space.

(b) Show that the Cauchy integral $f \mapsto C(f)$ gives the orthogonal projection from $L^2(\partial\mathcal{U}, d\beta)$ to the linear space of functions F_0 that arise as $\lim_{\epsilon\to 0} F_\epsilon$, for $F \in H^2(\mathcal{U})$.

[Hint: For (a), use the previous exercise.]

10 Problems

The problems below are not intended as exercises for the reader but are meant instead as a guide to further results in the subject. Sources in the literature for each of the problems can be found in the "Notes and References" section.

1.* Suppose $f = f(z_1, \ldots, z_n)$ is defined in a region $\Omega \subset \mathbb{C}^n$, and for each j, $1 \leq j \leq n$, the function f is holomorphic in z_j with the other variables fixed. Then f is holomorphic in Ω. This was shown at the start of the chapter when f is continuous, and the point of this problem is that no condition on f is required besides the analyticity in each separate variable.

An important ingredient in the proof of this result is an application of the Baire category theorem.

2.* Assume f is holomorphic in a neighborhood of the origin and $f(0) = 0$. Let $f(z) = \sum a_\alpha z^\alpha$ be the power series expansion of f valid near the origin. The order of the zero (at $z = 0$) is the integer k that is the smallest $|\alpha|$, for which $a_\alpha \neq 0$. Then, after a linear change of variables, we can write $f(z) = c(z)P(z)$ near the origin, where $P(z) = z_n^k + a_{k-1}(z')z_n^{k-1} + \cdots + a_0(z')$ with (z', z_n), and $c(z) \neq 0$ while $a_{k-1}(0) = \cdots = a_0(0) = 0$. This result is the **Weierstrass preparation** theorem.

[Hint: Assume that our coordinate system $(z', z_n) \in \mathbb{C}^{n-1} \times \mathbb{C}$ is such that $f(0, z_n) = z_n^k$. Then by Rouché's theorem we can choose $\epsilon, r > 0$, so that $z_k \to f(z', z_k)$ has k zeroes inside the disc $|z_k| \leq r$, but is non-vanishing on the boundary, for all $|z'| < \epsilon$. Let $\gamma_1(z')$, $\gamma_2(z')$, ... , $\gamma_k(z')$ be an arbitrary ordering of these zeroes. Then the symmetric functions $\sigma_1(z') = \sum_{\ell=1}^k \gamma_\ell(z')$, $\sigma_2(z') = \sum_{m<\ell} \gamma_\ell(z')\gamma_m(z'), \ldots$, are holomorphic in z', for $|z'| < \epsilon$. This follows since the sums $s_m(z') = \sum_{\ell=1}^k (\gamma_\ell(z'))^m$, $1 \leq m \leq k$, have this property because they are given by the formula

$$s_m(z') = \frac{1}{2\pi i} \int_{|w|=r} w^m \frac{(\partial f/\partial w)(z', w)}{f(z', w)}\, dw.$$

Now we need only take $a_{k-j}(z') = (-1)^j \sigma_j(z')$, and the result holds for $P(z) = z_n^k + a_{k-1}(z')z^{k-1} + \cdots + a_0(z')$.]

3.* The original proof of Theorem 4.1 represented F in terms of F_0 by Green's theorem via the "Bochner-Martinelli integral." The result then held for F_0 merely of class C^1.

4.* We are concerned with the problem

$$\overline{\partial} u = f, \qquad \text{on } \Omega, \tag{42}$$

where Ω is a bounded region in $\mathbb{C}^n$ with C^∞ boundary and f is given in Ω with $\overline{\partial} f = 0$ there.

(a) If Ω is pseudo-convex, and $f \in C^\infty(\overline{\Omega})$ then there is $u \in C^\infty(\overline{\Omega})$ that solves (42).

(b) The "normal" solution (if it exists) is defined as the (unique) solution u in $L^2(\Omega)$ for which

$$\int_\Omega u\overline{F}\, dm(z) = 0$$

for all F that are holomorphic in Ω and are in L^2 there. For Ω that are strongly pseudo-convex (and many other classes of Ω) whenever $f \in C^\infty(\overline{\Omega})$, the normal solution u also belongs to $C^\infty(\overline{\Omega})$. This results from the study of the "$\overline{\partial}$-Neumann problem."

5.* **Domains of holomorphy.** A domain of holomorphy is a region Ω with the property that there exists a holomorphic function F so that for every $z^0 \in \partial\Omega$, the function F cannot be continued into some ball centered at z^0. If Ω is a domain of holomorphy and has boundary of class C^2, then Ω is pseudo-convex, by Theorem 7.5. Conversely, it can be shown that if Ω is pseudo-convex, then it is a domain of holomorphy.

6.* The converse to Theorem 8.7 holds. If f is a distribution with compact support so that $C(f)(z)$ is analytic near $z = 0$ then $\overline{L}(U) = f$ is locally solvable near the origin.

This is proved by finding a kernel K so that the convolution operator $T(f) = f * K$ on the Heisenberg group is a relative inverse to $\overline{L}$ in the sense that $\overline{L}T(f) = f - C(f)$. Then write $f = f - C(f) + C(f) = f_1 + f_2$, with $f_1 = f - C(f)$ and $f_2 = C(f)$. We can solve $L(U_1) = f_1$ by what has just been asserted, and we can solve $L(Uu_2) = f_2$ locally by the Cauchy-Kowaleski theorem, since f_2 is real-analytic at the origin.

8 Oscillatory Integrals in Fourier Analysis

> The origin of my devotion to these problems is after I attended in 1839 Nichol's Senior Natural Philosophy class, I had become filled with the utmost admiration for the splendor and poetry of Fourier... I asked Nichol if he thought I could read Fourier. He replied 'perhaps.' He thought the book a work of most transcendent merit. So on the 1st of May... I took Fourier out of the University Library; and in a fortnight I had mastered it – gone right through it.
>
> *W. Thompson (Kelvin),* 1840

> This result might also have been obtained from the integral U in its original shape, namely, $\int_0^\infty \cos(x^3 - nx)\,dx$ ... If x_1 be the positive value of x which renders $x^3 - nx$ a minimum, we have $x_1 = 3^{-\frac{1}{2}}n^{\frac{1}{2}}$. Let the integral U be divided into three parts, by integrating separately from $x = 0$ to $x = x_1 - a$, from $x = x_1 - a$ to $x = x_1 + b$, and from $x = x_1 + b$ to $x = \infty$; then make n infinite...
>
> *G. G. Stokes,* 1850

The study of oscillatory integrals and their asymptotics has been a vital part of harmonic analysis from the beginnings of the subject. The Fourier transform and the attendant Bessel functions provided initial examples of such oscillatory integrals. One should also note the study of asymptotics in the early works of Airy, Lipschitz, Stokes, and Riemann. In the work of the last two, the principle of stationary phase appears, if only implicitly; for Stokes it was in a reexamination of Airy's integral and for Riemann it was in the calculation of certain Fourier series. This principle was then used more generally by Kelvin in an 1887 paper on water waves. The application of these ideas to number theory and lattice point problems was initiated in the first quarter of the next century by

Voronoi and van der Corput, among others.

Given this long history it is an interesting fact that only relatively recently (1967) did one realize the possibility of restriction theorems for the Fourier transform, and the relation of the above mentioned asymptotics to differentiation theory and maximal functions had to wait another ten years to come to light.

Here we present an introduction to the development of some of these ideas. Of importance to us is the bearing of certain geometric considerations (involving curvature) on the decay of the Fourier transform and these are explained by the behavior of oscillatory integrals.

Two pillars of the theory are: averaging operators, and restriction theorems for the Fourier transform. Once we have described some basic facts about these, we apply the results of the restriction theorems to partial differential equations of "dispersion" type. We also reexamine the Radon transform, emphasizing its common traits with the averaging operator. Finally, we turn to the problem of counting lattice points and see what the ideas of oscillatory integrals teach us.

1 An illustration

We begin with a simple example that hints at the role of curvature in harmonic analysis. The setting is $\mathbb{R}^d$ with $d = 3$, and we consider the **averaging operator** A that gives for each function f its average over the sphere of radius 1 centered at x. It can be written as

$$A(f)(x) = \frac{1}{4\pi} \int_{S^2} f(x-y)\, d\sigma(y),$$

with $d\sigma$ the induced Lebesgue measure on the sphere $S^2 = \{x \in \mathbb{R}^3 : |x| = 1\}$. (See Book III, Chapter 6 for the definition and properties of $d\sigma$.)

The unexpected fact about the operator A is that it smooths f in several senses, the simplest one being that when $f \in L^2(\mathbb{R}^3)$, then $A(f)$ will have first derivatives also in L^2. This is expressed in the inequality

$$(1) \qquad \left\| \frac{\partial}{\partial x_j} A(f) \right\|_{L^2} \leq c \|f\|_{L^2}, \qquad j = 1, 2, 3.$$

More precisely, this estimate states that for $f \in L^2$, the convolution $\frac{1}{4\pi}(f * d\sigma)$, which is itself an L^2 function (see for instance Exercise 17 in Chapter 1), has first derivatives taken in the sense of distributions that are L^2 functions and that satisfy (1).

Now these assertions are a direct consequence of a corresponding estimate for the Fourier transform $\widehat{d\sigma}$ of the measure $d\sigma$, namely

$$\widehat{d\sigma}(\xi) = \int_{S^2} e^{-2\pi i x \cdot \xi}\, d\sigma(x).$$

In the present case one knows $\widehat{d\sigma}$ explicitly:

$$\widehat{d\sigma}(\xi) = \frac{2\sin(2\pi|\xi|)}{|\xi|},$$

from which it is evident[1] that

$$(2) \qquad |\widehat{d\sigma}(\xi)| \leq c(1+|\xi|)^{-1}.$$

Now simple manipulations of distributions and their Fourier transforms (see Section 1.5 in Chapter 3) show that $(f * d\sigma)^\wedge = \hat{f}\widehat{d\sigma}$, and

$$\left(\frac{\partial}{\partial x_j}A(f)\right)^\wedge(\xi) = \frac{1}{4\pi}2\pi i\xi_j \hat{f}(\xi)\widehat{d\sigma}(\xi),$$

so (1) follows from (2) and Plancherel's theorem.

The results above have extensions to d dimensions for all $d > 1$. We define the **averaging operator** A in $\mathbb{R}^d$ by

$$A(f) = \frac{1}{\sigma(S^{d-1})}\int_{S^{d-1}} f(x-y)\, d\sigma(y),$$

with $d\sigma$ the induced measure on the unit sphere S^{d-1}. We also recall the Sobolev space L^2_k described in Section 3.1 of Chapter 1.

Proposition 1.1 *The mapping $f \mapsto A(f)$ is bounded from $L^2(\mathbb{R}^d)$ to $L^2_k(\mathbb{R}^d)$, with $k = \frac{d-1}{2}$.*

Note that if d is odd (and hence k is integral), this means

$$\sum_{|\alpha|\leq k} \|\partial_x^\alpha A(f)\|_{L^2} \leq c\|f\|_{L^2}.$$

The proof of that proposition relies on properties of Bessel functions which we do not prove here. However, these may be found in Book I,

---

[1]This formula follows by integrating over S^2, using polar coordinates; see Chapter 6 in Book III.

Chapter 6, Problem 2, and Book II, Appendix A. In any case, we will see below that these results can be deduced without the use of the theory of Bessel functions.

Proof. The proposition is a consequence of the identity

$$\widehat{d\sigma}(\xi) = 2\pi|\xi|^{-d/2+1}J_{d/2-1}(2\pi|\xi|), \tag{3}$$

where $\widehat{d\sigma}(\xi) = \int_{S^{d-1}} e^{-2\pi i x\cdot\xi}\,d\sigma(x)$, and J_m is the Bessel function of order m. In turn this is just another version of the formula for the Fourier transform of a radial function $f(x) = f_0(|x|)$, given by $\hat{f}(\xi) = F(|\xi|)$, with

$$F(\rho) = 2\pi\rho^{-d/2+1}\int_0^\infty J_{d/2-1}(2\pi\rho r)f_0(r)r^{d/2}\,dr, \tag{4}$$

from which (3) follows by a simple limiting argument. From (3) we obtain the key decay estimate

$$|\widehat{d\sigma}(\xi)| \leq O(|\xi|^{-\frac{d-1}{2}}) \quad \text{as } |\xi| \to \infty. \tag{5}$$

Indeed, (5) is deducible from (3) and the asymptotic behavior of the Bessel functions that guarantees that $J_m(r) = O(r^{-1/2})$ as $r \to \infty$.

Once (5) is established the proof of the proposition is finished via Plancherel's theorem as in the case $d = 3$.

The following comments may help put the result in perspective.

- It is natural to ask if it is some special feature of the sphere among hypersurfaces (for instance, its symmetry with respect to rotations) that guarantee the crucial decay estimate (5), or does that phenomenon hold in more general circumstances for hypersurfaces M? We will see below that the analog of (5) is true when an appropriate "curvature" of M is non-vanishing.

- Moreover, simple examples show that anything like (5) fails completely when M is "flat" (Exercise 2), and more generally, whatever decay one might hope for $\widehat{d\sigma}(\xi)$ is linked to the degree to which the curvature of M does not vanish.

- One can also observe that the degree of smoothing $k = (d-1)/2$ asserted in Proposition 1.1 can only happen in the context of L^2, and not for L^p, $p \neq 2$. (A result in this direction is outlined in Exercise 7.)

- Finally, it is interesting to remark that when $d = 3$ the averaging operator furnishes the solution of the wave equation $\triangle_x u(x,t) = \frac{\partial^2}{\partial t^2}u(x,t)$, $(x,t) \in \mathbb{R}^3 \times \mathbb{R}$ with $u(x,0) = 0$ and $\frac{\partial u}{\partial t}(x,0) = f(x)$. The solution for time $t = 1$ is given by $u(x,1) = A(f)(x)$, and for other times it can be obtained by rescaling. (See Chapter 6 in Book I, where A is denoted by M.)

2 Oscillatory integrals

Certain basic facts about oscillatory integrals will allow us to generalize the decay estimate (5) we have obtained for the sphere. What we have in mind are the integrals of the form

$$I(\lambda) = \int_{\mathbb{R}^d} e^{i\lambda\Phi(x)}\psi(x)\,dx, \tag{6}$$

and the question of their behavior for large λ.

The function Φ is called the **phase** and ψ the **amplitude**. In what follows we assume that both the phase Φ and the parameter λ are real-valued, but ψ may be allowed to be complex-valued.[2]

There is a basic principle underlying the analysis, that of **stationary phase**: in so far as the derivative (or gradient) of the phase is non-vanishing, the integral is rapidly decreasing in λ (and thus negligible); thus the main contribution of (6) comes from those points x where the gradient of Φ vanishes; so when $d = 1$ these are the x for which $\Phi'(x) = 0$.

The first observation along these lines is merely an extension of a simple estimate for the Fourier transform (effectively the case $\Phi(x) = 2\pi\frac{\xi}{|\xi|} \cdot x$, and $\lambda = |\xi|$). We assume here that Φ and ψ are C^∞ functions, and that ψ has compact support.

Proposition 2.1 *Suppose* $|\nabla\Phi(x)| \geq c > 0$ *for all* x *in the support of* ψ. *Then for every* $N \geq 0$

$$|I(\lambda)| \leq c_N \lambda^{-N}, \qquad \textit{whenever } \lambda > 0.$$

Proof. We consider the following vector field

$$L = \frac{1}{i\lambda}\sum_{k=1}^{d} a_k \frac{\partial}{\partial x_k} = \frac{1}{i\lambda}(a \cdot \nabla),$$

[2]However in some circumstances it is of interest to allow Φ or λ to be complex valued. This arises in particular when $d = 1$ and Φ (and ψ) are analytic and the integral (6) is treated by deforming contours of integration, as in Appendix A of Book II.

with $a = (a_1, \ldots, a_d) = \frac{\nabla\Phi}{|\nabla\Phi|^2}$. Then the transpose L^t of L is given by

$$L^t(f) = -\frac{1}{i\lambda}\sum_{k=1}^{d}\frac{\partial}{\partial x_k}(a_k f) = -\frac{1}{i\lambda}\nabla\cdot(af).$$

Because of our assumption on $\nabla\Phi$, the a_j and all their partial derivatives are each bounded on the support of ψ.

Now observe that $L(e^{i\lambda\Phi}) = e^{i\lambda\Phi}$, therefore $L^N(e^{i\lambda\Phi}) = e^{i\lambda\Phi}$ for every positive integer N. Thus

$$I(\lambda) = \int_{\mathbb{R}^d} L^N(e^{i\lambda\Phi})\psi\,dx = \int_{\mathbb{R}^d} e^{i\lambda\Phi}(L^t)^N(\psi)\,dx.$$

Taking absolute values in the last integral gives $|I(\lambda)| \leq c_N\lambda^{-N}$ for positive λ, thus proving the proposition.

The next two assertions are limited to dimension one, where we can obtain more precise conclusions with simpler hypotheses. In this situation it is appropriate to consider first the integral I_1 given by

$$I_1(\lambda) = \int_a^b e^{i\lambda\Phi(x)}\,dx, \tag{7}$$

where a and b are any real numbers. Thus in (7) there is no amplitude ψ present, (or put another way, $\psi(x) = \chi_{(a,b)}(x)$). Here we assume only that Φ is of class C^2, and $\Phi'(x)$ is monotonic (increasing or decreasing), while $|\Phi'(x)| \geq 1$ in the interval $[a,b]$.

Proposition 2.2 *In the above situation,* $|I_1(\lambda)| \leq c\lambda^{-1}$, *all* $\lambda > 0$, *with* $c = 3$.

What is important here is not the specific value of c, but that it is independent of the length of the interval $[a,b]$. Note that the order of decrease in λ cannot be improved, as the simple example $\Phi(x) = x$, and $I_1(\lambda) = \frac{1}{i\lambda}(e^{i\lambda b} - e^{i\lambda a})$ shows.

Proof. The proof uses the operator L that occurred in the previous proposition. We may assume $\Phi' > 0$ on $[a,b]$, because the case when $\Phi' < 0$ follows by taking complex conjugates. So $L = \frac{1}{i\lambda\Phi'(x)}\frac{d}{dx}$, and $L^t(f) = -\frac{1}{i\lambda}\frac{d}{dx}(f/\Phi')$, hence

$$I_1(\lambda) = \int_a^b L(e^{i\lambda\Phi})\,dx = \int_a^b e^{i\lambda\Phi}L^t(1)\,dx + \left[e^{i\lambda\Phi}\frac{1}{i\lambda\Phi'}\right]_a^b,$$

and now (because we do not have an amplitude ψ that vanishes at the end-points) there are boundary terms. Since $|\Phi'(x)| \geq 1$, these two terms contribute a total majorized by $2/\lambda$. But the integral on the right-hand side is clearly bounded by

$$\int_a^b |L^t(1)|\, dx = \frac{1}{\lambda}\int_a^b \left|\frac{d}{dx}\left(\frac{1}{\Phi'}\right)\right| dx.$$

However Φ' is monotonic and continuous while $|\Phi'(x)| \geq 1$, so $\frac{d}{dx}(1/\Phi')$ does not change sign in the interval $[a, b]$. Therefore

$$\int_a^b \left|\frac{d}{dx}\left(\frac{1}{\Phi'}\right)\right| dx = \left|\int_a^b \frac{d}{dx}\left(\frac{1}{\Phi'}\right) dx\right| = \left|\frac{1}{\Phi'(b)} - \frac{1}{\Phi'(a)}\right|.$$

Altogether then $|I_1(\lambda)| \leq 3/\lambda$ and the proposition is proved.

Remark. If in the above proposition we assumed that $|\Phi'(x)| \geq \mu$ (instead of $|\Phi'(x)| \geq 1$), then we could get $|I_1(\lambda)| \leq c(\lambda\mu)^{-1}$. This is obvious on replacing Φ by Φ/μ, and λ by $\lambda\mu$ in the proposition.

Next we ask what happens to $I_1(\lambda)$ when $\Phi'(x_0) = 0$ for some x_0, if we make the assumption that the **critical point** x_0 is **non-degenerate** in the sense that $\Phi''(x_0) \neq 0$. A good indication of what we may expect comes from the case $\Phi(x) = x^2$ (where the critical point is the origin). Here one has

$$\int e^{i\lambda x^2}\psi(x)\, dx = c_0\lambda^{-1/2} + O(|\lambda|^{-3/2}), \qquad \text{as } \lambda \to \infty,$$

and more generally

$$\int e^{i\lambda x^2}\psi(x)\, dx = \sum_{k=0}^{N} c_k\lambda^{-1/2-k} + O\left(|\lambda|^{-3/2-N}\right), \tag{8}$$

for every $N \geq 0$. To see (8) we start with the formula for the Fourier transform of the Gaussian that states

$$\int_{\mathbb{R}} e^{-\pi s x^2}\psi(x)\, dx = s^{-1/2}\int_{\mathbb{R}} e^{-\pi\xi^2/s}\hat{\psi}(\xi)\, d\xi.$$

Now since both sides have analytic continuations for $\mathrm{Re}(s) > 0$, the passing to the limit, $s = -i\lambda/\pi$ yields

$$\int e^{i\lambda x^2}\psi(x)\, dx = \left(\frac{\pi i}{\lambda}\right)^{1/2}\int e^{-i\pi^2\xi^2/\lambda}\hat{\psi}(\xi)\, d\xi.$$

So the expansion $e^{iu^2} = \sum_{k=0}^{N} \frac{(iu^2)^k}{k!} + O(|u|^{2N+2})$ gives us (8) with $c_k = (i\pi)^{1/2} \frac{i^k}{2^{2k}k!}\psi^{(2k)}(0)$. This indicates that a decrease of order $O(\lambda^{-1/2})$ can be expected when the phase has a critical point which is non-degenerate.

There is a version of Proposition 2.2 for the second derivative that takes this observation into account: it is the following estimate of van der Corput. Here Φ is again supposed to be of class C^2 in the interval $[a,b]$, but now we assume that $|\Phi''(x)| \geq 1$ throughout the interval.

Proposition 2.3 *Under the above assumptions, and with $I_1(\lambda)$ given by* (7) *we have*

$$(9) \qquad |I_1(\lambda)| \leq c'\lambda^{-1/2} \quad \textit{for all } \lambda > 0, \textit{ with } c' = 8.$$

Again, it is not the exact value of c' that matters, but that it is independent of $[a,b]$.

Proof. We may assume that $\Phi''(x) \geq 1$ throughout the interval, because the case $\Phi''(x) \leq -1$ follows from this by taking complex conjugates. Now $\Phi''(x) \geq 1$ implies that $\Phi'(x)$ is strictly increasing, so if Φ has a critical point in $[a,b]$, it can have only one. Assume x_0 is such a critical point and break the interval $[a,b]$ in three sub-intervals: the first is centered at x_0 and is $[x_0-\delta, x_0+\delta]$ with δ chosen momentarily. The other two make up the complement and are $[a, x_0-\delta]$ and $[x_0+\delta, b]$. Now the first interval has length 2δ, so trivially the integral taken over that interval contributes at most 2δ. On the interval $[x_0+\delta, b]$ we observe that $\Phi'(x) \geq \delta$ (because $\Phi'' \geq 1$) and so by Proposition 2.2 and the remark that follows it, the integral contributes at most $3/(\delta\lambda)$; similarly for the interval $[a, x_0-\delta]$. Thus altogether $I_1(\lambda)$ is majorized by $2\delta + 6/(\delta\lambda)$, and upon choosing $\delta = \lambda^{-1/2}$ we get (9). Note that if Φ has no critical points in $[a,b]$ and/or one of the three intervals is smaller than indicated, then each of the estimates holds *a fortiori*, and hence also the conclusion.

There is a similar conclusion when an amplitude ψ is present. We suppose ψ is of class C^1 in the interval $[a,b]$.

Corollary 2.4 *Assume Φ satisfies the hypotheses of Proposition* 2.3. *Then*

$$(10) \qquad \left| \int_a^b e^{i\lambda\Phi(x)}\psi(x)\,dx \right| \leq c_\psi \lambda^{-1/2},$$

where $c_\psi = 8\left(\int_a^b |\psi'(x)|\,dx + |\psi(b)|\right)$.

Proof. Let $J(x) = \int_a^x e^{i\lambda\Phi(u)}\,du$. We integrate by parts, using $J(a) = 0$. Then

$$\int_a^b e^{i\lambda\Phi(x)}\psi(x)\,dx = -\int_a^b J(x)\frac{d\psi}{dx}\,dx + J(b)\psi(b),$$

and the result follows, because $|J(x)| \leq 8\lambda^{-1/2}$ for each x, by the proposition.

As an illustration, we give a quick proof of the Bessel function estimate

$$J_m(r) = O(r^{-1/2}) \qquad \text{as } r \to \infty \tag{11}$$

when m is a fixed integer. We have (see, for instance, Section 4 in Chapter 6, Book I) that

$$J_m(r) = \frac{1}{2\pi}\int_0^{2\pi} e^{ir\sin x}e^{-imx}\,dx.$$

Here $\lambda = r$, $\Phi(x) = \sin x$, and $\psi(x) = \frac{1}{2\pi}e^{-imx}$. Now break the interval $[0, 2\pi]$ into two parts, according to whether $|\sin x| \geq 1/\sqrt{2}$ or $|\cos x| \geq 1/\sqrt{2}$. The first part consists of two sub-intervals to which we may apply the corollary, giving a contribution of $O(r^{-1/2})$. The second part is the sum of three sub-intervals to which one can apply a version of Proposition 2.2 (analogous to the corollary), and this gives a contribution of $O(r^{-1}) = O(r^{-1/2})$, as $r \to \infty$.

In dimension d greater than 1, the fact is that there are no analogs of the strict estimates given by Propositions 2.2 and 2.3. However, there is a workable version of the second derivative test of Proposition 2.3 that can be established. We now take this up and then apply it below.

We consider phase and amplitude functions Φ and ψ that are C^∞ and we suppose that ψ has compact support. We form the $d \times d$ **Hessian matrix** of Φ, given by $\left\{\frac{\partial^2\Phi}{\partial x_j \partial x_k}\right\}_{1\leq j,k\leq d}$, and abbreviated as $\nabla^2\Phi$.

The main assumption will be that

$$\det\{\nabla^2\Phi\} \neq 0 \qquad \text{on the support of } \psi. \tag{12}$$

Proposition 2.5 *Suppose* (12) *holds. Then*

$$I(\lambda) = \int_{\mathbb{R}^d} e^{i\lambda\Phi(x)}\psi(x)\,dx = O(\lambda^{-d/2}), \qquad \text{as } \lambda \to \infty. \tag{13}$$

We estimate $I(\lambda)$ via $|I(\lambda)|^2 = \overline{I(\lambda)}I(\lambda)$. This simple trick allows us to bring in the Hessian of Φ (that is, second derivatives) in terms of first derivatives of differences of Φ, an idea that has many variants.

Before we exploit this artifice we must take a precaution: we will assume that the support of ψ is sufficiently small, in particular, that it lies in a ball of fixed radius ϵ, where ϵ will be chosen in terms of Φ. Once the estimate (13) has been proved for such ψ, we can obtain (13) for general ψ as a finite sum of these estimates, by using a partition of unity to cover the support of the original ψ.

Now

$$\overline{I(\lambda)}I(\lambda) = \int_{\mathbb{R}^d}\int_{\mathbb{R}^d} e^{i\lambda[\Phi(y)-\Phi(x)]}\psi(y)\overline{\psi}(x)\,dx\,dy.$$

Here we make the change of variables $y = x + u$ (with x fixed), that is, $u = y - x$. Then the double integral becomes

$$\int_{\mathbb{R}^d}\int_{\mathbb{R}^d} e^{i\lambda[\Phi(x+u)-\Phi(x)]}\psi(x,u)\,dx\,du,$$

where $\psi(x,u) = \psi(x+u)\overline{\psi}(x)$ is a C^∞ function of compact support. Notice that $\psi(x,u)$ is supported where $|u| \leq 2\epsilon$, since both x and y are restricted to range in the same ball of radius ϵ. Therefore we have $|I(\lambda)|^2 = \int_{\mathbb{R}^d} J_\lambda(u)\,du$, where

$$J_\lambda(u) = \int_{\mathbb{R}^d} e^{i\lambda[\Phi(x+u)-\Phi(x)]}\psi(x,u)\,dx.$$

We claim that

$$|J_\lambda(u)| \leq c_N(\lambda|u|)^{-N}, \quad \text{for every } N \geq 0. \tag{14}$$

This is in the spirit of Proposition 2.1, and the proof of (14) follows the approach of that proposition.

We use the vector field

$$L = \frac{1}{i\lambda}(a \cdot \nabla)$$

and its transpose L^t given by $L^t(f) = -\frac{1}{i\lambda}\nabla \cdot (af)$. Here

$$a = \frac{\nabla_x(\Phi(x+u)-\Phi(x))}{|\nabla_x(\Phi(x+u)-\Phi(x))|^2} = \frac{b}{|b|^2},$$

with $b = \nabla_x(\Phi(x+u)-\Phi(x))$.

We have

$$|b| = |\nabla_x(\Phi(x+u) - \Phi(x))| \approx |u|, \tag{15}$$

if $|u|$ is sufficiently small, in particular if $|u| \leq 2\epsilon$.[3]

The upper estimate $|b| \lesssim |u|$ is clear since Φ is smooth. For the lower estimate observe that by Taylor's theorem, $\nabla_x(\Phi(x+u) - \Phi(x)) = \nabla^2\Phi(x) \cdot u + O(|u|^2)$. However our assumption (12) means that the linear transformation represented by $\nabla^2\Phi(x)$ is invertible, so $|\nabla^2\Phi(x) \cdot u| \geq c|u|$ for some $c > 0$. Therefore (15) is established if ϵ has been taken small enough. Observe also that $|\partial_x^\alpha b| \leq c_\alpha |u|$, for all α, and hence, using (15) we see that

$$|\partial_x^\alpha a| \leq c_\alpha |u|^{-1} \quad \text{for all } \alpha, \tag{16}$$

and as a result $|(L^t)^N(\psi(x,u))| \leq c_N(\lambda|u|)^{-N}$ for every positive integer N.

However,

$$\begin{aligned} J_\lambda(u) &= \int_{\mathbb{R}^d} L^N\left(e^{i\lambda[\Phi(x+u)-\Phi(x)]}\right)\psi(x,u)\,dx \\ &= \int_{\mathbb{R}^d} e^{i\lambda[\Phi(x+u)-\Phi(x)]}(L^t)^N(\psi(x,u))\,dx, \end{aligned}$$

and thus by (16), we have $|J_\lambda(u)| \leq c_N(\lambda|u|)^{-N}$, proving (14).

With this estimate established we take $N = 0$, and $N = d+1$ in (14) and see that

$$|I(\lambda)|^2 \leq \int_{\mathbb{R}^d} |J_\lambda(u)|\,du \leq c' \int_{\mathbb{R}^d} \frac{du}{(1+\lambda|u|)^{d+1}} = c\lambda^{-d},$$

as is evident by rescaling the last integral. This proves (13) and the proposition.

For later applications, it is of interest to elaborate some aspects of Proposition 2.5.

(i) The conclusion requires only that Φ is of class C^{d+2} and ψ of class C^{d+1}. In fact, as the patient reader may verify, in the estimate $|I(\lambda)| \leq A\lambda^{-d/2}$, the bound A depends only on the C^{d+2} norm of Φ, the C^{d+1} norm of ψ, the lower bound for $|\det\{\nabla^2\Phi\}|$, and the diameter of the support of ψ.

---

[3]Here we use the notation $X \lesssim Y$ and $X \approx Y$ to denote the fact that $X \leq cY$ and $c^{-1}Y \leq X \leq cY$ respectively, for appropriate constants c.

Similarly, the bound C_N appearing in Proposition 2.1 depends only on the C^{N+1} norm of Φ, the C^N norm of ψ, a lower bound for $|\nabla\Phi|$, and the diameter of the support of ψ.

(ii) There is a version of Proposition 2.5 in which we assume only that the rank of the Hessian of Φ is greater than or equal to m, $0 < m \leq d$, on the support of ψ. In that case the conclusion is

$$I(\lambda) = O(\lambda^{-m/2}), \qquad \text{as } \lambda \to \infty. \tag{17}$$

This may be deduced from the case $m = d$, already established. One proceeds as follows. For each x^0, the symmetric matrix $\nabla^2\Phi(x^0)$ can be diagonalized by introducing (via a rotation) a new coordinate system $x = (x', x'') \in \mathbb{R}^m \times \mathbb{R}^{d-m}$, so that $\nabla^2\Phi(x^0)$, when restricted to $\mathbb{R}^m$, has a non-vanishing determinant. Hence for a small open ball B centered at x^0, the same is true for $\nabla^2\Phi(x)$ when $x \in B$. Now for each fixed $x'' \in \mathbb{R}^{d-m}$ we use the proposition (where $d = m$) to obtain $|\int_{\mathbb{R}^m} e^{i\lambda\Phi(x',x'')}\psi_B(x', x'')\, dx'| \leq A\lambda^{-m/2}$, with ψ_B supported in B. After integrating in x'' and summing over finitely many such balls that cover the support of ψ, we obtain (17).

3 Fourier transform of surface-carried measures

We will now study surface-carried measures and their Fourier transforms. Our goal is a generalization of the estimate (5), which we had seen in the case of the sphere.

Recall from Section 4 of the previous chapter that given a point x^0 on a C^∞ hypersurface[4] M we dealt with a new coordinate system centered at x^0 (given via a translation and rotation of the initial coordinates), written as $x = (x', x_d) \in \mathbb{R}^{d-1} \times \mathbb{R}$, so that in a ball centered at x^0, the surface M is represented as

$$M = \{(x', x_d) \in \tilde{B} : \ x_d = \varphi(x')\} \tag{18}$$

where $\tilde{B}$ is the corresponding ball centered at the origin. We can also arrange matters so that the function φ, which is C^∞, satisfies $\varphi(0) = 0$, and $\nabla_{x'}\varphi(x')|_{x'=0}$.

Now this representation gives a defining function ρ_1 of M, with $\rho_1(x) = \varphi(x') - x_d$. Among the various possible defining functions of M near x^0, we now choose one, ρ, which is normalized by the condition $|\nabla\rho| = 1$ on M. This can be achieved by setting $\rho = \rho_1/|\nabla\rho_1|$ near M. With such

[4]The thrust of the C^∞ requirement is that M is of class C^k for sufficiently large k; later we will be more specific about how large k must be taken.

a normalized defining function, the **curvature form** of M at $x \in M$ (also known as the **second fundamental form**) is the form

$$\sum_{1 \le k,j \le d} \xi_k \xi_j \frac{\partial^2 \rho}{\partial x_k \partial x_j}(x), \tag{19}$$

restricted to vectors $\sum \xi_k \frac{\partial}{\partial x_k}$ that are tangent to M at x. The reader might note here the parallel between the curvature just described in terms of a quadratic form given by the defining function, and its complex analog (the Levi form) that was important in the previous chapter.

It is straightforward to verify that this form does not depend on the choice of a normalized defining function.

Now reverting to (18) and using $\nabla_{x'} \varphi(x')|_{x'=0} = 0$ we see that

$$\varphi(x') = \frac{1}{2} \sum_{1 \le k,j \le d-1} a_{kj} x_k x_j + O(|x'|^3),$$

and the curvature form is represented by the $(d-1) \times (d-1)$ matrix $\left\{ \frac{\partial^2 \varphi}{\partial x_k \partial x_j} \right\} = \{a_{kj}\}$, $1 \le k, j \le d-1$. Now if we make an appropriate rotation in the $x' \in \mathbb{R}^{d-1}$ space and relabel the coordinates accordingly we have

$$\varphi(x') = \frac{1}{2} \sum_{j=1}^{d-1} \lambda_j x_j^2 + O(|x'|^3).$$

The eigenvalues λ_j are called the **principal curvatures** of M (at x^0) and their product (the determinant of the matrix) is the **total curvature** or Gauss curvature of M.[5]

Notice that there is an implicit choice of signs (or "orientation") that has been made. The signs of the principal curvatures can be reversed if we use $-\rho$ instead of ρ as the defining function of M.

We mention briefly several examples.

EXAMPLE 1. The unit sphere in $\mathbb{R}^d$. If we start with $\rho_1 = |x|^2 - 1$ as a defining function, then $\rho = \frac{1}{2}\rho_1$ is "normalized." All the principal curvatures are equal to 1.

EXAMPLE 2. The parabolic hyperboloid $\{x_3 = x_1^2 - x_2^2\}$ in $\mathbb{R}^3$. This hypersurface has non-vanishing principal curvatures of opposite sign at each point.

[5]There is a neat geometric interpretation of the Gauss curvature in terms of "Gauss map," see Problem 1.

EXAMPLE 3. The circular cone $\{x_d^2 = |x'|^2,\ x_d \neq 0\}$ in $\mathbb{R}^d$. This hypersurface has $d-2$ identical non-vanishing principal curvatures at each point. The calculations involved are outlined in Exercise 9.

Next we consider the induced Lebesgue measure on M, the measure $d\sigma$ that has the following property: for any continuous function f on M with compact support

$$\int_M f\, d\sigma = \lim_{\epsilon\to 0} \frac{1}{2\epsilon} \int_{d(x,M)<\epsilon} F\, dx.$$

Here F is a continuous extension of f into a neighborhood of M and $\{x:\ d(x,M) < \epsilon\}$ is the "collar" of points at distance $< \epsilon$ from M. Now, as is well-known (see also Exercise 8), in our coordinate system $d\sigma = (1 + |\nabla_{x'}\varphi|^2)^{1/2}\, dx'$, in the sense that

$$\int_M f\, d\sigma = \int_{\mathbb{R}^{d-1}} f(x', \varphi(x'))(1 + |\nabla_{x'}\varphi|^2)^{1/2}\, dx'. \tag{20}$$

With this we can say that a measure $d\mu$ is a **surface-carried measure** on M with **smooth density** if $d\mu$ is of the form $d\mu = \psi d\sigma$, where ψ is a C^∞ function of compact support.

We now have all the ingredients necessary to state the main result concerning the Fourier transform of $d\mu$ defined by

$$\widehat{d\mu}(\xi) = \int_M e^{-2\pi i x\cdot\xi}\, d\mu.$$

Note that $\widehat{d\mu}(\xi)$ is bounded on $\mathbb{R}^d$ since the measure $d\mu$ is finite.

Theorem 3.1 *Suppose the hypersurface M has non-vanishing Gauss curvature at each point of the support of $d\mu$. Then*

$$|\widehat{d\mu}(\xi)| = O(|\xi|^{-(d-1)/2}) \quad \textit{as } |\xi| \to \infty. \tag{21}$$

Corollary 3.2 *If M has at least m non-vanishing principal curvatures at each point of the support of $d\mu$, then*

$$|\widehat{d\mu}(\xi)| = O(|\xi|^{-m/2}) \quad \textit{as } |\xi| \to \infty.$$

First some preliminary remarks. We can assume that the support of ψ is centered in a sufficiently small ball (so that in particular the representation (18) of M holds in it), because we can always write a given ψ as a finite sum of ψ_j of that type. Next, all our estimates can be made in the coordinate system used in (18) since the transformations of the x-space $\mathbb{R}^d$ used in that change of coordinates involves only a translation and a rotation. The Fourier transform $\widehat{d\mu}(\xi)$ then undergoes a multiplication by a factor of absolute value 1 (a character) and the same rotation in the ξ variable. Thus the estimate (21) is unchanged.

Now because of (20) we have

$$\widehat{d\mu}(\xi) = \int_{\mathbb{R}^{d-1}} e^{-2\pi i(x'\cdot\xi' + \varphi(x')\xi_d)} \tilde{\psi}(x')\, dx', \tag{22}$$

with $\xi = (\xi', \xi_d) \in \mathbb{R}^d$ and $\tilde{\psi}$ the C^∞ function with compact support given by

$$\tilde{\psi}(x') = \psi(x', \varphi(x'))(1 + |\nabla_{x'}\varphi|^2)^{1/2}.$$

We divide the ξ space into two parts: the "critical" region, the cone $|\xi_d| \geq c|\xi'|$, where c can be taken to be any fixed positive constant; and the subsidiary region, $|\xi_d| < c|\xi'|$, but here we need to assume that in fact c is small.

In the first region we may suppose that ξ_d is positive, since the case when ξ_d is negative follows by complex conjugation, or can be done similarly, and we write the exponent in the Fourier transform as

$$-2\pi i(x' \cdot \xi' + \varphi(x')\xi_d) = i\lambda\Phi(x'),$$

with the choice of $\lambda = 2\pi\xi_d$, and $\Phi(x') = -\varphi(x') - \frac{x'\cdot\xi'}{\xi_d}$. Observe that $\nabla^2_{x'}\Phi = -\nabla^2_{x'}\varphi$, and hence if the support of ψ is sufficiently small (which means we are sufficiently close to x^0), the determinant of the Hessian of Φ is non-vanishing. This is because of the corresponding property of φ that represents the non-vanishing of the curvature of M. Note also that Φ has, for any fixed N, a C^N norm that is uniformly bounded as ξ ranges over the set $|\xi_d| \geq c|\xi'|$. We can now apply Proposition 2.5 (with $\mathbb{R}^{d-1}$ in place of $\mathbb{R}^d$) and get

$$|\widehat{d\mu}(\xi)| = O(\lambda^{-\frac{d-1}{2}}) = O(\xi_d^{-\frac{d-1}{2}}) = O(|\xi|^{-\frac{d-1}{2}}),$$

since here $|\xi_d| \geq c|\xi'|$.

In the complementary region $|\xi_d| < c|\xi'|$ we write $\lambda = 2\pi|\xi'|$, and $\Phi(x') = -\varphi(x')\frac{\xi_d}{|\xi'|} - \frac{x'\cdot\xi'}{|\xi'|}$. Note that $|\nabla_{x'}\left(\frac{x'\cdot\xi'}{|\xi'|}\right)| = 1$, while $\frac{|\xi_d|}{|\xi'|}|\nabla_{x'}\varphi| \leq 1/2$ if c is so small that $c|\nabla_{x'}\varphi| \leq 1/2$ throughout the support of ψ. So if we invoke Proposition 2.1, the fact that $|\nabla_{x'}\Phi| \geq 1/2$ yields for each positive N,

$$|\widehat{d\mu}(\xi)| = O(\lambda^{-N}) = O(|\xi'|^{-N}) = O(|\xi|^{-N})$$

when ξ is in the second region. Taking $N \geq \frac{d-1}{2}$ completes the proof of the theorem.

The corollary can be proved by the same argument if one uses the estimate (17) instead of (13).

Suppose that Ω is a bounded region whose boundary $M = \partial\Omega$ satisfies the hypothesis of Theorem 3.1. If χ_Ω is the characteristic function of Ω, then its Fourier transform has a decay that is one order better than that of the corresponding surface-carried measure on its boundary.

Corollary 3.3 *If $M = \partial\Omega$ has non-vanishing Gauss curvature at each point, then*

$$\hat{\chi}_\Omega(\xi) = O(|\xi|^{-\frac{d+1}{2}}), \qquad \text{as } |\xi| \to \infty.$$

Proof. Using an appropriate partition of unity we can write

$$\chi_\Omega = \sum_{j=0}^{N} \psi_j \chi_\Omega,$$

with each ψ_j a C^∞ function of compact support; ψ_0 is supported in the interior of Ω, while each ψ_j, $1 \leq j \leq N$, is supported in a small neighborhood of the boundary in which the boundary is given as (18). Now since $\psi_0\chi_\Omega = \psi_0$, it is clear that $(\psi_0\chi_\Omega)^\wedge$ is rapidly decreasing. Next consider any $(\psi_j\chi_\Omega)^\wedge$ for $1 \leq j \leq N$. In analogy with (22), this has the form

$$\int_{x_d > \varphi(x')} e^{-2\pi i(x'\cdot\xi' + x_d\xi_d)} \psi_j(x', \xi_d)\, dx'\, dx_d,$$

which is, after changing variables so that $x_d = u + \varphi(x')$,

$$\int_{\mathbb{R}^{d-1}} e^{-2\pi i(x'\cdot\xi' + \varphi(x')\xi_d)} \Psi(x', \xi_d)\, dx' \tag{23}$$

where $\Psi(x', \xi_d) = \int_0^\infty e^{-2\pi i u\xi_d} \psi_j(x', u + \varphi(x'))\, du$. Note that $\Psi(x', \xi_d)$ is a C^∞ function in x' of compact support, uniformly in ξ_d. When $|\xi_d| < c|\xi'|$, the argument proceeds as before, giving an estimate $O(|\xi|^{-N})$ for each $N \geq 0$. To deal with the situation when $|\xi_d| \geq c|\xi'|$ write

$$\Psi(x', \xi_d) = -\frac{1}{2\pi i\xi_d} \int_0^\infty \frac{d}{du}(e^{-2\pi i u\xi_d}) \psi(x', u + \varphi(x))\, du.$$

and integrate by parts, giving us an additional decay of $O(1/|\xi_d|) = O(1/|\xi|)$ in (23). This proves the corollary.

Remark. In view of the comments following the proof of Proposition 2.5, we see that the results of this section hold if the C^∞ assumption we made about M is replaced by the requirement that M is only of class C^{d+2}.

4 Return to the averaging operator

We consider here a more general averaging operator. Given a hypersurface M in $\mathbb{R}^d$ and a surface-carried measure $d\mu = \psi d\sigma$ with smooth density of compact support, we set

$$A(f)(x) = \int_M f(x-y)\, d\mu(y).^6 \tag{24}$$

We shall prove that under the proper assumptions on M, the operator A regularizes f as a mapping from $L^2(\mathbb{R}^d)$ to $L^2_k(\mathbb{R}^d)$, and in addition that it "improves" f in the sense that it takes $L^p(\mathbb{R}^d)$ to $L^q(\mathbb{R}^q)$, for some $q > p$, if $1 < p < \infty$.

Theorem 4.1 *Suppose the Gauss curvature is non-vanishing at each point $x \in M$ in the support of $d\mu$. Then*

(a) *The map A given by (24) takes $L^2(\mathbb{R}^d)$ to $L^2_k(\mathbb{R}^d)$, with $k = \frac{d-1}{2}$.*

(b) *The map extends to a bounded linear transformation from $L^p(\mathbb{R}^d)$ to $L^q(\mathbb{R}^d)$ with $p = \frac{d+1}{d}$, and $q = d+1$.*

Corollary 4.2 *The Riesz diagram (see Section 2 in Chapter 2) of the map A is the closed triangle in the $(1/p, 1/q)$ plane whose vertices are $(0,0)$, $(1,1)$ and $(\frac{d}{d+1}, \frac{1}{d+1})$.*

In fact, the L^p, L^q boundedness asserted in this corollary is optimal, as is seen in Exercise 6.

Corollary 4.3 *If we only assume that M has at least m non-vanishing principal curvatures, then the same conclusions hold with $k = m/2$, and $p = \frac{m+2}{m+1}$, $q = m+2$.*

The proof of part (a) in the theorem is the same as that for the sphere once we invoke the decay (21), which implies that $(1+|\xi|^2)^{k/2}\widehat{d\mu}(\xi)$ is bounded. Hence

$$\begin{aligned}
\|A(f)\|_{L^2_k} &= \|(1+|\xi|^2)^{k/2}\widehat{Af}(\xi)\|_{L^2} \\
&= \|(1+|\xi|^2)^{k/2}\hat{f}(\xi)\widehat{d\mu}(\xi)\|_{L^2} \\
&\le c\|\hat{f}\|_{L^2} = c\|f\|_{L^2}.
\end{aligned}$$

[6]Here we have omitted a normalizing factor in the definition of A, since the density ψ is not necessarily positive.

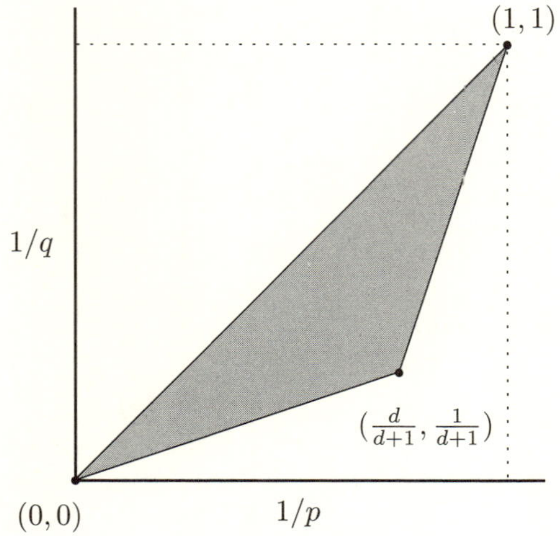

Figure 1. Riesz diagram of the map A in Corollary 4.2

The proof of part (b) combines two aspects of the operator A via interpolation, somewhat akin to the proof of the Hausdorff-Young theorem in Section 2 of Chapter 2. First, there is an $L^1 \to L^\infty$ estimate. The inequality involved is merely one of size, involving only the absolute value of our functions, but in order to get to it we have to "improve" the operator A by "integrating" it (of order 1). This estimate does not depend on the curvature of M.

Second, there is an $L^2 \to L^2$ estimate. It comes, like part (a) of the theorem, via Plancherel's theorem together with Theorem 3.1, and it allows us to "worsen" the operator A by essentially "differentiating" of degree $\frac{d-1}{2}$. The operator intermediate between the improved and the worsened operators is A itself, and the resulting intermediate estimate is then conclusion (b).

The scheme of the proof we have outlined in fact occurs in a number of situations. To carry it out we need a version of the Riesz interpolation theorem in which the operator in question is allowed to vary. The proper framework for this is an **analytic family of operators** defined as follows.[7]

For each s in the strip $S = \{a \leq \text{Re}(s) \leq b\}$ we assume we are given a linear mapping T_s taking simple functions on $\mathbb{R}^d$ to functions on $\mathbb{R}^d$ that are locally integrable. We also suppose that for any pair of simple

[7]Here we state the results for the space $\mathbb{R}^d$ with Lebesgue measure. The same ideas can be carried over to the setting of more general measure spaces as in Theorem 2.1 in Chapter 2.

functions f and g, the function

$$\Phi_0(s) = \int_{\mathbb{R}^d} T_s(f)g\,dx$$

is continuous and bounded in S and analytic in the interior of S. We further assume the two boundary estimates

$$\sup_{t\in\mathbb{R}} \|T_{a+it}(f)\|_{L^{q_0}} \leq M_0\|f\|_{L^{p_0}},$$

and

$$\sup_{t\in\mathbb{R}} \|T_{b+it}(f)\|_{L^{q_1}} \leq M_1\|f\|_{L^{p_1}}.$$

Proposition 4.4 *With the above assumptions,*

$$\|T_c\|_{L^q} \leq M\|f\|_{L^p},$$

for any c with $a \leq c \leq b$, where $c = (1-\theta)a + \theta b$ and $0 \leq \theta \leq 1$; and

$$\frac{1}{p} = \frac{1-\theta}{p_0} + \frac{\theta}{p_1} \quad \textit{and} \quad \frac{1}{q} = \frac{1-\theta}{q_0} + \frac{\theta}{q_1}.$$

Once we have formulated this result, we in fact observe that we can prove it by essentially the same argument as in Section 2 in Chapter 2.

We write $s = a(1-z) + bz$, so $z = \frac{s-a}{b-a}$, and the strip S is thereby transformed into the strip $0 \leq \mathrm{Re}(z) \leq 1$. For f and g given simple functions, we write $f_s = |f|^{\gamma(s)}\frac{f}{|f|}$ and $g_s = |g|^{\delta(s)}\frac{g}{|g|}$ where we define $\gamma(s) = p\left(\frac{1-s}{p_0} + \frac{s}{p_1}\right)$, and $\delta(s) = q\left(\frac{1-s}{q_0'} + \frac{s}{q_1'}\right)$. We then check that

$$\Phi(s) = \int_{\mathbb{R}^d} T_s(f_s)g_s\,dx$$

is continuous and bounded in the strip S and analytic in the interior. We then apply the three-lines lemma to $\Phi(s)$ and obtain the desired conclusion as in the proof of Theorem 2.1 in Chapter 2.

Returning to the averaging operator A, we shall assume (as we may) that the support of $d\mu$ has been chosen to lie in a ball for which M is given in coordinates by (18).

Now the operators T_s we will consider are convolution operators

$$T_s = f * K_s$$

defined initially for $\mathrm{Re}(s) > 0$, with

$$K_s = \gamma_s |x_d - \varphi(x')|_+^{s-1} \psi_0(x). \tag{25}$$

The following explains the several terms appearing in the definition of K_s.

- The factor γ_s equals $s(s+1)\cdots(s+N)e^{s^2}$.

  The purpose of the product $s(s+1)\cdots(s+N)$ will be clear momentarily, and the factor e^{s^2} is there to mitigate the growth of that polynomial as $\mathrm{Im}(s) \to \infty$. Here N is fixed with $N \geq \frac{d-1}{2}$.

- The function $|u|_+^{s-1}$ equals u^{s-1} when $u > 0$ and equals 0 when $u \leq 0$.

- $\psi_0(x) = \psi(x)(1 + |\nabla_{x'}\varphi(x')|^2)^{1/2}$, with ψ the density of $d\mu = \psi d\sigma$.

We note first that when $\mathrm{Re}(s) > 0$, the function K_s is integrable over $\mathbb{R}^d$. Our main claim is then the following.

Proposition 4.5 *The Fourier transform $\hat{K}_s(\xi)$ is analytically continuable into the half-plane $-\frac{d-1}{2} \leq \mathrm{Re}(s)$ and satisfies*

$$\sup_{\xi \in \mathbb{R}^d} |\hat{K}_s(\xi)| \leq M \qquad \text{in the strip } -\tfrac{d-1}{2} \leq \mathrm{Re}(s) \leq 1. \tag{26}$$

This is based on the following one-dimensional Fourier transform calculation. We suppose that F is a C^∞ function on $\mathbb{R}$ with compact support, and let

$$I_s(\rho) = s(s+1)\cdots(s+N)\int_0^\infty u^{s-1}F(u)e^{-2\pi i u\rho}\,du, \qquad \rho \in \mathbb{R}. \tag{27}$$

Lemma 4.6 *$I_s(\rho)$ initially given above for $\mathrm{Re}(s) > 0$, has an analytic continuation into the half-space $\mathrm{Re}(s) > -N-1$. Also*

(a) $|I_s(\rho)| \leq c_s(1+|\rho|)^{-\mathrm{Re}(s)}$, *when* $-N-1 < \mathrm{Re}(s) \leq 1$.

(b) $I_0(\rho) = N!F(0)$.

Here c_s is at most of polynomial growth in $\mathrm{Im}(s)$ and it depends only on the C^{N+1} norm of F and the support of F.

The reader should note that when when $\rho = 0$, we are dealing with the analytic continuation of a homogenous distribution, $|x|_+^{s-1}$, much in the same way as in Section 2.2 in Chapter 3.

Proof. Write $s(s+1)\cdots(s+N)u^{s-1} = \left(\frac{d}{du}\right)^{N+1} u^{s+N}$. Then an $(N+1)$-fold integration by parts yields

$$I_s(\rho) = (-1)^{N+1} \int_0^\infty u^{s+N} \left(\frac{d}{du}\right)^{N+1} (F(u)e^{-2\pi iu\rho})\, du,$$

from which the analytic continuation of I_s to the half-space $\mathrm{Re}(s) > -N-1$ is evident. It also proves the estimate (a) when ρ is bounded, for example when $|\rho| \leq 1$.

The proof of the size estimate (a) when $|\rho| > 1$ is similar but requires a little more care. We break the range of integration in (27) into two parts, essentially according to $u|\rho| \leq 1$ or $u|\rho| > 1$. We suppose η is a C^∞ cut-off function on $\mathbb{R}$ with $\eta(u) = 1$ if $|u| \leq 1/2$, and $\eta(u) = 0$ if $|u| \geq 1$, and insert $\eta(u\rho)$ or $1 - \eta(u\rho)$ in the integral (27).

When we insert $\eta(u\rho)$ we write the resulting integral as

$$(-1)^{N+1} \int_0^\infty u^{s+N} \left(\frac{d}{du}\right)^{N+1} (\eta(u\rho)e^{-2\pi iu\rho}F(u))\, du,$$

and so it is dominated by a constant multiple of

$$(1+|\rho|)^{N+1} \int_{0\leq u\leq 1/|\rho|} u^{\sigma+N}\, du, \qquad \text{with } \sigma = \mathrm{Re}(s).$$

Since $\sigma + N > -1$ this quantity is itself dominated by the product $(1+|\rho|)^{N+1}|\rho|^{-\sigma-N-1}$, which is $\lesssim (1+|\rho|)^{-\sigma}$, since we have assumed $|\rho| \geq 1$.

When we insert $1 - \eta(u\rho)$ we write the resulting integral as

$$s(s+1)\cdots(s+N)\frac{1}{(-2\pi i\rho)^k} \int_0^\infty u^{s-1}F(u)(1-\eta(u\rho))\left(\frac{d}{du}\right)^k (e^{-2\pi iu\rho})\, du$$

where k is chosen so that $\mathrm{Re}(s) < k$. Then, except for a factor that does not depend on ρ (and is a polynomial in s), the integral equals

$$\rho^{-k} \int_0^\infty e^{-2\pi iu\rho} \left(\frac{d}{du}\right)^k [u^{s-1}F(u)(1-\eta(\rho u))]\, du.$$

Since F has support in some interval $|u| \leq A$, it is easily verified that the above is dominated by a multiple of $\rho^{-k}\int_{1/(2|\rho|)}^A u^{\sigma-k-1}\, du$, which is $O(\rho^{-\sigma})$, because $\sigma = \mathrm{Re}(s) < k$; that yields the bound required in (a).

Finally, the integration by parts we have used also shows that

$$I_s(\rho) = -(s+1)\cdots(s+N) \int_0^\infty u^s \frac{d}{du}(F(u)e^{-2\pi iu\rho})\, du,$$

so setting $s = 0$ gives conclusion (b), since $F(0)$ is equal to the integral $-\int_0^\infty \frac{d}{du}(F(u)e^{-2\pi i u\rho})\,du$. The lemma is therefore proved.

We return to Proposition 4.5. Looking back at (25) we see that when $\text{Re}(s) > 0$, making the change of variables $u = x_d - \varphi(x')$ yields

$$\begin{aligned}\hat{K}_s(\xi) &= \gamma_s \int_{\mathbb{R}^d} |x_d - \varphi(x')|_+^{s-1} \psi_0(x) e^{-2\pi i(x'\cdot\xi' + x_d\xi_d)}\,dx \\ &= \gamma_s \int_0^\infty u^{s-1} e^{-2\pi i u\xi_d} \int_{\mathbb{R}^{d-1}} e^{-2\pi i(x'\cdot\xi' + \varphi(x')\xi_d)} \\ &\qquad\qquad \psi_0(x', u+\varphi(x'))\,dx'\,du \\ &= e^{s^2} I_s(\xi_d),\end{aligned} \tag{28}$$

with

$$F(u) = \int_{\mathbb{R}^{d-1}} e^{-2\pi i(x'\cdot\xi' + \varphi(x')\xi_d)} \psi_0(x', u+\varphi(x'))\,dx',$$

in the formula (27) for I_s.

However, by Theorem 3.1 (essentially the estimates we have for the integrals in (22)) it follows that $|F(u)| \le c(1+|\xi|)^{-\frac{d-1}{2}}$, with the same order of decay in $|\xi|$ for any derivative of F with respect to u. Therefore by conclusion (a) of the lemma we get that

$$|\hat{K}_s(\xi)| \le c_s |e^{s^2}| (1+|\xi_d|)^{-\text{Re}(s)} (1+|\xi|)^{-\frac{d-1}{2}},$$

which yields (26). Note that in the strip $-\frac{d-1}{2} \le \text{Re}(s) \le 1$, we have $|e^{s^2}| \le ce^{-(\text{Im}(s))^2}$ and c_s is at most of polynomial growth in $\text{Im}(s)$. Proposition 4.5 is therefore proved.

We now return to the operators T_s and apply our analysis of the kernels K_s.

Suppose f and g are a pair of simple functions on $\mathbb{R}^d$. The fact that these are in L^2 allows us to use the Fourier transform and Plancherel's theorem. So if we set $\Phi_0(s) = \int T_s(f)g\,dx$ for $\text{Re}(s) > 0$, then

$$\begin{aligned}\Phi_0(s) &= \int_{\mathbb{R}^d} (f * K_s)g\,dx = \int_{\mathbb{R}^d} (f*K_s)^\wedge \hat{g}(-\xi)\,d\xi \\ &= \int_{\mathbb{R}^d} \hat{K}_s(\xi)\hat{f}(\xi)\hat{g}(-\xi)\,d\xi.\end{aligned}$$

So the proposition and Schwarz's inequality show that $\Phi_0(s)$ is continuous and bounded on the strip $-\frac{d-1}{2} \le \text{Re}(s) \le 1$ and analytic in the interior. It is also apparent by the proposition that

$$\sup_t \|T_{-\frac{d-1}{2}+it}(f)\|_{L^2} \leq M\|f\|_{L^2}.$$

Next, clearly $\sup_x |K_s(x)| \leq M$, for $\mathrm{Re}(s) = 1$. Thus

$$\sup_t \|T_{1+it}(f)\|_{L^\infty} \leq M\|f\|_{L^1}.$$

However, by (28) and conclusion (b) of the lemma, $\hat{K}_0(\xi) = N!\widehat{d\mu}(\xi)$ and thus

$$T_0(f) = N!A(f).$$

We can therefore apply the interpolation theorem, Proposition 4.4. Here we have $a = -\frac{d-1}{2}$, $b = 1$, and $c = 0$. Also $p_0 = q_0 = 2$, $p_1 = 1$, $q_1 = \infty$. However $0 = (1-\theta)a + \theta b$ so $\theta = \frac{d-1}{d+1}$. Since $1/p = \frac{1-\theta}{2} + \theta$, we get $1/p = \frac{d}{d+1}$; similarly $1/q = \frac{1}{d+1}$, giving us the desired result for the operator A.

5 Restriction theorems

We come to a second significant application of oscillatory integrals. Here we focus on the possibility of restricting the Fourier transform of a function to a lower dimensional surface. The background for this is as follows.

5.1 Radial functions

To start with, the Fourier transform $\hat{f}$ of an L^1 function is continuous (see Section 4* in Chapter 2, Book III) while by the Hausdorff-Young theorem, $\hat{f}$ belongs to L^q if $f \in L^p$ for $1 \leq p \leq 2$, and $1/q + 1/p = 1$. Now L^q functions are in general determined only almost everywhere. Thus (without further examination) this suggests that the Fourier transform of an L^p function, $1 < p \leq 2$, cannot in general be meaningfully defined on a lower dimensional subset, and this is indeed the case when $p = 2$.

The first hint that things might in fact be quite different is the observation that for certain p, $1 < p < 2$, whenever f is radial and in L^p and $d \geq 2$, then its Fourier transform is continuous away from the origin.

Proposition 5.1 *Suppose $f \in L^p(\mathbb{R}^d)$ is a radial function. Then $\hat{f}$ is continuous for $\xi \neq 0$ whenever $1 \leq p < 2d/(d+1)$.*

Note the sequence of exponents $\frac{2d}{(d+1)}$: $1, \frac{4}{3}, \frac{3}{2}, \frac{8}{5}, \ldots$ that tends to 2 as $d \to \infty$.

Proof. Suppose $f(x) = f_0(|x|)$. Then $\hat{f}(\xi) = F(|\xi|)$ with F defined by (4), namely,

$$F(\rho) = 2\pi\rho^{-d/2+1} \int_0^\infty J_{d/2-1}(2\pi\rho r) f_0(r) r^{d/2}\, dr. \tag{29}$$

We can make the simplifying assumption that f vanishes in the unit ball (thus the integral above is taken for $r \geq 1$) because an L^p function supported in a ball is automatically in L^1 and its Fourier transform is then continuous.

We also restrict $\rho = |\xi|$ to a bounded interval excluding the origin, and note that then the integral in (29) converges absolutely and uniformly in ρ. In fact the integral is dominated by a constant multiple of

$$\int_1^\infty |f_0(r)| r^{d/2-1/2}\, dr, \tag{30}$$

since $|J_{d/2-1}(u)| \leq Au^{-1/2}$ if $u > 0$, as we have already seen. Now let q be the exponent dual to p, $(1/p + 1/q = 1)$, and write

$$r^{d/2-1/2} = r^{\frac{d-1}{p}} r^{\frac{d-1}{q}} r^{-\frac{d-1}{2}}.$$

Then by Hölder's inequality the integral (30) is majorized by the product of an L^p and an L^q norm. The L^p factor is

$$\left(\int_1^\infty |f_0(r)|^p r^{d-1}\, dr\right)^{1/p} = c\|f\|_{L^p(\mathbb{R}^d)},$$

while the second factor is

$$\left(\int_1^\infty r^{d-1-q\left(\frac{d-1}{2}\right)} dr\right)^{1/q},$$

and this is finite if $d - 1 - q\left(\frac{d-1}{2}\right) < -1$, which means $q > 2d/(d-1)$, and thus $p < 2d/(d+1)$. The asserted convergence of (30) therefore proves the continuity in ρ of F in (29) and establishes the proposition.

An examination of the proof shows the range $1 \leq p < 2d/(d+1)$ cannot be extended.

We now turn to the question of what happens when f is not assumed to be radial.

5.2 The problem

Let us fix a (local) hypersurface M in $\mathbb{R}^d$. One can then phrase the restriction problem for M as follows. Suppose $d\mu$ is a given surface-carried measure, $d\mu = \psi d\sigma$, with smooth non-negative density ψ of compact support. For a given $1 < p < 2$, does there exist a q (not necessarily the dual exponent to p) so that the *a priori* inequality below

$$(31) \qquad \left(\int_M |\hat{f}(\xi)|^q \, d\mu(\xi) \right)^{1/q} \leq c \|f\|_{L^p(\mathbb{R}^d)}$$

holds?

By this we mean the inequality (31) is to be valid for an appropriate dense class of functions f in L^p, with the bound c independent of f. If the answer to the question is affirmative we say that the (L^p, L^q) **restriction** holds for M.

Here is what we can assert about this problem.

1. Non-trivial results of the kind (31) are possible only if M has some degree of curvature.

2. Suppose M has non-zero Gauss curvature at each point (in particular when M is the sphere). Then one is led to guess that the correct range for (31) to be valid is $1 \leq p < 2d/(d+1)$ and $q \leq \left(\frac{d-1}{d+1}\right) p'$ with $1/p' + 1/p = 1$. Note the end-points of this relation, $q = \infty$ when $p = 1$, and $q \to 2d/(d+1)$ when $p \to 2d/(d+1)$. When $d = 2$ this guess is indeed correct; the proof is outlined in Problem 4.

3. For $d \geq 3$ it is still not known whether the expected result holds, but an interesting part, corresponding to $q = 2$ (and hence for $q \geq 2$) is settled. This is what we now address.

5.3 The theorem

Here we prove the following result.

Theorem 5.2 *Suppose M has non-zero Gauss curvature at each point of the support of $d\mu$. Then the restriction inequality* (31) *holds for $q = 2$ and $p = \frac{2d+2}{d+3}$.*

Note here that we have another sequence of exponents $\frac{2d+2}{d+3}$: 1, $\frac{6}{5}$, $\frac{4}{3}$, $\frac{10}{7}$, ..., tending to 2.

The proof starts with several quick observations. Let $\mathcal{R}$ denote the restriction operator

$$\mathcal{R}(f) = \hat{f}(\xi)\Big|_M = \int_{\mathbb{R}^d} e^{-2\pi i x\cdot\xi} f(x)\,dx\Big|_M,$$

which is initially defined to map continuous functions f of compact support on $\mathbb{R}^d$ to continuous functions on M. Consider also the "dual" $\mathcal{R}^*$, mapping continuous functions F on M to continuous functions on $\mathbb{R}^d$, defined by

$$\mathcal{R}^*(F)(x) = \int_M e^{2\pi i\xi\cdot x} F(\xi)\,d\mu(\xi).$$

We note that an interchange of integration proves the duality identity

$$(\mathcal{R}(f), F)_M = (f, \mathcal{R}^*(F))_{\mathbb{R}^d}, \tag{32}$$

where $(f,g)_{\mathbb{R}^d} = \int_{\mathbb{R}^d} f(x)\overline{g(x)}\,dx$ and $(F,G)_M = \int_M F(\xi)\overline{G(\xi)}\,d\mu(\xi)$.

Now we consider the composition $\mathcal{R}^*\mathcal{R}$. We have

$$\mathcal{R}^*\mathcal{R}(f)(x) = \int_M e^{2\pi i\xi\cdot x}\left\{\int_{\mathbb{R}^d} e^{-2\pi i y\cdot\xi} f(y)\,dy\right\} d\mu(\xi).$$

Hence

$$\mathcal{R}^*\mathcal{R}(f) = f * k, \qquad \text{with } k(x) = \widehat{d\mu}(-x). \tag{33}$$

There is then the following relation between bounds for $\mathcal{R}$, $\mathcal{R}^*$ and $\mathcal{R}^*\mathcal{R}$.

Proposition 5.3 *For a given p with $p \geq 1$, the three norm estimates below are equivalent:*

(i) $\|\mathcal{R}(f)\|_{L^2(M,d\mu)} \leq c\|f\|_{L^p(\mathbb{R}^d)}$.

(ii) $\|\mathcal{R}^*(F)\|_{L^{p'}(\mathbb{R}^d)} \leq c\|F\|_{L^2(M,d\mu)}$, *where* $1/p + 1/p' = 1$.

(iii) $\|\mathcal{R}^*\mathcal{R}(f)\|_{L^{p'}(\mathbb{R}^d)} \leq c^2\|f\|_{L^p(\mathbb{R}^d)}$.

The equivalence of (i) and (ii) follows directly from the duality of L^p spaces and the general duality theorem (Theorem 4.1 and Proposition 5.3 in Chapter 1).

We assume (i) (or (ii)) then this implies (iii) once we apply (ii) with $F = \mathcal{R}(f)$.

Conversely, we know by (32) that

$$(\mathcal{R}(f), \mathcal{R}(f))_M = (\mathcal{R}^*\mathcal{R}(f), f)_{\mathbb{R}^d}.$$

Hence if (iii) holds, then $(\mathcal{R}(f), \mathcal{R}(f))_M \leq c^2 \|f\|^2_{L^p(\mathbb{R}^d)}$ by Hölder's inequality. This gives (i) and the proposition is proved.

From this proposition, we see that to establish the theorem we have to show that the operator $\mathcal{R}^*\mathcal{R}$ is bounded from $L^p(\mathbb{R}^d)$ to $L^{p'}(\mathbb{R}^d)$, with $p = \frac{2d+2}{d+3}$. The argument is very much like that for the averaging operator A, except here inverted via the Fourier transform.

In fact, here the analytic family of operators we consider is $\{S_s\}$ given by

$$S_s(f) = f * k_s,$$

where k_s is defined by $k_s(x) = \hat{K}_s(-x)$, and K_s is given initially by (25), and with $\hat{K}_s$ extended in the strip $-\frac{d-1}{2} \leq \mathrm{Re}(s) \leq 1$ by Proposition 4.5.

Recall that $\hat{K}_0(\xi) = N!\widehat{d\mu}(\xi)$, so $S_0(f) = N!\mathcal{R}^*\mathcal{R}(f)$ by (33). But when $\mathrm{Re}(s) = 1$

$$\|S_s(f)\|_{L^2} \leq M\|f\|_{L^2},$$

since $K_{1+it} \in L^\infty$, and $\sup_t \|K_{1+it}\|_{L^\infty} \leq M$, we have

$$\hat{k}_{1+it}(\xi) = K_{1+it}(\xi).$$

Also, when $\mathrm{Re}(s) = -\frac{d-1}{2}$, then $k_s \in L^\infty$ by (26) in Proposition 4.5, since $k_s(x) = \hat{K}_s(-x)$. Thus

$$\sup_t \|S_{-\frac{d-1}{2}+it}(f)\|_{L^\infty} \leq M\|f\|_{L^1}.$$

Finally, it is easily verified (again using Proposition 4.5) that $\Phi_0(s) = \int_{\mathbb{R}^d} S_s(f)g\,dx$ is continuous, and is bounded in the strip $-\frac{d-1}{2} \leq \mathrm{Re}(s) \leq 1$ and analytic in the interior, whenever f and g are in $L^1(\mathbb{R}^d)$ (and in particular when f and g are simple). We therefore can apply the interpolation theorem (Proposition 4.4) to S_s. In this case $a = -\frac{d-1}{2}$, $b = 1$, and $c = 0$, so $0 = (1-\theta)a + \theta b$ implies that $\theta = \frac{d-1}{d+1}$. Also here $p_0 = 1$, $q_0 = \infty$, and $p_1 = 2$, $q_2 = 2$.

So $1/p = \frac{1-\theta}{p_0} + \frac{\theta}{p_1}$, gives $1/p = 1 - \theta + \theta/2 = 1 - \theta/2$ and as a result $1/p = \frac{d+3}{2d+2}$. Similarly $1/q = \frac{1-\theta}{q_0} + \frac{\theta}{q_1} = \theta/2$, and $1/q = 1 - 1/p = 1/p'$. Therefore $S_0 = N!\mathcal{R}^*\mathcal{R}$ maps L^p to $L^{p'}$ and by the equivalence guaranteed by Proposition 5.3, the theorem is proved.

Corollary 5.4 *Under the assumptions of the theorem, the restriction inequality* (31) *holds for* $1 \leq p \leq \frac{2d+2}{d+3}$ *and* $q \leq \left(\frac{d-1}{d+1}\right) p'$.

This follows by combining the critical case $p = \frac{2d+2}{d+3}$, $q \leq 2$ (a consequence of the theorem and Hölder's inequality) with the trivial case $p = 1$, $q = \infty$ via the Riesz interpolation theorem.

The key to the theorem is of course the decay of the Fourier transform of the surface-carried measure $d\mu$. This is highlighted by the following assertion which is clear upon reexamination of the proof of the theorem.

Here we deal with a hypersurface M, where we make no explicit assumptions about its curvature. The measures $d\mu$ considered will be of the form $\psi d\sigma$ as before.

Corollary 5.5 *Suppose that for some* $\delta > 0$, *we have*

$$|\widehat{d\mu}(\xi)| = O(|\xi|^{-\delta}), \qquad \text{as } |\xi| \to \infty, \text{ for all measures of the above form.}$$

Then the restriction property (31) *holds for* $p = \frac{2\delta+2}{\delta+2}$, $q = 2$.

In particular, if we assume M has m non-vanishing principal curvatures, then using the corollary in Section 3, we get this conclusion for $p = \frac{2m+4}{m+4}$.

6 Application to some dispersion equations

Dispersion equations have, broadly speaking, the property that as time varies, their solutions conserve some form of mass or energy (for example, the L^2 norm), yet these solutions disperse, in the sense that their sup-norms decay as time increases. In what follows we will see how the ideas we have discussed in this chapter apply to some equations of this kind, both linear and non-linear.

6.1 The Schrödinger equation

Typical of linear equations of the dispersion kind is the imaginary-time **Schrödinger equation**

$$\frac{1}{i}\frac{\partial u}{\partial t} = \triangle u, \tag{34}$$

for $u(x, t)$, and $(x, t) \in \mathbb{R}^d \times \mathbb{R} = \mathbb{R}^{d+1}$, with its Cauchy problem of determining a solution of (34) with initial data f, that is,

$$u(x, 0) = f(x). \tag{35}$$

Here $\triangle = \sum_{j=1}^{d} \frac{\partial^2}{\partial x_j^2}$ is the Laplacian on $\mathbb{R}^d$.

If we proceed formally, we are led to define the operators $e^{it\triangle}$ by

$$(36) \qquad (e^{it\triangle} f)^\wedge(\xi) = e^{-it4\pi^2|\xi|^2} \hat{f}(\xi),$$

where $^\wedge$ denotes the Fourier transform in the x-variable, and one expects that $u(x,t) = e^{it\triangle}(f)(x)$ is the solution of the problem (34) and (35). That this is so can be seen in two different contexts, the first of which is in the setting of the Schwartz space $\mathcal{S}$ of testing functions.

Proposition 6.1 *For each t:*

(i) $e^{it\triangle}$ *maps $\mathcal{S}$ to $\mathcal{S}$.*

(ii) *If we set $u(x,t) = e^{it\triangle}(f)(x)$, with $f \in \mathcal{S}$, then u is a C^∞ function of (x,t) that satisfies* (34) *and* (35).

(iii) $e^{it\triangle}(f) = f * K_t$, *if $t \neq 0$, where $K_t(x) = (4\pi i t)^{-d/2} e^{-|x|^2/(4it)}$.*

(iv) $\|e^{it\triangle}(f)\|_{L^\infty} \leq (4\pi|t|)^{-d/2}\|f\|_{L^1}$.

Proof. That $e^{it\triangle}$ maps $\mathcal{S}$ to $\mathcal{S}$ is clear because the multiplier $e^{-it4\pi^2|\xi|^2}$ has the property that each derivative in ξ is of at most polynomial increase. Next, the Fourier inversion formula gives

$$u(x,t) = \int_{\mathbb{R}^d} e^{-it4\pi^2|\xi|^2} e^{2\pi i x\cdot\xi} \hat{f}(\xi)\, d\xi.$$

The rapid decrease of $\hat{f}$ guarantees that the function u is C^∞ in the x and t variables. The fact that then u satisfies (34) is clear since the action of $\frac{1}{i}\frac{\partial}{\partial t}$ brings down a factor of $-4\pi^2|\xi|^2$, which is the same factor that results from the application of $\triangle$.

The conclusion (iii) is a consequence of the identity

$$(37) \qquad K_t^\wedge(\xi) = e^{-it4\pi^2|\xi|^2}, \qquad t \neq 0$$

when both bounded functions $K_t(x) = (4\pi i t)^{-d/2} e^{-|x|^2/(4it)}$ and $e^{-it4\pi^2|\xi|^2}$ are viewed as tempered distributions, and the usual relation between convolutions and Fourier transforms as in Chapter 3.

To prove (37) we start with the familiar identity for Gaussians

$$(u^{-d/2} e^{-\pi|x|^2/u})^\wedge(\xi) = e^{-u\pi|\xi|^2}, \qquad \text{when } u > 0.$$

Here we are dealing with rapidly decreasing functions and the Fourier transform is taken in, say, the L^1 sense. We now write $u = 4\pi s$, and we extend the above identity by analytic continuation to complex $s = \sigma + it$, with $\sigma > 0$, since the functions in question are still rapidly decreasing. Thus

$$((4\pi s)^{-d/2} e^{-|x|^2/(4s)})^\wedge = e^{-4\pi^2 s\,|\xi|^2}.$$

Finally, if t is fixed, $t \neq 0$, then letting $\sigma \to 0$, the functions on the left-hand side and right-hand side converge pointwise and boundedly (and hence in the sense of tempered distributions) to $K_t^\wedge(\xi)$ and $e^{-it4\pi^2|\xi|^2}$, respectively. Therefore (37) is established. Finally

$$\|f * K_t\|_{L^\infty} \leq \|K_t\|_{L^\infty} \|f\|_{L^1} = (4\pi|t|)^{-d/2} \|f\|_{L^1},$$

and the proposition is proved.

We look again at the operator $e^{it\triangle}$ given by (36), but now in the context of L^2.

Proposition 6.2 *For each t:*

(i) *The operator $e^{it\triangle}$ is unitary on $L^2(\mathbb{R}^d)$.*

(ii) *For every f, the mapping $t \mapsto e^{it\triangle}(f)$ is continuous in the $L^2(\mathbb{R}^d)$ norm.*

(iii) *If $f \in L^2(\mathbb{R}^d)$, then $u(x,t) = e^{it\triangle}(f)(x)$ satisfies (34) in the sense of distributions.*

Proof. Conclusion (i) is immediate from Plancherel's theorem, since the multiplier $e^{-it4\pi^2|\xi|^2}$ has absolute value one. Now if $\hat{f} \in L^2(\mathbb{R}^d)$, then clearly $e^{-it4\pi^2|\xi|^2}\hat{f}(\xi) \to e^{-it_0 4\pi^2|\xi|^2}\hat{f}(\xi)$ in the $L^2(\mathbb{R}^d)$ norm when $t \to t_0$, so (ii) follows again from Plancherel's theorem.

To prove the third conclusion we use the short-hand $\mathcal{L} = \frac{1}{i}\frac{\partial}{\partial t} - \triangle$, and $\mathcal{L}' = -\frac{1}{i}\frac{\partial}{\partial t} - \triangle$ for its transpose. Conclusion (iii) asserts that whenever φ is a C^∞ function on $\mathbb{R}^d \times \mathbb{R}$ of compact support, then

$$\iint_{\mathbb{R}^d \times \mathbb{R}} \mathcal{L}'(\varphi)(x,t)(e^{it\triangle} f)(x)\, dx\, dt = 0. \tag{38}$$

Now if $f \in \mathcal{S}$, then (38) holds for such f, because then $u(x,t) = e^{it\triangle}(f)(x)$ satisfies $\mathcal{L}(u) = 0$ in the usual sense, as we have seen. For general $f \in L^2$, approximate f in $L^2(\mathbb{R}^d)$ by a sequence $\{f_n\}$ with $f_n \in \mathcal{S}$. Then because of conclusion (i) we may pass to the limit and obtain (38) for any $f \in L^2(\mathbb{R}^d)$, finishing the proof of the proposition.

We remark that the decay estimate (iv) in the first proposition can be extended to read

$$(39) \qquad \|e^{it\triangle}f\|_{L^q(\mathbb{R}^d)} \leq c_p|t|^{-d(1/p-1/2)}\|f\|_{L^p(\mathbb{R}^d)},$$

if $1/q + 1/p = 1$, and $1 \leq p \leq 2$, with $c_p = (4\pi)^{-d(1/p-1/2)}$. This in fact is a direct consequence of the Riesz interpolation theorem (see Theorem 2.1 in Chapter 2) when we combine the cases corresponding to $p = 1$ and $p = 2$, in the propositions above. Another way to see (39) is to realize that the operator $e^{it\triangle}$ is a disguised version of a rescaled Fourier transform, and thus (39) is a restatement of the Hausdorff-Young theorem. This is outlined in Exercise 12.

Now the decay estimates (39) raise the question whether one can see any decrease for large time, when the initial data is merely assumed to be in L^2. Given the unitarity of $e^{it\triangle}$, the best one can hope for is an overall, or average, decay in both x and t. Thus one is led to ask whether an estimate of the kind

$$(40) \qquad \|u(x,t)\|_{L^q(\mathbb{R}^d\times\mathbb{R})} \leq c\|f\|_{L^2(\mathbb{R}^d)}$$

is possible (say for $q < \infty$).

By a simple scaling argument we can see that (40) can hold only with the exponent $q = \frac{2d+4}{d}$. Indeed, if $u(x,t) = e^{it\triangle}(f)(x)$, replace f by f_δ where $f_\delta(x) = f(\delta x)$, and u by u_δ, with $u_\delta(x,t) = u(\delta x, \delta^2 t)$, and $\delta > 0$. Then u_δ is a solution of (34) with corresponding initial data f_δ. That is, $u_\delta(x,t) = e^{it\triangle}(f_\delta)(x)$. Thus if (40) held, we would have $\|u_\delta\|_{L^q(\mathbb{R}^{d+1})} \leq c\|f_\delta\|_{L^2(\mathbb{R}^d)}$, for all $\delta > 0$, with c independent of δ. But $\|f_\delta\|_{L^2(\mathbb{R}^d)} = \delta^{-d/2}\|f\|_{L^2(\mathbb{R}^d)}$, while $\|u_\delta\|_{L^q(\mathbb{R}^{d+1})} = \delta^{-\frac{d+2}{q}}\|u\|_{L^q(\mathbb{R}^{d+1})}$, and so $\delta^{-\frac{d+2}{q}} \leq c'\delta^{-d/2}$ for all $\delta > 0$, which is possible only when $\frac{d+2}{q} = \frac{d}{2}$, that is, $q = \frac{2d+4}{d}$.

One should notice that $q = \frac{2d+4}{d}$ is exactly the (dual) exponent arising in the restriction result in Theorem 5.2 (that is, $1/p + 1/q = 1$, with $p = \frac{2d+4}{d+4}$) when we are in $\mathbb{R}^{d+1}$ instead of $\mathbb{R}^d$. This is no accident as we will now see.

Theorem 6.3 *If $u(x,t) = e^{it\triangle}(f)(x)$ with $f \in L^2(\mathbb{R}^d)$, then* (40) *holds when $q = \frac{2d+4}{d}$.*

Results of this kind are called **Strichartz estimates**. We will see that in fact this theorem is a direct consequence of the results in Section 5.

We consider the Fourier transform now on the space $\mathbb{R}^{d+1} = \mathbb{R}^d \times \mathbb{R} = \{(x, x_{d+1})\}$, relabeling the variable t as x_{d+1}. In the corresponding dual space (also $\mathbb{R}^{d+1}$) the dual variables are denoted by (ξ, ξ_{d+1}), with ξ dual to x and ξ_{d+1} dual to x_{d+1}. In this dual space we take M to be the paraboloid given by

$$M = \{(\xi, \xi_{d+1}) : \ \xi_{d+1} = -2\pi|\xi|^2\},$$

where $|\xi|^2 = \xi_1^2 + \cdots + \xi_d^2$.

On M we define the non-negative measure $d\mu = \psi d\sigma = \psi_0 d\xi$, where $d\xi$ is the Lebesgue measure on $\mathbb{R}^d$, and ψ_0 is a C^∞ function of compact support that equals 1 for $(\xi, \xi_{d+1}) \in M$ and $|\xi| \leq 1$. (As a result $\psi = \psi_0(1 + 16\pi^2|\xi|^2)^{1/2}$.)

Since the paraboloid M has a non-zero Gauss curvature, we can apply the restriction theorem, in particular its dual statement given in Proposition 5.3, with $\mathbb{R}^{d+1}$ in place of $\mathbb{R}^d$. This assertion deals with the operator

$$\mathcal{R}^*(F)(x) = \int_M e^{2\pi i(x\cdot\xi + x_{d+1}\xi_{d+1})} F(\xi, \xi_{d+1})\, d\mu$$

and then guarantees that

$$\|\mathcal{R}^*(F)\|_{L^q(\mathbb{R}^{d+1})} \leq c\|F\|_{L^2(M,d\mu)}.$$

Now let us take $F(\xi, \xi_{d+1}) = \hat{f}(\xi)$. Then we see that $\mathcal{R}^*(F) = e^{it\triangle}(f\psi_0)$, because we have set $x_{d+1} = t$, $d\mu = \psi_0 d\xi$, and on M we have $\xi_{d+1} = -2\pi|\xi|^2$. As a result

$$\|e^{it\triangle}(f)\|_{L^q(\mathbb{R}^{d+1})} \leq c\|f\|_{L^2(\mathbb{R}^d)}, \tag{41}$$

whenever $\hat{f}$ is supported in the unit ball. This is the essence of the result and from it the theorem follows easily.

In fact, if we replace f by $f_\delta(x) = f(\delta x)$, and u by $u_\delta(x) = u(\delta x, \delta^2 t)$ then, as we have seen above, (41) also holds with the same bound. However $(f_\delta)^\wedge(\xi) = \hat{f}(\xi/\delta)\delta^{-d}$, and now the support of $(f_\delta)^\wedge$ is the ball $|\xi| < \delta$. So allowing δ to be arbitrarily large shows that (41) is valid whenever f is in L^2 and $\hat{f}$ has compact support. Since such f are dense in L^2, a simple limiting argument establishes (41) for all $f \in L^2(\mathbb{R}^d)$, proving the theorem.

6.2 Another dispersion equation

We now digress briefly to touch on another dispersion equation and sketch certain aspects that are parallel with the Schrödinger equation.

We have in mind the cubic equation on $\mathbb{R} \times \mathbb{R}$

$$\frac{\partial u}{\partial t} = \frac{\partial^3 u}{\partial x^3},$$

with its initial value problem $u(x, 0) = f(x)$.

We can write the solution operator $f \mapsto e^{t\left(\frac{d}{dx}\right)^3}(f)$, with

$$\left(e^{t\left(\frac{d}{dx}\right)^3}(f)\right)^{\wedge}(\xi) = e^{t(2\pi i\xi)^3}\hat{f}(\xi).$$

Again this operator maps $\mathcal{S}$ to $\mathcal{S}$ for each t and is unitary on $L^2(\mathbb{R})$.

Note one difference with the Schrödinger equation: Here we can envisage solutions u that are real-valued, which is not possible for the equation (34), where the solutions need to be complex-valued because of the coefficient $1/i$.

When $t \neq 0$, we can write

$$e^{t\left(\frac{d}{dx}\right)^3}(f) = f * \tilde{K}_t, \qquad \text{for } f \in \mathcal{S},$$

where the kernel $\tilde{K}_t$ is given in terms of the Airy integral

$$\text{Ai}(u) = \frac{1}{2\pi}\int_{\mathbb{R}} e^{i\left(\frac{v^3}{3}+uv\right)}dv.$$ [8]

In fact, since $\tilde{K}_t(x) = \int_{\mathbb{R}} e^{t(2\pi i\xi)^3}e^{2\pi ix\xi}\,d\xi$, the change of variables $-(2\pi)^3 t\xi^3 = v^3/3$, $\xi = -v(3t)^{-1/3}(2\pi)^{-1}$ shows that

$$\tilde{K}_t(x) = (3t)^{-1/3}\text{Ai}(-x/(3t)^{1/3}).$$

Now one knows that

$$(42) \qquad \begin{cases} |\text{Ai}(u)| & \leq c \\ |\text{Ai}(u)| & \leq c|u|^{-1/4} \end{cases}$$

for all u. From the first of these inequalities we get the dispersion estimate

$$\|e^{t\left(\frac{d}{dx}\right)^3}(f)\|_{L^\infty} \leq c|t|^{-1/3}\|f\|_{L^1}.$$

There is also an analog of Theorem 6.3.

---

[8]The convergence of this integral and the estimates stated below can be found in Appendix A of Book II. There these are carried out using complex analysis. The results needed can also be obtained by the methods in Section 2 of this chapter, and are outlined in Exercise 13.

Theorem 6.4 *The solution* $e^{t\left(\frac{d}{dx}\right)^3}(f)$ *satisfies*

$$\|u\|_{L^q(\mathbb{R}^2)} \leq c\|f\|_{L^2(\mathbb{R})}, \qquad \text{with } q = 8.$$

The proof of this is result is parallel with that of the previous theorem and reduces to a restriction theorem on $\mathbb{R}^2$ for the cubic curve

$$\Gamma = \{(\xi_1, \xi_2) : \ \xi_2 = -4\pi^2\xi_1^3\}.$$

According to Corollary 5.5, what is needed is an estimate for $\widehat{d\mu}(\xi)$, where $d\mu$ is a smooth measure carried on the cubic curve Γ. The desired estimate can be rephrased as follows.

Lemma 6.5 *Let* $I(\xi) = \int_{\mathbb{R}} e^{2\pi i(\xi_1 t + \xi_2 t^3)}\psi(t)\,dt$, *where* ψ *is a* C^∞ *function of compact support. Then*

$$I(\xi) = O(|\xi|^{-1/3}), \qquad \text{as } |\xi| \to \infty.$$

Proof. First note that $I(\xi) = O(|\xi_2|^{-1/3})$. In fact

$$I(\xi) = \int_{|t|\leq|\xi_2|^{-1/3}} + \int_{|t|>|\xi_2|^{-1/3}}.$$

The first integral is obviously $O(|\xi_2|^{-1/3})$. For the second term we use the second derivative test (Proposition 2.3 and Corollary 2.4) noting that the second derivative of the phase exceeds $c|\xi_2||\xi_2|^{-1/3} = c|\xi_2|^{2/3}$, so this term is also $O(|\xi_2|^{-1/3})$, which proves that $I(\xi) = O(|\xi_2|^{-1/3})$. We apply this result when $|\xi_2| \geq c'|\xi_1|$, where c' is a suitably small constant, giving $I(\xi) = O(|\xi|^{-1/3})$ in this case.

In the case when $|\xi_1| > (1/c')|\xi_2|$, we apply the first derivative test (Proposition 2.1) noting that there the first derivative of the phase exceeds a multiple of $|\xi_1|$. Thus $I(\xi) = O(|\xi_1|^{-1}) = O(|\xi|^{-1/3})$. A combination of these two cases yields the lemma.

We can now invoke Corollary 5.5 with $\delta = 1/3$ and obtain

$$\|\mathcal{R}(f)\|_{L^2(\Gamma)} \leq c\|f\|_{L^p(\mathbb{R}^2)}$$

and

$$\|\mathcal{R}^*(F)\|_{L^q(\mathbb{R}^2)} \leq c\|F\|_{L^2(\Gamma)},$$

for $p = \frac{2\delta+2}{\delta+2} = \frac{8}{7}$, and $1/p + 1/q = 1$, so $q = 8$. The estimate for $\mathcal{R}^*$ then proves our theorem.

There are also corresponding space-time estimates for solutions of the wave equation in terms of its initial data. See Problem 5.

6.3 The non-homogeneous Schrödinger equation

We return to the imaginary-time Schrödinger equation and now consider the non-homogeneous problem

$$\frac{1}{i}\frac{\partial u}{\partial t} - \triangle u = F, \tag{43}$$

with F given. Here we require

$$u(x, 0) = 0. \tag{44}$$

It is easy to write down a formal solution to this problem by integrating the corresponding equation when $\triangle$ is replaced by a scalar. This leads to the solution operator

$$S(F)(x,t) = i\int_0^t e^{i(t-s)\triangle}F(\cdot, s)\,ds. \tag{45}$$

Here $e^{i(t-s)\triangle}F(\cdot, s)$ indicates that for each t and s the operator $e^{i(t-s)\triangle}$ has been applied to $F(x, s)$ as a function of x. The use of formula (45) can be justified in several different settings. The simplest is the following.

Proposition 6.6 *Suppose F is a C^∞ function on $\mathbb{R}^d \times \mathbb{R}$ of compact support. Then $S(F)$ is a C^∞ function that satisfies* (43) *and* (44).

Proof. Write $F = e^{it\triangle}G(\cdot, t)$ with $G(x,t) = i\int_0^t e^{-is\triangle}F(\cdot, s)\,ds$. Now $F(\cdot, s)$ is in the Schwartz space $\mathcal{S}(\mathbb{R}^d)$ for each s and depends smoothly on s. Thus the same is true for $G(\cdot, s)$ and then for $S(F)(\cdot, s)$, so this function is C^∞. Now differentiate both sides of the identity

$$e^{-it\triangle}(S(F))(\cdot, t) = i\int_0^t e^{-is\triangle}F(\cdot, s)\,ds,$$

with respect to t.

The left-hand side gives $e^{-it\triangle}\left(-i\triangle + \frac{\partial}{\partial t}\right)S(F)(\cdot, t)$. The right-hand side yields $ie^{-it\triangle}F(\cdot, t)$. After composing with $e^{it\triangle}$, we see that

$$\left(-i\triangle + \frac{\partial}{\partial t}\right)S(F)(\cdot, t) = iF(\cdot, t),$$

as was to be proved. Note that it is obvious that $S(F)(\cdot, 0) = 0$.

The corresponding result in the L^2 setting is detailed in Exercise 14.

We come to the key estimate for the operator S. It arises from the question of proving an estimate of the form

$$\|S(F)\|_{L^q(\mathbb{R}^d\times\mathbb{R})} \leq c\|F\|_{L^p(\mathbb{R}^d\times\mathbb{R})}, \tag{46}$$

where $q = \frac{2d+4}{d}$. Here q is the exponent for which $u \in L^q(\mathbb{R}^d \times \mathbb{R})$, whenever $u(x,t) = e^{it\triangle}(f)(x)$, with $f \in L^2$. Again, a simple scaling argument (which we leave to the reader) shows that (46) can hold only if $p = \frac{2d+4}{d+4}$, the dual exponent of q.

Theorem 6.7 *The estimate* (46) *holds if* $q = \frac{2d+4}{d}$ *and* $p = \frac{2d+4}{d+4}$.

This means that S, initially defined on C^∞ functions F of compact support, satisfies (46) with c independent of F, and hence has a unique extension to a bounded operator from $L^p(\mathbb{R}^d \times \mathbb{R})$ to $L^q(\mathbb{R}^d \times \mathbb{R})$ for which (46) is valid.

To prove the theorem we first make two simplifications. To begin with, we replace the operator S by S_+ given by

$$S_+(F)(x,t) = i\int_{-\infty}^{t} e^{i(t-s)\triangle}F(\cdot, s)\,ds,$$

and next, to avoid issues of convergence, we replace S_+ by S_ϵ, where

$$S_\epsilon(F)(x,t) = i\int_{-\infty}^{t} e^{i(t-s)\triangle}e^{-\epsilon(t-s)}F(\cdot, s)\,ds.$$

We will prove that

$$\|S_\epsilon(F)\|_{L^q(\mathbb{R}^d\times\mathbb{R})} \leq c\|F\|_{L^p(\mathbb{R}^d\times\mathbb{R})} \tag{47}$$

with c independent of ϵ. Once (47) is established then (46) will follow easily.

The advantage of S_+ (and S_ϵ) over S is that now we are dealing with convolutions on the space $\mathbb{R}^d \times \mathbb{R}$. For S_ϵ the kernel $\mathcal{K}(x,t)$ is formally $\frac{i}{(4\pi it)^{d/2}}e^{-\frac{|x|^2}{4it}}e^{-\epsilon t}$ when $t > 0$, and 0 when $t < 0$.

We prove (47) by the same method used in Theorem 4.1 and in the restriction theorem. We embed S_ϵ in an analytic family of operators, $\{T_z\}$, with the complex variable ranging over the half-plane $-1 \leq \mathrm{Re}(z)$. The operator will be first given when $d/2 - 1 < \mathrm{Re}(z)$ as a convolution, $T_z(f) = f * \mathcal{K}_z$, with the locally integrable kernel

$$\mathcal{K}_z(x,t) = \gamma(z)\frac{e^{-\frac{|x|^2}{4it}}}{(4\pi it)^{d/2}}e^{-\epsilon t}t_+^z. \tag{48}$$

Here $t_+^z = t^z$ when $t > 0$ and 0 otherwise, while $\gamma(z) = \frac{e^{z^2}}{\Gamma(z+1)}i$, and the factor $\gamma(z)$ is bounded in any strip $a \leq \text{Re}(z) \leq b$, because $\frac{1}{\Gamma(z+1)} = O(e^{|z|\log|z|})$ as $|z| \to \infty$, by Stirling's formula. We note that the Fourier transform of $\mathcal{K}_z$ on $\mathbb{R}^d \times \mathbb{R}$ (as a tempered distribution) is the function

$$\begin{aligned}\mathcal{K}_z^\wedge(\xi, \xi_{d+1}) &= \gamma(z) \int_0^\infty e^{-i4\pi^2 t|\xi|^2} e^{-\epsilon t} e^{-2\pi i t\xi_{d+1}} t^z \, dt \\ &= ie^{z^2}(\epsilon + i(4\pi^2|\xi|^2 + 2\pi\xi_{d+1}))^{-z-1}.\end{aligned}$$

This is because of (37) and the fact that

$$\int_0^\infty e^{-At} t^z \, dt = \Gamma(z+1)A^{-z-1} \qquad \text{whenever } \text{Re}(A) > 0,$$

as is seen by verifying the formula first when $A > 0$.

Next, if ϵ is fixed with $\epsilon > 0$, then $\mathcal{K}_z^\wedge$ is, by the above, a bounded function of $(\xi, \xi_{d+1}) \in \mathbb{R}^d \times \mathbb{R}$ as long as $-1 \leq \text{Re}(z)$. This Fourier multiplier defines T_z as a bounded operator on $L^2(\mathbb{R}^d \times \mathbb{R})$ whenever $-1 \leq \text{Re}(z)$, and gives a continuation of T_z, initially defined for $d/2 - 1 < \text{Re}(z)$. We also observe that $\mathcal{K}_z^\wedge$ is bounded independently of ϵ when $\text{Re}(z) = -1$, and therefore

$$\|T_z(F)\|_{L^2(\mathbb{R}^d\times\mathbb{R})} \leq c\|F\|_{L^2(\mathbb{R}^d\times\mathbb{R})} \qquad \text{when } \text{Re}(z) = -1, \tag{49}$$

with c independent of ϵ.

Now the kernel $\mathcal{K}_z$ given by (48) is clearly a bounded function on $\mathbb{R}^d \times \mathbb{R}$ when $\text{Re}(z) = d/2$, with a bound independent of ϵ. Thus

$$\|T_z(F)\|_{L^\infty} \leq c\|F\|_{L^1}, \qquad \text{when } \text{Re}(z) = d/2, \tag{50}$$

with c again independent of ϵ.

The interpolation theorem (Proposition 4.4) yields $\|T_0(F)\|_{L^q} \leq c\|F\|_{L^p}$, first for simple functions, and then by a passage to the limit for all F that are C^∞ of compact support. Again the bound is independent of ϵ. We also recognize that

$$T_0 = S_\epsilon \tag{51}$$

when acting on C^∞ functions of compact support.

In fact, by taking the Fourier transform in the x-variable we see that

$$S_\epsilon(F)^\wedge(\xi, t) = i \int_{-\infty}^t e^{-i(t-s)4\pi^2|\xi|^2} e^{-\epsilon(t-s)} \hat{F}(\xi, s) \, ds.$$

Then the Fourier transform in the t variable gives

$$\begin{aligned} S_\epsilon^\wedge(F)^\wedge(\xi,\xi_{d+1}) &= i\left(\int_0^\infty e^{-it4\pi^2|\xi|^2}e^{-\epsilon t}e^{-2\pi it\xi_{d+1}}\,dt\right)\hat{F}(\xi,\xi_{d+1}) \\ &= i(\epsilon - i(4\pi^2|\xi|^2+2\pi\xi_{d+1}))^{-1}\hat{F}(\xi,\xi_{d+1}), \end{aligned}$$

which establishes (51), and hence proves (47).

We now finish the proof by modifying F so that $F(x,s)=0$ when $s\leq 0$. Hence from (47), when we let $\epsilon\to 0$, we get

$$\left(\int_{\mathbb{R}^d}\int_0^\infty |S(F)(x,t)|^q\,dx\,dt\right)^{1/q} \leq c\|F\|_{L^p(\mathbb{R}^d\times\mathbb{R})}.$$

Changing t into $-t$ (and s into $-s$) gives us a parallel inequality, but with the integration in t now taken over $(-\infty,0)$. Adding these two finally yields (46) and the theorem is proved.

A final fact about the action of the solving operator S on the space $L^p(\mathbb{R}^d\times\mathbb{R})$ is as follows.

Proposition 6.8 *If $F\in L^p(\mathbb{R}^d\times\mathbb{R})$ then $S(F)$ can be corrected (that is, redefined on a set of measure zero) so that for each t, $S(F)(\cdot,t)$ belongs to $L^2(\mathbb{R}^d)$ and, moreover, the map $t\mapsto S(F)(\cdot,t)$ is continuous in the $L^2(\mathbb{R}^d)$ norm.*

This is based on the inequality

$$\|\int_\alpha^\beta e^{-is\triangle}F(\cdot,s)\,ds\|_{L^2(\mathbb{R}^d)} \leq c\|F\|_{L^p(\mathbb{R}^d\times\mathbb{R})}, \tag{52}$$

with c independent of the finite numbers α and β.

In fact, (52) is essentially the dual statement of (40) in Theorem 6.3. We let g be any element of $L^2(\mathbb{R}^d)$ with $\|g\|_{L^2(\mathbb{R}^d)}\leq 1$. Then by the unitarity of $e^{-is\triangle}$ we have

$$\int_\alpha^\beta\left(\int_{\mathbb{R}^d} e^{-is\triangle}F(x,s)\overline{g(x)}\,dx\right)ds = \int_\alpha^\beta\left(\int_{\mathbb{R}^d} F(x,s)\overline{v(x,s)}\,dx\right)ds,$$

where $v(x,s)=(e^{is\triangle}g)(x)$. So by (40), $\|v\|_{L^q(\mathbb{R}^d\times\mathbb{R})}\leq c$ and Hölder's inequality gives

$$\left|\int_{\mathbb{R}^d}\left(\int_\alpha^\beta e^{-is\triangle}F(\cdot,s)\,ds\right)\overline{g(x)}\,dx\right| \leq c\|F\|_{L^p(\mathbb{R}^d\times\mathbb{R})},$$

and since g was arbitrary, this suffices to establish (52).

Next, since $S(F)(x,t) = ie^{it\triangle}\int_0^t e^{-is\triangle}F(\cdot,s)\,ds$, taking $\alpha = 0$ and $\beta = t$ in (52), we see that for each t the function $S(F)(\cdot,s)$ belongs to $L^2(\mathbb{R}^d)$, and

$$\sup_t \|S(F)(\cdot,t)\|_{L^2(\mathbb{R}^d)} \leq c\|F\|_{L^p(\mathbb{R}^d\times\mathbb{R})}. \tag{53}$$

Finally, approximate F in the L^p norm by a sequence $\{F_n\}$ of C^∞ functions of compact support. Then for each n, $S(F_n)(\cdot,t)$ is clearly continuous in t in the $L^2(\mathbb{R}^d)$ norm. Since by (53)

$$\sup_t \|S(F)(\cdot,t) - S(F_n)(\cdot,t)\|_{L^2} \leq c\|F - F_n\|_{L^p} \to 0,$$

the continuity in t carries over to $S(F)(\cdot,t)$ and the proposition is proved.

6.4 A critical non-linear dispersion equation

We now consider the non-linear problem

$$\begin{cases} \dfrac{1}{i}\dfrac{\partial u}{\partial t} - \triangle u &= \sigma|u|^{\lambda-1}u \\ \\ u(x,0) &= f(x). \end{cases} \tag{54}$$

Here σ is a non-zero real number and the exponent λ is greater than 1. Besides its relative simplicity, the interest of the equation (54) is that its solution has two noteworthy conservation properties, namely that the "mass" $\int_{\mathbb{R}^d}|u|^2\,dx$, and the "energy" $\int_{\mathbb{R}^d}(\frac{1}{2}|\nabla u|^2 - \frac{\sigma}{\lambda+1}|u|^\lambda)\,dx$ are conserved over time. (See Exercise 15.)

We shall deal in particular with the initial-value problem for f in $L^2(\mathbb{R}^d)$. In this setting there is a "critical" exponent λ, the one for which the problem is scale-invariant. More precisely, suppose u is any solution of (54) with initial data f. Then we seek an exponent a so that $\delta^a u(\delta x, \delta^2 t)$ also solves the equation (54), (with initial data $\delta^a f(\delta x)$), for all $\delta > 0$. For the linear case $\sigma = 0$ of course any a will do, but in the present situation this requires $d + 2 = \lambda a$. Now if we also want the L^2 norm of the initial data to be invariant under these scalings then we need $a = d/2$, and as a result $\lambda = 1 + 4/d$.

We should observe a related significant fact about the critical exponent λ: we have $q = \lambda p$, where q and p are the dual exponents arising in our estimates (Theorems 6.3 and 6.7). This is the case because $q = \frac{2d+4}{d}$, $p = \frac{2d+4}{d+4}$, and $\lambda = \frac{d+4}{d}$.

Incidentally, one notices that the exact value of the coefficient σ in (54) is not significant; what matters is its sign, since it can be replaced by ± 1 via the fixed scaling $(x,t) \mapsto (|\sigma|^{1/2}x, |\sigma|t)$.

After these preliminaries we can now state the main result. Given an $f \in L^2(\mathbb{R}^d)$, we will say that a function u in $L^q(\mathbb{R}^d \times \mathbb{R})$ is a **strong solution** of (54) if

(i) u satisfies the differential equation in the sense of distributions.

(ii) For each t, the function $u(\cdot, t)$ belongs to $L^2(\mathbb{R}^d)$, the mapping $t \mapsto u(\cdot, t)$ is continuous in the $L^2(\mathbb{R}^d)$ norm, and $u(\cdot, 0) = f$.

We can also envisage solutions u that are given only for time t with $|t| < a$, for some fixed $0 < a < \infty$. In that case we assume u is in $L^q(\mathbb{R}^d \times \{|t| < a\})$ and consider u as a distribution on the open set $\mathbb{R}^d \times \{|t| < a\} \subset \mathbb{R}^d \times \mathbb{R}$, and define a strong solution in the same way as above.

The theorem below guarantees the solution of our problem under two scenarios. First for all times t, if the initial data is small enough. Second, for all initial data f, for a finite time interval.

Theorem 6.9 *Suppose λ, p and q are as above.*

(i) *There is an $\epsilon > 0$ so that whenever $\|f\|_{L^2(\mathbb{R}^d)} < \epsilon$ then there exists a strong solution of* (54).

(ii) *Given any $f \in L^2(\mathbb{R}^d)$, there is an $a > 0$, (depending on f), so that* (54) *has a strong solution for $|t| < a$.*

The proof exemplifies the use of fixed-point arguments in non-linear problems.

Suppose $u_0 = e^{it\triangle}(f)$. As will be seen, the problem reduces to finding u so that

$$u = \sigma S(|u|^{\lambda-1}u) + u_0. \tag{55}$$

The existence of u is obtained by a classical iteration argument, the existence of a fixed point of a suitable contraction mapping $\mathcal{M}$.

We consider first the alternative (i) of the theorem and here the mapping $\mathcal{M}$ will be defined on the underlying space

$$\mathcal{B} = \{u \in L^q(\mathbb{R}^d \times \mathbb{R}), \text{ with } \|u\|_{L^q} \leq \delta\},$$

with δ fixed below.

The mapping $\mathcal{M}$ will be given by

$$\mathcal{M}(u) = \sigma S(|u|^{\lambda-1}u) + u_0.$$

For an appropriate choice of δ, and then a choice of ϵ implying $\|f\|_{L^2} < \epsilon$, we will see that

(a) $\mathcal{M}$ maps $\mathcal{B}$ to itself.

(b) $\|\mathcal{M}(u) - \mathcal{M}(v)\|_{L^q} \leq \frac{1}{2}\|u - v\|_{L^q}$ for $u, v \in \mathcal{B}$.

In fact, $\|\mathcal{M}(u)\|_{L^q} \leq |\sigma|\|S(|u|^{\lambda-1}u)\|_{L^q} + \|u_0\|_{L^q}$. To estimate the first term we use Theorem 6.7, and this gives

$$\|S(|u|^{\lambda-1}u)\|_{L^q} \leq c\||u|^\lambda\|_{L^p} = c\|u\|^\lambda_{L^q},$$

since $q = p\lambda$. So if $\|u\|_{L^q} \leq \delta$, then $|\sigma|\|S(|u|^{\lambda-1}u)\|_{L^q} \leq \delta/2$, as long as $|\sigma|c\delta^\lambda \leq \delta/2$, which is the case if δ is small enough.

However by Theorem 6.3, $\|u_0\|_{L^q} \leq c\epsilon$, since $\|f\|_{L^2} < \epsilon$. Thus $\|u_0\|_{L^q} < \delta/2$, if $c\epsilon < \delta/2$, and with this choice of ϵ in terms of δ, property (a) is proved.

Next

$$\begin{aligned}\|\mathcal{M}(u) - \mathcal{M}(v)\|_{L^q} &= |\sigma|\|S(|u|^{\lambda-1}u - |v|^{\lambda-1}v)\|_{L^q} \\ &\leq c|\sigma|\||u|^{\lambda-1}u - |v|^{\lambda-1}v\|_{L^p}.\end{aligned}$$

However, as is easily verified

$$||u|^{\lambda-1}u - |v|^{\lambda-1}v| \leq c_\lambda|u - v|(|u| + |v|)^{\lambda-1}$$

for any pair of complex numbers u and v. Thus

$$\||u|^{\lambda-1}u - |v|^{\lambda-1}v\|_{L^p} \leq c_\lambda\|(u - v)(|u| + |v|)^{\lambda-1}\|_{L^p}.$$

Disregarding the constant c_λ, the pth power of the term on the right is $\int |u - v|^p(|u| + |v|)^{(\lambda-1)p}$. We estimate this by using Hölder's inequality with exponents λ and $\lambda' = \lambda/(\lambda - 1)$. Since $\lambda p = q$ and $\lambda'(\lambda - 1)p = q$ we see that this integral is majorized by

$$\left(\int |u - v|^q\right)^{1/\lambda}\left(\int (|u| + |v|)^q\right)^{1/\lambda'} = \|u - v\|^p_{L^q}\||u| + |v|\|^{(\lambda-1)p}_{L^q}.$$

Taking p^{th} root gives

$$\|\mathcal{M}(u) - \mathcal{M}(v)\|_{L^q} \leq c'_\lambda\|u - v\|_{L^q}\||u| + |v|\|^{(\lambda-1)}_{L^q},$$

and we only need to choose δ so that $c'_\lambda(2\delta)^{\lambda-1} \leq 1/2$ to obtain (b).

Next define $u_1, u_2, \ldots, u_k, \ldots$ successively according to $u_{k+1} = \mathcal{M}(u_k)$, $k = 0, 1, 2, \ldots$. Then, since $u_0 \in \mathcal{B}$, it follows from (a) that each $u_k \in \mathcal{B}$. Also by property (b) we have $\|u_{k+1} - u_k\|_{L^q} \le \frac{1}{2}\|u_k - u_{k-1}\|_{L^q}$ and hence $\|u_{k+1} - u_k\|_{L^q} \le \left(\frac{1}{2}\right)^k \|u_1 - u_0\|_{L^q}$.

Therefore the sequence $\{u_k\}$ converges in the L^q norm to a $u \in \mathcal{B}$, and hence $u = \mathcal{M}(u) = \sigma S(|u|^{\lambda-1}u) + u_0$, since $u_{k+1} = \mathcal{M}(u_k)$. To see that u is a distribution solution of (54) we must verify that

$$(56) \qquad \int_{\mathbb{R}^d \times \mathbb{R}} u\mathcal{L}'(\varphi)\, dx\, dt = \sigma \int_{\mathbb{R}^d \times \mathbb{R}} |u|^{\lambda-1} u\varphi\, dx\, dt,$$

for every φ that is C^∞ and has compact support, with $\mathcal{L}' = -\frac{1}{i}\frac{\partial}{\partial t} - \triangle$. However, by Proposition 6.6,

$$(57) \qquad \int S(F)\mathcal{L}'(\varphi)\, dx\, dt = \int F\varphi\, dx\, dt$$

if F is a C^∞ function of compact support. We now approximate an arbitrary F in $L^p(\mathbb{R}^d \times \mathbb{R})$ by a sequence $\{F_n\}$ of C^∞ functions of compact support. Since Theorem 6.7 implies that $S(F_n) \to S(F)$ in the L^q norm, the identity (57) for the F_n holds also for $F \in L^p$. Thus we may apply (57) to $F = \sigma|u|^{\lambda-1}u$ and use Proposition 6.2 part (iii) to conclude that (56) is valid, because $u = S(F) + u_0$.

Next, applying Proposition 6.8 to $F = \sigma|u|^{\lambda-1}u$ shows that for each t, the function $u(\cdot, t)$ belongs to $L^2(\mathbb{R}^d)$ and $t \mapsto u(\cdot, t)$ is continuous in the L^2 norm. Obviously $u(\cdot, 0) = f(\cdot)$ so the proof that u is a strong solution is complete.

In the second alternative, where we do not assume $\|f\| < \epsilon$, we instead choose a positive constant a so that

$$\left(\iint_{\mathbb{R}^d \times \{|t|<a\}} |e^{it\triangle}(f)(x,t)|^q\, dx\, dt\right)^{1/q} \le \delta/2.$$

Such a choice of a, which depends on f, is possible since $e^{it\triangle}f \in L^q(\mathbb{R}^d \times \mathbb{R})$. We then proceed as in the previous alternative with the understanding that now $\mathcal{B}$ consists of functions on $\mathbb{R}^d \times \{|t| < a\}$ (of norm $\le \delta$). Note that $S(F)(\cdot, t)$, for $|t| < a$, depends only on $F(\cdot, s)$ for $|s| < a$, so all inequalities used are still valid in this context, and the proof can be carried out as before.

The uniqueness of the solution of (54) and its continuous dependence on its initial data is outlined in Exercise 17.

7 A look back at the Radon transform

We now link the averaging operator studied in Section 4 with the Radon transform, pointing out certain striking affinities between these two, and formulating a common generalization.

Some elementary properties of the Radon transform were set down in Book I, where one can find an indication of its early interest. Of further significance is its role in the theory of Besicovitch-Kakeya sets. There, an L^2 smoothness property for $d \geq 3$, somewhat akin to that of averaging operators, is responsible for the continuity of measures of hyperplane sections asserted in Chapter 7 of Book III. Moreover the existence of Besicovitch sets may be said to be possible because when $d = 2$ the smoothing in L^2 is exactly of critical order $1/2$; in addition, this property of the Radon transform allowed one to see that Besicovitch sets in $\mathbb{R}^d$, $d = 2$, must have Hausdorff dimension 2.

7.1 A variant of the Radon transform

Recall that in $\mathbb{R}^d$ the **Radon transform** $\mathcal{R}$ is defined by

$$\mathcal{R}(f)(t,\gamma) = \int_{\mathcal{P}_{t,\gamma}} f$$

where $(t,\gamma) \in \mathbb{R} \times S^{d-1}$ and $\mathcal{P}_{t,\gamma}$ is the affine hyperplane $\{x : x \cdot \gamma = t\}$.

The smoothing property of $\mathcal{R}$ we have in mind can be stated most easily when $d = 3$ as the identity

$$\int_{S^2}\int_{\mathbb{R}} \left|\frac{d}{dt}\mathcal{R}(f)(t,\gamma)\right|^2 dt\, d\sigma(x) = 8\pi^2 \int_{\mathbb{R}^3} |f(x)|^2\, dx. \tag{58}$$

This is a direct consequence of the observation that $\hat{\mathcal{R}}(f)(\lambda,\gamma) = \hat{f}(\lambda\gamma)$, with $\hat{\mathcal{R}}(f)(\lambda,\gamma)$ denoting the Fourier transform in t of $\mathcal{R}(f)(t,\gamma)$ (with dual variable λ), and $\hat{f}$ denoting the usual 3-dimensional Fourier transform of f.

To pursue this point a little further we consider briefly a simple "linearized" variant of the Radon transform that, unlike $\mathcal{R}$, is directly given as a mapping of functions of $\mathbb{R}^d$ to functions of $\mathbb{R}^d$. This variant is determined once one fixes a non-degenerate bilinear form B on $\mathbb{R}^{d-1} \times \mathbb{R}^{d-1}$, and is denoted by

$$\mathcal{R}_B(f)(x) = \int_{\mathbb{R}^{d-1}} f(y', x_d - B(x',y'))\, dy',$$

where we have set $x = (x', x_d) \in \mathbb{R}^{d-1} \times \mathbb{R}$ and $y = (y', y_d) \in \mathbb{R}^{d-1} \times \mathbb{R}$. So $\mathcal{R}_B(f)(x)$ can be written as

$$\mathcal{R}_B(f)(x) = \int_{M_x} f,$$

with M_x denoting the affine hyperplane $\{(y', y_d) : \; y_d = x_d - B(x', y')\}$. The integration measure on M_x is taken to be dy', the Lebesgue measure on $\mathbb{R}^{d-1}$.

Note that the mapping $x \mapsto M_x$ is an injective mapping from $\mathbb{R}^d$ to the set of affine hyperplanes on $\mathbb{R}^d$, and this mapping is surjective on the collection of hyperplanes that are not perpendicular to the hyperplane M_0. Since the excepted collection of hyperplanes is a lower-dimensional subset, then, broadly speaking, $\mathcal{R}_B$ can be thought of as a substitute for $\mathcal{R}$.

Now let us revert to the simplest case, $d = 3$, where an analog of (58) is

$$\int_{\mathbb{R}^3} \left| \frac{\partial}{\partial x_3} \mathcal{R}_B(f)(x) \right|^2 dx = c_B \int_{\mathbb{R}^3} |f(x)|^2 \, dx, \tag{59}$$

which we prove when f is (say) a smooth function with compact support.

To see (59) consider the Fourier transform in the x_3-variable (with ξ_3 its dual variable), that is, $\hat{\mathcal{R}}_B(f)(x', \xi_3)$ is given by

$$\int_{\mathbb{R}^2} e^{-2\pi i \xi_3 B(x', y')} \hat{f}(y', \xi_3) \, dy',$$

where here $\hat{f}$ denotes the Fourier transform in the x_3-variable. Similarly, $\left(\frac{\partial}{\partial x_3} \mathcal{R}_B(f)\right)^{\wedge}(x', \xi_3)$ (also taking the Fourier transform in the x_3-variable) is given by

$$2\pi i \xi_3 \int_{\mathbb{R}^2} e^{-2\pi i \xi_3 B(x', y')} \hat{f}(y', \xi_3) \, dy'.$$

However, $B(x', y') = C(x') \cdot y'$ for some invertible linear transformation C on $\mathbb{R}^2$. Therefore, introducing the new variable $\xi_3 C(x') = u$, with $u \in \mathbb{R}^2$, we have $\xi_3 B(x', y') = u \cdot y'$ and $\xi_3^2 \, |\det(C)| \, dx' = du$. So an application of Plancherel's theorem in $\mathbb{R}^2$ leads to

$$\int_{\mathbb{R}^2} \left| \left(\frac{\partial}{\partial x_3} \mathcal{R}(f) \right)^{\wedge} (x', \xi_3) \right|^2 dx' = \frac{4\pi^2}{|\det(C)|} \int_{\mathbb{R}^2} |\hat{f}(y', \xi_3)|^2 \, dy'.$$

Hence, integrating in ξ_3 and applying Plancherel's theorem again, but this time in the x_3-variable, yields (59).

If we consider an appropriately localized version $\mathcal{R}'_B$ of $\mathcal{R}_B$, then using the above it is easy to see that

$$\|\mathcal{R}'_B(f)\|_{L^2_1(\mathbb{R}^3)} \leq c\|f\|_{L^2(\mathbb{R}^3)}.$$

Corresponding results for general d, when d is odd, giving L^2 smoothing of order $(d-1)/2$ can be obtained in the same way. The steps leading to these conclusions are outlined in Exercises 18 and 19.

7.2 Rotational curvature

We have learned from the above considerations that there seems to be a parallel between the averaging operator A and the Radon transform $\mathcal{R}_B$ in terms of their smoothing properties. Each of these operators is of the form

$$f \mapsto \int_{M_x} f(y)\, d\mu_x(y),$$

where for each $x \in \mathbb{R}^d$ we have a manifold M_x (that depends smoothly on x) over which we integrate. In the case of A it is $M_x = x + M$, and in the case of $\mathcal{R}_B$ it is $M_x = \{y = (y', x_d - B(x', y')),\ y' \in \mathbb{R}^{d-1}\}$. However, paradoxically, the key feature of A was the curvature of M, while in the case of $\mathcal{R}_B$ the corresponding manifolds M_x are hyperplanes and have no curvature. So how are we to see them as different manifestations of the same phenomenon? Another issue is the question of having a diffeomorphic-invariant formulation for the conclusions regarding these operators. This question arises naturally, because the spaces L^2, L^p, and L^2_k, are (at least locally) invariant under diffeomorphism.

What unifies the above examples is a common rotational curvature that takes into account not only the (possible) curvature of each fixed M_x, but how the M_x evolve (or "rotate") as x varies. This concept can be formulated as follows.

We start with a C^∞ function $\rho = \rho(x, y)$ given on a ball in $\mathbb{R}^d \times \mathbb{R}^d$ (a "double" defining function), and assign to it its **rotational matrix** $\mathcal{M}$, defined as the $(d+1) \times (d+1)$ matrix given by

$$\mathcal{M} = \begin{pmatrix} \rho & \frac{\partial \rho}{\partial y_1} & \cdots & \frac{\partial \rho}{\partial y_d} \\ \frac{\partial \rho}{\partial x_1} & & & \\ \vdots & & \frac{\partial^2 \rho}{\partial y_j \partial x_k} & \\ \frac{\partial \rho}{\partial x_d} & & & \end{pmatrix}.$$

We define the **rotational curvature** of ρ, denoted by rotcurv(ρ), as the determinant of the matrix $\mathcal{M}$,

$$\text{rotcurv}(\rho) = \det(\mathcal{M}).$$

Our basic condition is that where $\rho = 0$, then rotcurv$(\rho) \neq 0$. This clearly implies $\nabla_y \rho(x, y) \neq 0$ there. Hence if $M_x = \{y : \rho(x, y) = 0\}$, each M_x is a C^∞ hypersurface in $\mathbb{R}^d$, that in fact depends smoothly on x. We then note the following properties of rotational curvature that are straight-forward to verify.

1. If $\rho(x, y) = \rho(x - y)$, the translation-invariant case, then $M_x = x + M_0$. Here one also has the condition that rotcurv$(\rho) \neq 0$ is equivalent with the non-vanishing of the Gauss curvature of M_0.

2. In the case of $\mathcal{R}_B$, we take $\rho(x, y) = y_d - x_d + B(x', y')$, and then rotcurv$(\rho) \neq 0$ is equivalent to B being non-degenerate.

3. If $\rho'(x, y) = a(x, y)\rho(x, y)$, with $a(x, y) \neq 0$, then ρ' is another defining function for $\{M_x\}$ and rotcurv$(\rho') = a^{d+1}$rotcurv(ρ) whenever $\rho = 0$.

4. The invariance of rotational curvature under local diffeomorphisms can be stated this way: Suppose $x \mapsto \Psi_1(x)$ and $y \mapsto \Psi_2(y)$ are a pair of (local) diffeomorphisms on $\mathbb{R}^d$ and set $\rho'(x, y) = \rho(\Psi_1(x), \Psi_2(y))$. Then rotcurv$(\rho') = \mathcal{J}_1(x)\mathcal{J}_2(y)$rotcurv$(\rho)$ whenever $\rho'(x, y) = 0$, where $\mathcal{J}_1$ and $\mathcal{J}_2$ are the Jacobian determinants of Ψ_1 and Ψ_2 respectively.

With these notions in hand we can come to the regularity theorem for the general form of the Radon transform.

We assume we are given a double defining function ρ as above with rotcurv$(\rho) \neq 0$. We set $M_x = \{y : \ \rho(x, y) = 0\}$. For each x we let $d\sigma_x(y)$ be the induced Lebesgue measure on M_x, and define $d\mu_x(y) = \psi_0(x, y) d\sigma_x(y)$, where ψ_0 is some fixed C^∞ function on $\mathbb{R}^d \times \mathbb{R}^d$ of compact support. Given this, we define the general averaging operator $\mathcal{A}$ by

$$\mathcal{A}(f)(x) = \int_{M_x} f(y)\, d\mu_x(y), \tag{60}$$

initially for functions f on $\mathbb{R}^d$ that are (say) continuous with compact support.

Theorem 7.1 *The operator $\mathcal{A}$ extends to a bounded linear map of $L^2(\mathbb{R}^d)$ to $L^2_k(\mathbb{R}^d)$, with $k = \frac{d-1}{2}$.*

It should be pointed out that the averaging operator A of Section 4 is translation-invariant, and the Radon transform $\mathcal{R}_B$ is partially so; it is translation-invariant with respect to the x_3-variable. So in both cases the Fourier transform can be used. However in the general situation the Fourier transform is unavailable and we must proceed differently.

There will be two steps. The first will use an oscillatory integral operator that partly substitutes for the Fourier transform and Plancherel's theorem. The second is an L^2 estimate, obtained via a dyadic decomposition of "almost-orthogonal" parts, that further serves to implement this approach.

7.3 Oscillatory integrals

We turn to the first idea. We consider an operator T_λ (depending on a positive parameter λ) of the form

$$T_\lambda(f)(x) = \int_{\mathbb{R}^d} e^{i\lambda\Phi(x,y)}\psi(x,y)f(y)\,dy.$$

Here Φ and ψ are a pair of C^∞ functions on $\mathbb{R}^d \times \mathbb{R}^d$, the latter assumed to have compact support. The phase Φ is supposed to be real-valued, and the key assumption is that its mixed Hessian

$$\det\{\nabla^2_{x,y}\Phi\} = \det\{\frac{\partial^2\Phi}{\partial x_k \partial y_j}\}_{1\leq k,j\leq d}, \tag{61}$$

is non-vanishing on the support of ψ.

Proposition 7.2 *Under the above assumptions we have* $\|T_\lambda\| \leq c\lambda^{-d/2}$, $\lambda > 0$, *with* $\|\cdot\|$ *denoting the norm of the operator acting on* $L^2(\mathbb{R}^d)$.

For us the importance of this proposition is the consequence it has for a corresponding oscillatory integral that involves the defining function ρ. We set

$$S_\lambda(f)(x) = \int_{\mathbb{R}\times\mathbb{R}^d} e^{i\lambda y_0 \rho(x,y)}\psi(x,y_0,y)f(y)\,dy_0\,dy. \tag{62}$$

Here the integration is over $(y_0, y) \in \mathbb{R}\times\mathbb{R}^d$. The function ψ is again a C^∞ function with compact support in all variables, but the noteworthy further assumption is that ψ is supported away from $y_0 = 0$.

Corollary 7.3 *Assume that the double defining function* ρ *satisfies the condition* $\mathrm{rotcurv}(\rho) \neq 0$ *on the set where* $\rho = 0$. *Then*

$$\|S_\lambda\| \leq c\lambda^{-\frac{d+1}{2}}.$$

Note. We have an extra gain of $\lambda^{-1/2}$ over what can be said for T_λ.

The proof of the proposition is in many ways like that of the scalar version, Proposition 2.5 in Section 2, so we will be brief. As before, we begin by taking the precaution that ψ is supported in a small ball. Now if T is an operator on L^2, then $\|T^*T\| = \|T\|^2$, where T^* denotes the adjoint of T.[9]

However T_λ is given by the kernel $K(x,y) = e^{i\lambda\Phi(x,y)}\psi(x,y)$, that is, $T_\lambda(f)(x) = \int K(x,y)f(y)\,dy$, so T_λ^* is given by the kernel $\overline{K}(x,y)$ and $T_\lambda^* T_\lambda$ is given by the kernel

$$M(x,y) = \int_{\mathbb{R}^d} \overline{K}(z,x)K(z,y)\,dz = \int_{\mathbb{R}^d} e^{i\lambda[\Phi(z,y)-\Phi(z,x)]}\psi(x,y,z)\,dz,$$

with $\psi(x,y,z) = \overline{\psi}(z,x)\psi(z,y)$. The crucial point will be like (14), namely,

$$|M(x,y)| \leq c_N(\lambda|x-y|)^{-N} \quad \text{for every } N \geq 0.$$

Here, with $z = (z_1, \ldots, z_d) \in \mathbb{R}^d$, we use the vector field

$$L = \frac{1}{i\lambda}\sum_{j=1}^{d} a_j \frac{\partial}{\partial z_j} = a \cdot \nabla_z$$

and its transpose, $L^t(f) = -\frac{1}{i\lambda}\sum_{j=1}^{d} \frac{\partial(a_j f)}{\partial z_j}$, where

$$(a_j) = a = \frac{\nabla_z(\Phi(z,x) - \Phi(z,y))}{|\nabla_z(\Phi(z,x) - \Phi(z,y))|^2}.$$

Now because $u = x - y$ is sufficiently small in view of the support assumptions made on ψ, we see as before that $|a| \approx |x-y|^{-1}$ and $|\partial_x^\alpha a| \lesssim |x-y|^{-1}$, for all α. Thus

$$\begin{aligned}
|M(x,y)| &\leq \left| \int L^N\left(e^{i\lambda[\Phi(z,y)-\Phi(z,x)]}\right)\psi(x,y,z)\,dz \right| \\
&\leq \int |(L^t)^N \psi(x,y,z)|\,dz \\
&\leq c_N(\lambda|x-y|)^{-N}.
\end{aligned}$$

[9] In this connection, see for instance Exercise 19, in Chapter 4 of Book III.

However, then

$$\begin{aligned}|T_\lambda^* T_\lambda f(x)| &\le \int |M(x,y)||f(y)|\,dy \\ &\le \int M^0(x-y)|f(y)|\,dy \\ &= \int M^0(y)|f(x-y)|\,dy,\end{aligned}$$

where $M^0(u) = c'_N(1+\lambda|u|)^{-N}$, and by Minkowski's inequality

$$\|T_\lambda^* T_\lambda(f)\|_{L^2} \le \|f\|_{L^2} \int M^0(u)\,du.$$

However $\int M^0(u)\,du = c\lambda^{-d}$, if in the estimate for M^0 the N is taken to be greater than d. As a result $\|T_\lambda^* T_\lambda\| \le c\lambda^{-d}$, and the proposition is proved.

We turn now to the corollary. The link between the rotational curvature of ρ and the phase Φ in the proposition occurs in passing from $\mathbb{R}^d$ to $\mathbb{R}^{d+1}$. With $\overline{x} = (x_0, x) \in \mathbb{R}\times\mathbb{R}^d = \mathbb{R}^{d+1}$ and $\overline{y} = (y_0, y) \in \mathbb{R}\times\mathbb{R}^d = \mathbb{R}^{d+1}$, we set

$$\Phi(\overline{x}, \overline{y}) = x_0 y_0 \rho(x,y).$$

Then, as is evident,

$$\det(\nabla^2_{\overline{x},\overline{y}}\Phi) = (x_0y_0)^{d+1}\text{rotcurv}(\rho).$$

Now define $F_\lambda(x_0,x)$ by

$$\text{(63)}\quad \begin{aligned}F_\lambda(x_0,x) = F_\lambda(\overline{x}) &= \int_{\mathbb{R}^{d+1}} e^{i\lambda\Phi(\overline{x},\overline{y})}\psi_1(x_0,x,y_0,y)f(y)\,dy_0\,dy \\ &= \int_{\mathbb{R}^{d+1}} e^{i\lambda x_0 y_0 \rho(x,y)}\psi_1(x_0,x,y_0,y)f(y)\,dy_0\,dy\end{aligned}$$

with $\psi_1(1,x,y_0,y) = \psi(x,y_0,y)$, and ψ_1 having compact support that is disjoint from $x_0 = 0$ or $y_0 = 0$.

This means that $S_\lambda(f)(x) = F_\lambda(1,x)$.

To proceed we need the following little calculus lemma, valid for any function g which is of class C^1 in an interval I of length one. Suppose $u_0 \in I$, then

$$\text{(64)}\qquad |g(u_0)|^2 \le 2\left(\int_I |g(u)|^2\,du + \int_I |g'(u)|^2\,du\right).$$

Indeed, for any $u \in I$, one has $g(u_0) = g(u) + \int_u^{u_0} g'(r)\,dr$. So by Schwarz's inequality

$$|g(u_0)|^2 \leq 2\left(|g(u)|^2 + \int_I |g'(r)|^2\,dr\right),$$

and an integration in u ranging over I then yields (64).

We apply this inequality with $I = [1,2]$, $u_0 = 1$ and $g(u) = F_\lambda(u,x)$ (that is, u is the variable x_0). Since $F_\lambda(1,x) = S_\lambda(f)(x)$, we therefore have after an integration in $x \in \mathbb{R}^d$

$$\int_{\mathbb{R}^d} |S_\lambda(f)(x)|^2\,dx \leq 2\left(\int_{\mathbb{R}\times\mathbb{R}^d} |F_\lambda(x_0,x)|^2\,dx_0\,dx \right.$$
$$\left. + \int_{\mathbb{R}\times\mathbb{R}^d} |\frac{\partial}{\partial x_0} F_\lambda(x_0,x)|^2\,dx_0\,dx\right).$$

The first term on the right-hand side of the inequality is dominated by a multiple of $\lambda^{-(d+1)}\int_{\mathbb{R}^d}|f(y)|^2\,dy$, as we see by applying the proposition (with $\mathbb{R}^{d+1}$ in place of $\mathbb{R}^d$), since ψ_1 has compact support in y_0.

However the second term is more problematic, because differentiation in x_0 in (63) brings down an extra factor of λ. We get around this by observing that

$$\frac{\partial}{\partial x_0}\left(e^{i\lambda x_0 y_0 \rho(x,y)}\right) = \frac{\partial}{\partial y_0}\left(e^{i\lambda x_0 y_0 \rho(x,y)}\right)\frac{y_0}{x_0}$$

and then integrating by parts in the y_0 variable in (63). We note that because of the support property of ψ_1, the variable y_0 is bounded away from 0, and the differentiation in y_0 falls only on the smooth functions of the integrand, and not $f(y)$ since it is independent of y_0.

This shows that the second term also satisfies the desired estimate, establishing the corollary.

7.4 Dyadic decomposition

We now come to the dyadic decomposition of the operators $\mathcal{A}$. When we fix any Schwartz function h on $\mathbb{R}$ that is normalized by $\int_{\mathbb{R}} h(\rho)\,d\rho = 1$, then we know (see Exercise 8) that for any smooth hypersurface M in $\mathbb{R}^d$ with defining function ρ, and any continuous function f on $\mathbb{R}^d$ of compact support,

$$\lim_{\epsilon\to 0}\epsilon^{-1}\int_{\mathbb{R}^d} h(\rho(x)/\epsilon)f(x)\,dx = \int_M f\frac{d\sigma}{|\nabla\rho|},$$

with $d\sigma$ the induced Lebesgue measure on M.

As a result (see (60))

$$\mathcal{A}(f)(x) = \int_{M_x} f(y)\, d\mu_x(y) = \lim_{\epsilon \to 0} \epsilon^{-1} \int_{\mathbb{R}^d} h\left(\frac{\rho(x,y)}{\epsilon}\right) \psi(x,y) f(y)\, dy$$

where $\psi(x,y)$ is a C^∞ function of compact support given by $\psi(x,y) = \psi_0(x,y)|\nabla_y \rho|$ and $d\mu_x(y) = \psi_0(x,y)\, d\sigma_x(y)$.

Now choose $\gamma(u)$ to be a C^∞ function on $\mathbb{R}$ with γ supported in $|u| \leq 1$, and $\gamma(u) = 1$ if $|u| \leq 1/2$, and let $h(\rho) = \int_{\mathbb{R}} e^{2\pi i u \rho} \gamma(u)\, du$. Then by the Fourier inversion theorem $\int_{\mathbb{R}} h(\rho)\, d\rho = 1$, and also $\int e^{2\pi i u \rho} \gamma(\epsilon u)\, du = \epsilon^{-1} h(\rho/\epsilon)$.

Next write $\epsilon = 2^{-r}$, with r a positive integer, and $\gamma(2^{-r}u) = \gamma(u) + \sum_{k=1}^{r} (\gamma(2^{-k}u) - \gamma(2^{-k-1}u))$. Letting $r \to \infty$ we have

$$1 = \gamma(u) + \sum_{k=1}^{\infty} \eta(2^{-k}u)$$

with $\eta(u) = \gamma(u) - \gamma(u/2)$, and η is supported in $1/2 \leq |u| \leq 2$.

As a result of the above, we can write, whenever f is continuous,

$$\mathcal{A}(f)(x) = \sum_{k=0}^{\infty} \mathcal{A}_k(f)(x) = \lim_{r \to \infty} \sum_{k=0}^{r} \mathcal{A}_k(f)(x),$$

where

$$\mathcal{A}_k(f)(x) = \int_{\mathbb{R} \times \mathbb{R}^d} e^{2\pi i u \rho(x,y)} \eta(2^{-k}u) \psi(x,y) f(y)\, du\, dy \tag{65}$$

(with a similar formula for $\mathcal{A}_0$, but with $\eta(2^{-k}u)$ replaced by $\gamma(u)$). The limit here exists for each x.

We now make the following observations about the operators $\mathcal{A}_k(f)$, the first of which is self-evident.

(a) $\mathcal{A}_k(f)$ is a C^∞ function of compact support for each $f \in L^2(\mathbb{R}^d)$.

(b) We have the estimates

$$\|\mathcal{A}_k(f)\|_{L^2} \leq c 2^{-k\left(\frac{d-1}{2}\right)} \|f\|_{L^2}. \tag{66}$$

In fact, if we make the change of variables $2^{-k}u = y_0$, then

$$\mathcal{A}_k(f)(x) = 2^k \int_{\mathbb{R} \times \mathbb{R}^d} e^{2\pi i 2^k y_0 \rho(x,y)} \psi(x, y_0, y) f(y)\, dy_0\, dy, \tag{67}$$

with $\psi(x, x_0, y) = \psi(x, y)\eta(y_0)$, which, in light of (62), equals

$$2^k S_\lambda(f)(x),$$

where $\lambda = 2\pi 2^k$. Thus the inequality (66) is an immediate consequence of Corollary 7.3, since η is supported away from zero.

(c) We have the following strong "almost-orthogonality" of the collection $\{\mathcal{A}_k\}$: there is an integer $m > 0$ so that whenever $|k - j| \geq m$,

$$\|\mathcal{A}_k \mathcal{A}_j^*(f)\|_{L^2} \leq c_N 2^{-N \max(k,j)} \|f\|_{L^2}, \tag{68}$$

for each $N \geq 0$. A similar assertion holds for $\mathcal{A}_k^* \mathcal{A}_j$.

To verify (c) we make a simple estimate of the size of the kernel of the operator $\mathcal{A}_k \mathcal{A}_j^*$. A straight-forward calculation yields that its kernel is given by

(69)

$$K(x, y) = 2^k 2^j \int_{\mathbb{R}\times\mathbb{R}\times\mathbb{R}^d} e^{2\pi i(2^j v\rho(z,y) - 2^k u\rho(z,x))} \psi(z, x, y)\overline{\eta(u)}\eta(v)\, dz\, du\, dv$$

where $\psi(z, x, y) = \overline{\psi}(z, x)\psi(z, y)$. Now assume $j \geq k$ (the case $k \geq j$ is similar). Write the exponent in (69) as

$$2\pi i(2^j v\rho(z, y) - 2^k u\rho(z, x)) = i\lambda\Phi(z),$$

with $\lambda = 2\pi 2^j$ and $\Phi(z) = v\rho(z, y) - 2^{k-j} u\rho(z, x)$. Recall that because of the support properties of η we have $1/2 \leq |v| \leq 2$ and $1/2 \leq |u| \leq 2$. As a result $|\nabla_z \Phi(z)| \geq c' > 0$ if $j - k \geq m$, for some fixed m that is large enough, (because $|\nabla_z \rho(z, y)| \geq c$, while $|\nabla_z \rho(z, x)| \leq 1/c$ for a constant c that is small enough).

We now can invoke Proposition 2.1 to estimate $\int_{\mathbb{R}^d} e^{i\lambda\Phi(z)}\psi(z, x, y)\, dz$, and as a result obtain that for each $N \geq 0$

$$\begin{aligned} |K(x, y)| &\leq c_N 2^k 2^j 2^{-jN} \\ &\leq c_{N'} 2^{-N' \max(k,j)}, \qquad \text{with } N' = N - 2. \end{aligned}$$

Since K also has fixed compact support, the estimate (68) for $\mathcal{A}_k \mathcal{A}_j^*$ is therefore established. Of course a parallel argument works for $\mathcal{A}_k^* \mathcal{A}_j$, and property (c) is proved.

(d) Our last assertion concerns the operators $\left(\frac{\partial}{\partial x}\right)^\alpha \mathcal{A}_k = \partial_x^\alpha \mathcal{A}_k$, which we denote by $\mathcal{A}_k^{(\alpha)}$. Note that $\mathcal{A}_k^{(\alpha)}$, like $\mathcal{A}_k$, has a kernel that is C^∞ and has compact support. The $\{\mathcal{A}_k^{(\alpha)}\}$ satisfy estimates very similar to those for $\{\mathcal{A}_k\}$. In fact,

$$\|\mathcal{A}_k^{(\alpha)}\| \leq c_\alpha 2^{k|\alpha|} 2^{-k(\frac{d-1}{2})}, \tag{70}$$

and

$$\|\mathcal{A}_k^{(\alpha)}(\mathcal{A}_j^{(\alpha)})^*\| \leq c_{\alpha,N} 2^{-N\max(k,j)}, \qquad \text{if } |k-j| \geq m. \tag{71}$$

There is a parallel estimate for $(\mathcal{A}_k^{(\alpha)})^* \mathcal{A}_j^{(\alpha)}$. Here of course $\|\cdot\|$ denotes the operator norm on $L^2(\mathbb{R}^d)$.

Looking at (67) we see that carrying out the differentiation ∂_x^α on $\mathcal{A}_k(f)$ yields a finite sum of terms like $\mathcal{A}_k$ (but with modified ψ's) multiplied by factors that do not exceed $2^{k|\alpha|}$. Thus (70) and (71) are direct consequences of assertions (b) and (c) above.

7.5 Almost-orthogonal sums

Since we have appropriate control of the norms of the different pieces $\mathcal{A}_k$ making up $\mathcal{A}$, we now put these together by using a general almost-orthogonality principle.

We consider a sequence $\{T_k\}$ of bounded operators on $L^2(\mathbb{R}^d)$ and we assume we are given positive constants $a(k)$, with $-\infty < k < \infty$, so that the sum is finite, that is, $A = \sum_{k=-\infty}^{\infty} a(k) < \infty$.

Proposition 7.4 *Assume that*

$$\|T_k T_j^*\| \leq a^2(k-j) \quad \textit{and} \quad \|T_k^* T_j\| \leq a^2(k-j).$$

Then for every r,

$$\|\sum_{k=0}^{r} T_k\| \leq A. \tag{72}$$

The thrust of this proposition is of course that the bound A is independent of r.

Proof. We write $T = \sum_{k=0}^r T_k$ and recall that $\|T\|^2 = \|TT^*\|$. Since TT^* is self-adjoint we may use this identity repeatedly to obtain $\|T\|^{2n} = \|(TT^*)^n\|$, (at least when n is of the form $n = 2^s$ for some integer s). Now

$$(TT^*)^n = \sum_{i_1, i_2, \ldots, i_{2n}} T_{i_1} T_{i_2}^* \cdots T_{i_{2n-1}} T_{i_{2n}}^*.$$

We make two estimates for the norm of each term in the above sum. First

$$\|T_{i_1}T_{i_2}^* \cdots T_{i_{2n-1}}T_{i_{2n}}^*\| \leq a^2(i_1 - i_2)a^2(i_3 - i_4) \cdots a^2(i_{2n-1} - i_{2n}),$$

which is obtained by associating the product as $(T_{i_1}T_{i_2}^*)\cdots(T_{i_{2n-1}}T_{i_{2n}}^*)$. Next

$$\|T_{i_1}T_{i_2}^* \cdots T_{i_{2n-1}}T_{i_{2n}}^*\| \leq A^2a^2(i_2 - i_3)a^2(i_4 - i_5) \cdots a^2(i_{2n-2} - i_{2n-1}),$$

which is obtained by associating the product as $T_{i_1}(T_{i_2}^*T_{i_3})\cdots(T_{i_{2n-2}}^*T_{i_{2n-1}})T_{i_{2n}}^*$, and using the fact that T_{i_1} and $T_{i_{2n}}^*$ are both bounded by A. Taking the geometric mean of these estimates yields

$$\|T_{i_1}T_{i_2}^* \cdots T_{i_{2n-1}}T_{i_{2n}}^*\| \leq Aa(i_1 - i_2)a(i_2 - i_3) \cdots a(i_{2n-1} - i_{2n}).$$

Now we sum this first in i_1, then i_2, and so on, until i_{2n-1}, obtaining a further factor of A each time, because $A = \sum a(k)$. When we sum in i_{2n} we use the fact that there are $r+1$ terms in the sum. The result is then $\|T\|^{2n} \leq A^{2n}(r+1)$. Taking the $(2n)^{\text{th}}$ root and letting $n \to \infty$ gives (72) and the proposition.

7.6 Proof of Theorem 7.1

We consider first the case when the dimension d is odd, and thus the fraction $(d-1)/2$ is integral. The case when d is even is slightly more complicated, and will be dealt with separately.

In this first case we must show that whenever $|\alpha| \leq (d-1)/2$, and $f \in L^2(\mathbb{R}^d)$, the derivative $\partial_x^\alpha \mathcal{A}(f)$ exists in the sense of distributions, is an L^2 function, and the mapping $f \mapsto \partial_x^\alpha \mathcal{A}(f)$ is bounded on L^2.

For each r we consider

$$\partial_x^\alpha \sum_{k=0}^{r} \mathcal{A}_k = \sum_{k=0}^{r} T_k, \qquad \text{where } T_k = \mathcal{A}_k^{(\alpha)} = \partial_x^\alpha \mathcal{A}_k.$$

Now because of (70) and (71) we see that the hypotheses in Proposition 7.4 are satisfied with in fact $a(k) = c_N 2^{-|k|N}$, (and in particular for $N = 1$). Thus

$$\|\partial_x^\alpha \sum_{k=0}^{r} \mathcal{A}_k(f)\|_{L^2} \leq A\|f\|_{L^2}, \qquad |\alpha| \leq \tfrac{d-1}{2}. \tag{73}$$

However, by (70) for $\alpha = 0$, the sum $\sum_{k=0}^{r} \mathcal{A}_k$ converges in the L^2 norm as $r \to \infty$ to $\mathcal{A}(f)$, (since the latter also converges pointwise to $\mathcal{A}(f)$), and hence in the sense of distributions. Thus $\partial_x^\alpha \sum_{k=0}^{r} \mathcal{A}_k(f)$ also converges in the sense of distributions as $r \to \infty$ but since this sum is uniformly in L^2 as r varies, the limit is also in L^2.

Finally then we have

$$\|\partial_x^\alpha \mathcal{A}(f)\|_{L^2} \leq A\|f\|_{L^2},$$

whenever f is continuous and of compact support and $|\alpha| \leq (d-1)/2$, with d odd. Hence Theorem 7.1 is proved in this case.

Now we consider the case when d is even. Here we need to involve the "fractional derivative" operator D^s, defined on the Schwartz space $\mathcal{S}$ by its action as a multiplier on the Fourier transform, namely

$$(D^s f)^\wedge(\xi) = (1 + |\xi|^2)^{s/2} \hat{f}(\xi).$$

Note that $\|D^s(f)\|_{L^2} = \|f\|_{L^2_\sigma}$, where $\sigma = \mathrm{Re}(s)$, whenever f is in $\mathcal{S}$. We also need to observe that if $\mathrm{Re}(s) = m$ is a positive integer, then

$$\|D^s(f)\|_{L^2} \leq c \sum_{|\alpha| \leq m} \|\partial_x^\alpha f\|_{L^2}. \tag{74}$$

Indeed, this follows directly from the inequality $(1 + |\xi|^2)^{m/2} \leq c' \sum_{|\alpha| \leq m} |\xi^\alpha|$, $\xi \in \mathbb{R}^d$, and Plancherel's theorem.

Now arguing as above for the case when d is odd, it will suffice to prove that

$$\|D^{\frac{d-1}{2}} \sum_{k=0}^{r} \mathcal{A}_k(f)\|_{L^2} \leq c\|f\|_{L^2}, \tag{75}$$

with the bound c independent of r. To this end, consider the family of operators T^s, depending on the complex parameter s, defined by

$$T^s(f) = D^{s + \frac{d-1}{2}} \sum_{k=0}^{r} 2^{-ks} \mathcal{A}_k(f), \tag{76}$$

for $f \in L^2(\mathbb{R}^d)$ (in particular for simple f). As we have already noted, for such f the $\mathcal{A}_k(f)$ are in $\mathcal{S}$ so (76) is well-defined and $T^s(f)$ is itself in $\mathcal{S}$. Moreover, whenever $g \in L^2$, (in particular, if it is a simple function) then by Plancherel's theorem

$$\begin{aligned}\Phi(s) &= \int_{\mathbb{R}^d} T^s(f)g\,dx \\ &= \sum_{k=0}^{r} 2^{-ks}\int_{\mathbb{R}^d}(1+|\xi|^2)^{s/2}F_k(\xi)\hat{g}(-\xi)\,d\xi,\end{aligned}$$

where each F_k belongs to $\mathcal{S}$. Hence Φ is analytic (in fact entire) in s, and by Schwarz's inequality, bounded in any strip $a \leq \mathrm{Re}(s) \leq b$.

Next

$$\sup_t \|T^{-\frac{1}{2}+it}(f)\|_{L^2} \leq M\|f\|_{L^2}. \tag{77}$$

In fact, by (74) and (76) it suffices to see that

$$\|\sum_{k=0}^{r} 2^{-k/2}\partial_x^\alpha \mathcal{A}_k(f)\|_{L^2} \leq M\|f\|_{L^2}, \qquad \text{for } |\alpha| \leq \tfrac{d-2}{2}.$$

But this is proved like (73) by using estimates (70) and (71) for $\mathcal{A}_k^{(\alpha)} = \left(\frac{\partial}{\partial x}\right)^\alpha \mathcal{A}_k$ together with the almost-orthogonality proposition in Section 7.5.

Similarly, one shows that

$$\sup_t \|T^{\frac{1}{2}+it}(f)\|_{L^2} \leq M\|f\|_{L^2}. \tag{78}$$

Finally, we apply the analytic interpolation given by Proposition 4.4. Here the strip is $a \leq \mathrm{Re}(s) \leq b$, with $a = -1/2$, $b = 1/2$ and $c = 0$, while $p_0 = q_0 = p_1 = q_1 = 2$. The result is then

$$\|T^0(f)\|_{L^2} \leq M\|f\|_{L^2},$$

which in view of the definition (76) is the estimate (75). This completes the proof of the theorem.

Remark. The L^p, L^q boundedness result of Theorem 4.1 (b) and Corollary 4.2 extends to this setting. The proof is outlined in Exercise 20.

8 Counting lattice points

In this last section we will see the relevance of oscillatory integrals to some questions related to number theory.

8.1 Averages of arithmetic functions

The arithmetic functions we have in mind are $r_2(k)$, the number of representations of k as the sum of two squares, and $d(k)$, the number of divisors of k. Even a cursory examination of the size of these functions as $k \to \infty$ reveals a high degree of irregularity, so that it is not possible to capture by simple analytic expressions the essential behavior of these functions for large k.

In fact, it is an elementary observation that $r_2(k) = 0$ and $d(k) = 2$, each for infinitely many k, while given any $A > 0$, one has $r_2(k) \geq (\log k)^A$ for infinitely many k, and the same is true for $d(k)$.[10]

In this context an inspired idea was to inquire instead as to the *average* behavior of these arithmetic functions. That this might be a fruitful question is already indicated by the observation of Gauss: the average value of $r_2(k)$ is π. This means that $\frac{1}{\mu}\sum_{k=1}^{\mu} r_2(k) \to \pi$, as $\mu \to \infty$.

In more detail, we have the following result.

Proposition 8.1 $\sum_{k=1}^{\mu} r_2(k) = \pi\mu + O(\mu^{1/2})$, *as* $\mu \to \infty$.

The proof depends on the realization that $\sum_{k=0}^{\mu} r_2(k)$ represents the number of lattice points in the disc of radius R with $R^2 = \mu$. In fact, with $\mathbb{Z}^2$ denoting the **lattice points** in $\mathbb{R}^2$, that is, the points in $\mathbb{R}^2$ with integral coordinates, then $r_2(k) = \#\{(n_1, n_2) \in \mathbb{Z}^2 : \ k = n_1^2 + n_2^2\}$, and hence

$$\sum_{k=0}^{\mu} r_2(k) = \#\{(n_1, n_2) \in \mathbb{Z}^2 : \ n_1^2 + n_2^2 \leq R^2\}.$$

So if $N(R)$ is the quantity above, then the proposition is equivalent to

$$N(R) = \pi R^2 + O(R), \qquad \text{as } R \to \infty. \tag{79}$$

To prove this we write D_R for the closed disc $\{x \in \mathbb{R}^2 : |x| \leq R\}$, and let $\tilde{D}_R$ be the rectangular region that is the union of unit squares centered at points $n \in \mathbb{Z}^2$ with $n \in D_R$, that is,

$$\tilde{D}_R = \bigcup_{|n| \leq R, \ n \in \mathbb{Z}^2} (S + n),$$

with $S = \{x = (x_1, x_2) : -1/2 \leq x_i < 1/2, \quad i = 1, 2\}$.

[10]For the elementary facts about $r_2(k)$ and $d(k)$ stated here, including the asymptotic formula (81), see, for example, Chapter 8 in Book I and Chapter 10 in Book II.

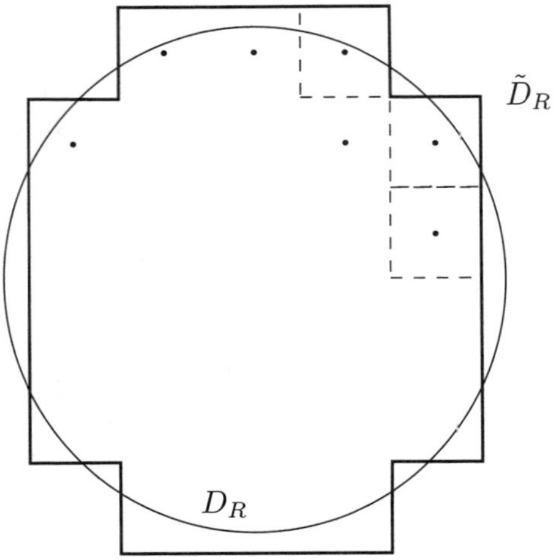

Figure 2. The region $\tilde{D}_R$

Since the squares $S + N$ are mutually disjoint and each has area 1, we see that $m(\tilde{D}_R) = N(R)$. However

$$D_{R-2^{-1/2}} \subset \tilde{D}_R \subset D_{R+2^{-1/2}}. \tag{80}$$

In fact if $x \in S + n$ with $|n| \leq R$, then $|x| \leq 2^{-1/2} + |n| \leq R + 2^{-1/2}$, so $\tilde{D}_R \subset D_{R+2^{-1/2}}$. The reverse inclusion can be proved the same way. It follows from (80) that

$$m(D_{R-2^{-1/2}}) \leq m(\tilde{D}_R) \leq m(D_{R+2^{-1/2}}),$$

and hence

$$\pi(R - 2^{-1/2})^2 \leq N(R) \leq \pi(R + 2^{-1/2})^2,$$

proving that $N(R) = \pi R^2 + O(R)$.

There is a similar but somewhat more intricate statement for the averages of the divisor function. Dirichlet's theorem asserts:

$$\sum_{k=1}^{\mu} d(k) = \mu \log \mu + (2\gamma - 1)\mu + O(\mu^{1/2}) \quad \text{as } \mu \to \infty, \tag{81}$$

where γ is Euler's constant.

Again this is a consequence of counting lattice points in the plane: the left-hand side of (81) is the number of lattice points (n_1, n_2), with $n_1, n_2 > 0$ that lie on or below the hyperbola $x_1 x_2 = \mu$.[11]

Both (79) and (81) raise the question of what are the true sizes of the error terms appearing in these asymptotic statements. Like other important questions of this kind in number theory, these problems have a long history involving much effort, but yet remain unsolved. It will be our purpose here to show only how the first results that go beyond (79) and (81) can be obtained by the help of ideas treated in this chapter.

8.2 Poisson summation formula

Indispensable for any further insight into these problems is the application of the Poisson summation formula. We state this identity here in the general context of $\mathbb{R}^d$, but with a restricted hypothesis sufficient for our applications.[12]

Proposition 8.2 *Suppose f belongs to the Schwartz space $\mathcal{S}(\mathbb{R}^d)$. Then*

$$\sum_{n\in\mathbb{Z}^d} f(n) = \sum_{n\in\mathbb{Z}^d} \hat{f}(n). \tag{82}$$

Here $\mathbb{Z}^d$ denotes the collection of **lattice points** in $\mathbb{R}^d$, the points with integral coordinates, and $\hat{f}$ is the Fourier transform of f.

For the proof consider two sums

$$\sum_{n\in\mathbb{Z}^d} f(x+n) \quad \text{and} \quad \sum_{n\in\mathbb{Z}^d} \hat{f}(n)e^{2\pi i n\cdot x}.$$

Both are rapidly converging series (since f and $\hat{f}$ are in $\mathcal{S}(\mathbb{R}^d)$), and hence both these sums are continuous functions. Moreover each is periodic, that is, each is unchanged when x is replaced by $x+m$, for any $m \in \mathbb{Z}^d$. For the sum $\sum_{n\in\mathbb{Z}^d} f(x+n)$ this is clear, because replacing x by $x+m$ merely reshuffles the sum. Also the second sum is unchanged, because of the periodicity of $e^{2\pi i n\cdot x}$ for each $n \in \mathbb{Z}^d$. Moreover both sums have the same Fourier coefficients. To see this, let Q be the fundamental cube

---

[11] That there might be some connection between the averages $r_2(k)$ and $d(k)$ is suggested by the fact that $r_2(k) = 4(d_1(k) - d_3(k))$, for $k \geq 1$, where d_1 and d_3 are respectively the number of divisors of $k \equiv 1 \mod 4$ or $\equiv 3 \mod 4$.

[12] Other settings for the formula can be found in Chapter 5 in Book I, and Chapter 4 in Book II.

$Q = \{x \in \mathbb{R}^d : 0 < x_j \leq 1, \ j = 1, \ldots, d\}$, and fix any $m \in \mathbb{Z}^d$. Then

$$\begin{aligned}\int_Q \left(\sum_n f(x+n)\right) e^{-2\pi i m\cdot x}\,dx &= \sum_n \int_{Q+n} f(x)e^{-2\pi i m\cdot x}\,dx \\ &= \int_{\mathbb{R}^d} f(x)e^{-2\pi i m\cdot x}\,dx \\ &= \hat{f}(m),\end{aligned}$$

since $\bigcup_{n\in\mathbb{Z}^d}(Q+n)$ is a partition of $\mathbb{R}^d$ into cubes $\{Q+n\}_{n\in\mathbb{Z}^d}$. Moreover

$$\int_Q \left(\sum_n \hat{f}(n)e^{2\pi i n\cdot x}\right) e^{-2\pi i m\cdot x}\,dx = \hat{f}(m),$$

because $\int_Q e^{2\pi i n\cdot x}e^{-2\pi i m\cdot x}\,dx = 1$ if $n = m$, and is 0 otherwise. Since $\sum_n f(x+n) = \sum_n \hat{f}(n)e^{2\pi i n x}$ have the same Fourier coefficients, these functions must be equal,[13]and setting $x = 0$ gives us (82).

Next let us see what happens to the summation formula (82) when we apply it first to the case of a radial function on $\mathbb{R}^2$, $f(x) = f_0(|x|)$, that is in $\mathcal{S}$, and then we try it with χ_R, the characteristic function of the disc D_R.

Using the formula (4) in Section 1 we obtain

$$\sum_{n\in\mathbb{Z}^2} f_0(|n|) = 2\pi\int_0^\infty f_0(r)r\,dr + \sum_{k=1}^\infty F_0(k^{1/2})r_2(k), \tag{83}$$

once we gather together the terms for which $|n|^2 = k$. Here $F_0(\rho) = 2\pi\int_0^\infty J_0(2\pi\rho r)f_0(r)r\,dr$, and we note that $J_0(0) = 1$.

If we could apply this formula to the case when f is χ_R, (the obstacle is that of course χ_R is not smooth), and use the fact that $rJ_1(r) = \int_0^r \sigma J_0(\sigma)\,d\sigma$, which is outlined in Exercise 23, this would give us Hardy's identity

$$N(R) = \pi R^2 + R\sum_{k=1}^\infty \frac{r_2(k)}{k^{1/2}} J_1(2\pi k^{1/2}R).$$

Note that since $J_1(u)$ is of order $u^{-1/2}$ as $u \to \infty$ (see (11)), the series does not converge absolutely, and this is the barrier in trying to apply (83), even if one is guaranteed the (conditional) convergence of the

[13]See, for instance, Exercise 16 in Chapter 6 of Book III.

series. Nevertheless, since each term of the series is $O(R^{-1/2})$, it might be hoped that the error term, $N(R) - \pi R^2$, is roughly of the order $O(R^{1/2})$, and this is what is conjectured.[14]

Here we prove the following weaker assertion that is, however, an improvement over (79).

Theorem 8.3 $N(R) = \pi R^2 + O(R^{2/3})$, *as* $R \to \infty$.

Proof. We replace the characteristic function χ_R by a regularized version as follows. We fix a non-negative "bump" function φ that is C^∞, is supported in the unit disc, and has $\int_{\mathbb{R}^2} \varphi(x)\, dx = 1$. We set $\varphi_\delta(x) = \delta^{-2}\varphi(x/\delta)$, and let

$$\chi_{R,\delta} = \chi_R * \varphi_\delta.$$

Then clearly $\chi_{R,\delta}$ is a C^∞ function of compact support and hence the summation formula (82) applies to it. Notice that $\hat{\chi}_{R,\delta}(\xi) = \hat{\chi}_R(\xi)\hat{\varphi}_\delta(\xi)$, and $\hat{\chi}_{R,\delta}(0) = \hat{\chi}_R(0)\hat{\varphi}_\delta(0) = \pi R^2$.

As a result, if we define $N_\delta(R) = \sum_{n\in\mathbb{Z}^2} \chi_{R,\delta}(n)$, then the summation formula yields

$$N_\delta(R) = \pi R^2 + \sum_{n\neq 0} \hat{\chi}_R(n)\hat{\varphi}(\delta n).$$

We now estimate the sum above by breaking it into two parts as

$$\sum_{0<|n|\leq 1/\delta} + \sum_{|n|>1/\delta}.$$

For the first sum we use the fact that

$$|\hat{\chi}_R(n)| = \frac{R}{|n|}|J_1(2\pi|n|R)| = O(R^{1/2}|n|^{-3/2}),$$

by what has been said above, and that $|\hat{\varphi}(n\delta)| = O(1)$. This gives

$$\begin{aligned}\sum_{0<|n|\leq 1/\delta} &= O\left(R^{1/2}\sum_{0<|n|\leq 1/\delta}|n|^{-3/2}\right)\\ &= O\left(R^{1/2}\int_{|x|\leq 1/\delta}|x|^{-3/2}\,dx\right)\\ &= O(R^{1/2}\delta^{-1/2}).\end{aligned}$$

[14]More precisely the guess is that the error term is $O(R^{1/2+\epsilon})$ for every $\epsilon > 0$. See also Problem 6.

Similarly,

$$\sum_{|n|>1/\delta} = O\left(R^{1/2}\delta^{-1} \sum_{|n|>1/\delta} |n|^{-5/2}\right),$$

since $|\hat{\varphi}(n\delta)| = O(|n|^{-1}\delta^{-1})$. (In fact $\hat{\varphi}(\xi)$ is rapidly decreasing.) Thus this sum is also $O(R^{1/2}\delta^{-1/2})$. We conclude therefore that

$$N_\delta(R) = \pi R^2 + O(R^{1/2}\delta^{-1/2}). \tag{84}$$

However there is a simple relation between $N_\delta(R)$ and $N(R)$, namely

$$N_\delta(R-\delta) \leq N(R) \leq N_\delta(R+\delta). \tag{85}$$

This in turn follows from the observation that

$$\chi_{R-\delta,\delta} \leq \chi_R \leq \chi_{R+\delta,\delta}.$$

The inequality on the right-hand side, $\chi_R(x) \leq \int \chi_{R+\delta}(x-y)\varphi_\delta(y)\,dy$, is clear because $x \in D_R$ and $|y| \leq \delta$ implies $x - y \in D_{R+\delta}$. Similarly for the inequality on the left-hand side.

Finally by (84), we have $N_\delta(R+\delta) = \pi R^2 + O(R^{1/2}\delta^{-1/2}) + O(R\delta)$ and analogously $N_\delta(R-\delta) = \pi R^2 + O(R^{1/2}\delta^{-1/2}) + O(R\delta)$. Altogether then, (85) yields that

$$N(R) = \pi R^2 + O(R^{1/2}\delta^{-1/2}) + O(R\delta).$$

By choosing $\delta = R^{-1/3}$ we make both O terms above equal, and this gives

$$N(R) = \pi R^2 + O(R^{2/3}).$$

The theorem is therefore proved.

The approach to Theorem 8.3, leads to a wide generalization in which the disc in $\mathbb{R}^2$ is replaced by an appropriate convex set in $\mathbb{R}^d$.

Recall that a set Ω is **convex** if whenever x and x' are in Ω, so is the line segment joining them. Suppose in addition that Ω is a bounded set with C^2 boundary (in the sense of Section 4 in Chapter 7). Then whenever ρ is a defining function for Ω, the second fundamental form (19) is positive semi-definite. (In fact, assuming the contrary, we can find a point on the boundary and coordinates $(x_1, \ldots, x_d)$ centered at this point, so that x_d is in the direction of the inward normal, and the quadratic form has an eigenvalue $\lambda_1 < 0$ in the direction x_1. Hence near the origin in this coordinate system, the intersection of Ω with the plane determined by the x_1 and x_d axes is then given by $\{x_d > \lambda_1 x_1^2 + o(x_1^2)\}$, which is clearly not convex, contradicting the convexity of Ω.)

With this in mind, we say that Ω is **strongly convex** when the quadratic form (19) is strictly positive definite at each point of the boundary of Ω. We denote by $R\Omega$ the dilated set $\{Rx : x \in \Omega\}$ and write $N_R = \#\{\text{lattice points in } R\Omega\}$.

Theorem 8.4 *Suppose Ω is a bounded domain in $\mathbb{R}^d$ with sufficiently smooth boundary.*[15] *Assume that Ω is strongly convex and $0 \in \Omega$. Then*

$$N_R = R^d m(\Omega) + O(R^{d-\frac{2d}{d+1}}) \quad \text{as } R \to \infty.$$

The proof follows closely the argument for Theorem 8.3.

Proof. Let χ denote the characteristic function of Ω and χ_R that of $R\Omega$, so $\chi_R(x) = \chi(x/R)$. With φ a non-negative C^∞ function supported in the unit ball that satisfies $\int \varphi(x)\,dx = 1$, we set $\varphi_\delta(x) = \delta^{-d}\varphi(x/\delta)$. We let $\chi_{R,\delta} = \chi_R * \varphi_\delta$, and set

$$N_{R,\delta} = \sum_{n\in\mathbb{Z}^d} \chi_{R,\delta}(n).$$

Now, by the summation formula (82)

$$N_{R,\delta} = R^\delta m(\Omega) + \sum_{n\neq 0} \hat{\chi}_{R,\delta}(n),$$

since $\hat{\chi}_{R,\delta}(0) = \hat{\chi}_R(0)\hat{\varphi}(0)$, $\hat{\chi}_R(0) = R^d\hat{\chi}(0) = R^d m(\Omega)$, and $\hat{\varphi}(0) = 1$.

However $\hat{\chi}(\xi) = O\left(|\xi|^{-\frac{d+1}{2}}\right)$ by Corollary 3.3. Thus

$$\hat{\chi}_R(n) = R^d\hat{\chi}(Rn) = O\left(R^{\frac{d-1}{2}}|n|^{-\frac{d+1}{2}}\right),$$

so

$$\hat{\chi}_{R,\delta}(n) = O\left(R^{\frac{d-1}{2}}|n|^{-\frac{d+1}{2}}\right)\hat{\varphi}(\delta n).$$

Now we break the sum $\sum_{n\neq 0}\hat{\chi}_{R,\delta}(n)$ as $\sum_{1\leq|n|\leq 1/\delta} + \sum_{1/\delta<|n|}$. For the first term we use the fact that $\hat{\chi}_{R,\delta} = O(R^{\frac{d-1}{2}}|n|^{-\frac{d+1}{2}})$, and we estimate that sum by $O(R^{\frac{d-1}{2}}\delta^{-\frac{d-1}{2}})$, (for example by comparing it with $R^{\frac{d-1}{2}}\int_{|x|\leq 1/\delta}|x|^{-\frac{d+1}{2}}\,dx$).

[15]The proof will show that C^{d+2} suffices. See the remark at the end of Section 3.

The second term leads to $R^{\frac{d-1}{2}} \sum_{|n|>1/\delta} |n|^{-\frac{d+1}{2}} (|n|\delta)^{-r}$ for any $r > 0$, since $\hat{\varphi}$ is rapidly decreasing. Choosing r sufficiently large, say $r = d/2$ gives in the same way the estimate $O\left(R^{\frac{d-1}{2}}\delta^{-\frac{d-1}{2}}\right)$ for this part of the sum. Hence

$$N_{R,\delta} = R^d m(\Omega) + O\left(R^{\frac{d-1}{2}}\delta^{-\frac{d-1}{2}}\right). \tag{86}$$

Next we observe that for an appropriate $c > 0$

$$N_{R-c\delta,\delta} \leq N_R \leq N_{R+c\delta,\delta}. \tag{87}$$

This inequality follows from

$$\chi_{R-c\delta,\delta} \leq \chi_R \leq \chi_{R+c\delta,\delta}.$$

The inequality on the right-hand side,

$$\chi_R(x) \leq \int \chi_{R+c\delta}(x-y)\varphi_\delta(y)\, dy,$$

is a consequence of the geometric observation that there is a $c > 0$, so that whenever $R \geq 1$, and $\delta \leq 1$,

$$x \text{ in } R\Omega \text{ and } |y| \leq \delta \text{ imply that } x - y \in (R + c\delta)\Omega. \tag{88}$$

The proof of this geometrical fact about the convexity of Ω is outlined in Exercise 21.

The inequality $\chi_{R-c\delta,\delta} \leq \chi_R$ is seen in the same way.

Now a combination of (86) and (87) show that

$$N_R = R^d m(\Omega) + O\left(R^{\frac{d-1}{2}}\delta^{-\frac{d-1}{2}}\right) + O(R^{d-1}\delta).$$

If we now choose $\delta = R^{-\frac{d-1}{d+1}}$, then both O terms are $O\left(R^{d-\frac{2d}{d+1}}\right)$, and the theorem is proved.

8.3 Hyperbolic measure

We turn to the improvement of (81) for the divisor problem that is analogous to Theorem 8.3.

Theorem 8.5

(89) $$\sum_{k=1}^{\mu} d(k) = \mu \log \mu + (2\gamma - 1)\mu + O(\mu^{1/3} \log \mu) \qquad \textit{as } \mu \to \infty.$$[16]

Now as much as we might wish to follow the lines of the proof of Theorem 8.3, there are serious obstacles that seem to stand in the way. In fact, if χ_μ is the characteristic function of the region

(90) $$\{(x_1, x_2) \in \mathbb{R}^2 : \; x_1 x_2 \leq \mu, \quad x_1 > 0, \text{ and } x_2 > 0\},$$

which consists of the area on or below the hyperbola $x_1 x_2 = \mu$, then indeed

$$\sum_{k=1}^{\mu} d(k) = \sum_{n \in \mathbb{Z}^2} \chi_\mu(n).$$

However the other side of the Poisson summation formula (82) for $f = \chi_\mu$ is problematic as it stands. In fact, $\hat{\chi}_\mu(0) = \int_{\mathbb{R}^2} \chi_\mu \, dx = \infty$, and for the same reason the integral giving each term $\hat{\chi}_\mu(n)$ is not well-defined.

A further issue is that the main term in (89) is $\mu \log \mu$, while a simple scaling of the region (90) would suggest rather a term linear in μ. Connected with this is the mysterious occurrence of Euler's constant γ in the subsidiary term.

Now the essence of our analysis of lattice points in D_R (and formulas like (83)) are the facts about Fourier transforms of radial functions in two dimensions, which in turn depend on the Fourier transform of the invariant measure of the circle. In parallel to this we seek the analog where instead of radial functions we consider functions invariant under "hyperbolic dilations" $(x_1, x_2) \to (\delta x_1, \delta^{-1} x_2)$, $\delta > 0$, and a corresponding invariant measure in $\mathbb{R}^2$ supported on the hyperbola $x_1 x_2 = 1$.

We begin with the **hyperbolic measure** on $\mathbb{R}^2$, denoted by $d\mathfrak{h}$, and which is defined by the integration formula

$$\int_{\mathbb{R}^2} f(x) \, d\mathfrak{h} = \int_0^\infty f(u, 1/u) \, \frac{du}{u},$$

valid for every continuous function f of compact support. Alternatively

$$\mathfrak{h}(E) = \int_0^\infty \chi_E(u, 1/u) \, \frac{du}{u}$$

[16] Recall the correspondence $\mu = R^2$ when comparing this with the result for lattice points in D_R.

for every Borel set E in $\mathbb{R}^2$, with the integral taken in the extended sense. Note that the measure $\mathfrak{h}$ is invariant under the scalings $(x_1, x_2) \to (\delta x_1, \delta^{-1} x_2)$, for $\delta > 0$.

Now the linear functional $f \mapsto \int_0^\infty f(u, 1/u)\,\frac{du}{u}$ is well-defined for $f \in \mathcal{S}$ in view of the rapid convergence of the integral, and moreover this convergence shows that the measure $\mathfrak{h}$ can be considered by this formula to be a tempered distribution. We seek to determine the Fourier transform of this distribution, and matters will depend on a pair of oscillatory integrals $\mathfrak{J}^+$ and $\mathfrak{J}^-$. These are given formally by

$$\mathfrak{J}^{\pm}(\lambda) = \int_0^\infty e^{i\lambda(u \pm 1/u)}\,\frac{du}{u}.$$

Since these integrals do not converge absolutely (either at 0 or at infinity) they must be considered as appropriate limits after truncation.

For this purpose we pick η to be a non-negative C^∞ function on $[0, \infty)$ with $\eta(u) = 0$ for small u, and $\eta(u) = 1$ if $u \geq 1$, and set $\eta_a(u) = \eta(u/a)$. We then define the convergent integral

$$\mathfrak{J}^+_{a,b}(\lambda) = \int_0^\infty e^{i\lambda(u + \frac{1}{u})}\eta_a(u)\eta_b(1/u)\,\frac{du}{u},$$

with a similar definition for $\mathfrak{J}^-_{a,b}(\lambda)$. To begin with we take $0 < a, b \leq 1/2$.

Proposition 8.6 *For each $\lambda \neq 0$, the limit $\mathfrak{J}^+(\lambda) = \lim_{a,b\to 0} \mathfrak{J}^+_{a,b}(\lambda)$ exists. Moreover, uniformly in a and b, we have:*

(i) $\mathfrak{J}^+_{a,b}(\lambda) = \left(\sum_{k=0}^N c_k \lambda^{-1/2-k}\right) e^{2i\lambda} + O\left(|\lambda|^{-3/2-N}\right)$, *for* $|\lambda| \geq 1/2$ *and for every $N \geq 0$, with $c_0, c_1, \ldots, c_k, \ldots$ appropriate constants.*

(ii) $\mathfrak{J}^+_{a,b}(\lambda) = O(\log 1/|\lambda|)$, *for* $|\lambda| \leq 1/2$.

Proof. We divide the integral $\mathfrak{J}^+_{a,b}$ into three parts as follows. Let α be a C^∞ function so that $\alpha(u) = 1$ when $3/4 \leq u \leq 4/3$, and α is supported in $[1/2, 2]$. Set $\beta = 1 - \alpha$ so β is supported where $u \leq 3/4$ or $u \geq 4/3$. Then split $\mathfrak{J}^+_{a,b}$ as $I + II + III$, where

$$II = \int_{1/2}^2 e^{i\lambda\Phi(u)}\alpha(u)\,\frac{du}{u}, \qquad I = \int_0^{3/4} e^{i\lambda\Phi(u)}\beta(u)\eta_a(u)\,\frac{du}{u},$$

and

$$III = \int_{4/3}^\infty e^{i\lambda\Phi(u)}\beta(u)\eta_b(u)\,\frac{du}{u}.$$

Here we have written $\Phi(u)$ for $u + 1/u$.

Now we observe that $\Phi'(1) = 0$, while $\Phi''(u) > 0$ for all u, so that $u = 1$ is the (only) critical point of Φ. Also, since $\Phi(1) = 2$, we are led to make the change of variables $\Phi(u) = u + 1/u = 2 + x^2$. Solving the quadratic equations involved gives

$$x = \frac{u-1}{u^{1/2}}, \qquad u = 1 + \frac{x^2}{2} + \frac{x(4+x^2)^{1/2}}{2},$$

which shows that $u \mapsto x$ is a smooth bijection of the intervals $[1/2, 2]$ with $[-2^{-1/2}, 2^{1/2}]$.

Making the indicated change of variables, we see that the integral II becomes

$$e^{2i\lambda} \int e^{i\lambda x^2} \tilde{\alpha}(x)\, dx,$$

with $\tilde{\alpha}$ a C^∞ function of compact support. We now invoke the asymptotic formula (8) to obtain

$$II = \left(\sum_{k=0}^{N} c_k \lambda^{-1/2-k} \right) e^{2i\lambda} + O\left(|\lambda|^{-3/2-N} \right),$$

for every $N \geq 0$.

Next, to deal with the integral I, we write

$$L = \frac{1}{i\lambda \Phi'(u)} \frac{d}{du}.$$

Then $L(e^{i\lambda\Phi}) = e^{i\lambda\Phi}$, and for every integer $N \geq 1$

(91) $$I = \int_0^{3/4} L^N(e^{i\lambda\Phi}) \beta(u) \eta_a(u) \frac{du}{u}.$$

Let us first consider the case $N = 1$. Since $\Phi'(u) = 1 - 1/u^2$, and $1/\Phi'(u) = u^2/(u^2-1)$, integration by parts shows

$$I = -\frac{1}{i\lambda} \int_0^{3/4} e^{i\lambda\Phi(u)} \frac{d}{du}(u\beta_1(u)\eta_a(u))\, du,$$

where $\beta_1(u) = \beta(u)/(u^2-1)$, and β_1 is smooth.

Carrying out the differentiation leads to two terms. First, if the derivative falls on $\beta_1(u)$, the resulting contribution to I is certainly $O(1/|\lambda|)$.

For the second, if the derivative falls on $\eta_a(u)$ the contribution is also $O(1/|\lambda|)$, since $(\eta_a(u))' = O(1/a)$, and $\eta_a'(u)$ is supported on $[0,a]$. This shows that $I = O(1/|\lambda|)$.

For $N > 1$ we use (91) again, and carry out the integration by parts N times. Now at each step we get a gain of a factor of u and a possible loss of a factor of a^{-1}, the latter occurring when η_a is differentiated. So altogether this shows that $I = O(|\lambda|^{-N})$ for each positive integer N. The integral III is similar to that of I, as can be seen by transforming it by the mapping $u \mapsto 1/u$. So we also have $III = O(|\lambda|^{-N})$, and hence conclusion (i) of the proposition is proved.

Next, when $|\lambda| \le 1/2$, since II is obviously bounded, we need only estimate I and III. Turning to I we write as before

$$\begin{aligned} I &= -\frac{1}{i\lambda}\int_0^{3/4} e^{i\lambda\Phi(u)}\frac{d}{du}(u\beta_1(u)\eta_a(u))\,du \\ &= -\frac{1}{i\lambda}\int_0^{|\lambda|} - \frac{1}{i\lambda}\int_{|\lambda|}^{3/4}. \end{aligned}$$

But the first term is majorized by a multiple of

$$\frac{1}{|\lambda|}\int_0^{|\lambda|}(1 + u|\eta_a'(u)|)\,du = O(1),$$

while the second term can be written

$$\int_{|\lambda|}^{3/4} e^{i\lambda\Phi(u)}\beta(u)\eta_a(u)\ \frac{du}{u} + O(1),$$

which is clearly $O\left(\int_{|\lambda|}^{3/4}\frac{du}{u}\right) + O(1) = O(\log 1/|\lambda|)$. The estimate for III is similar, so conclusion (ii) is established.

To prove the convergence of $\mathfrak{J}_{a,b}^{+}$ as $a, b \to 0$, note that the integral II is independent of a and b. Now consider I and recall that it depends only on a. We have

$$I_a - I_{a'} = \int e^{i\lambda\Phi(u)}(\eta_a(u) - \eta_{a'}(u))\beta(u)\,\frac{du}{u},$$

and the integrand is supported only on $(0, \max(a,a'))$. Now as before

$$I_a - I_{a'} = \frac{1}{i\lambda}\int \frac{d}{du}(e^{i\lambda\Phi(u)})(\eta_a(u) - \eta_{a'}(u))u\beta_1(u)\,du$$

and an integration by parts shows that this difference is $O\left(\frac{1}{|\lambda|}\max(a,a')\right)$. Since λ is fixed, $\lambda \neq 0$, this tends to zero with a and a', and so I_a tends to a limit as $a \to 0$. The term III is treated similarly, and hence $\mathfrak{J}^+_{a,b}$ tends to a limit, proving the proposition.

A similar result holds for $\mathfrak{J}^-_{a,b}$, except for one change.

Corollary 8.7 *The conclusions for $\mathfrak{J}^-_{a,b}$ are the same as those for $\mathfrak{J}^+_{a,b}$ stated in Proposition* 8.6, *except that* (i) *should be modified to read that uniformly in a, b,*

(i′) $\mathfrak{J}^-_{a,b} = O(|\lambda|^{-N})$ *for* $|\lambda| \geq 1/2$, *for every* $N \geq 0$.

The only change occurs in the treatment of II, namely $\int e^{i\lambda\Phi(u)}\alpha(u)\,\frac{du}{u}$, where now $\Phi(u) = u - 1/u$. In this case $\Phi'(u) = 1 + 1/u^2 > 1$, and there is no critical point. So Proposition 2.1 implies that $II = O(|\lambda|^{-N})$ for every $N \geq 0$, and then conclusion (i′) follows by the arguments we have used for I and III previously.

Remarks. Two further observations about $\mathfrak{J}^+_{a,b}$ are straight-forward consequences of the arguments given above.

1. $\mathfrak{J}^+$ and $\mathfrak{J}^-$ are both continuous in λ if $\lambda \neq 0$.

2. The uniformity of the estimates (i), (i′) and (ii) holds in the wider range $0 < a < \infty$, $0 < b < \infty$, with the only change being that in the asymptotic formula in (i) the constants c_k may depend on a and b, but are still uniformly bounded. For example, when $a \leq 1/2$ but now b is unrestricted, then in the term II, $\alpha(u)$ is replaced by $\alpha(u)\eta(1/(bu))$, which is still uniformly smooth when $b \geq 1/2$. In I, the function $\beta_1(u)$ is replaced by $\beta_1(u)\eta(1/(bu))$ with the same effect. This reasoning clearly applies when both a and b are large.

8.4 Fourier transforms

We now come to the Fourier transform of $\mathfrak{h}$. It is convenient at this point to change our notation slightly, so that a general point (x_1, x_2) of $\mathbb{R}^2$ will now be denoted instead by (x, y), and similarly the dual variable in $\mathbb{R}^2$ will be denoted by (ξ, η).[17]

We divide the plane $\mathbb{R}^2$ into its four proper quadrants Q_1, Q_2, Q_3 and Q_4, (together with the x and y axes) with $Q_1 = \{(x,y) :\ x > 0 \text{ and } y > 0\}$, $Q_2 = \{(x,y) :\ x < 0 \text{ and } y > 0\}$ and so on.

---

[17]This will reduce the burden of subscripts in some of our formulas.

Proposition 8.8 *The Fourier transform* $\hat{\mathfrak{h}}$ *(taken as a tempered distribution) is a continuous function when* $\xi\eta \neq 0$ *and is given by*

$$\begin{array}{ll} \mathfrak{J}^+(-2\pi|\xi\eta|^{1/2}) & \text{in } Q_1. \\ \mathfrak{J}^-(-2\pi|\xi\eta|^{1/2}) & \text{in } Q_2. \\ \mathfrak{J}^+(2\pi|\xi\eta|^{1/2}) & \text{in } Q_3. \\ \mathfrak{J}^-(2\pi|\xi\eta|^{1/2}) & \text{in } Q_4. \end{array}$$

Proof. We approximate $\mathfrak{h}$ by the finite measures $\mathfrak{h}_\epsilon$ given by

$$\int_{\mathbb{R}^2} f\, d\mathfrak{h}_\epsilon = \int_0^\infty f(u, 1/u)\eta_\epsilon(u)\eta_\epsilon(1/u)\, \frac{du}{u}.$$

Now clearly $\int f\, d\mathfrak{h}_\epsilon \to \int f d\mathfrak{h}$ as $\epsilon \to 0$, whenever $f \in \mathcal{S}$, so the measures $\mathfrak{h}_\epsilon$ converge to $\mathfrak{h}$ in the sense of tempered distributions. Now

$$\hat{\mathfrak{h}}_\epsilon(\xi, \eta) = \int_0^\infty e^{-2\pi i(\xi u + \eta/u)}\eta_\epsilon(u)\eta_\epsilon(1/u)\, \frac{du}{u}.$$

Suppose first (ξ, η) is in Q_1 and therefore $\xi > 0$ and $\eta > 0$. Keeping (ξ, η) fixed, we make the change of variables $u \mapsto (\eta/\xi)^{1/2}u$. Then $\xi u + \eta/u$ becomes $(\xi\eta)^{1/2}(u + 1/u)$, while $\eta_\epsilon(u) = \eta(u/\epsilon)$ is transformed to $\eta_a(u)$, with $a = \epsilon(\xi/\eta)^{1/2}$, while $\eta_\epsilon(1/u)$ becomes $\eta_b(1/u)$, with $b = \epsilon(\eta/\xi)^{1/2}$. Also, the measure $\frac{du}{u}$ is unchanged. So

$$\hat{\mathfrak{h}}_\epsilon = \mathfrak{J}^+_{a,b}(-2\pi|\xi\eta|^{1/2})$$

in the first quadrant, with analogous formulas in the other three quadrants.

Now the conclusions (i), (ii), and (i′) of Proposition 8.6 and its corollary show that

$$\begin{array}{lll} |\hat{\mathfrak{h}}_\epsilon(\xi,\eta)| & \leq A|\xi\eta|^{-1/2} & \text{for } |\xi\eta| \geq 1/2, \\ |\hat{\mathfrak{h}}_\epsilon(\xi,\eta)| & \leq A\log(1/|\xi\eta|) & \text{for } |\xi\eta| \leq 1/2, \end{array}$$

uniformly in ϵ. Moreover, for each (ξ, η) with $\xi\eta \neq 0$, $\hat{\mathfrak{h}}_\epsilon(\xi, \eta)$ converges to a limit as $\epsilon \to 0$. This suffices to show that $\hat{\mathfrak{h}}_\epsilon$ converges in the sense of tempered distributions to the function $\hat{\mathfrak{h}}$ given by $\lim_{\epsilon\to 0} \hat{\mathfrak{h}}_\epsilon(\xi, \eta)$. This is because the above estimates imply that

$$\int_{\mathbb{R}^2} \hat{\mathfrak{h}}_\epsilon g \to \int \hat{\mathfrak{h}} g, \qquad \text{for any } g \in \mathcal{S},$$

by the dominated convergence theorem. Thus the proposition is proved.

We next study the Fourier transform of functions in $\mathbb{R}^2$ that are invariant under the dilations $(x,y) \to (\delta x, \delta^{-1}y)$, with $\delta > 0$. We state the result for a restricted class of smooth functions of the type that is needed below, although the main identities hold for broader classes of functions. We will suppose that f is of the form $f(x,y) = f_0(xy)$ in the first quadrant, and vanishes in the other three quadrants. The function f_0 will be assumed to be a C^∞ function with compact support on $(0,\infty)$. Functions f of this form are never integrable on the whole of $\mathbb{R}^2$ (unless $f_0 = 0$) but since they are bounded, they are of course tempered distributions.

Theorem 8.9 *Let $\hat{f}$ be the Fourier transform of $f(x,y) = f_0(xy)$. Then $\hat{f}$ is a continuous function where $\xi\eta \neq 0$. It is given by*

$$\hat{f}(\xi,\eta) = 2\int_0^\infty \mathfrak{J}^+(-2\pi|\xi\eta|^{1/2}\rho) f_0(\rho^2)\rho\, d\rho \tag{92}$$

for $(\xi,\eta) \in Q_1$. In Q_2, Q_3, and Q_4 it is given by the analogous formulas, with $\mathfrak{J}^+(-\cdot)$ replaced by $\mathfrak{J}^-(-\cdot)$, $\mathfrak{J}^+(+\cdot)$ and $\mathfrak{J}^-(+\cdot)$, respectively.

Proof. We approximate f by f_ϵ, with $f_\epsilon(x,y) = f_0(xy)\eta_\epsilon(x)\eta_\epsilon(y)$. Then each f_ϵ is a C^∞ function of compact support, and clearly $f_\epsilon \to f$ in the sense of tempered distributions.

Now

$$\hat{f}_\epsilon(\xi,\eta) = \int e^{-2\pi i(\xi x + \eta y)} f_0(xy)\eta_\epsilon(x)\eta_\epsilon(y)\, dxdy.$$

We introduce new variables (u,ρ) in the first quadrant with $x = u\rho$, $y = \frac{\rho}{u}$, and observe that

$$\frac{\partial(x,y)}{\partial(u,\rho)} = \begin{pmatrix} \rho & u \\ -\frac{\rho}{u^2} & \frac{1}{u} \end{pmatrix},$$

which has a determinant equal to $2\rho/u$. Therefore $dx\, dy = 2\rho\frac{du}{u}d\rho$ and

$$\hat{f}_\epsilon(\xi,\eta) = 2\int_0^\infty\int_0^\infty e^{-2\pi i(\xi u\rho + \eta\rho/u)} f_0(\rho^2)\eta_\epsilon(\rho u)\eta_\epsilon(\rho/u)\rho\,\frac{du}{u}\, d\rho.$$

Again, if (ξ,η) is in the first quadrant and if we make the change of variables $u \mapsto (\eta/\xi)^{1/2}u$, then we have

$$\hat{f}_\epsilon(\xi,\eta) = 2\int_0^\infty \mathfrak{J}^+_{a,b}(-2\pi|\xi\eta|^{1/2}\rho) f_0(\rho^2)\rho\, d\rho, \tag{93}$$

with now $a = \frac{\epsilon}{\rho}\left(\frac{\xi}{\eta}\right)^{1/2}$ and $b = \frac{\epsilon}{\rho}\left(\frac{\eta}{\xi}\right)^{1/2}$.

The analogous formulas for $\hat{f}_\epsilon(\xi,\eta)$ hold when (ξ,η) are in the second, third and fourth quadrants. So the fact that $\hat{f}_\epsilon$ converges in the sense of tempered distributions to the limit $\hat{f}$ given by (92) then follows by the same reasoning used in the proof of Proposition 8.8.

Corollary 8.10 *The Fourier transforms $\hat{f}_\epsilon$ and $\hat{f}$ satisfy the following estimate, uniformly in ϵ:*

$$|\hat{f}_\epsilon(\xi,\eta)| \le A_N|\xi\eta|^{-N} \quad \text{when } |\xi\eta| \ge 1/2, \tag{94}$$

for every $N \ge 0$.

This is a consequence of the asymptotic behavior of $\mathfrak{J}^{\pm}(\lambda)$ for λ as given in Proposition 8.6 and its corollary together with the fact that $\int_0^\infty e^{-4\pi i\rho|\xi\eta|^{1/2}} f_0(\rho^2)\rho\, d\rho$ is $O(|\xi\eta|^{-N})$ for every $N \ge 0$, since $f_0(\rho^2)\rho$ is a C^∞ function with compact support in $(0,\infty)$.

8.5 A summation formula

Here we obtain the hyperbolic analog of the summation formula (83). It will be convenient now to put together the oscillatory integrals for the four quadrants and write $\mathfrak{J}$ for

$$\mathfrak{J}(\lambda) = 2\left(\mathfrak{J}^+(\lambda) + \mathfrak{J}^+(-\lambda) + \mathfrak{J}^-(\lambda) + \mathfrak{J}^-(-\lambda)\right).[18]$$

Again f_0 is a C^∞ function with compact support in $(0,\infty)$.

Theorem 8.11

$$\sum_{k=1}^{\infty} f_0(k)d(k) = \int_0^\infty (\log\rho + 2\gamma) f_0(\rho)\, d\rho + \sum_{k=1}^{\infty} F_0(k)d(k), \tag{95}$$

where

$$F_0(u) = \int_0^\infty \mathfrak{J}(2\pi u^{1/2}\rho) f_0(\rho^2)\rho\, d\rho.$$

Proof. We apply the Poisson summation formula

$$\sum_{\mathbb{Z}^2} f_\epsilon(m,n) = \sum_{\mathbb{Z}^2} \hat{f}_\epsilon(m,n),$$

to the approximating functions f_ϵ, and then pass to the limit as $\epsilon \to 0$. Now the sum on the left-hand side is clearly taken over a bounded set of

[18] The expression of $\mathfrak{J}$ in terms of Bessel-like functions is given in Problem 7.

lattice points since $f_0(u)$ has compact support in $(0, \infty)$. Thus, gathering together the (m, n) for which $mn = k$ gives the left-hand side of the formula.

Now divide the sum on the right-hand side in two parts. One part taken over those (m, n) for which $mn \neq 0$, and the other part taken over those (m, n) where either $m = 0$, or $n = 0$, or both, $m = n = 0$.

By the theorem and Corollary 8.10, we see first that

$$\lim_{\epsilon \to 0} \sum_{mn \neq 0} \hat{f}_\epsilon(m, n) = \sum_{mn \neq 0} \hat{f}(m, n),$$

since the series are dominated by the convergent series $\sum_{mn \neq 0} |mn|^{-2}$. Next, gathering together those (m, n) for which $|mn| = k$, gives us

$$\sum_{mn \neq 0} \hat{f}(m, n) = \sum_{k=1}^{\infty} F_0(k) d(k)$$

because of formula (92).

It remains to evaluate the limit as $\epsilon \to 0$ of

$$\sum_{mn=0} \hat{f}_\epsilon(m, n). \tag{96}$$

Now, one part of (96) is $\sum_m \hat{f}_\epsilon(m, 0)$, which, by the Poisson summation formula (this time in its one-dimensional form) equals

$$\sum_m \int_{\mathbb{R}} f_\epsilon(m, y)\, dy.$$

However $f_\epsilon(x, y) = f_0(xy)\eta_\epsilon(x)\eta_\epsilon(y)$ and f_ϵ is supported in the first quadrant, so this sum is

$$\sum_{m=1}^{\infty} \int_0^{\infty} f_0(my)\eta_\epsilon(m)\eta_\epsilon(y)\, dy.$$

Upon making the change of variables $my \to y$ in the integral and interchanging the summation and integration (which is easily justified), we see that the sum becomes

$$\int_0^{\infty} k_\epsilon(y) f_0(y)\, dy,$$

with $k_\epsilon(y) = \sum_{m=1}^{\infty} \eta_\epsilon(y/m)\frac{1}{m}$, when we take $0 < \epsilon \leq 1$. (Note that then $\eta_\epsilon(m) = 1$ if $m \geq 1$.)

We claim that if $c_0 = \int_0^1 \eta(x)\,\frac{dx}{x}$, then

$$k_\epsilon(y) = \log(y/\epsilon) + \gamma + c_0 + O(\epsilon/y) \quad \text{as } \epsilon \to 0, \tag{97}$$

and this estimate is uniform as long as y ranges over a compact subset of $(0, \alpha)$.

To see this we divide the sum $k_\epsilon(y)$ in two parts: where the summation is taken over m with $m \leq y/\epsilon$, and the complementary part. Since $\eta_\epsilon(y/m) = \eta(y/(\epsilon m)) = 1$ when $m \leq y/\epsilon$, that part of the sum is $\sum_{1\leq m\leq y/\epsilon} 1/m$ which equals $\log(y/\epsilon) + \gamma + O(\epsilon/y)$ by the defining property of Euler's γ.[19]

On the other hand,

$$\sum_{m\geq y/\epsilon} \eta(y/(\epsilon m))\frac{1}{m} - \int_{u\geq y/\epsilon} \eta(y/(\epsilon u))\,\frac{du}{u} = O\left(\int_{y/\epsilon}^{\infty} \frac{du}{u^2}\right) = O\left(\frac{\epsilon}{y}\right),$$

because $\frac{d}{du}\left(\eta\left(\frac{y}{\epsilon u}\right)\frac{1}{u}\right) = O(1/u^2)$, which in turn follows since $\eta'(u)$ is compactly supported in $(0, \infty)$. As a result (97) is established with

$$c_0 = \int_1^\infty \eta(1/u)\,\frac{du}{u} = \int_0^1 \eta(u)\frac{du}{u}.$$

By symmetry we also get

$$\sum_n \hat{f}_\epsilon(0, n) = \int_0^\infty k_\epsilon(y) f_0(y)\,dy,$$

with k_ϵ given by (97).

It remains to evaluate $\hat{f}_\epsilon(0,0)$, which is the excess of $\sum_m \hat{f}_\epsilon(m,0) + \sum_m \hat{f}_\epsilon(0,m)$ over $\sum_{mn=0} \hat{f}(m,n)$.

However,

$$\begin{aligned}
\hat{f}_\epsilon(0,0) &= \int_{\mathbb{R}^2} f_\epsilon(x,y)\,dx\,dy \\
&= \int_{\mathbb{R}^2} f_0(xy)\eta_\epsilon(x)\eta_\epsilon(y)\,dx\,dy \\
&= \int_0^\infty k'_\epsilon(y) f_0(y)\,dy,
\end{aligned}$$

with $k'_\epsilon(y) = \int_0^\infty \eta(x/\epsilon)\eta(y/(\epsilon x))\,\frac{dx}{x}$, as a simple change of variables shows.

[19] See, for instance, Proposition 3.10 in Chapter 8 of Book I.

Now divide the integration in x into four parts: where both x/ϵ and $y/(\epsilon x)$ are ≥ 1; where one is ≥ 1, but the other is < 1; and where both are < 1. The first part gives $\int_\epsilon^{y/\epsilon} \frac{dx}{x} = \log y - 2\log\epsilon$, since $\eta(x/\epsilon) = 1$ and $\eta(y/(\epsilon x)) = 1$ there. Next if $x/\epsilon \leq 1$ but $y/(\epsilon x) \geq 1$ the integral is $\int_0^\epsilon \eta(x/\epsilon)\,\frac{dx}{x} = \int_0^1 \eta(x)\,\frac{dx}{x} = c_0$. A similar evaluation holds when $y/(\epsilon x) \leq 1$ and $x/\epsilon > 1$. Finally the last range of x's is empty when ϵ is sufficiently small since $x < \epsilon$ implies $y/(\epsilon x) > 1$, whenever $\epsilon \leq y$ and y is bounded away from 0. Thus

$$k'_\epsilon(y) = \log y - \log 2\epsilon + 2c_0. \tag{98}$$

Altogether then

$$\sum_{mn=0} \hat{f}_\epsilon(m,n) = \int_0^\infty (2k_\epsilon - k'_\epsilon) f_0(y)\,dy,$$

and because of (97) and (98) this converges to $\int_0^\infty (\log y + 2\gamma) f_0(y)\,dy$ as $\epsilon \to 0$. Theorem 8.11 is therefore proved.

We come now to the proof of the main theorem, whose conclusion is stated in (89). Here we would like to apply the sum formula (95) to $f_0 = \chi_\mu$, the characteristic function of the interval $(0,\mu)$. However this function does not have the smoothness required for the validity of (95). We are guided instead by the reasoning used in the proofs of Theorems 8.3 and 8.4 that suggest we regularize χ_μ in an appropriate way.

To proceed, let us note that in the sense that Theorem 8.3 and (89) in Theorem 8.5 are parallel, we have to think of μ as playing the role of R^2. Indeed, setting $\mu = R^2$ will lead us to the proper choices below. With this in mind we want to replace χ_μ by a function $\chi_{\mu,\delta}$, which is defined so that effectively $\chi_{\mu,\delta}(t) = 1$ if $0 < t \leq \mu$, that is, $\chi_{\mu,\delta}(\rho^2) = 1$ if $0 \leq \rho \leq R = \mu^{1/2}$; and moreover so that $\chi_{\mu,\delta}(\rho^2)$ decreases smoothly to zero in $R \leq \rho \leq R + \delta$. Here δ is the quantity $R^{-1/3}$ that arises in the proof of Theorem 8.3.

To give the precise definition of $\chi_{\mu,\delta}$ we fix a C^∞ function ψ on $[0,1]$ so that $0 \leq \psi \leq 1$, with $\psi = 0$ near the origin and $\psi = 1$ near 1. We define

$$\chi_{\mu,\delta}(\rho^2) = \begin{cases} \psi(\rho) & \text{for } 0 \leq \rho \leq 1, \\ 1 & \text{for } 1 \leq \rho \leq R, \\ 1 - \psi\left(\frac{\rho - R}{\delta}\right) & \text{for } R \leq \rho \leq R + \delta. \end{cases}$$

Now consider the sum formula (95) with $f_0(u) = \chi_{\mu,\delta}(u)$. Then the integral term on the right-hand side is $\int_0^\infty (\log\rho + 2\gamma)\chi_{\mu,\delta}(\rho)\,d\rho$, which is equal to

$$\int_1^{\mu}(\log\rho+2\gamma)\,d\rho+O(1)+O\left(\int_{\mu}^{\mu+c\mu^{1/3}}\log\rho\,d\rho\right),$$

because $R^2=\mu$ and $(R+\delta)^2=(R+R^{-1/3})^2=\mu+O(\mu^{1/3})$. Thus the integral equals

$$\mu\log\mu+(2\gamma-1)\mu+O(\mu^{1/3}\log\mu). \tag{99}$$

We now estimate each term $\int_0^\infty \mathfrak{J}(2\pi k^{1/2}\rho)f_0(\rho^2)\rho\,d\rho$ that arises in the sum on the right-hand side of (95) with $f_0(\rho^2)=\chi_{\mu,\delta}(\rho^2)$. We make two estimates for this term, with $R=\mu^{1/2}$:

(a) $O(R^{1/2}/k^{3/4})$ and

(b) $O(R^{1/2}\delta^{-1}/k^{5/4})$.

To see this consider the main contribution to $\mathfrak{J}(\lambda)$ for large λ as given via (i) and (i′) in Proposition 8.6 and its corollary. This is the term $c_0\lambda^{-1/2}e^{2i\lambda}$. Thus for its contribution we need to estimate

$$\sigma^{-1/2}\int_0^\infty e^{i\sigma\rho}\chi_{\mu,\delta}(\rho^2)\rho^{1/2}\,d\rho, \tag{100}$$

where we have set $\sigma=\pm2\cdot2\pi k^{1/2}$.

First since $e^{i\sigma\rho}=\frac{1}{i\sigma}\frac{d}{d\rho}(e^{i\sigma\rho})$, we may integrate by parts in (100) and see that (100) is majorized by a multiple of

$$\sigma^{-3/2}\left(\int_0^R\rho^{-1/2}\,d\rho+\int_R^{R+\delta}\rho^{1/2}\,d\rho\right),$$

because $\chi_{\mu,\delta}(\rho^2)=1$ for $1\le\rho\le R$, and $\frac{d}{d\rho}\chi_{\mu,\delta}(\rho^2)=O(1/\delta)$ for $R\le\rho\le R+\delta$. This gives us the estimate $O(\sigma^{-3/2}R^{1/2})=O(k^{-3/4}R^{1/2})$ and this is (a) above. If instead we integrate by parts twice we see that (100) is majorized by a multiple of

$$\sigma^{-5/2}\int_0^\infty\left|\left(\frac{d}{d\rho}\right)^2(\chi_{\mu,\delta}(\rho^2)\rho^{1/2})\right|\,d\rho.$$

However $\left(\frac{d}{d\rho}\right)^2(\chi_{\mu,\delta}(\rho^2)\rho^{1/2})=O(1)$ when $0\le\rho\le1$; it is $c\rho^{-5/2}$ when $1\le\rho\le R$; and $O(R^{1/2}\delta^{-2})$ when $R\le\rho\le R-\delta$. So we obtain the bound of the form $\sigma^{-5/2}\left(O(1)+R^{1/2}\delta^{-1}\right)=O(\sigma^{-5/2}R^{1/2}\delta^{-1})$ for (100). Thus we have established the bounds (a) and (b) for the main contribution coming from the first term in (i) of Proposition 8.6. The other terms in the asymptotic series give obviously smaller contributions, and we need only go as far as $N=1$ in the formula (i), because then the error term will contribute less then either (a) or (b). Thus the estimates (a) and (b) have been established for the individual terms of the series on the right-hand side of (95).

Our conclusion is then that modulo an error term that is $O(\mu^{1/3}\log\mu)$ we have

$$(101)\quad \sum \chi_{\mu,\delta}(m,n)=\mu\log\mu+(2\gamma-1)\mu+ \\ +O\left(R^{1/2}\sum_{1\le k\le 1/\delta^2} d(k)k^{-3/4}+R^{1/2}\delta^{-1}\sum_{k>1/\delta^2} d(k)k^{-5/4}\right).$$

Now it is a simple fact that

$$\sum_{1\le k\le r} d(k)k^{\alpha}=O\left(r^{\alpha+1}\log r\right)\qquad \text{as } r\to\infty, \text{ if } \alpha>-1,$$

and

$$\sum_{r<k} d(k)k^{\alpha}=O\left(r^{\alpha+1}\log r\right)\qquad \text{as } r\to\infty, \text{ if } \alpha<-1.$$

(The proof of this is outlined in Exercise 22.) Taking $r=1/\delta^2=R^{2/3}$ and $\alpha=-3/4$ or $\alpha=-5/4$, the above shows that the O term in (101) is majorized by a multiple of

$$(R^{1/2}R^{2/3\cdot 1/4}+R^{1/2}R^{1/3}R^{-2/3\cdot 1/4})\log R=2R^{2/3}\log R.$$

Now if we set $N_\delta(R)=\sum_{m,n}\chi_{\mu,\delta}(m,n)$, with $\mu=R^2$, then (101) states that

$$(102)\qquad N_\delta(R)=R^2\log R^2+(2\gamma-1)R^2+O(R^{2/3}\log R).$$

However by the way $\chi_{\mu,\delta}$ has been defined it is clear that

$$\chi_{(R-\delta)^2,\delta}\le\chi_\mu\le\chi_{(R+\delta)^2,\delta},$$

with $\mu=R^2$. Thus

$$N_\delta(R-\delta)\le\sum_{1\le k\le\mu} d(k)\le N_\delta(R+\delta).$$

If we look back at (102) we see this implies

$$\sum_{1\leq k\leq \mu} d(k) = \mu \log \mu + (2\gamma - 1)\mu + O(\mu^{1/3}\log \mu),$$

since $\mu = R^2$ and $\delta = R^{-1/3}$. Therefore our main result is now established.

9 Exercises

1. Use spherical coordinates to show that in $\mathbb{R}^d$

$$\int_{S^{d-1}} e^{-2\pi i x\cdot\xi}\, d\sigma = c_d \int_{-1}^{1} e^{-2\pi i|\xi|u}(1-u^2)^{\frac{d-3}{2}}\, du,$$

with c_d the area of the unit sphere S^{d-2} in $\mathbb{R}^{d-1}$. Then deduce formula (3) from Problem 2 in Chapter 6, Book I.

2. Let the hypersurface M contain a neighborhood of a hyperplane (for example $\{x_d = 0\}$). Show that in this case $\widehat{d\mu}(\xi) \neq O(|\xi|^{-\epsilon})$, as $|\xi| \to \infty$ for any $\epsilon > 0$.

3. Principle of stationary phase when $d = 1$. Consider

$$I(\lambda) = \int_{-\infty}^{\infty} e^{i\lambda\Phi(x)}\psi(x)\, dx,$$

where ψ is a C^∞ function of compact support and $x = 0$ is the only critical point of Φ in the support of ψ, while $\Phi''(0) \neq 0$. Then for every positive integer N,

$$I(\lambda) = \frac{e^{i\lambda\Phi(0)}}{\lambda^{1/2}}\left(a_0 + a_1\lambda^{-1} + \cdots + a_N\lambda^{-N}\right) + O(\lambda^{-N-1/2}), \qquad \text{as } \lambda \to \infty.$$

The a_k are determined by $\Phi''(0), \dots, \Phi^{(2k+2)}(0)$, and $\psi(0), \dots, \psi^{(2k)}(0)$. In particular $a_0 = \left(\frac{2\pi}{-i\Phi''(0)}\right)^{1/2}\psi(0)$.

Prove this in two steps:

(a) Consider first the special case when $\varphi(x) = x^2$ dealt with by (8).

(b) Pass to the case of general φ by a change of variables that brings $\varphi(x)$ to x^2 or $-x^2$.

4. Suppose Φ is of class C^k in an interval $[a, b]$ with $k \geq 2$. Assume that $|\Phi^{(k)}(x)| \geq 1$ throughout the interval. Prove the following generalization of Proposition 2.3

$$\left|\int_a^b e^{i\lambda\Phi(x)}\, dx\right| \leq c_k\lambda^{-1/k}.$$

[Hint: Suppose $\Phi^{(k-1)}(x_0) = 0$, and argue by induction as in the proof of Proposition 2.3.]

5. Consider the curve $\gamma(t) = (t, t^k)$ in $\mathbb{R}^2$, with k an integer ≥ 2. Its curvature vanishes nowhere when $k = 2$ and only at the origin, of order $k-2$, when $k > 2$. Let $d\mu$ be defined by $\int_{\mathbb{R}^2} f\, d\mu = \int_{\mathbb{R}} f(t, t^k)\psi(t)\, dt$, where ψ is a C^∞ function of compact support, and $\psi(0) \neq 0$. Then prove:

(a) $|\widehat{d\mu}(\xi)| = O(|\xi|^{-1/k})$.

(b) However, this decay estimate is optimal, that is, $|\widehat{d\mu}(0, \xi_2)| \geq c|\xi_2|^{-1/k}$ if ξ_2 is large.

[Hint: For (a) use Exercise 4. For (b) consider for example the case when k is even and verify that $\int_{-\infty}^{\infty} e^{i\lambda x^k} e^{-x^k}\, dx = c_\lambda (1 - i\lambda)^{-1/k}$.]

6. Show that the (L^p, L^q) results for the averaging operator A given by Corollary 4.2 are optimal by proving the following (in, say, the case of the sphere in $\mathbb{R}^3$):

(a) Suppose $f(x)$ vanishes for small x and $f(x) \geq |x|^{-r}$, for $|x| \geq 1$. Then observe $A(f)(x) \geq c|x|^{-r}$ and thus we must always have $q \geq p$. This restriction corresponds to the side of the triangle joining $(0,0)$ and $(1,1)$.

(b) Next let $f = \chi_{B_\delta}$, where B_δ is the ball of radius δ. Note that if δ is small $A(\chi_{B_\delta}) \geq c\delta^2$ for $|1 - |x|| < \delta/2$. So $\|f\|_{L^p} \approx \delta^{3/p}$ while $\|A(f)\|_{L^q} \gtrsim \delta^2 \delta^{1/q}$. Hence the inequality $\|A(f)\|_{L^q} \leq c\|f\|_{L^p}$ implies $2 + 1/q \geq 3/p$, which corresponds to the side of the triangle joining $(3/4, 1/4)$ and $(1,1)$.

(c) For the third inequality, use duality and (b).

7. By refining the argument given in Exercise 6 (b) one can show that the smoothing of degree $(d-1)/2$ asserted in Proposition 1.1 fails when $p \neq 2$.

In the case $p < 2$ and $d = 3$, this can be seen by taking $\delta > 0$ small and setting $f = \varphi_\delta$, where $\varphi_\delta = \varphi(x/\delta)$ and φ is a non-negative smooth function of compact support. Here $\|\varphi_\delta\|_{L^p} \approx c\delta^{3/p}$, while $\|\nabla A(\varphi_\delta)\|_{L^p} \gtrsim \delta\delta^{1/p}$. Hence the inequality $\|A(\varphi_\delta)\|_{L^p_1(\mathbb{R}^3)} \leq C\|\varphi_\delta\|_{L^p(\mathbb{R}^3)}$ fails for small delta when $p < 2$.

[Hint: If $c_1 > 0$ is sufficiently small, then $\delta^2 \lesssim A(\varphi_\delta)$ and $|\nabla A(\varphi_\delta)| \gtrsim \delta$, whenever $|1 - |x|| \leq c_1\delta$.]

8. Let M be a (local) hypersurface given in coordinates $(x', x_d) \in \mathbb{R}^{d-1} \times \mathbb{R}$ as $\{x_d = \varphi(x')\}$. Suppose F is any continuous function of small support defined in a a neighborhood of M and set $f = F|_M$.

(a) Show that $\lim_{\epsilon \to 0} \frac{1}{2\epsilon} \int_{d(x,M) < \epsilon} F\, dx$ exists and equals $\int_{\mathbb{R}^{d-1}} f(x', \varphi(x'))(1 + |\nabla_{x'}\varphi|^2)^{1/2}\, dx'$. This limit defines the induced Lebesgue measure $d\sigma$ and equals $\int_M f\, d\sigma$.

(b) Suppose ρ is any defining function of M. Show that

$$\lim_{\epsilon\to 0}\frac{1}{2\epsilon}\int_{|\rho|<\epsilon} F\,dx = \int_M f\frac{d\sigma}{|\nabla\rho|}.$$

(c) Suppose h is a Schwartz function on $\mathbb{R}$ with $\int_{\mathbb{R}} h(u)\,du = 1$. Then

$$\lim_{\epsilon\to 0}\epsilon^{-1}\int_{\mathbb{R}^d} h(\rho/\epsilon)F\,dx = \int_M f\frac{d\sigma}{|\nabla\rho|}.$$

[Hint: For (c), assume h is even and let $I(t) = \int_{|\rho(x)|<t} F(x)\,dx$. Then

$$\epsilon^{-1}\int h(\rho/\epsilon)F\,dx = \epsilon^{-1}\int_0^\infty h(u/\epsilon)\frac{dI_u}{du}\,du = -\epsilon^{-1}\int_0^\infty (u/\epsilon)h'(u/\epsilon)\left(\frac{1}{u}I_u\right)du.$$

Now use the fact that $-\int_0^\infty uh'(u)\,du = 1/2$ and $\frac{I_u}{2u}\to\int_M f\frac{d\sigma}{|\nabla\rho|}$, as $u\to 0$.]

9. Observe the following Euclidean-invariance properties of the principal curvatures of a hypersurface M in $\mathbb{R}^d$. For each $h\in\mathbb{R}^d$ consider the translate $M+h$ of M; also for each rotation r of $\mathbb{R}^d$, the rotated surface $r(M)$; and for each $\delta\in\mathbb{R}$, $\delta\neq 0$, the dilated surface δM. Denote by $\{\lambda_j(x)\}$ the principal curvatures of M at x.

(a) Show that $\{\lambda_j(x-h)\}$, $\{\lambda_j(r^{-1}(x))\}$, and $\{\delta^{-2}\lambda_j(x/\delta)\}$ are the principal curvatures of $M+h$, $r(M)$, δM at the points $x+h$, $r(x)$ and δx, respectively.

(b) Consider the cone $\{x_d^2 = |x'|^2,\ x\neq 0\}$ with defining function $\rho = |x'|^2 - x_d^2$. Using (a), show that at x there are $d-2$ principal curvatures that equal x_d^{-2}, and one that vanishes.

10. Let $f_0(r) = r^{-1/2}(\log r)^{-\delta}$, $0<\delta<1$, when $r\geq 2$, and $f_0(r)=0$ otherwise.

(a) Prove that $\int |J_k(2\pi\rho r)|f_0(r)\,dr = \infty$ for every $\rho>0$.

(b) Show as a result that, if $p\geq 2d/(d+1)$, then (31) cannot hold for any q when M is the sphere.

11. One can prove that the conjectured condition $q\leq\left(\frac{d-1}{d+1}\right)p'$ for (L^p,L^q) restriction cannot hold in a larger range, by the following argument given in the case $d=2$.

(a) Suppose the inequality (31) holds for some p and q. Show that as a result

$$\int_{1-\delta\leq|\xi|\leq 1}|\hat{f}(\xi)|^q\,d\xi \leq c'\delta\|f\|_{L^p}^q \qquad \text{for small } \delta.$$

(b) Next choose $\hat{f}(\xi_1, \xi_2) = \eta((\xi_1 - 1)/\delta)\eta(\xi_2/\delta)$ when $\eta(u) = 1$ if $|u| \geq 1$. That is, $\hat{f}(\xi)$ dominates the characteristic function of a rectangle of approximate side lengths δ and $\delta^{1/2}$ that fits inside the annulus $1 - \delta \leq |\xi| \leq 1$. Use this to obtain a contradiction $q > \left(\frac{d-1}{d+1}\right) p'$ by letting $\delta \to 0$.

12. Connect the operator $e^{it\triangle}$ and the Fourier transform as follows. Let m_t be the multiplication operator $m_t : f(x) \mapsto \frac{1}{(4\pi it)^d} e^{-\frac{i|x|^2}{4t}} f(x)$.

(a) Show that $e^{it\triangle}(f) = i^{-d} m_t (f m_t)^\wedge$ when $t = 1/4\pi$.

(b) Generalize this identity to any $t \neq 0$ by rescaling.

13. Let $\text{Ai}(u) = \lim_{N\to\infty} \frac{1}{2\pi} \int_{-N}^{N} e^{i\left(\frac{v^3}{3} + uv\right)} dv$.

(a) Show that this limit exists for every $u \in \mathbb{R}$.

(b) Prove that $|\text{Ai}(u)| \leq c(1 + |u|)^{-1/4}$.

(c) Moreover, show that $\text{Ai}(u)$ is rapidly decreasing as $u \to \infty$, for $u > 0$.

[Hint: Write $\Phi(r) = \frac{r^3}{3} + ru$, and apply the estimates in Section 2. For (a) use the fact that $\Phi'(r) \to \infty$ as $|r| \to \infty$. For (b), use the fact that $|\Phi'(r)| \geq |u|/2$, when $|r| \leq (\frac{1}{2}|u|)^{1/2}$, while $|\Phi''(r)| \geq 2|r|$ when $|r| > (\frac{1}{2}|u|)^{1/2}$.]

14. Suppose $F \in L^2(\mathbb{R}^d \times \mathbb{R})$ and $S(F)(x,t) = i\int_0^t e^{i(t-s)\triangle} F(\cdot, s)\, ds$. Prove that:

(a) For each t, $S(F)(\cdot, t) \in L^2(\mathbb{R}^d)$, and

$$\|S(F)(\cdot,t)\|_{L^2(\mathbb{R}^d)} \leq |t|^{1/2} \|F\|_{L^2(\mathbb{R}^d\times\mathbb{R})}.$$

(b) If $F(\cdot, t) = e^{it\triangle} G(\cdot, t)$, then

$$\|G(0,t_1) - G(0,t_2)\|_{L^2(\mathbb{R}^d)} \leq |t_1 - t_2|^{1/2} \|G\|_{L^2(\mathbb{R}^d\times\mathbb{R})}.$$

(c) As a result, $t \mapsto F(0,t)$ is continuous in the $L^2(\mathbb{R}^d)$ norm.

[Hint: For (a) and (b) use the unitarity of $e^{it\triangle}$ and Schwarz's inequality. For (c), approximate F by C^∞ functions of compact support, using (b) and (c).]

15. Suppose u is a smooth solution of (54) that decays sufficiently quickly as $|x| \to \infty$. Show that both $\int_{\mathbb{R}^d} |u|^2\, dx$, and $\int_{\mathbb{R}^d} (\frac{1}{2}|\nabla u|^2 - \frac{\sigma}{\lambda+1}|u|^\lambda)\, dx$ are independent of t.

[Hint: For the first, note that $\int_{\mathbb{R}^d} \triangle u\, v\, dx = \int_{\mathbb{R}^d} u \triangle v\, dx$. For the second, observe that $\frac{\partial}{\partial t} \int_{\mathbb{R}^d} |\nabla u|^2\, dx = -\int_{\mathbb{R}^d} \left(\frac{\partial u}{\partial t} \triangle \overline{u} + \frac{\partial \overline{u}}{\partial t} \triangle u\right) dx$.]

16. The following is a converse of Propositions 6.6 and 6.8. Suppose $u(\cdot, t)$ is in $L^2(\mathbb{R}^d)$ for each t, with $t \mapsto u(\cdot, t)$ continuous in the L^2 norm, and $u(\cdot, 0) = 0$. Assume that $\frac{1}{i}\frac{\partial u}{\partial t} - \triangle u = F$ as distributions, with $F \in L^2(\mathbb{R}^d \times \mathbb{R})$. Then show that $u = S(F)$.

[Hint: Use the following fact. If $H(\cdot, t)$ is in $L^2(\mathbb{R}^d)$ for each t, with $t \mapsto H(\cdot, t)$ continuous in the L^2 norm, $H(\cdot, 0) = 0$, and $\frac{\partial H}{\partial t} = 0$ in the sense of distributions, then $H = 0$. Apply this to $H(\cdot, t) = e^{-it\triangle}(u(\cdot, t) - S(F)(\cdot, t))$.]

17. A solution u of the non-linear Schrödinger equation (54) is uniquely determined by its initial data f. Moreover the solution depends continuously on this data. These are two features of the "well-posedness" of the problem and can be stated as follows. Assume $\lambda = \frac{d+4}{d}$ and $q = \frac{2d+4}{d}$.

(a) Suppose u and v are two strong solutions defined for $|t| < a$, having the same initial data $f \in L^2(\mathbb{R}^d)$. Show that $u = v$.

(b) Given $f \in L^2(\mathbb{R}^d)$, prove that there are $\epsilon > 0$ and $a > 0$ (depending on f) so that if $\|f - g\|_{L^2} < \epsilon$, and u and v are strong solutions of (54) with initial data f and g respectively, then

$$\|u - v\|_{L^q} \leq c\|f - g\|_{L^2(\mathbb{R}^d)}.$$

Here $L^q = L^q(\mathbb{R}^d \times \{|t| < a\})$.

[Hint: Adapt the argument in Theorem 6.9, and for (a) proceed as follows: note that for small $\ell > 0$

$$\|u\|_{L^q(\mathbb{R}^d \times I)} < \delta \qquad \text{and} \qquad \|v\|_{L^q(\mathbb{R} \times I)} < \delta$$

for all intervals I of length $\leq 2\ell$. Thus

$$\|u - v\|_{L^q} \leq \|\mathcal{M}(u) - \mathcal{M}(v)\|_{L^q} \leq \frac{1}{2}\|u - v\|_{L^q},$$

with $L^q = L^q(\mathbb{R}^d \times \{|t| < \ell\})$, and so $u = v$ for $0 \leq t \leq \ell$. Now use the t-translation invariance to apply the same argument for $u(\cdot, t + \ell)$ and $v(\cdot, t + \ell)$, and so on.

For (b) note that by choosing a and ϵ sufficiently small $\|e^{it\triangle} f\|_{L^q} < \delta/4$ and then $\|e^{it\triangle} g\|_{L^q} < \delta/2$, where $L^q = L^q(\mathbb{R}^d \times \{|t| < a\})$. Now the iteration argument shows that the solutions u and v satisfy $\|u\|_{L^q}, \|v\|_{L^q} < \delta$. Also $\|u - v\|_{L^q} \leq \|S(|u|^{\lambda-1}u - |v|^{\lambda-1}v)\|_{L^q} + c\|f - g\|_{L^2}$. But $\|S(|u|^{\lambda-1}u - |v|^{\lambda-1}v)\|_{L^q} \leq \frac{1}{2}\|u - v\|_{L^q}$, so this proves (b).]

18. Consider the Radon transform $\mathcal{R}_B$ defined by

$$\mathcal{R}_B(f)(x', x_d) = \int_{\mathbb{R}^{d-1}} f(y', x_d - B(x', y'))\, dy'$$

$x = (x', x_d) \in \mathbb{R}^{d-1} \times \mathbb{R}$, where B is a fixed non-degenerate bilinear form on $\mathbb{R}^{d-1} \times \mathbb{R}^{d-1}$. We write $B(x', y') = C(x') \cdot y'$, and assume that the dimension d is odd.

Verify that:

(a) $\| \left(\frac{\partial}{\partial x_d}\right)^{\frac{d-1}{2}} \mathcal{R}_B(f)\|^2_{L^2(\mathbb{R}^d)} = c_B\|f\|^2_{L^2}$ for every $f \in \mathcal{S}$, with $c_B = \frac{2(2\pi)^{d-1}}{|\det(C)|}$.

(b) If $(\mathcal{R}_B)^*$ is the (formal) adjoint of $\mathcal{R}_B$, then $\mathcal{R}_B^* = \mathcal{R}_{B^*}$, with $B^*(x,y) = -B(x,y)$. Also $\frac{\partial}{\partial x_d}\mathcal{R}_B = \mathcal{R}_B\frac{\partial}{\partial x_d}$.

(c) Deduce from (a) and (b) the inversion formula

$$\left(i\frac{\partial}{\partial x_d}\right)^{d-1} \mathcal{R}_B^*\mathcal{R}_B(f) = c_B f.$$

19. We take the Radon transform $\mathcal{R}_B$ as in the previous exercise (with the dimension d odd) and consider a localized version of it, $\mathcal{R}'_B$, given by

$$\mathcal{R}'_B = \eta'\mathcal{R}_B(\eta f)$$

where η and η' are a pair of C^∞ functions of compact support. Show that:

(a) $\|\mathcal{R}'_B(f)\|_{L^2} \le c\|f\|_{L^2}$.

(b) $\left(\frac{\partial}{\partial x}\right)^\alpha \mathcal{R}'_B(f)$ is a finite linear combination of terms of the form $\left(\frac{\partial}{\partial x_d}\right)^\ell (\eta'_\ell\mathcal{R}_B(\eta_\ell(f)))$ with $0 \le \ell \le |\alpha|$.

(c) Deduce from the above and part (a) of the previous exercise that $f \mapsto \mathcal{R}'_B(f)$ is a bounded linear transformation from L^2 to $L^2_{\frac{d-1}{2}}$.

20. The averaging operator from Section 7 satisfies the L^p, L^q conclusions stated for the operator A in Corollary 4.2. Prove this by proceeding according to the following steps.

First, recall that $\mathcal{A} = \sum_{k=0}^\infty \mathcal{A}_k$, with $\mathcal{A}_k$ given by (65) in Section 7.4, with the sum convergent in the L^2 norm. Now fix r and consider

$$T_s = (1 - 2^{1-s})e^{s^2}\sum_{k=0}^r 2^{-ks}\mathcal{A}_k.$$

Note that $T_0 = -\sum_{k=0}^r \mathcal{A}_k$, and so it will suffice to make $L^p \to L^q$ estimates for T_0 that are independent of r. Now prove:

(a) $\|T_s(f)\|_{L^2(\mathbb{R}^d)} \le M\|f\|_{L^2(\mathbb{R}^d)}$ if $\mathrm{Re}(s) = -\frac{d-1}{2}$.

(b) $\|T_s(f)\|_{L^\infty(\mathbb{R}^d)} \le M\|f\|_{L^1(\mathbb{R}^d)}$ if $\mathrm{Re}(s) = 1$.

Once (a) and (b) have been established, an interpolation via Proposition 4.4 yields

$$\|T_0(f)\|_{L^q} \le M\|f\|_{L^p},$$

with $p = \frac{d+1}{d}$ and $q = d+1$, and this leads to the desired conclusion.

[Hint: Part (a) follows from the estimates (70) and (71) for $\alpha = 0$, and the almost-orthogonality argument, Proposition 7.4. To prove (b) note that it suffices to prove that $(1 - 2^{1-s})e^{s^2} \sum_{k=0}^{r} 2^{-ks}\eta(2^{-k}u)$ has a bounded Fourier transform if $\mathrm{Re}(s) = 1$. Let v be the dual variable to u. We assume first $|v| \leq 1$. Let k_0 be the integer for which $2^{k_0} \leq 1/|v| \leq 2^{k_0+1}$. Now

$$\sum_{k=1}^{r} 2^{-ks} \int \eta(2^{-k}u)e^{2\pi iuv}\, du = \sum_{k\leq k_0} + \sum_{k>k_0} .$$

In the first sum, write $e^{2\pi iuv} = 1 + O(|u||v|)$, and recall that $\eta(\gamma)$ is supported in $1/2 \leq |\gamma| \leq 2$, thus

$$\sum_{k\leq k_0} = O\left(c \sum_{k\leq k_0} 2^{-ks}2^k\right) + O\left(\sum_{k\leq k_0} 2^{-ks} \int \eta(2^{-k}u)\, |v|\, |u|\, du\right),$$

where $c = \int \eta$. However $\sum_{k\leq k_0} 2^{-ks}2^k$ is $O(1/|1 - 2^{1-s}|)$ if $\mathrm{Re}(s) = 1$, while the second term above is (when $\mathrm{Re}(s) = 1$)

$$= O(|v|)\left(\sum_{k\leq k_0} 2^{-k} \int |\eta(2^{-k}u)||u|\, du\right) = O(|v|) \sum_{k\leq k_0} 2^k = O(1).$$

Finally for the second sum, $\sum_{k>k_0}$, integrate by parts, writing $e^{2\pi iuv}$ as $\frac{1}{2\pi iv}\frac{d}{du}(e^{2\pi iuv})$ to obtain a sum that is $O\left(\frac{1}{|v|}\sum_{k>k_0} 2^{-k}\right) = O(2^{-k_0}/|v|) = O(1)$.

If $|v| > 1$, take $k_0 = 0$, and argue similarly.]

21. Suppose Ω is a bounded open convex set with $0 \in \Omega$ and with C^2 boundary. Then there is a constant $c > 0$ so that whenever $R \geq 1$ and $\delta \leq 1$, then $x \in R\Omega$ and $|y| \leq \delta$ implies $x + y \in (R + c\delta)\Omega$.

[Hint: One may reduce to the case $R = 1$ by rescaling. To see, for example, that there is a μ so that $x + y = (1 + \mu\delta)\Omega$ whenever $x \in \partial\Omega$ and $|y| < \delta$ for δ sufficiently small, proceed as follows. By a Euclidean change of variables, introduce new coordinates so that x has been moved to $(0, 0) \in \mathbb{R}^{d-1} \times \mathbb{R}$, and near that point Ω is given by $x_d > \varphi(x')$, with $\phi(0) = 0$ and $\nabla_{x'}\varphi(0) = 0$. Then by convexity of Ω, the point corresponding to the initial origin is given by (z', z_d) with $z_d \geq c_1 > 0$. Also $x + y \in (1 + \mu\delta)\Omega$ is equivalent with

$$\frac{y_d + \mu\delta z_d}{1 + \mu\delta} > \varphi\left(\frac{y' + \mu\delta z'}{1 + \mu\delta}\right).$$

Since $|y_d| < \delta$, the left-hand side is $\geq \frac{c_1}{2}\frac{\mu\delta}{1+\mu\delta}$, as soon as $\mu \geq 2/c_1$. Fix such a μ. Now the right-hand side is dominated by

$$A\left|\frac{y' + \mu\delta z'}{1 + \mu\delta}\right|^2 \leq A'\left(\frac{\delta^2 + (\mu\delta)^2}{1 + \mu\delta}\right),$$

and we need only choose $\delta \leq c_2/\mu$, for appropriately small c_2.]

22. Prove the following two estimates for $r \to \infty$:

(a) $\sum_{1\leq k\leq r} d(k)k^\alpha = O\left(r^{\alpha+1}\log r\right)$ if $\alpha > -1$.

(b) $\sum_{r<k} d(k)k^\alpha = O\left(r^{\alpha+1}\log r\right)$ if $\alpha < -1$.

[Hint: Write

$$\sum_{k>r} d(k)k^\alpha = \sum\sum_{mn>r}(mn)^\alpha = \sum_n n^\alpha\left(\sum_{m>r/n} m^\alpha\right)$$
$$= O\left(\sum_n n^\alpha \min(1,(r/n)^{\alpha+1})\right).$$

23. Prove that $rJ_1(r) = \int_0^r \sigma J_0(\sigma)\,d\sigma$, by verifying the following:

(a) $J_1'(r) = \frac{1}{2}\left(J_0(r) - J_2(r)\right)$.

(b) $J_1(r) = \frac{r}{2}\left(J_0(r) + J_2(r)\right)$.

The above shows that $rJ_1'(r) + J_1(r) = rJ_0(r)$, so $\frac{d}{dr}\left(rJ_1(r)\right) = rJ_0(r)$, proving the assertion.

[Hint: Recall that $J_m(r) = \frac{1}{2\pi}\int_0^{2\pi} e^{ir\sin\theta}e^{-im\theta}\,d\theta$. For (a), differentiate in r under the integral sign. For (b), write $e^{i\theta} = -\frac{1}{i}\frac{d}{d\theta}(e^{-i\theta})$ and integrate by parts.]

10 Problems

The problems below are not intended as exercises for the reader but are meant instead as a guide to further results in the subject. Sources in the literature for each of the problems can be found in the "Notes and References" section.

1.* Suppose M is a local hypersurface in $\mathbb{R}^d$. In a neighborhood of a point $x_0 \in M$ one can choose a smooth vector field ν, defined in this neighborhood restricted to M, so that $\nu(x)$ is a unit normal vector of M at each $x \in M$. (There are two choices of this vector field, determined up to a sign.) The map $x \mapsto \nu(x)$ from M to S^{d-1} (with S^{d-1} the unit sphere in $\mathbb{R}^d$) is called the **Gauss map**.

One can prove that the Gauss curvature of M near x_0 is non-vanishing if and only if the Gauss map is a diffeomorphism near x_0. Moreover, if $d\sigma_M$ and $d\sigma_{S^{d-1}}$ are the induced Lebesgue measures of M and S^{d-1}, and $(d\sigma_{S^{d-1}})^*$ the pull-back of $d\sigma_{S^{d-1}}$ to M defined by

$$\int_M f\,(d\sigma_{S^{d-1}})^* = \int_{S^{d-1}} f(\nu^{-1}(x))\,d\sigma_{S^{d-1}}(x),$$

then $Kd\sigma_M = (d\sigma_{S^{d-1}})^*$, with K the absolute value of the Gauss curvature.

2.* The spherical maximal function. Define

$$A_t(f)(x) = \frac{1}{\sigma(S^d)} \int_{S^d} f(x - ty)\, d\sigma(y)$$

for each $t \neq 0$, and $A^*(f)(x) = \sup_{t\neq 0} |A_t(f)(x)|$. Then

$$\|A^*(f)\|_{L^p} \leq c_p \|f\|_{L^p}, \qquad \text{if } p > d/(d-1) \text{ and } d \geq 2.$$

As a result, if $f \in L^p$, $p > d/(d-1)$, then $\lim_{t\to 0} A_t(f)(x) = f(x)$ a.e. Simple examples show that this fails if $p \leq d/(d-1)$.

A hint that there may be estimates for $\sup_t |A_t(f)|$ (and in particular that the result holds for $p = 2$ and $d \geq 3$) is the following simple observation for $d \geq 3$:

$$\| \sup_{1\leq t\leq 2} |A_t(f)| \, \|_{L^2} \leq c \|f\|_{L^2}.$$

To establish this, one notes that

$$\int_1^2 \left| \frac{\partial A_s(f)(x)}{\partial s} \right|^2 dx\, ds \leq c' \|f\|^2_{L^2}$$

by using Theorem 3.1. However $\sup_{1\leq t\leq 2} |A_t(f)| \leq \int_1^2 \left|\frac{\partial A_s(f)}{\partial s}\right| ds + |A_1(f)(x)|$, hence the assertion follows by using Schwartz's inequality.

Refinements of this argument prove the result for $\sup_{t>0} |A_t(f)|$, $p = 2$ and $d \geq 3$, and then also for $p > d/(d-1)$. Further ideas are needed for the case $d = 2$.

3.* There is a variant of Problem 2 that applies to the wave equation.

Suppose u solves $\triangle_x u = \frac{\partial^2 u}{\partial t^2}$ for $(x,t) \in \mathbb{R}^d \times \mathbb{R}$, with $u(x,0) = 0$, and $\frac{\partial u}{\partial t}(x,0) = f(x)$. If $f \in L^2$ we observe that $\frac{u(x,t)}{t} \to f(x)$ in the $L^2(\mathbb{R}^d)$ norm as $t \to 0$. One can show that $\lim_{t\to 0} \frac{u(x,t)}{t}$ exists and equals $f(x)$ a.e. if $f \in L^p$, $p > 2d/(d+1)$.

4.* The restriction phenomenon (inequality (31)) is valid in $\mathbb{R}^2$, for the full range $1 \leq p < 4/3$.

[Hint: One may dualize the assertion as in the proof of Theorem 5.2. Consider the operator $\mathcal{R}^*$ defined by

$$\mathcal{R}^*(F)(x) = \int_M e^{2\pi i x\cdot\xi} F(\xi)\, d\mu(\xi).$$

The desired result then becomes the inequality

$$\|\mathcal{R}^*(F)\|_{L^q(\mathbb{R}^2)} \leq A \|F\|_{L^p(d\mu)}$$

when $q = 3p'$ and $1 \le p < 4$. Now the key point is that if we consider the singular measure $d\nu = F d\mu$, then the convolution $\nu * \nu$ is actually an absolutely continuous measure $f\,dx$ with density f, a locally integrable function on $\mathbb{R}^2$. This fact reflects the assumed curvature of M. Indeed, it can be shown that $f \in L^p(\mathbb{R}^2)$, with $\frac{3}{r} = \frac{2}{p} + 1$, whenever $F \in L^p(d\mu)$ and $1 \le p \le 4$, and $\|f\|_{L^r(\mathbb{R}^2)} \le c\|F\|^2_{L^p(d\mu)}$ and $1 \le p < 4$. Now if this is so, then

$$\mathcal{R}^*(F)^2 = (\hat{\nu}(-x))^2 = (\nu * \nu)\hat{}(-x) = \hat{f}(x),$$

and by the Hausdorff-Young inequality,

$$\|\mathcal{R}^*(F)\|^2_{L^{2r'}} \le \|(\mathcal{R}^*(F))^2\|_{L^{r'}} = \|\hat{f}\|_{L^{r'}} \le c\|F\|^2_{L^p},$$

and this proves the assertion since $2r' = 3p'$.]

5.* An analog of Theorem 6.3 for the wave equation is as follows. Let $u(x,t)$ be the solution of the wave equation $\frac{\partial^2 u}{\partial t^2} = \triangle u$ for $(x,t) \in \mathbb{R}^d \times \mathbb{R}$, with initial data

$$\begin{cases} u(x,0) &= 0 \\ \frac{\partial u}{\partial t}(x,0) &= f(x). \end{cases}$$

Then $\|u(x,t)\|_{L^q(\mathbb{R}^d\times\mathbb{R})} \le c\|f\|_{L^2(\mathbb{R}^d)}$ if $q = \frac{2d+2}{d-2}$ and $d \ge 3$.

6.* The following further results are known about $E(R) = N(R) - \pi R^2$, the error term appearing in Theorem 8.3.

(a) The Hardy series $R\sum_{k=1}^{\infty} \frac{r_2(k)}{k^{1/2}} J_1(2\pi k^{1/2} R)$ converges for each $R \ge 0$, and its sum equals $E(R)$ whenever $R \ne k^{1/2}$, for any positive integer k.

(b) The error $E(R)$ is on the average a multiple of $R^{1/2}$ in the sense that

$$\int_0^r E(R)^2 R\,dR = cr^3 + O(r^{2+\epsilon}),$$

for some $c > 0$ and every $\epsilon > 0$.

(c) However, $E(R)$ is not exactly $O(R^{1/2})$ since

$$\limsup_{R\to\infty} \frac{|E(R)|}{R^{1/2}} = \infty.$$

(d) It has been proved that $E(R) = O(R^{\alpha+\epsilon})$, for certain α, $1/2 < \alpha < 2/3$. A relatively recent result of this kind is for $\alpha = 131/208$.

7.* The oscillatory integral $\mathfrak{I}(\lambda)$ can be identified in terms of Bessel functions of the second and third kind. One has that

$$\mathfrak{I}(\lambda) = 4K_0(2\lambda) - 2\pi Y_0(2\lambda),$$

where Y_m and K_m are respectively Neumann and Macdonald functions.

8.* Consider the error term in the divisor problem

$$\Delta(\mu) = \sum_{k=1}^{\mu} d(k) - \mu \log \mu - (2\gamma - 1)\mu - 1/4.$$

It is given for μ not an integer by the convergent series

$$\frac{-2}{\pi}\mu^{1/2} \sum_{k=1}^{\infty} \frac{d(k)}{k^{1/2}} \left[K_1(4\pi k^{1/2}\mu^{1/2}) + \frac{\pi}{2} Y_1(4\pi k^{1/2}\mu^{1/2}) \right].$$

For Δ there are estimates, analogous to those for E in Problem 6, with $\Delta(\mu) = O(\mu^{\beta+\epsilon})$ and $\beta = \alpha/2$.

Notes and References

Chapter 1

The first citation is taken from the article [40] by F. Riesz, while the second is a translation from an excerpt of Banach's book [3].

General sources for topics in this chapter are Hewitt and Stromberg [23], Yosida [59], and Folland [18].

For Problem 7*, we refer, for instance, to the book [9] by Carothers, while results related to the Clarkson inequalities in Problem 6* can be found in Chapter 4 of Hewitt and Stromberg [23]. For a treatment of Orlicz spaces, see Rao and Ren [39]. Finally, in Wagon [57] the reader will find further information on the ideas described in Problems 8* and 9.

Chapter 2

The first citation is taken from Young's article [60]. The second citation, translated from the French, is an extract of a letter from M. Riesz to Hardy. The last citation is an extract from a letter from Hardy to M. Riesz. Both are cited in Cartwright [10]. In addition, this reference also contains the M. Riesz citation in the text in Section 1.

For the theory of the conjugate function on the circle, analogous to the Hilbert transform on the real line, see Chapter VII of Zygmund [61], and Katznelson [31]. The theory of $\mathbf{H}_r^1$ and BMO is treated in Stein [45] where other sources in the literature can be found.

For Problem 6* see for example Chapter III in Stein [45].

The proof of the result in Problem 7 can be carried out by complex methods using Blaschke products. For the details of this approach in the analogous situation when the upper half-plane is replaced by the unit disc, see Chapter VII in Zygmund [61]. An alternate approach by real methods is, for example, in Chapter III of Stein and Weiss [47].

Problem 9* is a result of Jones and Journé, which can be found in [28], while the reader can consult Coifman *et al.* [38] for results related to Problem 10*.

Chapter 3

The first citation is taken from Bochner [7], while the second comes from the preface of Zygmund [61].

The foundations of distribution theory can be found in the work of Schwartz [41].

A further in depth source for distribution theory is Gelfand and Shilov [20], which is the first volume of a series of books on the topic.

Formulations of Theorem 3.2 that are more general, because they require less regularity of the kernels of the operators, may be found in Stein [44], Chapter 2, and Stein [45], Chapter 1.

For Problems 5* and 6* see Bernstein and Gelfand [4], and Atiyah [1]. In fact, Hörmander [26] is also relevant for Problems 6* and 7*.

Finally, for Problem 8*, see for instance Folland [17], where other references may be found, in particular the original work of M. Riesz, Methée, and others.

Chapter 4

The citation is a translation taken from the original work of Baire [2].

The proof of the existence of Besicovitch sets using the Baire category theorem was originally given in Körner [34].

The concept of a universal element defined in Exercise 14 and also discussed in Problem 7* comes originally from ergodic theory and the study of dynamical systems. For a good survey regarding universality, and also the related hypercyclic operators see Grosse-Erdmann's article [21].

Chapter 5

The first citation is taken from an article by Shiryaev on Kolmogorov that appears in *Kolmogorov in Perspective*, History of Mathematics, Volume 20, American Mathematical Society, 2000. The second citation is an excerpt of a translation from [29].

There are many good texts for general probability theory and stochastic processes. For instance, the reader may consult Doob [13], Durrett [14] and Koralov and Sinai [33].

For more information on the Walsh-Paley functions in Exercise 16 and Problem 2*, the reader may turn to Schipp *et al.* [42]. The reader will also find some information on lacunary series relevant for Problem 2*, in Sections 6 to 8, Chapter V in Volume 1 of Zygmund [61].

Chapter 6

Doobs' citation is from a review of Masani's book, *Norbert Wiener*. This review appeared in the *Bulletin of the American Mathematical Society*, Volume 27, Number 2, October 1992.

The following are general sources for material on Brownian motion: Billingsley [5] and [6], Durrett [14], Karatzas and Shreve [30], Stroock [52], Koralov and Sinai [33], and Çinlar [11].

For problems 4* and 7*, see Durrett [14] or Karatzas and Shreve [30].

Chapter 7

Lewy's citation is from [37].

Relevant references for the topics discussed in this chapter, as well as the general theory of several complex variables, are Gunning and Rossi [22], Hörmander [25], and Krantz [35].

The approximation result in Theorem 7.1 can be found, for example in Boggess [8], Baouendi *et al.* [15] or Treves [56].

For further information on the theory of Cauchy-Riemann equations and extensions of some results discussed in this chapter, the reader may turn to Boggess [8].

More about analysis on the upper half-space $\mathcal{U}$ treated in the Appendix and its relation to the Heisenberg group can be found in Stein [45], Chapters XII and XIII.

For Problems 1 and 2, see for instance Gunning and Rossi [22] or Krantz [35].

Problem 3* is in Chapter 2 of Chen and Shaw [12], while the theory of the $\overline{\partial}$-Neumann equation in Problem 4* can be found in Folland and Kohn [19], and Chen and Shaw [12].

Finally, for domains of holomorphy in Problem 5* see, for instance, Chapter 2 in Hörmander [25], or Chapters 3 and 4 in Chen and Shaw [12], while for Problem 6*, see for instance Chapter XIII in Stein [45].

Chapter 8

The epigraph (1840) from Kelvin is taken from [54], while the epigraph of Stokes is taken from [48].

Some general references for topics covered in Sections 1 to 5 and 7 of this chapter are Sogge [43] and Stein [45], Chapters 8–11. We have omitted any discussion of the important topic of Fourier integral operators. An introduction to this subject is in Sogge [43], Chapter 6, where further references may be found.

Early work on dispersion equations was by done by Segal, Strichartz [51], Ginibre and Velo, and Strauss [49]. A systematic survey and exposition of the subject is in Tao [53], where further references to the literature may be found.

Sources for the results on lattice points in Section 8 are Landau [36], Part 8; Titchmarsh [55], Chapter 12; Hlawka [24]; and Iwaniec and Kowalski [27], Chapter 4.

For more about the Gauss map discussed in Problem 1* see, for example, Kobayashi and Nomizu [32], Sections 2 and 3.

A treatment of the spherical maximal function can be found in Stein and Wainger [46] for $d \geq 3$ and Sogge [43] for $d = 2$.

For Problem 4*, the restriction theorem when $d = 2$, see Stein [45], Section 5 in Chapter 9.

The result in Problem 5*, in a more general form, is in Strichartz [51].

For the results (a)–(c) in Problem 6* concerning $r_2(k)$, see Landau [36]. The exponent $\alpha = 131/208$ is due to M. N. Huxley.

The identification of $\mathfrak{J}$ with Bessel-type functions in Problem 7* can be deduced from formulas (15) and (25) in Erdélyi [16], and Sections 6.21 and 6.22 in Watson [58]. With the aid of these formulas one can connect Proposition 8.8 and Theorem 8.9 in this chapter with Theorem 1 in Strichartz [50], and formulas in Sections 2.6–2.9 in Gelfand and Shilov [20].

The identity for $\Delta(\mu)$ in Problem 8* goes back to Voronoi and in fact predates Hardy's identity for $r_2(k)$.

Bibliography

[1] M. F. Atiyah. Resolution of singularities and division of distributions. *Comm. Pure. Appl. Math*, 23:145–150, 1970.

[2] R. Baire. Sur les fonctions de variables réelles. *Annali. Mat. Pura ed Appl*, III(3):1–123, 1899.

[3] S. Banach. Théorie des opérations linéaires. *Monografje Matematyczne, Warsawa*, 1, 1932.

[4] I. N. Bernstein and S. J. Gelfand. The polynomial p^λ is meromorphic. *Funct. Anal. Appl*, 3:68–69, 1969.

[5] P. Billingsley. *Convergence of Probability Measures.* John Wiley & Sons, 1968.

[6] P. Billingsley. *Probability and Measure.* John Wiley & Sons, 1995.

[7] S. Bochner. "The rise of functions" in complex analysis. *Rice University Studies*, 56(2), 1970.

[8] A. Boggess. *CR Manifolds and the Tangential Cauchy-Riemann Complex.* CRC Press, Boca Raton, 1991.

[9] N. L. Carothers. *A Short Course on Banach Space Theory.* Cambridge University Press, 2005.

[10] M. L. Cartwright. Manuscripts of Hardy, Littlewood, Marcel Riesz and Titchmarsh. *Bull. London Math. Soc*, 14(6):472–532, 1982.

[11] E. Çinlar. *Probability and Statistics*, volume 261 of *Graduate texts in mathematics.* Springer Verlag, 2011.

[12] S-C Chen and M-C Shaw. *Partial Differential Equations in Several Complex Variables*, volume 19 of *Studies in Advanced Mathematics.* American Mathematical Society, 2001.

[13] J. L. Doob. *Stochastic Processes.* John Wiley & Sons, New York, 1953.

[14] R. Durrett. *Probability: Theory and Examples.* Duxbury Press, Belmont, CA, 1991.

[15] M. S. Baouendi, P. Ebenfelt, and L. P. Rothschild. *Real Submanifolds in Complex Space and Their Mappings.* Princeton University Press, Princeton, NJ, 1999.

[16] A. Erdélyi *et al.* *Higher Transcendental Functions.* Bateman Manuscript Project, Volume 2. McGraw-Hill, 1953.

[17] G. B. Folland. Fundamental solutions for the wave operator. *Expo. Math*, 15:25–52, 1997.

[18] G. B. Folland. *Real Analysis.* John Wiley & Sons, 1999.

[19] G. B. Folland and J. J. Kohn. *The Neumann Problem for the Cauchy-Riemann Complex.* Ann. Math Studies 75. Princeton University Press, Princeton, NJ, 1972.

[20] I. M. Gelfand and G. E. Shilov. *Generalized Functions*, volume 1. Academic Press, New York, 1964.

[21] K-G. Grosse-Erdmann. Universal families and hypercyclic operators. *Bull. Amer. Math. Soc*, 36(3):345–381, 1999.

[22] R. C. Gunning and H. Rossi. *Analytic Functions of Several Complex Variables.* Prentice-Hall, Englewood Cliffs, NJ, 1965.

[23] E. Hewitt and K. Stromberg. *Real and Abstract Analysis.* Springer, New York, 1965.

[24] E. Hlawka. Uber Integrale auf konvexen Körpern I. *Monatsh. Math.*, 54:1–36, 1950.

[25] L. Hörmander. *An Introduction to Complex Analysis in Several Variables.* D. Van Nostrand Company, Princeton, NJ, 1966.

[26] L. Hörmander. *The Analysis of Linear Partial Differential Operators II.* Springer, Berlin Heidelberg, 1985.

[27] H. Iwaniec and E. Kowalski. *Analytic Number Theory*, volume 53. American Mathematical Society Colloquium Publications, 2004.

[28] P. W. Jones and J-L. Journé. On weak convergence in $H^1(\mathbb{R}^d)$. *Proc. Amer. Math. Soc*, 120:137–138, 1994.

[29] M. Kac. Sur les fonctions independantes I. *Studia Math*, pages 46–58, 1936.

[30] I. Karatzas and S. E. Shreve. *Brownian Motion and Sochastic Calculus.* Springer, 2000.

[31] Y. Katznelson. *An Introduction to Harmonic Analysis.* John Wiley & Sons, 1968.

[32] S. Kobayashi and K. Nomizu. *Foundations of Differential Geometry*, volume 2. Wiley, 1996.

[33] L. B. Koralov and Y. G. Sinai. *Theory of Probability and Random Processes.* Springer, 2007.

[34] T.W. Körner. Besicovitch via Baire. *Studia Math*, 158:65–78, 2003.

[35] S. G. Krantz. *Function Theory of Several Complex Variables.* Wadsworth & Brooks/Cole, Pacific Grove, CA, second edition, 1992.

[36] E. Landau. *Vorlesungen über Zahlentheorie*, volume II. AMS Chelsea, New York, 1947.

[37] H. Lewy. An example of a smooth linear partial differential equation without solution. *Ann. of Math.*, 66(1):155–158, 1966.

[38] R. R. Coifman, P. L. Lions, Y. Meyer, and S. Semmes. Compacité par compensation et espaces de Hardy. *C. R. Acad. Sci. Paris*, 309:945–949, 1989.

[39] M. M. Rao and Z. D. Ren. *Theory of Orlicz Spaces.* Marcel Dekker, New York, 1991.

[40] F. Riesz. Untersuchungen über Systeme integrierbarer Funktionen. *Mathematische Annalen*, 69, 1910.

[41] L. Schwartz. *Théorie des distributions*, volume I and II. Hermann, Paris, 1950-1951.

[42] F. Schipp, W. R. Wade, P. Simon and J. Pál. *Walsh Series: An Introduction to Dyadic Harmonic Analysis.* Adam Hilger, Bristol, UK, 1990.

[43] C. D. Sogge. *Fourier Integrals in Classical Analysis.* Cambridge University Press, 1993.

[44] E. M. Stein. *Singular Integrals and Differentiability Properties of Functions.* Princeton University Press, Princeton, NJ, 1970.

[45] E. M. Stein. *Harmonic Analysis: Real-Variable Methods, Orthogonality, and Oscillatory Integrals.* Princeton University Press, Princeton, NJ, 1993.

[46] E. M. Stein and S. Wainger. Problems in harmonic analysis related to curvature. *Bull. Amer. Math. Soc*, 84:1239–1295, 1978.

[47] E. M. Stein and G. Weiss. *Introduction to Fourier Analysis on Euclidean Spaces.* Princeton University Press, Princeton, NJ, 1971.

[48] G. G. Stokes. On the numerical calculations of a class of definite integrals and infinite series. *Camb. Phil. Trans*, ix, 1850.

[49] W. Strauss. *Nonlinear Wave Equations*, volume 73 of *CBMS*. American Mathematical Society, 1978.

[50] R. S. Strichartz. Fourier transforms and non-compact rotation groups. *Ind. Univ. Math. Journal*, 24:499–526, 1974.

[51] R. S. Strichartz. Restriction of the Fourier transform to quadratic surfaces and decay of solutions of the wave equations. *Duke Math Journal*, 44:705–714, 1977.

[52] D. W. Stroock. *Probability Theory: An Analytic View.* Cambridge University Press, 1993.

[53] T. Tao. *Nonlinear Dispersive Equations*, volume 106 of *CBMS*. American Mathematical Society, 2006.

[54] S. P. Thompson. *Life of Lord Kelvin*, volume 1. Chelsea reprint, New York, 1976.

[55] E. C. Titchmarsh. *The Theory of the Riemann Zeta-function.* Oxford University Press, 1951.

[56] F. Treves. *Hypo-Analytic Structures.* Princeton University Press, Princeton, NJ, 1992.

[57] S. Wagon. *The Banach-Tarski Paradox.* Cambridge University Press, 1986.

[58] G. N. Watson. *A Treatise on the Theory of Bessel Functions.* Cambridge University Press, 1945.

[59] K. Yosida. *Functional Analysis.* Springer, Berlin, 1965.

[60] W. H. Young. On the determination of the summability of a function by means of its Fourier constants. *Proc. London Math. Soc*, 2-12:71–88, 1913.

[61] A. Zygmund. *Trigonometric Series*, volume I and II. Cambridge University Press, Cambridge, 1959. Reprinted 1993.

Symbol Glossary

The page numbers on the right indicate the first time the symbol or notation is defined or used. As usual, $\mathbb{Z}$, $\mathbb{Q}$, $\mathbb{R}$, and $\mathbb{C}$ denote the integers, the rationals, the reals, and the complex numbers respectively.

$L^p(X,\mathcal{F},\mu)$, $L^p(X,\mu)$, $L^p(X)$	L^p space	2
$\|\cdot\|_{L^p(X)}$, $\|\cdot\|_{L^p}$, $\|\cdot\|_p$	L^p norm	2
$L^\infty(X,\mathcal{F},\mu)$	L^∞ space	7
$\|\cdot\|_{L^\infty}$	L^∞ norm or essential-supremum	8
$C(X)$	Continuous functions on X with the sup-norm	9
Λ^α	Hölder space of exponent α	10
L^p_k	Sobolev space	11
$\mathcal{B}^*$	Dual space of $\mathcal{B}$	12
$\mathcal{B}_X$	Borel sets of X	29
$M(X)$	Finite signed Borel measures on X	29
$C_b(X)$	Bounded functions in $C(X)$	33
$L^{p_0}+L^{p_1}$	Sum of L^{p_0} and L^{p_1}	36
$A\triangle B$	Symmetric difference of A and B	36
$L^{p,r}$, $\|\cdot\|_{L^{p,r}}$	Mixed space and mixed norm	38
L^Φ	Orlicz space	40
$C^{k,\alpha}$	Functions whose k^{th} derivative are in Λ^α	42
$\mathbb{R}^2_+$	Upper half-plane	61
$H(f)$	Hilbert transform of f	62
$\mathcal{P}_y$, $\mathcal{Q}_y$	Poisson and conjugate Poisson kernels	63
$O(\cdots)$	O notation	64
$C_0^\infty(\mathbb{R})$	Space of indefinitely differentiable functions with compact support on $\mathbb{R}$	66
$\lambda_F(\alpha)$, $\lambda(\alpha)$	Distribution function of F	72
$\mathbf{H}^1_r(\mathbb{R}^d)$	Real Hardy space	75
$\|\cdot\|_{\mathbf{H}^1_r}$	$\mathbf{H}^1_r(\mathbb{R}^d)$ norm	75
$f^\dagger$	Truncated maximal function	76
$\|\cdot\|_{\mathrm{BMO}}$	Bounded mean oscillation (or BMO) norm	86
$C_0^\infty(\Omega)$, $\mathcal{D}(\Omega)$	Smooth functions with compact support in Ω, or test functions	100
∂_x^α, $\|\alpha\|$, $\alpha!$	Partial derivatives and related functions	100

$\mathcal{D}^*(\Omega)$	Space of distributions on Ω	100
$\delta(\cdot)$	Dirac delta	101
C^k, $C^k(\Omega)$	Functions of class C^k on Ω	101
$\mathcal{S}(\mathbb{R}^d)$, $\mathcal{S}$	Schwartz space, or test functions	105
$\|\cdot\|_N$	Sup-norm for derivatives up to order N	105
$\mathcal{S}^*$	Space of tempered distributions	106
$\mathrm{pv}(\frac{1}{x})$	Principal value	111
$\triangle$	Laplacian operator	119
A_d	Area of unit sphere in $\mathbb{R}^d$	123
$\partial_{\overline{z}}$, $\frac{\partial}{\partial \overline{z}}$, ∂_z, $\frac{\partial}{\partial z}$	Derivative with respect to $\overline{z}$ and z	148, 277
$\square$	Wave operator	155
A^δ	δ-neighborhood of A	177
$\mathbb{Z}_2^N$	N-fold product of $\mathbb{Z}_2$	189
$\mathbb{Z}_2^\infty$	Infinite product of $\mathbb{Z}_2$	191
r_n	Rademacher functions	192
m_0, σ^2	Mean or expectation, and variance	196
ν_{σ^2}	Gaussian distribution with mean zero and variance σ^2	196
$\mathbb{E}_{\mathcal{A}}(f)$, $\mathbb{E}(f\|\mathcal{A})$, $\mathbb{E}$	Conditional expectation of f with respect to $\mathcal{A}$	209
$\mathcal{P}$	Continuous paths in $\mathbb{R}^d$ starting at the origin	240
$\tau(\omega)$	Stopping time	254
$\mathbb{P}_r(z_0)$	Polydisc in $\mathbb{C}^n$	277
$C_r(z_0)$	Boundary circles of $\mathbb{P}_r(z_0)$	277
$\overline{\partial}$	Cauchy-Riemann operator	291
L_j	Tangential Cauchy-Riemann operator	300
$\mathcal{U}$	Upper half-space in $\mathbb{C}^n$	307
$H^2(\mathcal{U})$	Hardy space on $\mathcal{U}$	308
$X \lesssim Y$, $X \approx Y$	$X \leq cY$ and $c^{-1}y \leq X \leq cY$ for some $c > 0$	331
$\mathrm{rotcurv}(\rho)$	Rotational curvature	366
$r_2(k)$	Number of ways k is a sum of two squares	377
$d(k)$	Number of divisors of k	385
$\mathfrak{h}$	Hyperbolic measure	385

Index

Relevant items that also arose in Book I, Book II or Book III are listed in this index, preceded by the numerals I, II or III, respectively.